FERROMAGNETIC MATERIALS
VOLUME 2

FERROMAGNETIC MATERIALS

A handbook on the properties
of magnetically ordered substances

VOLUME 2

EDITED BY

E.P. WOHLFARTH

Department of Mathematics
Imperial College of Science and Technology
London

ELSEVIER
Amsterdam – Lausanne – New York – Oxford – Shannon – Singapore – Tokyo

ELSEVIER SCIENCE B.V.
Sara Burgerhartstraat 25
P.O. Box 211, 1000 AE Amsterdam, The Netherlands

First edition 1980
Second impression 1986
Third impression 1999
Transferred to digital printing 2005

Library of Congress Cataloging in Publication Data (Revised)
Main entry under title:

Ferromagnetic materials.

   Includes indexes.
   1. Ferromagnetic materials - Handbooks, manuals, etc.
I. Wohlfarth, E.P.
QC761.5.F47   538'.44   79-9308

ISBN: 0 444 85312 X

♾ The paper used in this publication meets the requirements of ANSI/NISO Z39.48-1992 (Permanence of Paper).

Printed and bound by Antony Rowe Ltd, Eastbourne

PREFACE

This Handbook on the Properties of Magnetically Ordered Substances, *Ferromagnetic Materials*, is intended as a comprehensive work of reference and textbook at the same time. As such it aims to encompass the achievements both of earlier compilations of tables and of earlier monographs. In fact, one aim of those who have helped to prepare this work has been to produce a worthy successor to Bozorth's classical and monumental book on Ferromagnetism, published some 30 years ago. This older book contained a mass of information, some of which is still valuable and which has been used very widely as a work of reference. It also contained in its text a remarkably broad coverage of the scientific and technological background.

One man can no longer prepare a work of this nature and the only possibility was to produce several edited volumes containing review articles. The authors of these articles were intended to be those who are still active in research and development and sufficiently devoted to their calling and to their fellow scientists and technologists to be prepared to engage in the heavy tasks facing them. The reader and user of the Handbook will have to judge as to the success of the choice made.

One drawback of producing edited volumes is clearly the impossibility of having all the articles ready within a time span sufficiently short for the whole work, once ready, to be up-to-date. This is an effect occurring every time an editor of such a work engages in his task and has been found to be particularly marked in the present case, as might have been expected. Hence a decision was made to edit the first two volumes of the projected four volume work on the basis of the articles available about the end of 1978. Again, the reader must judge whether on balance the lack of a complete logical order of the articles in these two volumes is outweighed by their immediacy and topicality. The future of the work is in the hands of the remaining authors. The projected remaining two volumes will complete a broad and comprehensive coverage of the whole field.

Each author had before him the task of producing a description of material properties in graphical and tabular form in a broad background of discussion of

v

the physics, chemistry, metallurgy, structure and, to a lesser extent, engineering aspects of these properties. In this way, it was hoped to produce the required combined comprehensive work of reference and textbook. The success of the work will be judged perhaps more on the former than on the latter aspect. Ferromagnetic materials are used in remarkably many technological fields, but those engaged on research and development in this fascinating subject often feel themselves as if in a strife for superiority against an opposition based on other physical phenomena such as semiconductivity. Let the present Handbook be a suitable and effective weapon in this strife!

I have to thank many people and it gives me great pleasure to do so. I have had nothing but kindness and cooperation from the North-Holland Publishing Company, and in referring only to P.S.H. Bolman by name I do not wish to detract from the other members of this institution who have also helped. In the same way, I am deeply grateful to many people at the Philips Research Laboratory, Eindhoven, who gave me such useful advice in the early stages. Again, I wish to mention in particular H.P.J. Wijn and A.R. Miedema, without prejudice. Finally, I would like to thank all the authors of this Handbook, particularly those who submitted their articles on time!

E.P. Wohlfarth
Imperial College.

TABLE OF CONTENTS

LIST OF CONTRIBUTORS

G. Bate, Verbatim Corporation, Sunny Vale, CA 94086, USA.

S.W. Charles, School of Physical and Molecular Sciences, University College of North-Wales, Bangor, Gwynedd, UK.

G.Y. Chin, Bell Laboratories, Murray Hill, NJ 07974, USA.

A.H. Eschenfelder, IBM Corporation, San Jose, CA 95193, USA.

M.A. Gilleo, Allied Chemical Corporation, Morristown, NJ 07960, USA.

J. Nicolas, Thomson-CSF, Laboratoire Central de Recherches, 91401 Orsay, France.

J. Popplewell, School of Physical and Molecular Sciences, University College of North-Wales, Bangor, Gwynedd, UK.

P.I. Slick, Bell Laboratories, Allentown, PA 18103, USA.

J.H. Wernick, Bell Laboratories, Murray Hill, NJ 07974, USA.

chapter 1

FERROMAGNETIC INSULATORS: GARNETS

M. A. GILLEO

Allied Chemical Corporation
Morristown, New Jersey 07960
U.S.A.

Ferromagnetic Materials, Vol. 2
Edited by E. P. Wohlfarth
© North-Holland Publishing Company, 1980

CONTENTS

1. Introduction

The silicate mineral garnet, which occurs fairly commonly in nature, has long been known as a source of abrasive grit and has served as a semiprecious stone. Since the synthesis of crystalline silicates is generally difficult, it was only quite recently that garnets of this type have been made (e.g., Skinner 1956) and even more recently that nonsilicate garnet-structure materials have been produced (Yoder and Keith 1951).

The crystal structure of garnet is rather complex even though the symmetry is cubic, the most probable space group being $O_h^{10} - Ia3d$. There are, moreover, eight formula units $A_3B_2C_3O_{12}$ in a unit cell for a total of 160 atoms, as was shown by Menzer (1929) in a valuable X-ray diffraction study of natural garnets by powder techniques.

Despite this extensive body of knowledge concerning garnets and their structure as well as of the nature and origin of ferrimagnetism, it was not until 1956 that ferrimagnetic garnets, exemplified by yttrium–iron garnet (YIG), were discovered (Bertaut and Forrat 1956, Geller and Gilleo 1957a).

Since that time, a tremendous body of research results has been published. Furthermore, at least two new technological areas have grown enormously with the aid of YIG-based devices: (i) tunable filters, circulators, and gyrators for use in the microwave region and (ii) magnetic-bubble-domain-type digital memories. As the bulk of the material used in these applications consists of thin films or discs which are produced by epitaxial growth, the material properties depend on the growth process which is tailored to the specific application. Consequently, these specialized materials will be treated in other sections of this book.

However, the fundamental properties of the garnet-structure ferrimagnetic material will be discussed in this section to lay a basis for further, more specialized, discussions to follow. Therefore, topics relating to the crystalline structure, to the stereochemistry, and to the effects of substituent ions of both the nonmagnetic and magnetic kind on the magnetic properties of ferrimagnetic garnets will be taken up. In general, YIG will be treated as the prototype material with the effects of the substitution of other rare earths for yttrium being considered first. Thereafter, the consequences of replacing iron in YIG with other cations will be examined. In some cases, the use of partial substitution for both yttrium and iron by other ions will be examined, though to a limited extent as magnetic-bubble-domain films are the principal province for such work. In particular, detailed data for uniaxial anisotropy in garnets will be omitted as being a subject specific to magnetic-bubble-domain material and most appropriately taken up in sections dealing with this material.

There have been a number of works in which various aspects of the ferrimagnetic garnets have been reviewed and summarized in excellent fashion. These works have contributed valuably to this article and include the following: Geller (1967, 1977), Nielsen (1971), Von Aulock (1965), Wang (1973), Wolf (1961).

2. Crystalline and magnetic structures and properties

2.1. Crystalline structure

2.1.1. Morphology

The growth habit in the garnet structure in nature, as well as from the flux, shows a predominance of {110} faces (Nielsen and Dearborn 1958, Belov 1958). These {110} faces favor layer growth. At faster growth rates Nielsen and Dearborn (1958) found that {211} faces begin to dominate. These {211} faces are parallel to a close packed chain of octahedra along ⟨111⟩ directions. On the other hand {111} faces are rarely observed on garnets and feature no close packed chains (Vertoprahkov et al. 1968) (see also a discussion by Wang 1973).

The morphology of growth is most important in the event that more than one type of cation is competing for a given site in the growth process, since in the complex garnet structure the aspect of a site presented in a given plane is changed as the plane changes. Thus the crystal being grown may no longer exhibit nearly cubic symmetry, i.e., it may show uniaxial anisotropy.

2.1.2. X-ray diffraction

An early and very detailed X-ray diffraction analysis of the crystal structure of YIG by Geller and Gilleo (1957b, 1959) was made possible by single crystals of YIG grown by Nielsen and Dearborn (1958). The neutron diffraction studies on powdered YIG (Bertaut et al. 1956, Prince 1957) provide confirmation of the O^{2-} position parameters and of the iron moment orientation expected for a ferrimagnetic material of the garnet structure.

The X-ray diffraction data show that YIG belongs to space group O_h^{10}–Ia3d (more properly to a Shubnikov group below the Curie temperature when a spontaneous magnetization is present). The unit cell contains 8 formula units $\{Y_3\}[Fe_2](Fe_3)O_{12}$ where { } designates a 24(c) site dodecahedrally coordinated with the O^{2-} in 96(h), [ ] a 16(a) octahedrally coordinated site, and ( ) a 24(d) tetrahedrally coordinated site (table 1). Each of the formula unit octants has a threefold axis along one of its body diagonals; the octants are related one to the other by 180° rotations.

The distribution of the cations in the garnet structure can best be visualized by

TABLE 1
Point positions in the garnet structure, space group O_h^{10}–Ia3d

O_h^{10}–Ia3d point positions:

$(000; \tfrac{1}{2}\tfrac{1}{2}\tfrac{1}{2})+$

16(a) $000; 0\tfrac{1}{2}\tfrac{1}{2}; \tfrac{1}{2}0\tfrac{1}{2}; \tfrac{1}{2}\tfrac{1}{2}0; \tfrac{1}{4}\tfrac{1}{4}\tfrac{1}{4}; \tfrac{1}{4}\tfrac{3}{4}\tfrac{3}{4}; \tfrac{3}{4}\tfrac{1}{4}\tfrac{3}{4}; \tfrac{3}{4}\tfrac{3}{4}\tfrac{1}{4}.$

24(c) $\tfrac{1}{8}0\tfrac{1}{4}; \tfrac{1}{4}\tfrac{1}{8}0; 0\tfrac{1}{4}\tfrac{1}{8}; \tfrac{3}{8}0\tfrac{3}{4}; \tfrac{3}{4}\tfrac{3}{8}0; 0\tfrac{3}{4}\tfrac{3}{8};$
$\tfrac{5}{8}0\tfrac{1}{4}; \tfrac{1}{4}\tfrac{5}{8}0; 0\tfrac{1}{4}\tfrac{5}{8}; \tfrac{7}{8}0\tfrac{3}{4}; \tfrac{3}{4}\tfrac{7}{8}0; 0\tfrac{3}{4}\tfrac{7}{8}.$

24(d) $\tfrac{3}{8}0\tfrac{1}{4}; \tfrac{1}{4}\tfrac{3}{8}0; 0\tfrac{1}{4}\tfrac{3}{8}; \tfrac{1}{8}0\tfrac{3}{4}; \tfrac{3}{4}\tfrac{1}{8}0; 0\tfrac{3}{4}\tfrac{1}{8};$
$\tfrac{7}{8}0\tfrac{1}{4}; \tfrac{1}{4}\tfrac{7}{8}0; 0\tfrac{1}{4}\tfrac{7}{8}; \tfrac{5}{8}0\tfrac{3}{4}; \tfrac{3}{4}\tfrac{5}{8}0; 0\tfrac{3}{4}\tfrac{5}{8}.$

96(h) $xyz; zxy; yzx;$
$\tfrac{1}{2}+x,\tfrac{1}{2}-y,\bar{z}; \tfrac{1}{2}+z,\tfrac{1}{2}-x,\bar{y}; \tfrac{1}{2}+y,\tfrac{1}{2}-z,\bar{x};$
$\bar{x},\tfrac{1}{2}+y,\tfrac{1}{2}-z; \bar{z},\tfrac{1}{2}+x,\tfrac{1}{2}-y; \bar{y},\tfrac{1}{2}+z,\tfrac{1}{2}-x;$
$\tfrac{1}{2}-x,\bar{y},\tfrac{1}{2}+z; \tfrac{1}{2}-z,\bar{x},\tfrac{1}{2}+y; \tfrac{1}{2}-y,\bar{z},\tfrac{1}{2}+x;$
$\bar{x}\bar{y}\bar{z}; \bar{z}\bar{x}\bar{y}; \bar{y}\bar{z}\bar{x};$
$\tfrac{1}{2}-x,\tfrac{1}{2}+y,z; \tfrac{1}{2}-z,\tfrac{1}{2}+x,y; \tfrac{1}{2}-y,\tfrac{1}{2}+z,x;$
$x,\tfrac{1}{2}-y,\tfrac{1}{2}+z; z,\tfrac{1}{2}-x,\tfrac{1}{2}+y; y,\tfrac{1}{2}-z,\tfrac{1}{2}+x;$
$\tfrac{1}{2}+x,y,\tfrac{1}{2}-z; \tfrac{1}{2}+z,x,\tfrac{1}{2}-y; \tfrac{1}{2}+y,z,\tfrac{1}{2}-x;$
$\tfrac{1}{4}+y,\tfrac{1}{4}+x,\tfrac{1}{4}+z; \tfrac{1}{4}+z,\tfrac{1}{4}+y,\tfrac{1}{4}+x; \tfrac{1}{4}+x,\tfrac{1}{4}+z,\tfrac{1}{4}+y;$
$\tfrac{3}{4}+y,\tfrac{1}{4}-x,\tfrac{3}{4}-z; \tfrac{3}{4}+z,\tfrac{1}{4}-y,\tfrac{3}{4}-x; \tfrac{3}{4}+x,\tfrac{1}{4}-z,\tfrac{3}{4}+y;$
$\tfrac{3}{4}-y,\tfrac{3}{4}+x,\tfrac{1}{4}-z; \tfrac{3}{4}-z,\tfrac{3}{4}+y,\tfrac{1}{4}-x; \tfrac{3}{4}-x,\tfrac{3}{4}+z,\tfrac{1}{4}-y;$
$\tfrac{1}{4}-y,\tfrac{3}{4}-x,\tfrac{3}{4}+z; \tfrac{1}{4}-z,\tfrac{3}{4}-y,\tfrac{3}{4}+x; \tfrac{1}{4}-x,\tfrac{3}{4}-z,\tfrac{3}{4}+y;$
$\tfrac{1}{4}-y,\tfrac{1}{4}-x,\tfrac{1}{4}-z; \tfrac{1}{4}-z,\tfrac{1}{4}-y,\tfrac{1}{4}-x; \tfrac{1}{4}-x,\tfrac{1}{4}-z,\tfrac{1}{4}-y;$
$\tfrac{3}{4}-y,\tfrac{1}{4}+x,\tfrac{3}{4}+z; \tfrac{3}{4}-z,\tfrac{1}{4}+y,\tfrac{3}{4}+x; \tfrac{3}{4}-x,\tfrac{1}{4}+z,\tfrac{3}{4}+y;$
$\tfrac{3}{4}+y,\tfrac{3}{4}-x,\tfrac{1}{4}+z; \tfrac{3}{4}+z,\tfrac{3}{4}-y,\tfrac{1}{4}+x; \tfrac{3}{4}+x,\tfrac{3}{4}-z,\tfrac{1}{4}+y;$
$\tfrac{1}{4}+y,\tfrac{3}{4}+x,\tfrac{3}{4}-z; \tfrac{1}{4}+z,\tfrac{3}{4}+y,\tfrac{3}{4}-x; \tfrac{1}{4}+x,\tfrac{3}{4}+z,\tfrac{3}{4}-y.$

noting that each octant of the unit cell has the (16)a ions octahedrally coordinated with oxygen at each corner and in the center in body centered cubic structure. The 24(d) and 24(c) ions, which are tetrahedrally and dodecahedrally coordinated with oxygen ions, each take positions along a bisector of a cube face and are spaced $\tfrac{1}{2}$ of the octant edge apart while being centered along the bisector (see fig. 2 in section 2.1.3).

The polyhedra share edges and corners as follows (fig. 1): The dodecahedron shares two edges with tetrahedra, four with octahedra, and four with other dodecahedra. Each octahedron shares six edges and each tetrahedron shares two edges with dodecahedra. The octahedra and tetrahedra share only corners.

Effectively, as pointed out by Geller and Gilleo (1957b), the garnet structure is a rather loose one with a volume of 236.9 Å^3 per formula unit. In comparison, the perovskite-like structure of YFeO$_3$ with the Y:Fe ratio of 1:1, instead of 3:5 as in the garnet structure, has a volume of 225.9 Å^3 per formula unit despite the greater proportion of larger ions.

Actually, as will be seen, the looseness of structure provides a great technical advantage in that it is possible to accommodate a very wide variety of cations in the garnet structure. Thus, it is feasible to achieve an enormous range of control

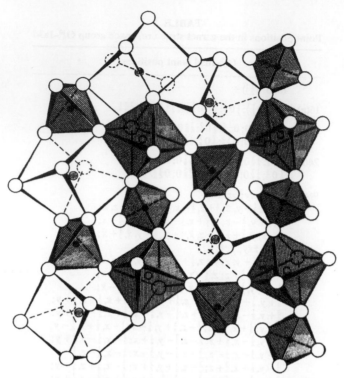

Fig. 1. The garnet structure is built-up of tetrahedra, octahedra, and dodecahedra of oxygen ions coordinated with the metal cations (Novak and Gibbs 1971).

of the magnetic properties in the garnet structure system as will be seen below.

The lattice constants of the rare earth–iron garnets show a regular decrease as the size of the rare earth ion in the dodecahedral site decreases (table 2). The largest of the rare earth–iron garnets that can be formed is $Sm_3Fe_5O_{12}$ with a lattice constant of 12.529 Å and the smallest is $Lu_3Fe_5O_{12}$ with a lattice constant of 12.283 Å. Partial substitution of the lighter and larger rare earth ions from La^{3+} to Pm^{3+} can be achieved as discussed by Geller (1967) with 12.538 Å being the upper limit of attainable lattice constant. Consequently, the limit of substitution is smaller for the larger or lighter rare earth ions in the $Pm^{3+} \rightarrow La^{3+}$ sequence.

2.1.3. Neutron diffraction

Although the X-ray diffraction method has been brought to a high degree of refinement, there are certain respects in which the diffraction of monoenergetic neutrons of near thermal energy (~ 1 Å wavelength) is advantageous: (i) the neutron scattering cross-section for lighter ions, such as O^{2-}, may be comparable to that of heavier ions, (ii) magnetic scattering occurs from the magnetic moments of ions, such as Fe^{3+}, so that magnetic ordering and its direction may

TABLE 2

Lattice constants of rare earth–iron garnets. The lattice constant in bold-face type has been chosen to represent a value of good accuracy and which is part of a consistent set; other values are given for reference. The unit cell contains eight formula units. The temperature is nominally 25°C

Formula unit	Lattice constant	Formula unit	Lattice constant
$Y_3Fe_5O_{12}$	**12.376**[a,b]	$Tb_3Fe_5O_{12}$	**12.436**[k]
	12.3452[c]		12.4275[c]
	12.360[d]		12.447[b]
	12.373[e]		12.452[h]
	12.374[f,g]	$Dy_3Fe_5O_{12}$	**12.405**[k]
	12.378[h]		12.3472[c]
	12.379[i]		12.385[l]
	12.380[j]		12.403[h]
$Sm_3Fe_5O_{12}$	**12.529**[k]		12.414[b]
	12.505[l]	$Ho_3Fe_5O_{12}$	**12.375**[k]
	12.524[c]		12.350[l]
	12.525[e]		12.3545[c]
	12.528[m]		12.380[b]
	12.530[n,o]		12.381[h]
	12.533[p]	$Er_3Fe_5O_{12}$	**12.347**[n]
	12.534[b]		12.3248[c]
$Eu_3Fe_5O_{12}$	**12.498**[k]		12.330[l]
	12.518[b,h]		12.349[b]
$Gd_3Fe_5O_{12}$	**12.471**[k]		12.360[h]
	12.4445[c]	$Tm_3Fe_5O_{12}$	**12.323**[k]
	12.445[d]		12.2849[c]
	12.463[q]		12.321[h]
	12.465[h]		12.325[b]
	12.469[i]	$Yb_3Fe_5O_{12}$	**12.302**[k]
	12.470[r]		12.283[l]
	12.472[n,o]		12.291[b,h]
	12.479[b]		12.3393[c]
		$Lu_3Fe_5O_{12}$	**12.283**[k]
			12.2451[c]
			12.277[b,h]

[a] Geller and Gilleo (1957a, b)
[b] Bertaut and Forrat (1957)
[c] Telesnin et al. (1962)
[d] Wolf and Rodrigue (1958)
[e] Loriers and Villers (1961)
[f] Ramsey et al. (1962)
[g] Anderson (1959)
[h] Pauthenet (1958a)
[i] Ancker-Johnson and Rawley (1958)
[j] Schieber and Kalman (1961)
[k] Espinosa (1962)
[l] Rodrigue et al. (1958)
[m] Geller and Mitchell (1959)
[n] Geller et al. (1961)
[o] Sirvetz and Zneimer (1958)
[p] Cunningham and Anderson (1960)
[q] Anderson and Cunningham (1960)
[r] Villers et al. (1962b)

be observed, (iii) magnetic scattering is removed if the moments of the ions are aligned parallel or antiparallel to the scattering vector, as by an externally applied field, so that magnetic and nonmagnetic scattering may be distinguished.

By means of neutron diffraction techniques applied to powdered samples of yttrium–iron garnet, it was determined quite soon after the discovery of ferrimagnetism in YIG that the model of Néel (1948), in which the spins of the Fe_a^{3+} ions (Fe^{3+} ions in 16(a) sites; also Fe^{3+}(a)) would be parallel to one another and antiparallel to the spins of the Fe_d^{3+} ions (Fe^{3+} ions in 24(d) sites; also Fe^{3+}(d)) is correct (Bertaut et al. 1956, Prince 1957) (fig. 2). The oxygen ion parameters of YIG were examined by Fischer et al. (1966) with powdered specimens and found to agree fairly well with those of Euler and Bruce (1965) which were quite close to those of Geller and Gilleo (1957b, 1959) with a correction from Geller (1977): $x = -0.0274$, $y = 0.0572$, $z = 0.1495$.

However, not only is it possible to verify the orientation of the magnetic moments of the Fe^{3+} ions on the (a) and (d) sublattices but, as Prince (1965) has shown, it is possible to determine the temperature dependence of the magnetic moment of each sublattice and the net moment by means of neutron diffraction

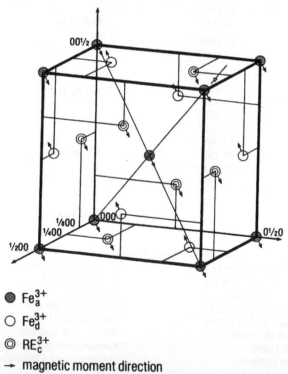

● Fe_a^{3+}

○ Fe_d^{3+}

◎ RE_c^{3+}

→ magnetic moment direction

Fig. 2. Orientation in one octant of a unit cell of a rare earth–iron garnet of the magnetic moments of Fe_a^{3+}, Fe_d^{3+}, and RE_c^{3+} where the easy axis of magnetization is $\langle 111 \rangle$ and RE^{3+} is Gd^{3+} or heavier. Canting of the RE^{3+} magnetic moment has been neglected and the oxygen ions have been omitted.

data obtained with a single crystal of YIG. The data of Prince (1965) cover the temperature range from 4.2 to 598 K, i.e., from complete magnetic saturation to complete paramagnetism ($T_C = 559 \pm 5$ K was observed). The sublattice and net magnetization data thus obtained were compared with the precise data of Anderson (1964a,b). Further, Anderson's analysis of his data in terms of the molecular-field model of Néel (1948) provided calculated curves of sublattice magnetization vs temperature against which Prince (1965) could compare his neutron diffraction data. The conclusion of Prince was that without a correction of the kind based on the approximate application of the biquadratic exchange hypothesis of Rodbell et al. (1963) to the temperature scale of the molecular-field model, the agreement between his data and Anderson's analysis was not as close as desirable. The neutron diffraction data for sublattice magnetic moments were as much as 20% above the curve provided by the molecular-field model.

However, with the correction to the temperature scale given by

$$T/T_C = (T'/T_C)[1 + \beta^2 S^2 M_a(T') M_d(T') / M_a(0) M_d(0)]$$

where $M_d(T')$ and $M_a(T')$ are the sublattice magnetizations calculated at room temperature T' from the molecular-field equations which include intersublattice interaction N_{ad} only ($N_{aa}, N_{dd} = 0$), S is the spin of the magnetic ions (Fe^{3+}), and β is an adjustable parameter approximately given by the ratio of biquadratic to linear exchange coefficients. For the data of Prince, β was found to be about 0.04.

Both neutron and X-ray diffraction methods were employed by Fischer et al. (1966) to determine the fraction of Ga^{3+} ions which were present on the tetrahedral (d) site in a formula unit of Ga-substituted YIG $Y_3Fe_{5-x}Ga_xO_{12}$. Their results were about equally accurate by both methods and showed an increasing preference of Ga^{3+}, an ion smaller than Fe^{3+}, for the smaller tetrahedral site consistent with the results deduced from magnetization data by Gilleo and Geller (1958) who found that for small x over 90% of the Ga entered the (d) sites. As will be discussed later, the Ga distribution is actually a function of the heat treatment of the sample or of the temperature of growth of a substituted-YIG crystal.

2.1.4. Stereochemistry

a. Site symmetry and coordination

In sections 2.1.1 and 2.1.2 the full symmetry of the garnet structure was discussed as it affects macroscopic morphology and as it determines the smallest building block of the structure, the unit cell. However, the fundamental magnetic properties of the iron garnets have their origin in the magnetic ions (principally Fe^{3+}) and their relationship to the surrounding oxygen ions. These oxygen ions influence the electronic configuration of the enclosed iron ions and provide the superexchange interaction between the iron ions in unlike sites. This interaction is the basis of ferrimagnetism (Néel 1948).

Therefore, close attention will be given to the symmetry of the cation sites, to the coordination of cations in one site with those in another, to the geometry of Fe^{3+}–O^{2-}–Fe^{3+} linkages involved in the superexchange interactions, and to the size and electronic configuration which affect the likelihood of occupancy of a given site by these cations. For the purpose of this discussion, YIG will be taken as the prototype.

The largest cation sites in the garnet structure are designated 24(c) and have orthorhombic point group symmetry D_2–222. In this site eight oxygen ions in positions 96(h) form the corners of a dodecahedral configuration which amounts to a cube with the faces slightly bent along one diagonal of each (fig. 3) with four different cube-edge lengths. In a unit cell there are twenty-four dodecahedral sites of which there are six which are nonequivalent with regard to crystal field orientation. The average oxygen distance to the 24(c) ion is about 2.40 Å (table 3).

The next largest cation sites of which there are sixteen in a unit cell in the garnet structure are designated 16(a) with rhombohedral point group symmetry C_{3i}–$\bar{3}$. The six oxygen ions 96(h) surrounding the site form an octahedron stretched along one three-fold axis (fig. 4). The edges of this octahedron are of two different lengths with all oxygen ions being equidistant from the Fe_a^{3+} at 2.01 Å (table 3). There are two octahedra which are nonequivalent in regard to crystalline field orientation and which have their threefold axes along the ⟨111⟩ directions but which are rotated by equal and opposite angles $\alpha \approx 28.6°$ (Geschwind 1961).

The smallest cation sites of which there are twenty-four in a unit cell are designated 24(d) with tetrahedral point group symmetry S_4–$\bar{4}$. The four oxygen ions 96(h) forming a tetrahedron occupy the corners of a cube which has been stretched along the direction of one parallel set of edges (fig. 5). This stretching is along one of the unit cell axes ⟨100⟩ about which the tetrahedra are rotated by equal and opposite angles $\beta \approx 15.6°$ in the unit cell (Geschwind 1961). Thus, there are two tetrahedral-type sites which are nonequivalent with regard to crystalline

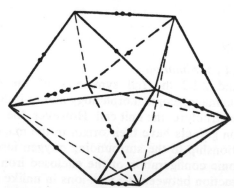

Fig. 3. Dodecahedral coordination of oxygen ions about the yttrium ion (· 2.68, ·· 2.81, ··· 2.87, ···· 2.96 Å) (Geller and Gilleo 1957b, with correction from Geller in private communication).

TABLE 3

Nearest-neighbor interionic distances in yttrium–iron garnet (Geller and Gilleo 1957b, with corrections from Geller 1977)

Ion	Interionic distances (Å)	
Y^{3+}	$4Fe^{3+}(a)$	at 3.46
	$6Fe^{3+}(d)$	at 3.09(2), 3.79(4)
	$8O^{2-}$	at 2.37(4), 2.43(4)
$Fe^{3+}(a)$	$2Y^{3+}$	at 3.46
	$6Fe^{3+}(d)$	at 3.46
	$6O^{2-}$	at 2.01
$Fe^{3+}(d)$	$6Y^{3+}$	at 3.09(2), 3.79(4)
	$4Fe^{3+}(a)$	at 3.46
	$4Fe^{3+}(d)$	at 3.79
	$4O^{2-}$	at 1.87
O^{2-}	$2Y^{3+}$	at 2.37, 2.43
	$1Fe^{3+}(a)$	at 2.01
	$1Fe^{3+}(d)$	at 1.87
	$9O^{2-}$	at 2.68(2), 2.81, 2.87, 2.96, 2.99(2), 3.16(2)

field orientation in the garnet structure. The edge lengths have two values while the oxygen ions are all at 1.87 Å from the Fe_d^{3+} ion (table 3).

The cation site symmetries and sizes are sufficiently different so that the occupancy of these sites by ions of different sizes and different electronic configurations is quite rigidly determined. Nevertheless, as will be seen, the

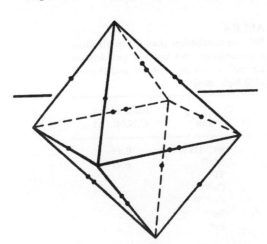

Fig. 4. Octahedral coordination of oxygen ions about an Fe_a^{3+} ion ($\cdot$ 2.68, $\cdot\cdot$ 2.99 Å) (Geller and Gilleo 1957b).

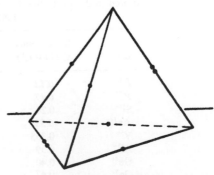

Fig. 5. Tetrahedral coordination of oxygen ions about an Fe_d^{3+} ion ($\cdot$ 3.16, $\cdot\cdot$ 2.87 Å) (Geller and Gilleo 1957b).

variety of sites is so large that a very wide range of cations can be incorporated in the garnet structure.

b. Substitution of cations

The substitution of a wide range of cations in YIG and other garnets has been treated in exhaustive detail, principally by Geller et al. (1964a) and Geller (1967). Therefore, the results of this work as they are relevant to substitutions of cations in YIG will be summarized here.

The first site to be considered for substituent ions will be the tetrahedral one which is the smallest and most restrictive in the number of possible substituent ions. In this case, and all others, the problem of charge compensation as a result of the substitution of ions of valence other than 3+ for Fe^{3+} will not be considered in any detail as this problem is one of crystal chemistry which is not being considered here. The tetrahedral site substituent ions will be tabulated in two groups. The first (table 4) consists of those ions which have been shown to substitute for Fe^{3+} on tetrahedral sites and the second (table 4, in italics) those ions which it should, in principle, be possible to substitute for Fe_d^{3+} in view of the size of the ion. The tetrahedral substituent ions are almost entirely of the spherically-symmetrical, closed-shell type.

The octahedral site is larger than the tetrahedral site in YIG with the distance to the surrounding oxygen ions being 2.01 Å vs 1.87 Å. Accordingly, there is a large variety of ions which have been observed to substitute for Fe^{3+} on the octahedral site of YIG (table 5) and others which might be expected to do so (table 5, in italics). Furthermore, the octahedral symmetry is favorable to certain ions, such as Cr^{3+}, with an electronic configuration of compatible symmetry. In fact, the octahedral site in YIG can actually accommodate some of the smaller rare earth ions (Suchow and Kokta 1971, 1972) (table 6). Many but not all, of the ions which

TABLE 4

Cations known to substitute for Fe^{3+} on tetrahedral sites (d) in yttrium–iron garnet and (*in italics*) those presumably able but not yet demonstrated to do so (Geller 1967). The radii are from Shannon and Prewitt (1969) [(HS) high spin]

Ion	Ionic radius C.N.(4) (Å)	Ion	Ionic radius C.N.(4) (Å)
B^{3+}	0.12	Fe^{3+} (HS)	0.49
P^{5+}	0.17	Mg^{2+}	0.49
Si^{4+}	0.26	Ti^{4+}	–
As^{5+}	0.335	Co^{2+} [a]	–
V^{5+}	0.355	Co^{3+} [b]	–
Al^{3+}	0.39	Sn^{4+}	–
Ge^{4+}	0.40	Fe^{4+}	–
Ga^{3+}	0.47		

[a] Geller et al. (1964b)
[b] Sturge et al. (1969)

TABLE 5

Cations known to substitute for Fe^{3+} on octahedral sites (a) in yttrium–iron garnet and (*in italics*) those presumably able but not yet demonstrated to do so (Geller 1967). The radii are from Shannon and Prewitt (1969). [(HS) high spin, (LS) low spin]

Ion	Ionic radius C.N.(6) (Å)	Ion	Ionic radius C.N.(6) (Å)
Al^{3+}	0.530	Ni^{2+}	0.70
Ge^{4+}	0.540	Hf^{4+}	0.71
Co^{3+} (LS)	*0.525*	Mg^{2+}	0.720
Fe^{3+} (LS)	*0.55*	Zr^{4+}	0.72
Ti^{4+}	0.605	*Cu^{2+}*	*0.73*
Co^{3+} (HS)	0.61	Sc^{3+}	0.730
Sb^{5+}	0.61	Co^{2+} (HS)	0.735
Fe^{2+} (LS)	0.61	*Li^{+}*	*0.74*
Cr^{3+}	0.615	*Zn^{2+}*	*0.745*
Rh^{4+}	*0.615*	Fe^{2+} (HS)	0.770
Ga^{3+}	0.62	In^{3+}	0.790
Ru^{4+}	0.620	Mn^{2+}(HS)	0.820
Nb^{5+}	*0.64*	Lu^{3+}	0.848
Ta^{5+}	*0.64*	Yb^{3+}	0.858
V^{3+}	*0.64*	Tm^{3+}	0.869
Fe^{3+} (HS)	0.645	Er^{3+}	0.881
Co^{2+} (LS)	0.65	Y^{3+}	0.892
Mn^{3+}	0.65	Ho^{3+}	0.894
Rh^{3+}	*0.665*	Dy^{3+}	0.908
Mn^{2+} (LS)	*0.67*	Tb^{3+}	0.923
Sn^{4+}	0.69	Gd^{3+}	0.938

replace Fe^{3+} in the tetrahedral site will also do so in the octahedral site (cf. tables 4 and 5).

The dodecahedral site is, of course, the largest of the three cation sites with the average distance to the oxygen ions being about 2.40 Å. Accordingly, a very large selection of ions, including all of the rare earths and alkaline earths, will occupy this site (table 6). Some other cations which occupy the dodecahedral site in some garnets might also do so in YIG (table 6, in italics).

The substitution of various ions for Fe^{3+} on the tetrahedral and octahedral sites and for Y^{3+} on the dodecahedral site has been considered above mainly from the stereochemical point of view. In some instances the preference of an ion for one site is not very strong so that it may occur in two sites at one time. Two good examples of this ambivalence are Ga^{3+} and Al^{3+} in YIG where these ions occupy both tetrahedral and octahedral sites simultaneously. This occurrence is not surprising as $Y_3Ga_2Ga_3O_{12}$ and $Y_3Al_2Al_3O_{12}$ exist. Mn^{2+} and Fe^{2+} show the tendency to occupy the octahedral and dodecahedral sites at the same time. These tendencies for some ions to occupy two different sites simultaneously and their consequences will be taken up in the next sections.

TABLE 6

Cations which are known to substitute for Y^{3+} on the dodecahedral site (c) in yttrium–iron garnet and (*in italics*) those presumably able but not yet demonstrated to do so (Geller 1967). The radii are from Shannon and Prewitt (1969)

Ion	Ionic radius C.N.(8) (Å)	Ion	Ionic radius C.N.(8) (Å)
Hf⁴⁺	*0.83*	Sm³⁺	1.09
Zr⁴⁺	*0.84*	Bi³⁺	1.11
Mn²⁺	0.93	Ca²⁺	1.12
Lu³⁺	0.97	Nd³⁺	1.12
Yb³⁺	0.98	Ce³⁺	1.14
Tm³⁺	0.99	Pr³⁺	1.14
Er³⁺	1.00	Na⁺	1.16
Y³⁺	1.015	La³⁺	1.18
Ho³⁺	1.02	Sr²⁺	1.25
Dy³⁺	1.03	Pb²⁺	1.29
Gd³⁺	1.06	*Co²⁺*	–
Cd²⁺	1.07	*Cu²⁺*	–
Eu³⁺	1.07	Fe²⁺	–

c. Site selectivity

As was mentioned in section 2.1.4a, in the growth of a garnet crystal from a flux, for example, there will be site selectivity present not only with respect to site size and coordination but also with respect to the aspect presented by a given site in the growth face. The selectivity which arises purely with respect to the site size and symmetry was first reported by Gilleo and Geller (1958) when it was observed that Ga^{3+} and Al^{3+} showed a strong preference for tetrahedral sites while In^{3+}, Sc^{3+}, and Cr^{3+} had a clearly predominant preference for the octahedrally coordinated site. The site preference observed was readily explained on the basis of size alone, except for Cr^{3+} for which the electronic configuration is the determining factor. The statistical theory of Gilleo (1960) which related the Curie temperature and the spontaneous magnetization at 0 K to the fraction of tetrahedral and octahedral Fe^{3+} ions which were replaced by nonmagnetic ions provided a great deal of aid in obtaining an estimate of the site selectivity of substituent ions.

The question of site selectivity of nonmagnetic substituent ions for Fe^{3+} in YIG has been examined and reviewed in considerable detail by Geller et al. (1964a) and Geller (1967). In the case of Al^{3+} it was found that the fraction f_t of the Al^{3+} ions which enter tetrahedral sites decreased smoothly from unity for $x = 0$, where x is defined in the formula unit $Y_3Fe_{5-x}Al_xO_{12}$, to 0.6 for $x = 5$ (fig. 6). The case of Ga^{3+} was found to be similar except that, surprisingly as Ga^{3+} has a larger radius than Al^{3+}, f_t was larger for Ga^{3+} than for Al^{3+} (fig. 7) for $x \leqslant 2.8$.

The spontaneous magnetization at 0 K of $Y_3Fe_{5-x}Ga_xO_{12}$ as a function of x has

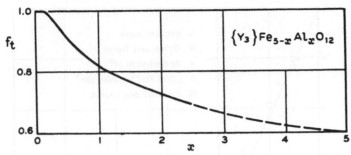

Fig. 6. The fraction f_t of Al^{3+} ions in the tetrahedral site of $Y_3Fe_{5-x}Al_xO_{12}$ vs x (Geller et al. 1964a).

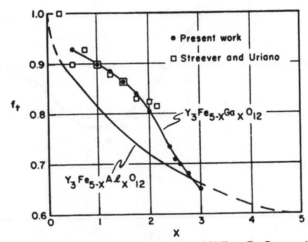

Fig. 7. The fraction f_t of Ga^{3+} ions in the tetrahedral site of $Y_3Fe_{5-x}Ga_xO_{12}$ vs x (Geller et al. 1966a).

been measured with good agreement by five sets of workers (Geller et al. 1966a) (fig. 8). However, this work was all carried out with ceramic specimens which were fired at higher temperatures (Geller et al. 1962). The work of Cohen and Chegwidden (1966) showed that heat treatment of a ceramic (polycrystalline) garnet of the formula $Y_3Al_{1.25}Fe_{3.75}O_{12}$ can cause its magnetization at room temperature to vary over a range of 60%. They concluded that Al^{3+} ions must be redistributed among the octahedral and tetrahedral sites by this heat treatment. With increased temperature of heating prior to quenching, the magnetization of the sample increased so that Al^{3+} ions must be remaining on the octahedral sites in higher concentration. That is, at higher temperatures the Al^{3+} ion distribution becomes more nearly random.

Leo et al. (1966) carried out experiments similar to those of Cohen and Chegwidden (1966) except that Leo et al. used flux-grown single crystals of $Y_3Fe_{5-x}M_xO_{12}$ with $x = 1.26$ for $M = Ga$ and $x = 1.35$ for $M = Al$. They observed

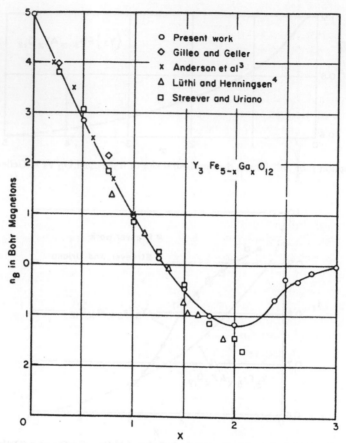

Fig. 8. The magnetic moment n_B in Bohr magnetons per formula unit of $Y_3Fe_{5-x}Ga_xO_{12}$ vs x (Geller et al. 1966a).

that quenching these materials from 1275°C instead of slowly cooling them from 1000°C resulted in an increase of spontaneous magnetization in both cases while for $M = Al$ there was an increase in T_C and for $M = Ga$ there was a decrease in T_C. It was known that in both cases the values of f_t became larger with slow cooling under which circumstance $f_t = 0.89$ for $M = Ga$ and $x = 1.26$ (Streever and Uriano 1965), and $f_t = 0.75$ for $M = Al$ and $x = 1.35$ (Geller et al. 1964a). With the aid of the theory of Gilleo (1960), Leo et al. (1966) were able to show that the like changes in spontaneous magnetization and the opposite changes in T_C which were observed as a consequence of quenching were consistent with the known values of f_t and changes in them resulting from different heat treatment.

The activation energy for the redistribution of Ga^{3+} ions among the octahedral and tetrahedral sites when Ga^{3+} has been substituted for Fe^{3+} has been estimated to be about 1 eV by Kurtzig and Dixon (1972) for both bulk crystals and films grown from a PbO flux. A similar result had been obtained by Euler et al. (1967) for both Al- and Ga-substituted YIG single crystals except that the activation energy for Ga motion was about 0.67 eV and for Al about 0.78 eV.

A different sort of site selectivity has been observed in the case of rare earth–iron garnet crystals when during crystal growth more than one rare earth or another type of ion is present which may occupy the dodecahedral sites. The pair-ordering process which would be involved in the site selection by the rare earth ions during growth was first considered by LeCraw et al. (1971) and attempts were made in theoretical interpretation by Rosencwaig and co-workers (1971). However, Callen (1971) was able to give a more successful treatment in which the interaction of the rare earth ions with second-nearest-neighbor tetrahedral iron ions was involved.

Akselrad and Callen (1971) found that uniaxial anisotropy could be achieved in iron garnets which contained only one type of magnetic rare earth ion or which contained nonmagnetic ions in the dodecahedral site. In these cases, however, there was a nonmagnetic substituent ion on the tetrahedral site. A theoretical treatment by them showed that the uniaxial anisotropy observed could be explained by a distortion of the local symmetry around the tetrahedral ions as a consequence of preferential site ordering among the tetrahedral neighbors to this site. According to Akselrad and Callen further distortion of the tetrahedral site could be caused by ordering of dodecahedral ions.

Nevertheless, there remains unambiguous evidence for site selectivity by rare earth ions as was shown by Wolfe et al. (1971) in the case of Nd^{3+}- and Yb^{3+}-doped, flux-grown yttrium–aluminum garnet crystals by means of spin resonance measurements. Heilner and Grodkiewicz (1973) showed that a maximum in uniaxial anisotropy was achieved in flux-grown bulk crystals of $Y_{3-x}Sm_xFe_5O_{12}$ and $Gd_{3-x}Eu_xFe_5O_{12}$ for $x = 1.5$, i.e., for the maximum product of the concentrations of the rare earth ions, which confirms the site selectivity effects for rare earth ions. This effect was further supported by the experiments of Gyorgy et al. (1974) for $Y_{3-x}Lu_xFe_5O_{12}$ in which case site selectivity by the rare earth ions would only be on a size basis.

The energy required to bring about a random distribution of rare earth ions among the dodecahedral sites in a flux grown crystal appears to be considerably larger than the values of approximately 0.7 to 0.8 eV for Ga^{3+} or Al^{3+}. The growth-induced anisotropy attributable to site selectivity of rare earth ions is little affected by annealing for 16 h at 1150°C in air (Hagedorn et al. 1973) (fig. 9), for example, whereas annealing in a forming gas atmosphere 1 h at 600°C (fig. 10) can lead to almost a 50% reduction in magnetization in a $Eu_2Er_1Fe_{4.3}Ga_{0.7}O_{12}$ composition.

An important difference in these two cases is, however, that the annealing out of the rare earth ion order is essentially irreversible since the ordering achieved with growth temperatures which are ~1000°C can not be retrieved by annealing at a temperature as low as the growth temperature as the rate of ordering there is far too slow.

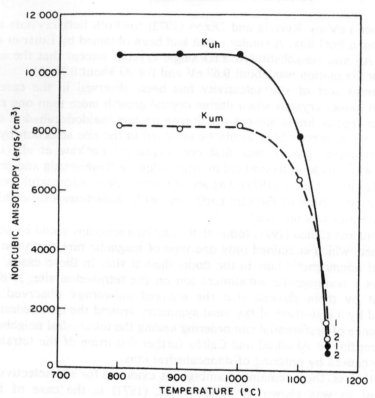

Fig. 9. Noncubic anisotropy as a function of annealing temperature for garnet composition $Y_{1.9}Eu_{0.2}Gd_{0.5}Tb_{0.4}Al_{0.6}Fe_{4.4}O_{12}$. The annealing period at each temperature was 6 h, except at 1150°C at which two 16 h anneals were applied (Hagedorn et al. 1973).

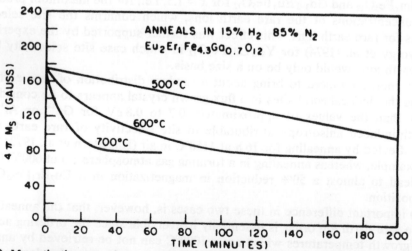

Fig. 10. Magnetization vs time of anneal at three different annealing temperatures for $Eu_2Er_1Fe_{4.3}Ga_{0.7}O_{12}$ in 15%H_2 + 85%N_2 (Kurtzig and Dixon 1972).

d. Solid solutions

Previously the treatment of substitutions in YIG has been on the basis of one ion for another. However, in many cases it is advantageous to consider the problem from the point of view of a solid solution of a garnet of one composition with another. The advantage of this approach is greatest when the end members of the solid-solution system exist. Perhaps the most familiar and most thoroughly studied solid-solution systems are $Y_3Fe_2Fe_3O_{12}$–$Y_3Al_2Al_3O_{12}$ and $Y_3Fe_2Fe_3O_{12}$–$Y_3Ga_2Ga_3O_{12}$. In these instances iron can be entirely replaced by Al^{3+} or Ga^{3+}, respectively. In both cases the lattice constant decreases approximately linearly as the proportion of YIG decreases. However, for $Y_3Fe_{5-x}Ga_xO_{12}$ the lattice constant for $0 < x < 5$ is always larger than would be calculated by a linear interpolation between the lattice constants of the end members (fig. 11). For $Y_3Fe_{5-x}Al_xO_{12}$ the lattice constant again is greater than that given by linear interpolation, though for $x \leqslant 2$ the variation is quite accurately linear (fig. 12).

There are other examples of solid solutions in which the end members exist but for which all Fe^{3+} may not be replaced. One interesting early example is $(1-y)Y_3Fe_2Fe_3O_{12} \cdot yCa_3Fe_2Sn_3O_{12}$ reported by Geller et al. (1959). This system is interesting in that for small values of y most of the octahedral iron and part of the tetrahedral iron are replaced with Sn. Again, the lattice constant is always greater than that obtained by linear interpolation, though in this case the lattice constant for the end member with $y = 1$ is greater than that of YIG.

Two other cases were reported by Geller et al. (1964a) which behave differently in that the replacement of iron is almost wholly on the tetrahedral site. These cases are $(1-y)Y_3Fe_2Fe_3O_{12} \cdot yCa_3Fe_2Si_3O_{12}$ and $(1-y)Y_3Fe_2Fe_3O_{12} \cdot yCa_3Fe_2Ge_3O_{12}$. In both cases the lattice constant is always larger than that obtained by linear interpolation (fig. 13). The instance of $(1-y)Y_3Fe_2Fe_3O_{12} \cdot yCa_3Fe_2Ge_3O_{12}$ is interesting in that the lattice constant has decreased only about 0.004 Å for $y = \frac{1}{3}$ and about 0.056 Å for $y = 1$.

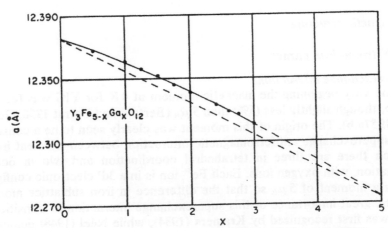

Fig. 11. Lattice constant a vs x for $Y_3Fe_{5-x}Ga_xO_{12}$ (Geller et al. 1966a).

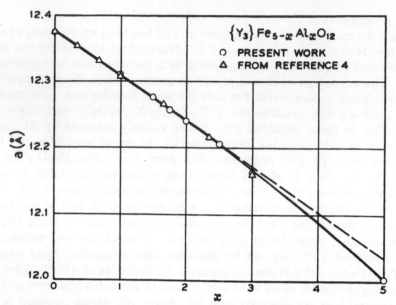

Fig. 12. Lattice constant a vs x for $Y_3Fe_{5-x}Al_xO_{12}$ (Geller et al. 1964a).

Of course, solid solutions of one rare earth–iron garnet in another proceed readily when the end members exist, as they do for all rare earths from Sm to higher atomic numbers plus Y. The rare earth–aluminum garnets also exist for Gd and higher atomic numbers plus Y while the rare earth–gallium garnets exist for Pr and higher atomic numbers plus Y (e.g., Geller 1967).

However, despite the absence of end member iron garnets, the lighter rare earths all the way to La may be substituted to some extent for Y in YIG (Geller 1960, Aharoni and Schieber 1961).

2.2. Magnetic structure

2.2.1. Yttrium–iron garnet

a. Superexchange interactions

From the very beginning the magnetic moment at 0 K for YIG was found to be close to, though slightly less (5%) than 5 μ_B (Bertaut and Forrat 1956, Geller and Gilleo 1957a,b). The origin of this moment was clearly seen to be a consequence of the superexchange antiferromagnetic interaction between trivalent iron ions, of which there are three in tetrahedral coordination and two in octahedral coordination, with oxygen ions. Each Fe^{3+} ion is in a $3d^5$ electronic configuration and has a moment of 5 μ_B so that the difference in iron sublattice moments is 5 μ_B. The great importance of this superexchange interaction in transition-metal oxides was first recognized by Kramers (1934), while Néel (1948) demonstrated that ferrimagnetism results from this superexchange interaction.

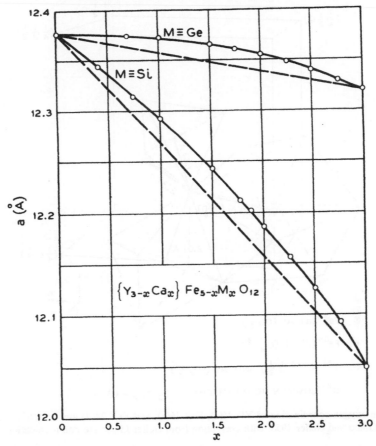

Fig. 13. Lattice constant a vs x for $\{Y_{3-x}Ca_x\}Fe_{5-x}M_xO_{12}$ where $M =$ Si and Ge (Geller et al. 1964a).

Subsequent work (Gilleo and Geller 1958, Geller et al. 1962) placed the magnetic moment of YIG within a fraction of a percent of $5\,\mu_B$. The ferrimagnetism of YIG in which the octahedral iron moments are antiparallel to the tetrahedral moments was conclusively demonstrated by the neutron diffraction experiments performed by Prince (1957).

The Fe_a^{3+}–O^{2-}–Fe_d^{3+} superexchange linkage geometry is clearly exhibited by fig. 14. The included angle in this linkage is 125.9° (table 7) while the Fe_a^{3+}–O^{2-} distance is 2.01 Å and the Fe_d^{3+}–O^{2-} distance is 1.87 Å (table 3). Although this (a)–(d) linkage is by far the strongest for superexchange interaction, there are one other (a)–(a) linkage and four other (d)–(d) linkages (table 7) which will lead to antiferromagnetic interaction in the (a) and (d) sublattices. These intrasublattice interactions can become important when the tetrahedral or octahedral iron is substantially depleted by nonmagnetic ion substitution.

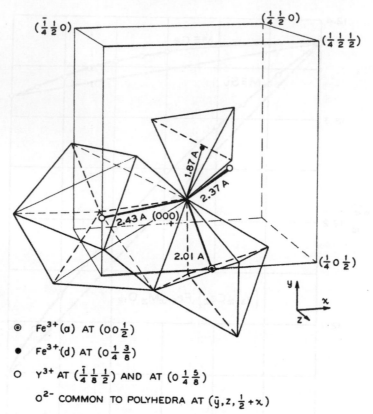

$\odot$ $Fe^{3+}(a)$ AT $(00\frac{1}{2})$

$\bullet$ $Fe^{3+}(d)$ AT $(0\frac{1}{4}\frac{3}{8})$

$\circ$ Y^{3+} AT $(\bar{\frac{1}{4}}\frac{1}{2}\frac{1}{2})$ AND AT $(0\frac{1}{4}\frac{5}{8})$

O^{2-} COMMON TO POLYHEDRA AT $(\bar{y},z,\frac{1}{2}+x)$

Fig. 14. The oxygen coordination of Y^{3+} 24(c), Fe^{3+} 16(a), and Fe^{3+} 24(d) ions in yttrium–iron garnet (Gilleo and Geller 1958, with correction from Geller in private communication).

TABLE 7

Angles in yttrium–oxygen–iron and iron–oxygen–iron linkages in yttrium–iron garnet (Geller and Gilleo 1957b, with corrections from Geller 1977)

Ions	Angles (°)	Ions	Angles (°)
$Fe^{3+}(a)-O^{2-}-Fe^{3+}(d)$	125.9	$Fe^{3+}(a)-O^{2-}-Fe^{3+}(a)$ (4.41) *	147.2
$Fe^{3+}(a)-O^{2-}-Y^{3+}$ **	102.8	$Fe^{3+}(d)-O^{2-}-Fe^{3+}(d)$ (3.41)	86.6
$Fe^{3+}(a)-O^{2-}-Y^{3+}$ †	104.7	$Fe^{3+}(d)-O^{2-}-Fe^{3+}(d)$ (3.68)	78.8
$Fe^{3+}(d)-O^{2-}-Y^{3+}$ **	123.0	$Fe^{3+}(d)-O^{2-}-Fe^{3+}(d)$ (3.83)	74.7
$Fe^{3+}(d)-O^{2-}-Y^{3+}$ †	92.2	$Fe^{3+}(d)-O^{2-}-Fe^{3+}(d)$ (3.83)	74.6
$Y^{3+}-O^{2-}-Y^{3+}$	104.7		

*Numbers in parentheses are the longer Fe^{3+}(a or d)–O^{2-} distances. The shorter distances (table 3) are Fe^{3+}(a)–O^{2-} = 2.01, Fe^{3+}(d)–O^{2-} = 1.87 Å

**Y^{3+}–O^{2-} distance = 2.43 Å

†Y^{3+}–O^{2-} distance = 2.37 Å

b. Magnetization

The most accurate representation of the magnetization of YIG as a function of temperature has been achieved by Anderson (1964a,b) on the basis of the two-sublattice model of Néel (1948). The data for evaluation of the molecular-field coefficients were obtained by extremely careful measurements made on YIG samples of high purity over the temperature range of 4.2 to 650 K. A similar analysis was later made by Dionne (1970a) for data which had been published by Gilleo and Geller (1958) for YIG which had been made with substitutions of Al^{3+} and Ga^{3+} for Fe_a^{3+} and Fe_d^{3+}, of In^{3+}, and Sc^{3+} for Fe_a^{3+}, and of Mg_a^{2+}, Si_d^{4+} for Fe_a^{3+} and Fe_d^{3+}, respectively (Geller et al. 1964a). Therefore, the analysis and the results will be given in terms of the work of Dionne (1970a) which will later be extended to the case of substituted YIG.

In the Néel theory of ferrimagnetism the temperature dependence of the magnetic moment per mole of each sublattice may be expressed in terms of Brillouin functions

$$M_i(T) = M_i(0)B_{S_i}(x_i)$$

in which the subscript i refers to the ith sublattice. In the garnet-structure ferrimagnets to be considered here, only the octahedral (a) and tetrahedral (d) sublattices matter; these sublattices contain Fe^{3+} as the only magnetic ions.

Then the net magnetic moment per mole (formula unit $Y_3Fe_5O_{12}$) is given by

$$M(T) = M_d(T) - M_a(T)$$

where

$$M_d(T) = M_d(0)B_{S_d}(x_d) \qquad M_a(T) = M_a(0)B_{S_a}(x_a).$$

The Brillouin functions are

$$B_{S_d} = [(2S_d + 1)/2S_d] \coth [(2S_d + 1)/2S_d]x_a - (2S_d)^{-1} \coth (2S_d)^{-1}x_d$$

$$B_{S_a} = [(2S_a + 1)/2S_a] \coth [(2S_a + 1)/2S_a]x_a - (2S_a)^{-1} \coth (2S_a)^{-1}x_a$$

where

$$x_d = (S_d g \mu_B/kT)(N_{dd}M_d + N_{da}M_a) \qquad x_a = (S_a g \mu_B/kT)(N_{ad}M_d + N_{aa}M_a).$$

In these equations N_{aa}, N_{dd}, and $N_{ad} = N_{da}$ are the molecular-field coefficients, S_a and S_d are the spin quantum numbers ($\frac{5}{2}$ for Fe^{3+} $3d^5$), g, the spectroscopic splitting factor, is equal to 2, μ_B is the Bohr magneton, and k is the Boltzmann constant.

At absolute zero of temperature the moments per mole for the two sublattices are

$$M_d(0) = 3g_{J_d}J_d\mu_B N \qquad M_a(0) = 2g_{J_a}J_a\mu_B N$$

in which N is Avogadro's number while the factors 3 for (d) and 2 for (a) are the multiplicity of Fe^{3+} ions per formula unit in the respective sites; g_{J_a} and g_{J_d} are the spectroscopic splitting factors and are equal to 2 for Fe^{3+}; J_a and J_d are the angular momenta which for Fe^{3+} are equal to S_a and S_d, respectively.

When the data of Anderson (1964a,b), of Gilleo and Geller (1958), and of Geller et al. (1964a) were fitted as well as possible, Dionne (1970a) found $N_{ad} = 97.0$ mole cm^{-3}, $N_{aa} = -65.0$ mole cm^{-3}, and $N_{dd} = -30.4$ mole cm^{-3}.

For YIG the sublattice magnetizations at absolute zero are (in emu mole^{-1})

$$M_d(0) = 8.374 \times 10^4 \qquad M_a(0) = 5.583 \times 10^4.$$

For convenience, the molecular fields are (in gauss)

$$H_d = N_{dd}M_d(T) + N_{da}M_a(T) \qquad H_a = N_{ad}M_d(T) + N_{aa}M_a(T).$$

The equations for the magnetization of YIG as a function of temperature then are

$$M(T) = M_d(T) - M_a(T)$$
$$M_d(T) = 8.374 \times 10^4 B_{5/2}(3.358 \times 10^{-4}/T)H_d$$
$$M_a(T) = 5.583 \times 10^4 B_{5/2}(3.358 \times 10^{-4}/T)H_a$$
$$H_d = -30.4\, M_d(T) + 97.0\, M_a(T)$$
$$H_a = 97.0\, M_d(T) - 65.0\, M_a(T).$$

The fit obtained by Anderson (1964a,b) is accurate to a few tenths of a percent except near the Curie point (fig. 15). The temperature dependences of the

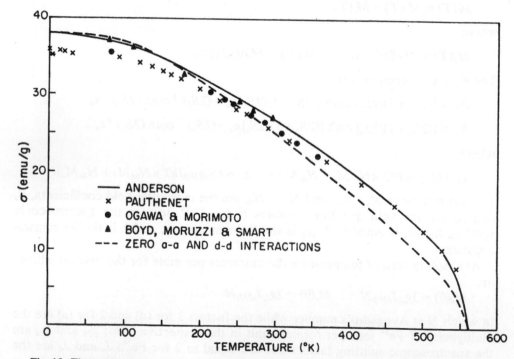

Fig. 15. The net magnetic moment σ(emu/g) for $Y_3Fe_5O_{12}$ vs temperature T (Anderson 1964a).

magnetization of the octahedral M_a iron sublattice (fig. 16) and the tetrahedral M_d sublattice (fig. 17) are not quite alike with M_a decreasing less rapidly with increasing temperature than M_d in the region $100\,K < T < 500\,K$ so that the net magnetization M decreases less rapidly than M_a or M_d in that temperature region. The calculated data for M_a and M_d (figs. 16 and 17) are compared with NMR data. Good agreement is obtained between the two sets of data below about 430 K, as might be expected on the basis of the suggestion by Sternheimer (1952) that the spin of the aligned $3d^5$ electrons of Fe^{3+} polarizes the core electrons to produce a net unpaired spin density at the nucleus.

Although the data of Anderson (1964a) for YIG are well fitted by the molecular-field model with his coefficients, it may be noted that the experimental points given for other authors (figs. 15–17) seem not to agree as well. Part of the reason for this disparity is that Anderson chose data from authors who also had made molecular-field calculations of sublattice magnetization (Pauthenet 1958b) or who had obtained the sublattice magnetizations from nuclear magnetic resonance measurements (Ogawa and Morimoto 1962, Boyd et al. (1963). However, agreement of the magnetization data of Anderson (1964a) with, for example, that of Gilleo and Geller (1958) on single crystals of YIG is within 1% at 0 K and 2% at 300 K.

The model of Dionne (1970a) will later be extended to YIG in which very substantial portions of Fe_a^{3+} and Fe_d^{3+} have been replaced by nonmagnetic ions.

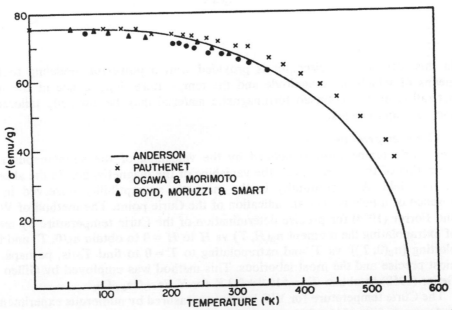

Fig. 16. The magnetic moment σ(emu/g) for the octahedral iron $[Fe_a^{3+}]$ in $\{Y_3\}[Fe_2](Fe_3)O_{12}$ (Anderson 1964a).

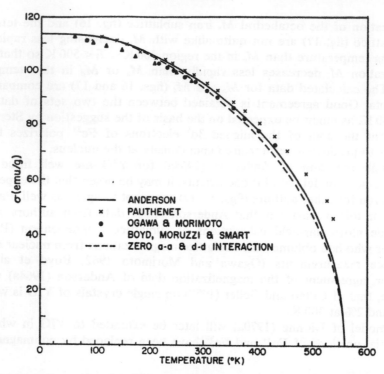

Fig. 17. The magnetic moment σ(emu/g) for the tetrahedral iron (Fe_3^{3+}) in $\{Y_3\}[Fe_2](Fe_3)O_{12}$ (Anderson 1964a).

In this way the experimenter is provided with a powerful modeling tool by means of which the magnitude and the temperature dependence of the magnetization of a YIG-based ferrimagnetic material may be carefully tailored to specific requirements.

c. Curie temperature

The Curie temperature is defined by the vanishing of the spontaneous magnetization of the material, i.e., the vanishing of the magnetization in the absence of any field. A discontinuity in the magnetic susceptibility observed in the absence of a field is also an indication of the Curie point. The method of Weiss and Forrer (1924) for precise determination of the Curie temperature by means of extrapolating the moment $n_B(H, T)$ vs H to $H = 0$ to obtain $n_B(0, T)$ and then plotting $[n_B(0, T)]^2$ vs T and extrapolating to $T = 0$ to find T_C is, perhaps, the most precise and the most laborious. This method was employed by Gilleo and Geller (1958) to obtain $T_C = 519.0 \pm 0.3$ K for $Y_3Fe_{4.75}Ga_{0.25}O_{12}$.

The Curie temperature for YIG has been measured by numerous experimenters and ranges from 545 ± 3 K reported by Gilleo and Geller (1958) to 560 K by Pauthenet (1958a). However, as the molecular-field coefficients have been obtained by Anderson (1964a,b) from a set of data very carefully measured, his $T_C = 559$ K will be taken as the best value for YIG.

The value of the Curie temperature of antiferromagnetic ($Fe_a^{3+} = Fe_b^{3+}$) and ferrimagnetic ($Fe_a^{3+} \neq Fe_b^{3+}$) oxides, where subscripts a and b refer to sublattices based on Fe^{3+} as the only magnetic ion, has been shown by Gilleo (1958) to depend on the number n of Fe_a^{3+}–O^{2-}–Fe_b^{3+} linkages per formula unit. In YIG there are $n = 24/5$ such linkages per formula unit $Y_3Fe_2Fe_3O_{12}$ so that $T_C/n = 559/(24/5) = 116.5$ K which is nearly equal to an average value of 115 K obtained for eight different oxides of Fe^{3+} by Gilleo (1958).

As will later be shown, this counting of the number of antiferromagnetic superexchange interaction linkages can provide by means of a statistical model quite an accurate estimate of the Curie temperature for YIG in which appreciable portions of Fe_a^{3+} and Fe_d^{3+} have been replaced by nonmagnetic ions. In this way it is also possible to estimate the composition of a substituted YIG by means of a measured Curie temperature and the saturation magnetization.

2.2.2. Rare earth–iron garnets

a. Rare earth–iron interaction

When a rare earth ion, such as Gd^{3+} through Yb^{3+}, is present in the dodecahedral site instead of a diamagnetic ion, such as Y^{3+} in YIG, it is found that the magnetization as a function of temperature shows compensation points, i.e., temperatures at which the spontaneous magnetization reaches or passes through zero (fig. 18) (Pauthenet 1958b, Bertaut and Pauthenet 1957). The explanation in these cases turns out to be quite simple in that it is only necessary to add

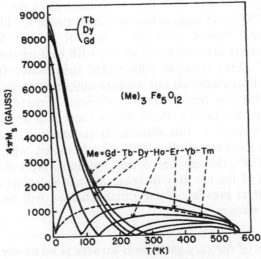

Fig. 18. Magnetization vs temperature for rare earth–iron garnets for rare earths Gd, Tb, Dy, Ho, Er, Tm, and Yb (Bertaut and Pauthenet 1957).

additional negative exchange interaction terms to the molecular-field equations of Dionne (1970a,b, 1976), viz.

$$M(T) = M_d(T) - M_a(T) - M_c(T)$$
$$M_c(T) = M_c(0)B_{S_c}(x_c)$$
$$B_{S_c} = [(2S_c + 1)/2S_c] \coth [(2S_c + 1)/2S_c]x_c - (2S_c)^{-1} \coth (2S_c)^{-1}x_c$$

where $M_c(0) = 3g_{J_c}J_c'\mu_B N$ in which J_c' is the effective angular momentum of the rare earth ion in the crystal. However, the equations for x_d and x_a must be expanded to include molecular-field contributions from M_c and an equation for x_c must be added

$$x_d = (S_d g\mu_B/kT)(N_{dd}M_d + N_{da}M_a + N_{dc}M_c)$$
$$x_a = (S_a g\mu_B/kT)(N_{ad}M_d + N_{aa}M_a + N_{ac}M_c)$$
$$x_c = (S_c g\mu_B/kT)(N_{cd}M_d + N_{ca}M_a + N_{cc}M_c).$$

The complexity of these equations is considerably greater and would present a substantial problem. However, for almost all practical purposes, the effect of M_c on the M_d and M_a through N_{cd} and N_{ca} is negligible as was shown by the work of Litster and Benedek (1966) who employed NMR techniques. Moreover, only N_{cd} proves to be important and it is an order of magnitude smaller than N_{ad}, while N_{ca} is smaller than N_{cd} though of opposite sign and N_{cc} is an order of magnitude smaller than N_{cd}. The modeling of Dionne (1970a,b, 1971, 1976) and of Brandle and Blank (1976) with the above eqs. will be presented in section 2.2.3b together with the molecular-field coefficients for their models.

The size and variation of N_{cd} among the rare earth–iron garnets have been made apparent by Pauthenet (1958b) whose data have been reworked by Wolf and Van Vleck (1960) and augmented by others (table 8). The magnitude of the exchange field for a typical exchange energy of 20 K is 3×10^5 Oe. The conclusion is that there is little variation in N_{cd} with rare earths in the sequence of Gd^{3+} through Yb^{3+}. Data obtained with partial substitution for Y by lighter rare earths in yttrium–iron garnet do not indicate any great variation of N_{cd} from the values obtained for the heavier rare earth–iron garnets. Unfortunately, iron garnets do not exist for La, Ce, Pr, or Nd as these ions are too large; Pm garnet has not been attempted as this element is radioactive and scarce as well as presumably being too large (Geller 1967). Furthermore, with trivalent rare earth ions lighter than Gd^{3+}, the spin and orbital moments are not additive so that the magnetic moment of the ion may increase the net magnetic moment of the iron garnet in which it is present. These circumstances will be discussed in more detail in the next section.

b. Magnetization

The magnetization of the rare earth–iron garnets is relatively simply understood for the heavier ions from Gd^{3+} through Yb^{3+}; for Lu^{3+} there is little difference from YIG. However, for the lighter rare earths the problems are quite different,

TABLE 8

Exchange energy $H_c\mu_B/k$ (K) at the dodecahedral site in rare earth–iron garnets

Rare earth ion	Source of $H_c\mu_B/k$	
	Room temp. data[c] (K)	Comp.temp. data[c] (K)
Eu^{3+}	22[a]	–
Gd^{3+}	25	25
Tb^{3+}	23	20
Dy^{3+}	22	22
Ho^{3+}	21	16
Er^{3+}	19	16
Tm^{3+}	19[b]	14
Yb^{3+}	20	–

[a]Wolf and van Vleck (1960) with the data of Pauthenet (1958b) found 24 K, while with the more precise data of Geller et al. (1963a) the value of 22 K is found
[b]Caspari et al. (1964)
[c]Pauthenet (1958b)

particularly for Sm and Eu iron garnets which exist. For the rare earths lighter than Sm, the garnets do not exist so that the magnetic properties of the end member garnets in substitutional systems such as (Nd, Y) or (Nd, Lu) etc.–iron garnets are hypothetical and must be obtained by extrapolation. Therefore, the heavier rare earth–iron garnets will be considered first. After that, the lighter rare earth–iron garnets will be considered case by case.

The simplicity of the behavior of the rare earth–iron garnets for Gd^{3+} through Yb^{3+} derives from Hund's rule which requires that $J = L + S$ when the electron shell is half-filled or more, as it is in this sequence. Since the exchange field acts on the spin and since the spin and orbital contributions to the moments are parallel in this series, it turns out that in all cases $M_c(0) > M_d(0) - M_a(0)$ so that at some temperature T_c, $0 < T_c < T_C$, there must be a compensation point, i.e., a temperature at which $M(T_c) = 0$.

Moreover, the moments of the rare earth ions in this series are quite independent of temperature as any excited states are well above the ground state. However, the moment of the rare earth ion in the dodecahedral site differs appreciably from that of a free ion as a consequence of crystalline field and magnetic exchange field effects (White and Andelin 1959). The calculations of White and Andelin are in quite good agreement with the values subsequently determined by Guillot and Pauthenet (1964) from susceptibility measurements on rare earth–gallium garnets and by Pauthenet (1958b), Villers and Loriers (1958), Geller et al. (1961, 1963a, 1965) from magnetization data for rare earth–iron garnets (table 9).

TABLE 9

Rare earth ions Pr³⁺ through Yb³⁺ and their calculated and observed moments together with values of L, S, J, g and gJ

	Pr³⁺	Nd³⁺	Pm³⁺	Sm³⁺	Eu³⁺[b]	Gd³⁺	Tb³⁺	Dy³⁺	Ho³⁺	Er³⁺	Tm³⁺	Yb³⁺
L	5	6	6	5	3	0	3	5	6	6	5	3
S	1	3/2	2	5/2	3	7/2	3	5/2	2	3/2	1	1/2
J	4	9/2	4	5/2	0	7/2	6	15/2	8	15/2	6	7/2
g	4/5	8/11	3/5	2/7	–	2	3/2	4/3	5/4	6/5	7/6	8/7
gJ	16/5	36/11	12/5	5/7	–	7	9	10	10	9	7	4
M$_z$[111][a]$_{calc.}$ (μ_B)	–	–	–	–0.43	–	7	8.4	10	9.5	8.3	5.9	2.2
M$_z$[110][a]$_{calc.}$ (μ_B)	–	–	–	0.14	–	7	5.8	6.4	7.2	6.3	4.2	2.1
M[b] (μ_B)	–	–	–	–	–	6.8	7.93	6.8	6.8	5.1	–	1.7
M$_{meas.}$ (μ_B)	1.6[c]	~1.2[c]	–	0.14[c]	0.74[c]	7.00[e]	7.73[e]	7.30[d,f]	6.73[e]	5.07[e]	2.07[e]	1.69[e,f]
M[g]$_{meas.}$ (μ_B)	–	–	–	–	–	6.72	6.90	7.08	6.25	5.52	2.00	1.67
M[b]$_{meas.}$ (μ_B)	–	–	–	–	–	6.92	7.75	7.28	6.30	4.87	–	–

[a]White and Andelin (1959)
[b]Guillot and Pauthenet (1964)
[c]Geller et al. (1963a)
[d]Geller et al. (1961)

[e]Geller et al. (1965)
[f]Clark and Callen (1968)
[g]Pauthenet (1958b)
[h]Villers and Loriers (1958)

[i]Ground state values applicable only at 0 K are given; nearby excited states become important at higher temperatures for Sm³⁺ and Eu³⁺ as discussed in section 2.2.2b

An independent indication of the exchange field at the rare earth nucleus has been made by using the $\gamma-\gamma$ angular correlation method which is feasible in a few cases by use of a radioactively decaying rare earth nucleus (Caspari et al. 1960, 1962).

With the lighter rare earth elements, those with 4f shells less than half-filled, the orbital angular momentum is antiparallel to the spin, i.e., $J = |L - S|$, while the exchange field acts only on the spin. Further, of course, there is the influence of lattice interaction on the orbital moment. The rare earth ions with $L > 2S$ might be expected to contribute a moment additively to that of the iron system. Such a contribution does occur in the case of all of these lighter rare earth ions, except for Eu^{3+} and for Sm^{3+} at higher temperatures. The ions Eu^{3+} and Sm^{3+} are cases which require special attention.

Magnetization data for europium–iron garnet (Pauthenet 1958a, Geller et al. 1963a,b) show that at 0 K the moment of Eu^{3+} is antiparallel to that of the iron system and is $0.73 \mu_B$ at high concentration – for $x > 2.5$ in, e.g., $\{Eu_xCa_{3-x}\}$ $[Fe_2](Fe_xSi_{3-x})O_{12}$. The value of Geller et al. (1963a) for the moment of Eu^{3+} at 0 K has been chosen as more precise; the choice does not affect the following discussion. At lower concentrations ($x < 2.5$) the Eu^{3+} moment decreases becoming as low as $0.2 \mu_B$ for $x = 1.2$, probably as a consequence of the reduced exchange field from Fe_d^{3+} ions (Geller et al. 1966b).

The explanation of the magnetic moment of Eu^{3+} was given by Wolf and Van Vleck (1960). The ground state of Eu^{3+} has $J = 0$. Even so, Eu^{3+} is paramagnetic as a consequence of the induced moment associated with the second-order Zeeman effect. Furthermore, the $J = 1$ state is only about 480 K above the ground state for a free ion so that the magnetic contribution of this state can not be ignored at higher temperatures. The calculations of Wolf and Van Vleck which took these matters into account were in excellent agreement with the data of Pauthenet (1958a,b) and yielded an exchange energy parameter of 24 K. With the more precise data of Geller et al. (1963a), an exchange energy parameter of 22 K is obtained (cf. table 8).

A very detailed determination of the sublattice magnetizations of europium–iron garnet by NMR and Mössbauer techniques has been made by Myers et al. (1968). It was shown that $|N_{ac}/N_{ad}| < 0.03$. Atzmony et al. (1968) by Mössbauer studies were able to show that $|N_{cc}/N_{ad}| \simeq 0.006$.

A unique consequence of the $J = 0$ electronic configuration of Eu^{3+} is that a magnetic body may be produced with a $g \to \pm \infty$ at 0 K or with very large $|g|$ values at room temperature by adjusting the moment of the iron system to pass through zero with the introduction of nonmagnetic substituents for iron in the correct amount (LeCraw et al. 1965) (See further discussion in section 2.2.3d).

The case of samarium–iron garnet is also a special one involving a second-order Zeeman effect for the susceptibility of Sm^{3+} (White and Van Vleck 1961). The free Sm^{3+} ion has $L = 5$ and $S = \frac{5}{2}$ so that $L = 2S$. With the $4f^5$ configuration of Sm^{3+} the 4f shell is less than half filled and $J = L - S = \frac{5}{2}$ according to Hund's rule while $L - 2S = 0$ at 0 K. The exchange field acts only on the spin so the net moment of the ion adds to or subtracts from that of the iron depending on whether

$L > 2S$ or $< 2S$. In addition, the lowest multiplet component of Sm^{3+} must be taken into account as it is about 1500 K above the ground state. On the basis of these considerations, White and Van Vleck (1961) estimated that the net moment of Sm^{3+} would cross over from orbital dominance at low temperature, at which the Sm^{3+} moment adds to the net iron moment, to spin dominance at about 300 K above which temperature the Sm^{3+} moment subtracts from that of iron. However, a sign reversal of the rare earth contribution to the Knight shift in $SmAl_2$ was observed (Jaccarino et al. 1960) to take place at 150 K.

A detailed analysis of the magnetization vs temperature data of several authors was carried out by Van Loef (1968) with the conclusion that the cross-over of the contribution of the magnetic moment of samarium in samarium–iron garnet takes place at 150 ± 15 K. Below the cross-over point, the Sm^{3+} moment adds to the net moment of the iron system at 0 K about 0.15 μ_B per Sm^{3+} ion (Van Loef 1968, Geller et al. 1963a) which is 0.20 of the moment of a free ion (Nowik and Ofer 1968, White and Andelin 1959). The cross-over temperature for the Sm^{3+} in samarium iron garnet appears to be about 230 K from a curve published by Nowik and Ofer (1968). The deduction of Geller et al. (1963a) that the cross-over would occur at about 330 K is in the best agreement with the calculation of White and Van Vleck (1961) which yields $\frac{5}{24}h\nu_{7/2, 5/2}/k = 312\,K$ where $\nu_{7/2, 5/2}$ is the lowest multiplet separation for which $h\nu_{7/2, 5/2}/k = 1500$ K (Geller 1977).

The substitution of Nd^{3+} for Y^{3+} in YIG is limited because of the larger size of Nd^{3+} to about 1.5 out of 3 (Geller et al. 1961). Therefore, while this is not the first case considered of a rare earth ion with $L > J$, it is the first case without the complications present for Eu^{3+} with $J = 0$ and Sm^{3+} with $L - 2S = 0$, so that it is unfortunate that a pure neodymium–iron garnet does not exist. However, the data of Geller et al. (1961) show that the moment of Nd^{3+} adds to that of the iron system in $Y_{3-x}Nd_xFe_2Fe_3O_{12}$ and is about 1.2 μ_B with considerable uncertainty which arises from the very high anisotropy present in iron garnets containing Nd^{3+}. For a free Nd^{3+} ion, $M_z = L - 2S = 3\,\mu_B$ (table 9) so that the effect of crystalline surroundings has appreciably reduced the orbital contribution ($L = 6$) to the Nd^{3+} moment.

The contribution of Pr^{3+} to the magnetic moment of YIG is a case entirely parallel to that of Nd^{3+}. Geller et al. (1961) have shown that by means of substitution for Y^{3+} – a maximum of $x = 1.33$ was found by Espinosa (1962) – that Pr^{3+} adds to the magnetization of YIG in the amount of 1.6 μ_B per ion. Again the anisotropy introduced by Pr^{3+} is very high and $M_z = L - 2S = 3\,\mu_B$ (table 9) so that the orbital contribution ($L = 5$) to the moment has been substantially reduced by the crystalline surroundings.

The substitution of La^{3+} for Y^{3+} is limited to $x \simeq 0.45$. (Loriers and Villers 1961, unpublished work of Gilleo referred to in Geller et al. 1962; Espinosa 1962) because of the very large size of La^{3+} in comparison with Y^{3+}. There is no effect on the magnetic moment at $T = 0\,K$ (Geller et al. 1963a).

c. Curie temperature

The Curie temperatures T_C of rare earth–iron garnets seem to be virtually independent of any variable other than lattice constant while the compensation temperatures T_c decrease with decreasing moment of the rare earth ion (table 10). For example, the expansion of the lattice constant to 12.430 Å from

TABLE 10

The Curie, T_C, and compensation, T_c, temperatures for rare earth–iron garnets

Formula unit	T_C (K)	T_c (K)	Ref.
$Y_3Fe_5O_{12}$	545	None	a,b,c,d)
	548		e)
	550		f)
	552.5		g)
	556		h)
	559		i)
	560		k)
	570		j)
$Sm_3Fe_5O_{12}$	562	None	g)
	565		d)
	578		k)
$Eu_3Fe_5O_{12}$	565	None	l)
	566		k)
$Gd_3Fe_5O_{12}$	562		h)
	564	290	k)
	–	286	m,x)
	–	290	n,o,p)
	566	287	q)
$Tb_3Fe_5O_{12}$	568	246	k)
	–	246	m,o,p)
	–	244	r)
$Dy_3Fe_5O_{12}$	563	220	k)
	–	215	r)
	–	220	o,p)
	–	221	s)
	–	226	m)
	552	226	r)
$Ho_3Fe_5O_{12}$	558	144	w)
	567	136	k)
	–	130	r)
	–	136	o,p)
	–	137	m)
$Er_3Fe_5O_{12}$	556	84	k)
	–	83	m,r)
	–	84	o,p)

TABLE 10 (cont.)

Formula unit	T_C(K)	T_c(K)	Ref.
$Tm_3Fe_5O_{12}$	549	$0 \leqslant T_c \leqslant 20.4$	[k]
	–	$0 \leqslant T_c \leqslant 20.4$	[o]
	–	None observed	[m]
$Yb_3Fe_5O_{12}$	548	$\simeq 0$	[k]
		7.6	[t,u]
$Lu_3Fe_5O_{12}$	549	None	[k]

[a] Geller and Gilleo (1957a)
[b] Geller and Gilleo (1957b)
[c] Gilleo and Geller (1958)
[d] Loriers and Villers (1961)
[e] Ramsey et al. (1962)
[f] Dillon (1958)
[g] Aharoni et al. (1961)
[h] Vassiliev et al. (1961a)
[i] Anderson (1964a)
[j] Cunningham and Anderson (1960)
[k] Pauthenet (1958a)
[l] Miyadai (1960)

[m] Geller et al. (1965)
[n] Maguire and Green (1962)
[o] Harrison et al. (1965)
[p] Villers and Loriers (1958)
[q] Anderson (1964b)
[r] Pauthenet (1958b)
[s] Pearson (1962)
[t] Henderson and White (1961)
[u] Clark and Callen (1968)
[v] Vassiliev et al. (1961b)
[w] Vassiliev et al. (1961c)
[x] Gilleo (quoted in Geschwind and Walker 1959)

12.376 Å by the substitution of $La^{3+}_{0.45}$ for $Y^{3+}_{0.45}$ in YIG causes the Curie temperature to increase by about 2 to 3 K (Loriers and Villers 1961, Aharoni et al. 1961) for a rate change of T_C with lattice constant of 37 to 56 K/Å. The data of Loriers and Villers (1961) for the system $Y_{3-x}Sm_xFe_2Fe_3O_{12}$ for which the lattice constant increases with x show a rate of change of T_C with lattice constant of about 86 K/Å while the data of Villers et al. (1962a) for the system $Y_{3-x}Eu_xFe_2Fe_3O_{12}$ for which the lattice constant increases with x yield a rate of change of T_C of about 87 K/Å. For the system $Lu_{3-x}Y_xFe_2Fe_3O_{12}$ in which the lattice constant increases with x the data of Pauthenet (1958a) show a rate of change of T_C of about 100 K/Å. However, in the system $Y_{3-x}Bi_xFe_2Fe_3O_{12}$ which was examined by Geller et al. (1963b), the rate of change of T_C with lattice constant which increases with x is 460 K/Å, which is vastly greater than in the other cases.

The substitution of Bi^{3+} for Y^{3+} has been shown to have the profound effect of increasing the Faraday rotation of YIG as discovered by Buhrer (1969). Akselrad (1971) was the first to identify correctly the origin of the effect as lying in the $Bi^{3+}_c-O^{2-}-Fe^{3+}_d$ complex. The most likely explanation was given by Wittekoek and Lacklison (1972) who showed that an important effect can result from the mixing of the 6p orbitals of the Bi^{3+} ion, which is rather covalent, with the oxygen orbitals. Since the spin–orbit coupling for the 6p electrons of bismuth is large (about 17 000 cm^{-1} for the free ion (McLay and Crawford 1933)), this mixing will lead to a strong increase in the spin–orbit coupling of a 2p electron in O^{2-}. The main effect observed is associated with a great strengthening of one or two transitions arising from the tetrahedral iron complex. As a consequence, the $Fe_a^{3+}-O^{2-}-Fe_d^{3+}$ interaction might be increased as well.

2.2.3. Substituted magnetic garnets

a. Magnetic-ion interaction in YIG with nonmagnetic substitutions for iron and magnetic rare earths for yttrium

The magnetic garnets thus far discussed have been of the pure type in that each crystallographic site has been wholly occupied by one type of ion. In particular, the tetrahedral and octahedral sites have been entirely occupied by Fe^{3+} ions and the dodecahedral sites by a single nonmagnetic ion (La, Y, Lu, Bi) or by a magnetic rare earth ion.

The behavior of YIG with nonmagnetic ion substitutions in the tetrahedral site, the octahedral site, or partially in both sites has received extensive theoretical treatment of the statistical type (Gilleo 1960, Borghese 1967), of the molecular-field type (Nowik 1968, Dionne 1970a,b, 1976) and of the phenomenological type (Geller et al. 1964a, Geller 1966). While each of these approaches, as typified by the references cited, which are representative rather than exhaustive, has merit, none of them are adequate under all circumstances. The statistical theory of Gilleo (1960) provides the only means to calculate T_C with reasonable accuracy as a function of nonmagnetic substituent concentration while at the same time and on the same basis providing a reasonable estimate of the spontaneous magnetization at 0 K only. The treatment of Borghese (1967) provides an improved estimate of $M(0)$ but not of T_C.

In the theory of Gilleo the basic and unifying assumptions are (i) that T_C is proportional to the number of $Fe_a^{3+}-O^{2-}-Fe_d^{3+}$ linkages present per magnetic ion per formula unit and (ii) that an Fe_a^{3+} or Fe_d^{3+} ion contributes to the magnetic moment and to countable linkages only when it is linked to at least two Fe^{3+} ions in the opposite lattice. With these two assumptions, $M(0)$ and T_C are calculable.

Borghese (1967) makes a slightly different assumption that all Fe^{3+} ions with one or more linkages to an Fe^{3+} ion in the opposite lattice contribute to the magnetic moment. With this assumption good success is achieved for $M(0)$ but no attempt is made to calculate T_C; this calculation on the basis of the results of Gilleo (1960) would not succeed. Explanations for this situation are unknown.

Of the molecular-field type treatments, that of Dionne (1970a,b), while somewhat empirical, provides an excellent model for the magnetization $M(T)$ and for the sublattice magnetizations $M_c(T)$, $M_d(T)$, and $M_a(T)$ as a function of nonmagnetic substituent ion concentrations.

The phenomenological approach of Geller et al. (1964a) and Geller (1966) provides a good basis for estimating $M_d(0)$, $M_a(0)$, and $M(0)$ under all circumstances of nonmagnetic ion substitution for iron. In this treatment a set of curves (fig. 19) is provided (Geller 1966) with which the effective moments of Fe^{3+} in the tetrahedral and octahedral sublattices may be determined on the basis of the nonmagnetic substitution in the opposite lattice. The effective moment of the Fe^{3+} in one lattice begins to decrease for a nonmagnetic ion concentration in the

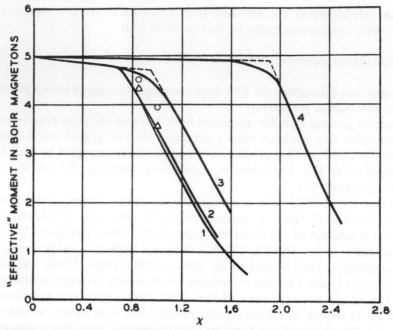

Fig. 19. Effective moment per Fe^{3+} ion in (1) tetrahedral sites for $\{Y_{3-x}Ca_x\}[Fe_{2-x}Zr_x](Fe_3)O_{12}$ system; in (2) tetrahedral sites for $\{Y_3\}[Fe_{2-x}Sc_x](Fe_3)O_{12}$ system; in (3) tetrahedral sites for $\{Y_3\}[Fe_{2-x}Mg_x](Fe_{3-x}Si_x)O_{12}$ system; in (4) octahedral sites for $\{Y_{3-x}Ca_x\}[Fe_2](Fe_{3-x}Si_x)O_{12}$ system. Circles are for $\{Y_{3-x}Ca_x\}[Fe_{2-x}Sc_x](Fe_{3-x}Si_x)O_{12}$ specimens and triangles are for $\{Y_{3-2x}Ca_{2x}\}$ $[Fe_{2-x}Zr_x](Fe_{3-x}Si_x)O_{12}$ specimens (Geller et al. 1964a).

opposite lattice above a certain value. The cause for the decrease in effective moment is the canting of the Fe^{3+} ion moments in the lattice as a consequence of the antiferromagnetic interaction between ions in the same lattice. This tendency is enhanced by a weakening of the antiferromagnetic interaction with the opposite lattice.

The effective moment of Fe^{3+} in (a) sites begins to decrease when more than 1.9 out of 3 Fe^{3+} ions on (d) sites have been replaced. For iron on the (d) sites, however, the decrease in effective moment starts immediately, though slowly, with substitution for Fe^{3+} on (a) sites and quite sharply after $Fe^{3+}_{0.7}$ has been replaced on (a). In the event that Fe^{3+} on (d) and (a) sites is replaced equally, the rapid decrease of the effective moment of Fe^{3+} on these sites does not begin to occur until $Fe^{3+}_{0.95}$ has been replaced on both sites.

The moment which arises from the rare earth ions is generally taken as a simple matter to estimate since virtually the entire exchange field at the (c) site arises from the (d) site Fe^{3+} ions with moment $M_d(T)$. Therefore, in a rare earth garnet of formula $\{RE_{3-x}Y_x\}[Fe_{2-y}M_y](Fe_{3-z}N_z)O_{12}$ the moment M_c would be given by

$$M_c(T) = M_c(0)\{[(2S_c + 1)/2S_c] \coth [(2S_c + 1)/2S_c]x_c - (2S_c)^{-1} \coth (2S_c)^{-1}x_c\}$$
$$x_c \approx (S_c g \mu_B / kT)(N_{cd} M_d + N_{ca} M_a)$$

and $M_c(0) = (3 - x)M_{RE}^{(c)} N$ with $M_{RE}^{(c)}$ being the effective moment in emu ion^{-1} of the RE ions in the (c) site (table 9) and N is Avogadro's number (cf. section 2.2.2a). In some cases the fact that the exchange field acts only on the spin of the rare earth ion (Wolf and Van Vleck 1960) and B_{J_c} rather than B_{S_c} is used (Dionne 1976, 1979).

b. Magnetization

The treatment of Dionne (1970a,b, 1971, 1976) of the magnetization of yttrium–iron garnet was presented in section 2.2.1b. His extension of this treatment to yttrium–iron garnet in which nonmagnetic ions have been substituted for Fe^{3+} ions on the (a) and (d) sites and magnetic RE ions for Y^{3+} on the (c) sites is a simple expansion of the treatment given earlier. A similar treatment with emphasis on the room temperature region was given by Brandle and Blank (1976).

The fractions of the Fe^{3+} ions on the (a) and (d) sites which have been replaced by nonmagnetic ions are k_a and k_d and the fraction of RE ions which has been replaced by Y^{3+} is k_c. Dionne found empirically that

$$M_d(0) = 3gS_d\mu_B N(1 - k_d)(1 - 0.1\ k_a)$$
$$M_a(0) = 2gS_a\mu_B N(1 - k_a)(1 - k_d^{5.4})$$
$$M_c(0) = 3g_{J_c}J'_c\mu_B N(1 - k_c).$$

These equations are valid for k_i values which are small enough that none of the sublattices makes a transition to an antiferromagnetic state. There was no obvious explanation as to why a linear factor $(1 - 0.1\ k_a)$ served in the equation for $M_d(0)$ and the exponential factor $(1 - k_d^{5.4})$ was appropriate in the equation for $M_a(0)$. The symbol g_{J_c} denotes the g value for the ion on the (c) site (table 9).

On the other hand, Gilleo (1960) earlier proposed, on the basis that a magnetic ion should be linked to at least two other magnetic ions in the other lattice in order to participate in ferromagnetism, that

$$M_d(0) = 3gS_d\mu_B N(1 - k_d)(1 - 4k_a^3 + 3k_a^4) \qquad M_a(0) = 2gS_a\mu_B N(1 - k_a)(1 - 6k_d^5 + 5k_d^6)$$

which is closely similar. Again, k_a and k_d may not become too large, say, not greater than 0.7, for these relations to depart seriously from experimental observations. The same assumptions will be shown in section 2.2.3c to provide a good estimate of the dependence of T_C on k_a and k_d.

However, the equations of Gilleo (1960), while reasonably good for small k_a and k_d, do not serve well for larger values. Therefore, the remaining treatment will be that of Dionne.

As a consequence of the substitution of nonmagnetic ions for Fe^{3+}, the molecular-field coefficients are also functions of k_a and k_d while N_{cd} and N_{aa} remain approximately constant and $N_{cc} \approx 0$. The equations for the N_{ij} in units of mole cm^{-3} are

$$N_{dd} = -30.4(1 - 0.87\,k_a) \qquad\qquad N_{cd} = \text{see tables 11 and 12}$$
$$N_{aa} = -65.0(1 - 1.26\,k_d) \qquad\qquad N_{ac} = \text{see tables 11 and 12}$$
$$N_{ad} = 97.0(1 - 0.25\,k_a - 0.38\,k_d) \qquad N_{cc} = \text{see tables 11 and 12}$$

for $k_a \leqslant 0.35$ and $k_d \leqslant 0.65$. An antiferromagnetic transition occurs in both the (a) and (d) sublattices when N_{dd}, $N_{aa} \gtrsim -20$ mole cm^{-3}.

These equations represent a material with a formula unit

$$\{RE_{3-x}Y_x\}[Fe_{2-y}M_y](Fe_{3-z}N_z)O_{12}$$

where $x = 3k_c$, $y = 2k_a$, and $z = 3k_d$.

The relations above provide a very good representation of the magnetizations of the sublattices (a), (d), and (c) for $RE = Gd^{3+}$ to Yb^{3+} as well as of the net magnetization (figs. 20 and 21). The equally good agreement of the case of $Ga^{3+}_{a,d}$ substitution (fig. 22) with data from Gilleo and Geller (1958) and of Mg^{2+}_a, Si^{4+}_d substitution (fig. 23) from the data of Geller et al. (1964a) is also remarkable.

The molecular-field coefficients N_{ac} and N_{cd} and the J'_c and $g_{J_c}J'_c$ values employed by Dionne (1976, 1978) for the rare earth–iron garnets (table 11) emphasize a fit of the calculated saturation magnetization at $T = 0\,K$. In these calculations $B_{J'_c}[x_c(g_{J_c}J'_c)]$ is used instead of $B_{S_c}[x_c(S_c)]$ (Dionne 1979). On the other hand the molecular-field coefficients of Brandle and Blank (1976) have

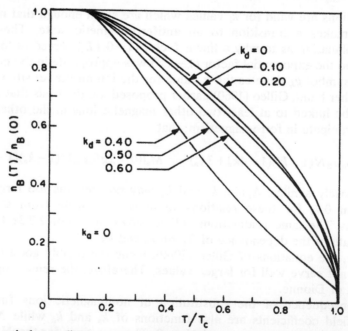

Fig. 20. Theoretical normalized variations with temperature of magnetic moment per formula unit $\{Y_3\}[Fe_2](Fe_{3-x}M_x)O_{12}$ for tetrahedral site substitutions $k_d = \tfrac{1}{3}x$ (Dionne 1970a).

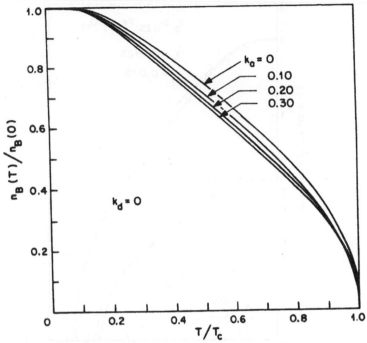

Fig. 21. Theoretical normalized variations with temperature of magnetic moment per formula unit $\{Y_3\}[Fe_{2-x}M_x](Fe_3)O_{12}$ for octahedral site substitutions $k_a = \frac{1}{2}x$ (Dionne 1970a).

been chosen for a good fit of T_c and T_C and the magnetization at room temperature (table 12). In both cases the values of Dionne (1970a,b) for N_{aa}, N_{ad}, N_{dd} and the coefficients for k_a and k_d have been retained. It will also be noted that Brandle and Blank employed the J_c for a free atom and took the effect of the crystalline environment into account by means of a multiplier $\mathcal{M}$ applied to

TABLE 11

Molecular-field coefficients, J'_c, and $g_{J_c}J'_c$ values for rare earth–iron garnets. $N_{cc} = 0$ in all cases (Dionne 1976, 1979)

c-site ion	J'_c	$g_{J_c}J'_c$	N_{ca} (mol cm^{-3})	N_{cd} (mol cm^{-3})
Gd^{3+}	3.5	7.00	-3.44	6.0
Tb^{3+}	4.60	6.90	-4.2	6.5
Dy^{3+}	5.30	7.07	-4.0	6.0
Ho^{3+}	4.98	6.22	-2.1	4.0
Er^{3+}	4.62	5.54	-0.2	2.2
Tm^{3+}	1.085	1.27	-1.0	17.0
Yb^{3+}	1.49	1.70	-4.0	8.0

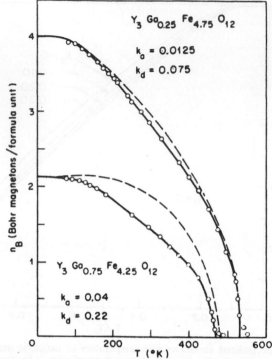

Fig. 22. Comparison of theory (solid lines) and experiment for magnetic moment in Bohr magnetons per formula unit vs temperature for $Y_3Ga_{0.25}Fe_{4.75}O_{12}$ and $Y_3Ga_{0.75}Fe_{4.25}O_{12}$. Experimental data (points) are from Gilleo and Geller (1958). Dashed lines represent the results calculated by using the molecular-field coefficients for $Y_3Fe_5O_{12}$ (Dionne 1970a).

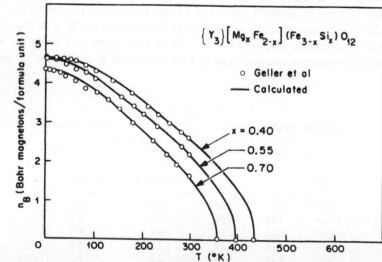

Fig. 23. Comparison of theory (solid lines) with experiment (points) for three compositions $\{Y_3\}$ $[Mg_xFe_{2-x}](Fe_{3-x}Si_x)O_{12}$ which feature substitution in both sublattices according to Geller et al. (1964a) (Dionne 1970a).

TABLE 12

Molecular-field coefficients, $g_{J_c}J_c$ values, and multiplying factor $\mathcal{M}$ for rare earth–iron garnets (Brandle and Blank 1976)

c-site ion	$g_{J_c}J_c$	N_{ca} (mol cm^{-3})	N_{cd} (mol cm^{-3})	N_{cc} (mol cm^{-3})	$\mathcal{M}$
Sm^{3+}	0	0	0	0	0.94
Eu^{3+}	0	0	0	0	0.518
Gd^{3+}	7	−3.44	6.00	0	1
Tb^{3+}	9	−1.80	3.40	0	1
Dy^{3+}	10	−3.35	3.95	0.10	1
Ho^{3+}	10	−0.75	1.50	0.25	1
Er^{3+}	9	−0.75	1.25	0	1
Tm^{3+}	7	−1.00	8.00	0	0.760
Yb^{3+}	4	−1.70	2.00	0	0.967
Lu^{3+}	0	0	0	0	0.862
Y^{3+}	0	0	0	0	1

the overall magnetization calculated. In this way they were able to handle the cases of samarium θ and europium θ iron garnets. Thus their coefficients are of particular value for the calculation of the room-temperature properties of magnetic-bubble-domain films, as was their intent.

The magnetization of the rare earth–iron garnets with Sm^{3+} through Lu^{3+} and Y^{3+} in the dodecahedral site has been determined in detail by numerous authors over temperatures ranging from 4.2 K and below to well above the Curie temperature. The data of most interest are generally for the magnetization in Bohr magnetons per formula unit and in gauss at 0 K and at room temperature or 300 K. These data have been provided separately for reference (table 13) from those which could be obtained by calculation with the constants of tables 11 and 12 as the original data are often of fundamental value. However, there is not complete agreement between the 0 K magnetization for the rare earth–iron garnets among these tables. The data of Dionne (1976) (table 11) were taken from Pauthenet (1958a) and Vassiliev et al. (1961a,b) while those from table 13 are from more recent sources.

c. Curie temperature

The molecular field treatment of the magnetization curves of garnets by Dionne (1970a,b, 1971, 1976) also yields a Curie temperature for the composition in question, though rather laboriously. The method of Gilleo (1960) is quite simple and effective over a range of substitutions approaching that of Dionne. This method, as mentioned earlier in section 2.2.1c, amounts to counting the number of exchange linkages per magnetic ion per formula unit with the assumption that only those Fe^{3+} ions which are linked to two or more Fe^{3+} ions in the other sublattice participate in the ferrimagnetic interaction. On this basis T_C is given by

TABLE 13

Magnetization of rare earth–iron garnets in Bohr magnetons n_B per formula unit ($RE_3Fe_5O_{12}$) at 0 K and in gauss and at 0 and 300 K

Rare earth–iron garnet	n_B(0 K) (μ_B)	$4\pi M$(0 K) (gauss)	$4\pi M$(300 K) (gauss)	Ref.
$Y_3Fe_5O_{12}$	5.00	2462	1758	[a]
$Sm_3Fe_5O_{12}$	5.43	2574	1762	[b]
$Eu_3Fe_5O_{12}$	2.78	1328	1172	[b]
$Gd_3Fe_5O_{12}$	16.00	7691	122	[c]
$Tb_3Fe_5O_{12}$	18.2	8824	331	[c]
$Dy_3Fe_5O_{12}$	16.9	8256	433	[c]
$Ho_3Fe_5O_{12}$	15.2	7479	822	[c]
$Er_3Fe_5O_{12}$	10.2	5053	1154	[c]
$Tm_3Fe_5O_{12}$	1.2	598	1431	[c]
$Yb_3Fe_5O_{12}$	0.069	35	1554	[d]
$Lu_3Fe_5O_{12}$	5.07	2551	1807	[c]

[a] Anderson (1964a, b)
[b] Geller et al. (1963a)
[c] Geller et al. (1965)
[d] Clark and Callen (1968)

$$T_C(k_d, k_a) = \frac{5n(k_d, k_a)}{24N(k_d, k_a)} T_C(0, 0)$$

where

$$n(k_d, k_a) = 24(1 - k_a)(1 - 6k_d^3 + 5k_d^6)(1 - k_d)(1 - 4k_a^3 + 3k_a^4)$$

which is the number of active interactions per O_{12} formula unit and

$$N(k_d, k_a) = 2(1 - k_a)(1 - 6k_d^3 + 5k_d^6) + 3(1 - k_d)(1 - 4k_a^3 + 3k_a^4)$$

is the number of magnetic ions which participate in ferrimagnetism. $T_C(0, 0)$ is the Curie temperature of the unsubstituted ferrimagnetic material which in this case is YIG for which $T_C = 559$ K has been taken (section 2.2.1c).

The introduction of rare earths in the dodecahedral site has little effect on T_C except through change of lattice constant while another nonmagnetic ion, such as Bi^{3+}, may have a much larger effect (section 2.2.1c).

d. The effective g-factor

The ferrimagnetic rare earth–iron garnets, because of their three sublattices of magnetization, provide an interesting question concerning the effective g-factor (g_{eff}) for the net magnetic moment of the system. The first recognition of this problem was by Tsuya (1952) and Wangsness (1953) who derived the formula

$$g_{eff} = (M_{Fe} + M_{RE})/(M_{Fe}/g_{Fe} + M_{RE}/g_{RE}).$$

In this equation M_{Fe}, the net moment of the two Fe^{3+} sublattices, could be treated as a unit since the coupling of these sublattices is very much greater than that of either of them to the rare earth ion sublattice (see section 2.2.2b).

Later Kittel (1959) examined the equations of motion for the magnetic sublattices in the case of a high relaxation frequency for the RE sublattice. Here the term high means in comparison with the exchange frequency of the RE and Fe sublattice interactions and with the relaxation frequency of the Fe sublattices. The relaxation term in this model is of the Landau–Lifshitz form, i.e., proportional to $M \times (M \times H)$. In these respects, Kittel (1959) acknowledged having considered a special case of a very general one Wangsness (1958) had investigated. These conditions obtain at room temperature, except for gadolinium–iron garnet in which the Gd^{3+} ion in the S-state has very small interaction with the lattice. The result was that

$$g_{eff} = (M_{Fe} + M_{RE})/(M_{Fe}/g_{Fe}).$$

In the limit of a low relaxation frequency for M_{RE}, the Tsuya–Wangsness equation for g_{eff} given previously resulted.

Van Vleck (1961) reviewed the derivations of the results for g_{eff} given above and found that a treatment from the point of view of crystalline field effects and/or spin–orbit interaction for the rare earth ions would yield valuable results which could be compared to the equation of Kittel (1959). Furthermore, it was not necessary to invoke high damping of the rare earth ion magnetic moment motion by spin–lattice interaction to obtain the results. The correspondence between the results of Van Vleck and those of Kittel would be particularly strong for rare earth ions with an even number of electrons. Van Vleck neglected the effect of anisotropy except to the extent that it could be handled by the artifice of an anisotropy field.

The g_{eff} factors for a large number of rare earth–iron garnets have been measured. The finding has been that at temperatures high enough for the rapid relaxation of the rare earth lattice to apply, the data are in reasonable agreement in most cases with the equation of Kittel (1959) or with the refinements of Van Vleck (1961). For the cases in which M_{Fe} has the full value of $5 \mu_B$ per formula unit, the g_{eff} for the rare earth–iron garnet is $\leqslant 2$ since $g_{RE} \leqslant 2$ for all the rare earth ions (table 9). Only for Gd^{3+} does $g_{RE} = 2$ and, therefore, $g_{eff} = 2$ also.

The samarium θ and europium–iron garnets are exceptions because of the presence of excited states of these rare earth ions not too many multiples of kT away from the ground state at room temperature. Further complications arise from the $J = 0$ ground state of europium.

Furthermore, examination of the equations of Tsuya–Wangsness and of Kittel reveals cases in which g_{eff} could become infinite. In fact, such infinities are observed.

Perhaps the first case is that of gadolinium–iron garnet with a low relaxation frequency for the gadolinium magnetic sublattice which would be better described by the Tsuya–Wangsness equation than by that of Kittel. Accordingly, when $M_{Gd}/g_{Gd} + M_{Fe}/g_{Fe} = 0$, as it does at the compensation temperature T_c (see section 2.2.2c), an infinite g_{eff} should occur. Such behavior of g_{eff} for gadolinium–iron garnet was reported by Calhoun et al. (1953).

Another case unique with Eu^{3+}, which can have a very large g-value because of its $J = 0$ ground state, has been observed in resonance experiments by LeCraw et al. (1965). The garnet composition was $Eu_3Fe_{5-x}Ga_xO_{12}$ in which x is chosen to make $M_{Fe} = 0$. The g_{eff} by the Tsuya–Wangsness equation is then g_{Eu}. The same result was later obtained by LeCraw et al. (1975) for a composition $Eu_{1.45}Y_{0.45}Ca_{1.1}Fe_{3.9}Ge_{1.1}O_{12}$ again with a composition chosen to make $M_{Fe} = 0$.

However, it should be noted that g_{eff} would also become large according to the equation of Kittel (1959) by virtue of $M_{Fe} = 0$ and independently of a very large value for g_{Eu}. In fact, most recently Ohta et al. (1977) have observed large values of g_{eff} for $(La, Tm, Ca)_3(Fe, Ge)_5O_{12}$, in which Tm^{3+} is an even-electron rare earth ion, and in $(Y, Er, Ca)_3(Fe, Ge)_5O_{12}$ in which Er^{3+} is an odd-electron rare earth ion.

It is evident, therefore, that the full details of the manifestation of high g_{eff} factors in rare earth–iron garnets have not yet been fully worked out.

2.2.4. Magnetocrystalline anisotropy

As has been pointed out by Landau and Lifshitz (1960), spontaneous magnetization arises from the exchange interaction which is essentially independent of the orientation of the magnetic moment with respect to the lattice. However, there are also crystalline field effects in which the moments of the magnetic ions interact with the lattice. These effects will be of the order of $(v/c)^2$, where v is the velocity of the electrons and c is the velocity of light, and may be called relativistic. Therefore, the ratio of the anisotropy energy per unit volume to the exchange energy is of the order of 10^{-4} to 10^{-5}.

The anisotropy energy which arises from spin–spin and spin–orbit interactions is invariant under time reversal so that it must be an even function of the direction cosines of the magnetization vector M. Accordingly, in a cubic crystal, such as the garnet, the direction dependent portion of the anisotropy energy (any constant component is omitted) is given by

$$E_a = K_1(\alpha_1^2\alpha_2^2 + \alpha_2^2\alpha_3^2 + \alpha_3^2\alpha_1^2) + K_2(\alpha_1^2\alpha_2^2\alpha_3^2) + \cdots$$

where α_i are the direction cosines measured with respect to the cube axes $\langle 100 \rangle$; the higher order terms are very rarely necessary.

As Landau and Lifshitz (1960) indicated, in a cubic crystal the first term in the anisotropy energy is of fourth order in M so that this energy is relatively small in comparison with the exchange energy. However, in a uniaxial crystal the anisotropy energy is of second order in M so that much higher values may be expected and are observed.

In practice, the anisotropy energy difference between any two crystallographic directions is the difference in the energies A_{hkl} required to magnetize the crystal to saturation in these two directions (Bozorth 1951). The A_{hkl} are given by

$$A_{hkl} = \int_0^{M_s} H \, dM.$$

Then it is found that

$$K_1 = 4(A_{110} - A_{100})$$
$$K_2 = 27(A_{111} - A_{100}) - 36(A_{110} - A_{100}) = 27(A_{111} - A_{100}) - 9K_1.$$

From the equation for E_a it can be seen that for $K_2 = 0$ and $K_1 < 0$ the easy direction of magnetization is $\langle 111 \rangle$ and for $K_1 > 0$ it is $\langle 100 \rangle$. With $K_1, K_2 < 0$ the easy direction remains $\langle 111 \rangle$. However, with $K_1 < 0$ and $K_1/K_2 \leqslant -3$ so that $E_a\langle 111 \rangle \geqslant 0$, it can be seen that $E_a\langle 110 \rangle < 0$, and $E_a\langle 100 \rangle = 0$, so that $\langle 110 \rangle$ becomes an easy direction. Finally, with $K_1, K_2 > 0$ the easy direction is again $\langle 100 \rangle$.

For yttrium–iron garnet, $K_1 < 0$ for all temperatures below the Curie temperature. The origin of K_1 has been shown (Wolf 1957) to be accessible through the crystalline field effect in the spin Hamiltonian for the individual paramagnetic ions,

$$\mathcal{H} = \mu_B H \cdot g \cdot S + \tfrac{1}{6}a(S_x^4 + S_y^4 + S_z^4) + DS_\alpha^2 + FS_\alpha^4$$

where α is the direction of the axial distortion and $S \leqslant \tfrac{5}{2}$. Then it was shown that

$$K_1 = K_d Y_d(T) + K_a Y_a(T)$$

where K_a and K_d are related to the crystal field parameters a and F in the (d) and (a) sites as follows (Rimai and Kushida 1966):

$$K_a = 5.01a_a - 4.15F_a$$
$$K_d = 18.76a_d + 9.3F_d$$

$Y_d(0)$ and $Y_a(0) = -\tfrac{5}{2}$ at 0 K.

The substitution of nonmagnetic ions for Fe^{3+} on the (d) or (a) sites was found by Pearson and Annis (1968) to reduce $Y_a(T)$ and $Y_d(T)$, respectively, nearly linearly for tetrahedral (d) substitution and more rapidly for octahedral (a) substitution as a consequence the reduction of the number of ferrimagnetic ions on the (d) sites by heavy substitution in the (a) sites as calculated by Gilleo (1960). The crystalline field parameters found by Rimai and Kushida (1966) are as listed below (in cm^{-1}/unit cell):

$$a_a = 170 \qquad a_d = 50 \qquad F_a = 130 \qquad F_d = -30.$$

In some cases these were best estimates based on the data. These values for a and F yielded for K_a and K_d, 0.032 and 0.068 cm^{-1}/unit cell, respectively, while the magnetic data for YIG yielded -0.006 and 0.101 cm^{-1}/unit cell, respectively.

The single-ion method of calculating K_1 has not been entirely successful, partly because of the difficulty of obtaining the a and F crystalline field parameters in different nonmagnetic hosts such as (Y–Al), (Lu–Al), and (Lu–Ga) garnets. However, the results are good enough to justify the method and to conclude that the effects of nonmagnetic ion substitution for Fe^{3+} in YIG may be estimated as discussed above.

The contributions to K_1 and K_2 for ions with numbers of different (3d)^n

electronic configurations ($4 \leqslant n \leqslant 9$) have been estimated for partial substitutions in YIG at 0 K by Sturge et al. (1969) (table 14).

The contributions to the magnetocrystalline anisotropy by pure rare earth ion substitution for Y are much more substantial in most cases then the magnetocrystalline anisotropy of YIG alone. Furthermore, K_2 may become important at low temperatures. Again, as in the case of the substitution of nonmagnetic ions or other transition-metal ions for Fe^{3+}, the anisotropy contributions are essentially purely additive on an ion per formula unit basis (table 15).

With the exception of Sm, all of the rare earths either have no effect on K_1 for YIG at 300 K or increase the negative value of K_1 as much as by doubling it in the case of Tm (table 15). At lower temperatures (78 to 4 K region) the influence of the rare earths on K_1 and K_2 is more striking. K_1 is invariably taken to a larger negative value than for YIG and in the case of Dy and Ho by almost 3 orders of magnitude, and for Tb and Tm by an order of magnitude. K_2 on the other hand either becomes large and positive (Sm, Dy, Tm) or large and negative (Tb, Ho) while for Eu and Er there is little effect.

TABLE 14

Estimated single-ion contributions to the anisotropy constants K_1 and K_2 of YIG on the basis of one ion per formula unit at 0 K (Sturge et al. 1969)

Configuration	Site	Cubic field[a]	Typical ion	Anisotropy (10^6 erg cm^{-3})	
				K_1	K_2
$(3d)^4$	(a)	W	Mn^{3+}	-0.34	-0.084
	(a)	S	Fe^{4+}	$+2.5$	$+7.5$
	(d)	W	Fe^{4+}	-71	-210
$(3d)^5$	(a)	W	Fe^{3+}	-0.008	0
	(a)	S	Co^{4+}	$+28$	-105
	(d)	W	Fe^{3+}	-0.008	0
	(d)	W	Co^{4+}	-0.025	0
$(3d)^6$	(a)	W	Fe^{2+}	$+5.4$	-251
				$+2.5$	$+5.0$
	(a)	S	Co^{3+}	0	0
	(d)	W	Co^{3+}	-5.9	-4.2
$(3d)^7$	(a)	W	Co^{2+}	$+23$	-69
	(a)	S	Ni^{3+}	0	0
	(d)	–	Co^{2+}	-3.4	-1.7
$(3d)^8$	(a)	–	Ni^{2+}	$+0.034$	-0.008
	(d)	–	Ni^{2+}	-13	$+14$
$(3d)^9$	(a)	–	Cu^{2+}	0	0
	(d)	–	Cu^{2+}	-29	-92

[a] W means a weak field (Hund's rule, maximum spin, ground term), S means strong field (low-spin ground term)

TABLE 15

Changes in the anisotropy constants K_1 and K_2 from those of YIG by nonmagnetic ion substitutions for Fe^{3+} or by replacement of Y^{3+} by RE^{3+}. For YIG, $K_1(300 \text{ K}) = -5.0 \times 10^3 \text{ erg cm}^{-3}$ [a], $K_1(78 \text{ K}) = -22.4 \times 10^3 \text{ erg cm}^{-3}$ [b], $K_1(4 \text{ K}) = -24.8 \times 10^3 \text{ erg cm}^{-3}$ [a]; $K_2 = 0$

Formula unit	Temp. (K)	$K_1 - K_1$(YIG) (10^3 erg cm^{-3})	$K_2 - K_2$(YIG) (10^3 erg cm^{-3})	Ref.
$Sm_3Fe_2Fe_3O_{12}$	300	$+3.0$	–	[c]
	80	-11.8×10^2	$+10 \times 10^2$	[c]
	77	–	$K_2/K_1 = -1.25$	[d]
	<77	–	$K_2/K_1 = -4$	[d]
$Eu_3Fe_2Fe_3O_{12}$	230	-4.0	0	[e]
	170	-3.7	$+9$	[e]
$Gd_3Fe_2Fe_3O_{12}$	300	-1.0	–	[f]
	78	-7.6	–	[b]
	4	-2.0	–	[f]
$Tb_3Fe_2Fe_3O_{12}$	300	~ 0	0	[e]
	80	-7.38×10^2	-7.6×10^3	[e]
	78	$>1.0 \times 10^3$	–	[b]
$Dy_3Fe_2Fe_3O_{12}$	300	~ 0	–	[e]
	80	-9.38×10^2	$+2.14 \times 10^2$	[e]
	78	-9.48×10^3	–	[b]
$Ho_3Fe_2Fe_3O_{12}$	300	~ 0	–	[e]
	80	-7.78×10^2	-2.70×10^2	[e]
	78	-7.78×10^2	–	[b]
	4	-11.8×10^3	–	[e]
$Er_3Fe_2Fe_3O_{12}$	300	-0.5	–	[e]
	78	~ 0	–	[b]
$Tm_3Fe_2Fe_3O_{12}$	300	-6.5	0	[e]
	77	-1.88×10^2	$+1 \times 10^2$	[e]
$Yb_3Fe_2Fe_3O_{12}$	300	-2	–	[e]
	78	-16	–	[b]
	4	-6.18×10^3	$+6.8$	[e]
$Y_3[Sc_{0.4}Fe_{1.6}]Fe_3O_{12}$	300	$+3.2$	–	[e]
	4	$+4.3$	–	[e]
$Y_3[Sc_{0.9}Fe_{1.1}]Fe_3O_{12}$	300	$+4.7$	–	[e]
	4	$+1.88$	–	[e]

[a] Von Aulock (1965)
[b] Iida (1967)
[c] Pearson (1962)
[d] Harrison et al. (1967)
[e] Miyadai (1960)
[f] Rodrigue et al. (1960)

The case of $Sm_3Fe_2Fe_3O_{12}$ is unusual in that the direction of easy magnetization changes from $\langle 111 \rangle$ to $\langle 110 \rangle$ at a temperature somewhat below 77 K at which point $K_2/K_1 = -4$ (Harrison et al. 1967). It is possible that a similar occurrence might take place at low enough temperatures for EuIG, DyIG or TmIG, though only in the case of EuIG does this show much likelihood of happening.

2.2.5. *Magnetoelastic effect*

Magnetostriction is the deformation of a ferrimagnetic body caused by a change
of the magnetization in a magnetic field. According to Landau and Lifshitz
(1960), magnetostriction may be due either to exchange interactions or to the
relativistic interactions which in section 2.2.4 were pointed out as being the
source of magnetic anisotropy.

In a ferrimagnet of cubic symmetry a fairly weak applied field changes the
direction of M as the anisotropy energy is relatively small being of fourth order in
M. Furthermore, any change in the magnitude of M is negligible. The remaining
effects are relativistic. It is found that the first terms of the magnetoelastic energy
which do not vanish are linear in stress and quadratic in the components of M
because of the symmetry of M with respect to time reversal. In a cubic crystal,
therefore, it is found that the dimensional change δl associated with the saturation
values of longitudinal magnetostriction is (see, for example, Kittel 1956, Chikazumi
1964)

$$\delta l/l = \tfrac{3}{2}\lambda_{100}(\alpha_1^2\beta_1^2 + \alpha_2^2\beta_2^2 + \alpha_3^2\beta_3^2 - \tfrac{1}{3}) + 3\lambda_{111}(\alpha_1\alpha_2\beta_1\beta_2 + \alpha_2\alpha_3\beta_2\beta_3 + \alpha_3\alpha_1\beta_3\beta_1)$$

where the α_i are the direction cosines of M referred to the cubic axes, the β_i are
the cosines for the direction in which δl is measured, while λ_{100} and λ_{111} are the
saturation magnetostriction constants in the $\langle 100 \rangle$ and $\langle 111 \rangle$ directions, respectively. The λ_{100} and λ_{111} may be expressed in terms of the magnetoelastic
coupling constants B_1 and B_2 as follows:

$$\lambda_{100} = -\tfrac{2}{3}B_1/(c_{11} - c_{12}) \qquad \lambda_{111} = -\tfrac{1}{3}B_2/c_{44}$$

where the c_{ij} are the elastic stiffness constants. However, the elongation in the
$\langle 110 \rangle$ direction is not independent but is formed by a combination of λ_{100} and λ_{111}

$$\lambda_{110} = \tfrac{1}{4}\lambda_{100} + \tfrac{3}{4}\lambda_{111}.$$

There is, of course, an inverse magnetostrictive effect which may be expressed
in terms of the direction cosines γ_i of the tension $\sigma_{ij} = \sigma\gamma_i\gamma_j$ so that the strain e_{ij}
is given by

$$e_{xx} = \sigma[s_{11}\gamma_1^2 + s_{12}(\gamma_2^2 + \gamma_3^2) \cdots] \qquad e_{xy} = \sigma s_{44}\gamma_1\gamma_2 \cdots$$

where the s_{ij} are the elastic constants. Then the magnetoelastic energy is

$$E_\sigma = B_1\sigma(s_{11} - s_{12})(\alpha_1^2\gamma_1^2 + \alpha_2^2\gamma_2^2 + \alpha_3^2\gamma_3^2 - \tfrac{1}{3}) + B_2\sigma s_{44}(\alpha_1\alpha_2\gamma_1\gamma_2 + \alpha_2\alpha_3\gamma_2\gamma_3 \\ + \alpha_3\alpha_1\gamma_3\gamma_1).$$

The c_{ij} and s_{ij} are related as follows:

$$c_{11} = (s_{11} + s_{12})/(s_{11} - s_{12})(s_{11} + 2s_{12})$$

$$c_{12} = -s_{12}/(s_{11} - s_{12})(s_{11} + 2s_{12})$$

$$c_{44} = 1/s_{44}.$$

Thus, E_σ may be rewritten in terms of λ_{100} and λ_{111}

$$E_\sigma = -\tfrac{3}{2}\lambda_{100}\sigma(\alpha_1^2\gamma_1^2 + \alpha_2^2\gamma_2^2 + \alpha_3^2\gamma_3^2) - 3\lambda_{111}\sigma(\alpha_1\alpha_2\gamma_1\gamma_2 + \alpha_2\alpha_3\gamma_2\gamma_3 + \alpha_3\alpha_1\gamma_3\gamma_1).$$

Consequently, with $K_1 > 0$ and M in the [100] direction,

$$E_\sigma[100] = -\tfrac{3}{2}\lambda_{100}\sigma\gamma_1^2$$

and so on. With $K_1 < 0$ and M in the [111] direction,

$$E_\sigma[111] = -\lambda_{111}\sigma(\gamma_1\gamma_2 + \gamma_2\gamma_3 + \gamma_3\gamma_1)$$

so that if ϕ is the angle between [111] and σ

$$E_\sigma[111] = -\tfrac{3}{2}\lambda_{111}\sigma\cos^2\phi.$$

Accordingly, it can be seen that the application of tension to a magnetic crystal, such as cubic ferrimagnetic garnet, will result in the addition of a uniaxial anisotropy energy term which is negative when $\lambda_{100}\sigma$ or $\lambda_{111}\sigma$ is positive and σ is along an easy axis, $\langle 100 \rangle$ or $\langle 111 \rangle$, respectively.

In the rare earth– and yttrium–iron garnets, λ_{111} is usually <0 and λ_{100} is usually >0 (table 16).

The elastic stiffness constants for the rare earth– and yttrium–iron garnets show small elastic anisotropy and are much alike (table 17).

TABLE 16

Saturation values of the longitudinal magnetostriction constants for rare earth– and yttrium–iron garnets at room temperature (Iida 1967)

RE^{3+}	$\lambda_{111} \times 10^6$	$\lambda_{100} \times 10^6$	RE^{3+}	$\lambda_{111} \times 10^6$	$\lambda_{100} \times 10^6$
Y	-2.4	-1.4	Ho	-4.0	-3.4
Sm	-8.5	$+21.0$	Er	-4.9	$+2.0$
Eu	$+1.8$	$+21.0$	Tm	-5.2	$+1.4$
Gd	-3.1	0	Yb	-4.5	$+1.4$
Tb	$+12.0$	-3.3	Lu	-2.4	-1.4
Dy	-5.9	$+12.5$			

TABLE 17

The elastic stiffness constants for yttrium θ, europium θ and gadolinium θ iron garnets at room temperature

Garnet	c_{11}	c_{12}	c_{44}	$A = \dfrac{2c_{44}}{c_{11} - c_{12}}$	Ref.
	\(10^{11} dyne cm^{-2}\)				
YIG	26.8	11.06	7.66	0.973	[a]
	26.9	10.77	7.64	0.947	[b]
EuIG	25.10	10.70	7.62	1.051	[a]
GdIG	27.31	12.50	7.41	1.0007	[c]

[a] Bateman (1966)
[b] Clark and Strakna (1961)
[c] Comstock et al. (1966)

The magnetostrictive properties of mixed rare earth–iron garnets turn out to be linearly additive, i.e., a linear combination of the contributions from each rare earth–iron garnet in proportion to its concentration (Clark et al. 1968). A quantum-mechanical theoretical treatment of the problem (Callen and Callen 1963, 1965) yields a good fit of the experimental results for the rare earth–iron garnet system.

3. Summary

It has been shown how the discovery of ferrimagnetism in yttrium–iron garnet has brought a wide range of new, highly technological applications to materials of this structure. The looseness of the garnet structure and the variety of site size and coordination for metallic cations allow the accommodation of nearly half of the elements of the periodic table and most of the metals. For this reason garnet is often referred to as a scavenger mineral.

The advantages of the garnet structure have been exhibited for ferrimagnetic materials by the display of the wide range of magnetization achievable and the variation of temperature dependence and anisotropy that can be accomplished through substitutions on rare earth and iron sites. Moreover, the magnetoelastic properties may also be varied in magnitude and sign as may the effective g-factor. At the same time the structural symmetry of the garnet remains cubic.

The existence of three magnetic sublattices within one structure of high symmetry is an important factor to these accomplishments. Somewhat independent control of the magnetization and single-ion anisotropy on each of these sublattices is the main reason for the exceptional flexibility of ferrimagnetic properties which may be achieved in these materials.

Only the bulk properties of essentially isotropic garnets have been treated in this chapter. An even wider range of properties can be attained, as will be seen, (in Eschenfelder's chs. 5 and 6 of this volume), in liquid phase epitaxial films which have multiple substitutions on each site and which are no longer cubic as a consequence of growth-induced ordering of cations, particularly in the dodecahedral sites.

Acknowledgements

The author is much indebted to Drs. S. Geller and J.W. Nielsen for knowledge gained during years of association, from papers published by them in the following years, and for their reviews and comments on this manuscript. Particularly, the manuscript (Geller 1977) received in advance of publication was helpful in apprising the author of the results of many years of consideration given to some problems by Dr. Geller.

References

Aharoni, A. and M. Scheiber, 1961, J. Phys. Chem. Solids 19, 304.

Aharoni, A., E.H. Frei, Z. Scheidlinger and M. Schieber, 1961, J. Appl. Phys. 32, 1851.

Akselrad, A., 1971, AIP Conf. Proc. 5, 249.

Akselrad, A. and H. Callen, 1971, Appl. Phys. Lett. 19, 464.

Ancker-Johnson, B. and J.J. Rawley, 1958, Proc. IRE 46, 1421.

Anderson, E.E., 1959, J. Appl. Phys. 30, 299S.

Anderson, E.E., 1964a, Phys. Rev. 134, A1581.

Anderson, E.E., 1964b, The magnetizations of yttrium– and gadolinium–iron garnets, Ph.D. thesis (Physics Department, University of Maryland).

Anderson, E.E. and J.R. Cunningham Jr., 1960, J. Appl. Phys. 31, 1687.

Atzmony, U., E.R. Bauminger, B. Einhorn, J. Hess, A. Mustachi and S. Ofer, 1968, J. Appl. Phys. 39, 1250.

Bateman, T.B., 1966, J. Appl. Phys. 37, 2194.

Belov, N.V., 1958, Sov. Phys. Crystallogr. 3, 222.

Bertaut, F. and F. Forrat, 1956, Compt. Rend. 242, 382.

Bertaut, F. and F. Forrat, 1957, Compt. Rend. 244, 96.

Bertaut, F., F. Forrat, A. Herpin and P. Mériel, 1956, Compt. Rend 243, 898.

Bertaut, F. and R. Pauthenet, 1957, Proc. Inst. Elec. Eng. B104, 261.

Borghese, C., 1967, J. Phys. Chem. Solids 28, 2225.

Boyd, E.L., V.L. Moruzzi and J.S. Smart, 1963, J. Appl. Phys. 34, 3049.

Bozorth, R.M., 1951, Ferromagnetism (Van Nostrand, New York).

Brandle, C.D. and S.L. Blank, 1976, IEEE Trans. Magn. MAG-12, 14.

Buhrer, C.F., 1969, J. Appl. Phys. 40, 4500.

Calhoun, B.A., J. Overmeyer and W.V. Smith, 1953, Phys. Rev. 91, 1085.

Callen, E.R. and H.B. Callen, 1963, Phys. Rev. 129, 578.

Callen, E.R. and H.B. Callen, 1965, Phys. Rev. 139, A455.

Callen, H., 1971, Appl. Phys. Lett. 18, 311.

Caspari, M.E., S. Frankel and M.A. Gilleo, 1960, J. Appl. Phys. 31, 320S.

Caspari, M.E., S. Frankel and G.T. Wood, 1962, Phys. Rev. 127, 1519.

Caspari, M.E., A. Koicki, S. Koicki and G.T. Wood, 1964, Phys. Lett. 11, 195.

Chikazumi, S., 1964, Physics of magnetism (Wiley, New York).

Clark, A.E. and E. Callen, 1968, J. Appl. Phys. 39, 5972.

Clark, A.E. and R.E. Strakna, 1961, J. Appl. Phys. 32, 1172.

Clark, A.E., J.J. Rhyne and E.R. Callen, 1968, J. Appl. Phys. 39, 573.

Cohen, H.M. and R.A. Chegwidden, 1966, J. Appl. Phys. 37, 1081.

Comstock, R.L., J.J. Raymond, W.G. Nilsen and J.P. Remeika, 1966, Appl. Phys. Lett. 9, 274.

Cunningham, J.R. Jr. and E.E. Anderson, 1960, J. Appl. Phys. 31, 45S.

Dillon, J.F. Jr., 1958, J. Appl. Phys. 29, 1286.

Dionne, G.F., 1970a, J. Appl. Phys. 41, 4874.

Dionne, G.F., 1970b, Lincoln Laboratory (Mass Inst. Tech., TR-480, ESD-TR-70-150 ZZ1694).

Dionne, G.G., 1971, J. Appl. Phys. 42, 2142.

Dionne, G.F., 1976, J. Appl. Phys. 47, 4220.

Dionne, G.F., 1978, Lincoln Laboratory (Mass. Inst. Tech., TR-534).

Espinosa, G.P., 1962, J. Chem. Phys. 37, 2344.

Euler, F. and J.A. Bruce, 1965, Acta Crystallogr. 19, 971.

Euler, F., B.R. Capone and E.R. Czerlinsky, 1967, IEEE Trans. Magn. MAG-3, 509.

Fischer, P., W. Hälg, E. Stoll and A. Segmüller, 1966, Acta Crystallogr. 21, 765.

Geller, S., 1960, J. Appl. Phys. 31, 305.

Geller, S., 1966, J. Appl. Phys. 37, 1408.

Geller, S., 1967, Z. Kristallogr. 125, 1.

Geller, S., 1977, Proc. Int. School of Physics, Enrico Fermi.

Geller, S. and M.A. Gilleo, 1957a, Acta Crystallogr. 10, 239.

Geller, S. and M.A. Gilleo, 1957b, J. Phys. Chem. Solids 3, 30.

Geller, S. and M.A. Gilleo, 1959, J. Phys. Chem. Solids 9, 235.

Geller, S. and D.W. Mitchell, 1959, Acta Crystallogr. 12, 936.

Geller, S., R.M. Bozorth, M.A. Gilleo and C.E. Miller, 1959, J. Phys. Chem. Solids 12, 111.

Geller, S., H.J. Williams and R.C. Sherwood, 1961, Phys. Rev. 123, 1692.

Geller, S., H.J. Williams, R.C. Sherwood and G.P. Espinosa, 1962, J. Phys. Chem. Solids 23, 1525.

Geller, S., H.J. Williams, R.C. Sherwood, J.P. Remeika and G.P. Espinosa, 1963a, Phys. Rev. 131, 1080.

Geller, S., H.J. Williams, G.P. Espinosa, R.C. Sherwood and M.A. Gilleo, 1963b, Appl. Phys. Lett. 3, 21.

Geller, S., H.J. Williams, G.P. Espinosa and R.C. Sherwood, 1964a, Bell Syst. Tech. J. 43, 565.

Geller, S., H.J. Williams, G.P. Espinosa and R.C. Sherwood, 1964b, Phys. Rev. 136, A1650.

Geller, S., J.P. Remeika, R.C. Sherwood, H.J. Williams and G.P. Espinosa, 1965, Phys. Rev. 137, A1034.

Geller, S., J.A. Cape, G.P. Espinosa and D.H. Leslie, 1966a, Phys. Rev. 148, 522.

Geller, S., J.A. Cape, G.P. Espinosa and D.H. Leslie, 1966b, Phys. Lett. 21, 495.

Geschwind, S., 1961, Phys. Rev. 121, 363.

Geschwind, S. and L.R. Walker, 1959, J. Appl. Phys. 30, 163S.

Gilleo, M.A., 1958, Phys. Rev. 109, 777.

Gilleo, M.A., 1960, J. Phys. Chem. Solids 13, 33.

Gilleo, M.A. and S. Geller, 1958, Phys. Rev. 110, 73.

Guillot, M. and R. Pauthenet, 1964, Compt. Rend. 259, 1303.

Gyorgy, E.M., M.D. Sturge and L.G. Van Uitert, 1974, AIP Conf. Proc. 18, 70.

Hagedorn, F.B., W.J. Tabor and L.G. Van Uitert, 1973, J. Appl. Phys. 44, 432.

Harrison, F.W., J.F.A. Thompson and G.K. Lang, 1965, J. Appl. Phys. 36, 1014.

Harrison, F.W., R.F. Pearson and K. Tweedale, 1967, Philips Tech. Rev. 28, 135.

Heilner, E.J. and W.H. Grodkiewicz, 1973, J. Appl. Phys. 44, 4218.

Henderson, J.W. and R.L. White, 1961, Phys. Rev. 123, 1627.

Iida, S., 1967, J. Phys. Soc. Japan 22, 1201.

Jaccarino, U., B.T. Matthias, M. Peter, A. Suhl and J.H. Wernick, 1960, Phys. Rev. Lett. 5, 251.

Kittel, C., 1956, Introduction to solid state physics (Wiley, New York).

Kittel, C., 1959, Phys. Rev. 115, 1587.

Kramers, H.A., 1934, Physica 1, 182.

Kurtzig, A.J. and M. Dixon, 1972, J. Appl. Phys. 43, 2883.

Landau, L.D. and E.M. Lifshitz, 1960, Electrodynamics of continuous media (Pergamon, New York).

LeCraw, R.C., J.P. Remeika and H. Matthews, 1965, J. Appl. Phys. 36, 901.

LeCraw, R.C., R. Wolfe, A.H. Bobeck, R.D. Pierce and L.G. Van Uitert, 1971, J. Appl. Phys. 42, 1643.

LeCraw, R.C., S.L. Blank and G.P. Vella-Coleiro, 1975, Appl. Phys. Lett. 26, 402.

Leo, D.C., D.A. Lepore and J.W. Nielsen, 1966, J. Appl. Phys. 37, 1083.

Litster, J.D. and G.B. Benedek, 1966, J. Appl. Phys. 37, 1320.

Loriers, J. and G. Villers, 1961, Compt. Rend. 252, 1590.

Maguire, E.A. and J.J. Green, 1962, J. Appl. Phys. 33, 1373S.

McLay, A.D. and M.F. Crawford, 1933, Phys. Rev. 44, 986.

Menzer, S., 1929, Z. Kristallogr. 69, 300.

Miyadai, T., 1960, J. Phys. Soc. Japan 15, 2205.

Myers, S.M., J.P. Remeika and H. Meyer, 1968, Phys. Rev. 170, 520.

Néel, L., 1948, Annls. de Phys. 3, 137.

Nielsen, J.W., 1971, Metall. Trans. 2, 625.

Nielsen, J.W. and E.F. Dearborn, 1958, J. Phys. Chem. Solids 5, 202.

Novak, G.A. and G.V. Gibbs, 1971, Am. Mineral. 56, 791.

Nowik, I., 1968, Phys. Rev. 171, 550.

Nowik, I. and S. Ofer, 1968, J. Appl. Phys. 39, 1252.

Ogawa, S. and S. Morimoto, 1962, J. Phys. Soc. Japan 17, 654.

Ohta, N., T. Ikeda, F. Ishida and Y. Sugita, 1977, J. Phys. Soc. Japan 43, 705.

Pauthenet, R., 1958a, Thesis (University of Grenoble).

Pauthenet, R., 1958b, Ann. Chim. Phys. 3, 424.

Pearson, R.F., 1962, J. Appl. Phys. 33, 1236S.

Pearson, R.F. and A.D. Annis, 1968, J. Appl. Phys. 39, 1338.

Prince, E., 1957, Acta Crystallogr. 10, 787.

Prince, E., 1965, J. Appl. Phys. 36, 1845.

Ramsey Jr., T.H., H. Steinfink and E.J. Weiss, 1962, J. Phys. Chem. Solids 23, 1105.

Rimai, L. and T. Kushida, 1966, Phys. Rev. 143, 160.

Rodbell, D.S., I.S. Jacobs, J. Owen and E.A. Harris, 1963, Phys. Rev. Lett. 11, 10.

Rodrigue, G.P., J.E. Pippin, W.P. Wolf and C.L. Hogen, 1958, IRE Trans. MTT-6, 83.

Rodrigue, G.P., H. Meyer and R.V. Jones, 1960, J. Appl. Phys. 31, 376S.

Rosencwaig, A., 1971, Solid State Commun. 9, 1899.

Rosencwaig, A. and W.J. Tabor, 1971a, Solid State Commun. 9, 1691.

Rosencwaig, A. and W.J. Tabor, 1971b, J. Appl. Phys. 42, 1643.

Rosencwaig, A., W.J. Tabor, F.B. Hagedorn and L.G. Van Uitert, 1971a, Phys. Rev. Lett. 26, 775.

Rosencwaig, A., W.J. Tabor and R.D. Pierce, 1971b, Phys. Rev. Lett. 26, 779.

Schieber, M. and Z.H. Kalman, 1961, Acta Crystallogr. 14, 1221.

Shannon, R.D. and C.T. Prewitt, 1969, Acta Crystallogr. B25, 925.

Sirvetz, M.H. and J.E. Zneimer, 1958, J. Appl. Phys. 29, 431.

Skinner, B.J., 1956, Am. Mineral. 41, 428.

Sternheimer, R., 1952, Phys. Rev. 86, 316.

Streever, R.L. and G.A. Uriano, 1965, Phys. Rev. 139, A305.

Sturge, M.D., E.M. Gyorgy, R.C. LeCraw and J.P. Remeika, 1969, Phys. Rev. 180, 413.

Suchow, L. and M. Kokta, 1971, AIP Conf. Proc. 5, 700.

Suchow, L. and M. Kokta, 1972, J. Solid State Chem. 5, 85.

Telesnin, R.V., A.M. Efimova and R.A. Yus'kaev, 1962, Sov. Phys. Solid State 4, 259.

Tsuya, N., 1952, Prog. Theoret. Phys. (Kyoto) 7, 263.

Van Loef, J.V., 1968, Solid State Commun. 6, 541.

Van Vleck, J.H., 1961, Phys. Rev. 123, 58.

Vassiliev, A., J. Nicolas and M. Hildebrandt, 1961a, Compt. Rend. 252, 2529.

Vassiliev, A., J. Nicolas and M. Hildebrandt, 1961b, Compt. Rend. 252, 2681.

Vassiliev, A., J. Nicolas and M. Hildebrandt, 1961c, Compt. Rend. 253, 242.

Vertoprahkov, V.N., V.O. Zamozhskii and P.V. Klevtsov, 1968, Sov. Phys. Crystallogr. 13, 113.

Villers, G. and J. Loriers, 1958, Compt. Rend. 247, 1101.

Villers, G., J. Loriers and F. Clerc, 1962a, Compt. Rend. 255, 1196.

Villers, G., M. Paulus and J. Loriers, 1962b, J. Phys. Rad. 23, 466.

Von Aulock, W.H., ed., 1965, Handbook of microwave ferrite materials (Academic, New York).

Wang, F.F.Y., 1973, Treatise on materials science and technology, Vol. 2 (ed. H. Hermann) (Academic, New York) pp. 279–398.

Wangsness, R.K., 1953, Phys. Rev. 91, 1085.

Wangsness, R.K., 1958, Phys. Rev. 111, 813.

Weiss, P. and R. Forrer, 1924, Compt. Rend. 178, 1670.

White, J.A. and J.H. Van Vleck, 1961, Phys. Rev. Lett. 6, 412.

White, R.L. and J.P. Andelin Jr., 1959, Phys. Rev. 115, 1435.

Wittekoek, S. and D.E. Lacklison, 1972, Phys. Rev. Lett. 28, 740.

Wolf, W.P., 1957, Phys. Rev. 108, 1152.

Wolf, W.P., 1961, Rep. Prog. Phys. XXIV, 212.

Wolf, W.P. and G.P. Rodrigue, 1958, J. Appl. Phys. 29, 105.

Wolf, W.P. and J.H. Van Vleck, 1960, Phys. Rev. 118, 1490.

Wolfe, R., M.D. Sturge, F.R. Merritt and L.G. Van Uitert, 1971, Phys. Rev. Lett. 26, 1570.

Yoder, H.S. and M.L. Keith, 1951, Am. Mineral. 36, 519.

chapter 2

SOFT MAGNETIC METALLIC MATERIALS

G. Y. CHIN AND J. H. WERNICK

Bell Laboratories
Murray Hill, NJ 07974
USA

Ferromagnetic Materials, Vol. 2
Edited by E. P. Wohlfarth
© North-Holland Publishing Company, 1980

CONTENTS

1. General introduction

Metals and alloys referred to as magnetically soft enjoy wide use in such diverse applications as pole pieces in electric motors and generators, laminations in transformer cores, relays for switching circuits for communication systems, and other electromagnetic equipment and devices. Because of the need for special combinations of properties for a given application, often conflicting in a given material, a number of different magnetic materials have been developed over the last 40–50 years. These soft magnetic alloys range from iron and low carbon steel through silicon–iron alloys to high nickel–iron and cobalt–iron alloys. Usually, the most important properties these alloys must possess, in addition to the requisite saturation magnetization, are: (a) low loss, (b) low coercive force and (c) high permeability, both initial and/or maximum, depending on the application. Some special applications may require a high constant permeability at low field strengths or a linear relation between permeability and temperature over a given temperature range.

This chapter is organized into five major sections: iron and low carbon steels, iron–silicon alloys, iron–aluminum and iron–aluminum–silicon alloys, nickel–iron alloys and iron–cobalt alloys. Emphasis is placed on these alloys as they are currently manufactured and in use, and on the fundamental relationship between processing, structure and technical magnetic behavior. For information on numerous other soft magnetic alloys studied over the years, the reader is referred to Bozorth (1951). (The reader may also find the Appendix listing the trade names of some common soft magnetic alloys useful.)[*]

The preferred magnetic units used in this chapter are in the cgs system since most of the original data reported to date are in cgs units and for practical purposes the cgs units are still the most widely used. Since the SI system is increasingly being adopted, however, the newer units are often found throughout the text. This should pose no problem to the reader as the conversion from one system to another, as noted below, is relatively simple for the properties

[*] References in the text, tables and figures do not only appear as authors' names plus years of publication, but also as book titles or company names, etc, with or without publication years. Full information on these references can be found in the reference list at the end of this chapter.

discussed here. Both units may also be quoted from time to time to facilitate the conversion.

Induction: 10 000 gausses (G) = 1 tesla (T)
Magnetic field strength: 1 oersted (Oe) = 79.58 amperes/meter (A/m)

Permeability values are generally understood to be relative to those in free space, and are thus the same in both systems. Core loss is usually quoted in watts/pound (W/lb) at 60 Hz in the US and watts/kilogram force (W/kg) at 50 Hz in Japan and Europe. Other than the 1 kg = 2.2 lb conversion, the loss conversion from 50 to 60 Hz depends on the induction level and the quality of the material. The following conversion table is generally adequate for silicon steel:

Non-oriented grades: 1 W/kg, 50 Hz = 1.74 W/lb, 60 Hz at 1.5 T
Oriented grades: 1 W/kg, 50 Hz = 1.68 W/lb, 60 Hz at 1.5 T
Oriented grades: 1 W/kg, 50 Hz = 1.67 W/lb, 60 Hz at 1.7 T
High induction grades: 1 W/kg, 50 Hz = 1.64 W/lb, 60 Hz at 1.7 T

Unless otherwise noted, all compositions are in weight percent. Also quotes may be made of original data in units such as inches, pounds per square inches, kilogram (force) per square millimeters, etc. for practical reasons. Standard conversions are found in many texts.

2. Iron and low carbon steels

2.1. Introduction

Iron and iron–silicon alloys are the principal soft ferromagnetic materials in use today. Improvements in magnetic properties have been achieved primarily by minimizing chemical impurities and controlling crystal orientation. In this section, we confine our attention to commercially pure irons and several iron-base alloys other than the important iron–silicon alloys.

The term "pure iron" generally refers to irons of 99.9 + % purity containing small amounts of C, Mn, Si, P, S, N, and O. Impurities that have the greatest effects on magnetic properties are the nonmetallic elements C, O, S, and N which enter the lattice interstitially. Commercial irons with maximum permeabilities of 10 000 to 20 000 can be increased to 100 000 or more by annealing in H_2 and in vacuum (Cioffi et al. 1937). The process removes much of the carbon, oxygen, sulfur, and nitrogen. Highly purified (and expensive) iron annealed for long periods in hydrogen and in vacuum can display maximum permeabilities as high as 10^6.

The metallic impurity content of a very pure research-grade iron (cost in excess of $ 700/kg is shown in table 1. Nitrogen and carbon contents initially are of the order of 200 ppm and 17 ppm respectively and can be reduced to <0.5 ppm by annealing in purified H_2 for 100 h at 800°C followed by an anneal under vacuum at 750°C for 3 h at $<10^{-9}$ Torr (Da Silva and McLellan 1976).

TABLE 1
Metallic impurities in a Johnson–Matthey high purity iron

Impurity	Ni	Mn	Si	Ca	Cu	Cr	Mg	Ag
Quantity (ppm)	10	3	3	2	2	1	1	<1

Trade names for several special commercially purified irons are Ferrovac-E, Puron, Vacuum-Melted Electrolytic Iron and Plast-Iron. Table 2 lists several pure irons, including zone refined and other highly purified irons obtained by laboratory purification methods, methods of preparation, and estimates of typical impurity contents. All are compared to Armco iron, a large volume low carbon bearing material used for soft magnetic applications. A typical analysis of an Armco iron (ingot iron) is shown in table 3.

Annealing of hot or cold rolled sheets of Armco iron improves permeability and coercive force by recrystallization and grain growth. Anneals are generally

TABLE 2
Several pure irons and preparation methods*

Iron and method of preparation	Metallic (ppm)	Non-metallic (ppm)	Purity (%)
Ingot iron, basic open hearth (Armco)**	1280	1180	99.75
Ingot, H_2-purified	1280	175	99.85
Ingot, H_2-purified, selected skin	–	–	to 99.9
Carbonyl, H_2-purified	840	100	99.90
Puron (Westinghouse)	46	530	99.94
Electrolytic, vacuum remelt (Ferrovac E) (Vacuum Metals Corp.)	<640	320	99.90
AISI, arc melted (Battelle)	55–218	50–179	99.96
Nitrate process (National Bureau of Standards)	32	65	99.990
Chloride process (National Bureau of Standards)	10–15	10–20	99.997
Zone refined (Franklin Institute)	–	–	>99.99

*Metals handbook (1961)
**Ingot iron (Armco) is shown for comparison

TABLE 3
Analysis of a magnetic ingot iron (Armco iron, mill sheet)*

Element	C	Mn	Si	P	S	O	N	Al
Quantity (weight %)	0.015	0.030	0.003	0.005	0.025	0.15	0.007	0.003

*Littmann (1971)

conducted at approximately 800°C. Carbon is the most important impurity; it results in magnetic aging by precipitation of iron carbide. By annealing in moist hydrogen, the carbon content can be reduced to as low as 0.002%. This treatment must be followed by annealing in dry hydrogen to avoid brittleness.

Magnetic aging effects can also be eliminated by reducing the carbon and nitrogen content through treatments of the molten iron with titanium and aluminum thus ensuring that the remaining carbon and nitrogen are present as second phase compounds of titanium and aluminum (Littmann 1971).

2.2. Magnetic properties

Typical magnetic properties of ingot iron and stabilized ingot iron (also referred to as electromagnet or non-aging iron) are shown in table 4. For comparison, selected magnetic properties of several high purity irons are shown in table 5. It should be borne in mind that cold working, while not altering saturation magnetization, will alter coercivity, permeability, and resistivity.

TABLE 4

Typical magnetic properties of ingot iron and stabilized (electromagnet, non-aging) iron*

	B_s (G)	H_c (Oe) ($B_m = 10$ kG)	Permeability ($H = 1$)	($H = 10$)
Cast magnetic ingot iron	21 500	0.85	3500	1500
Magnetic ingot iron, 0.08 in sheet	21 500	1.11	1800	1575
Electromagnet iron, 0.08 in sheet	21 500	1.02	2750	1575

*Littmann (1971)

TABLE 5

Magnetic properties of several high purity irons*

	B_s (G)	H_c (Oe)	Permeability initial	max.
Ingot Iron (vacuum-melted)	–	0.28–0.34		17 000
Electrolytic (annealed)	–	0.15–0.32	44–60	21 000 8 100
Electrolytic (vacuum-melted and annealed)	–	0.09	1 150	41 500 61 000
Puron (H$_2$-treated)	21 600	0.05		100 000

*Metals handbook (1961)

The behavior of the magnetization of single crystal iron is shown in fig. 1. These data show that saturation can be achieved at quite low fields and that the ⟨100⟩ direction is the "easy direction" of magnetization. Commercial grades of iron for magnetic purposes are not grain oriented because most applications require magnetic flux in more than one direction. Dunn and Walker (1961) and Kohler (1967) have described methods of producing cube-on-edge texture in pure iron by secondary recrystallization.

The variation of the crystal anisotropy constants K_1 and K_2, with temperature for iron is shown in fig. 2. The data for iron, which are quite old, are in poor agreement with the accepted room temperature values [$K_1 = 4.8 \times 10^5$, $K_2 =$

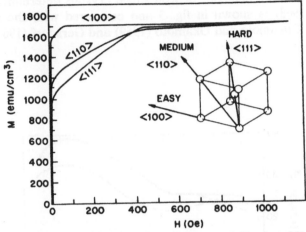

Fig. 1. Magnetization curves for an iron single crystal (Honda and Kaya 1926).

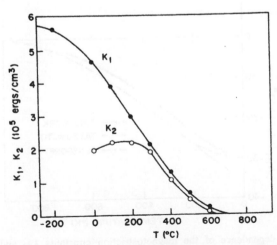

Fig. 2. Variation of the crystal anisotropy constants for iron (Bozorth 1937, from measurements of Honda et al. 1928).

$\pm 0.5 \times 10^5$ ergs/cm^3 (Graham 1958, Cullity 1972)]. Graham (1958) determined K_1 to be 5.05×10^5 and 5.20×10^5 ergs/cm^3 at 195 K and 77 K respectively and K_2 to be constant ($\pm 0.5 \times 10^5$ ergs/cm^3) in this temperature range (including room temperature). The values of K_2 determined by Sato and Chandrasekhar (1957) at 300 K and 77 K, 0.71×10^5 and 0.90×10^5 respectively, are positive and slightly higher than the results given by Graham. Although crystal anisotropy is considered a structure insensitive property because it is due to spin–orbit coupling, measurements of various samples suggest that crystal imperfections (residual strain, lattice vacancies, inclusions, porosity, etc.) interfere with its accurate measurement in such a way that crystal anisotropy becomes effectively structure sensitive (Graham 1958, Cullity 1972).

The magnetostriction behavior of iron single crystals determined by Williams and Pavlovic (1968) is shown in fig. 3 and compared with the results of the earlier work of Tatsumoto and Okamoto (1959) and Gersdorf (1961).

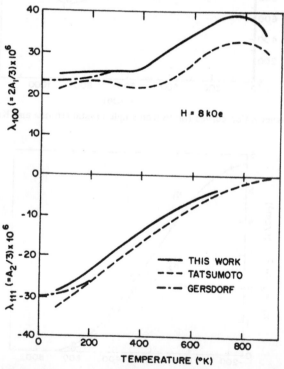

Fig. 3. Temperature dependence of the magnetostriction constants λ_{100} and λ_{111} (Williams and Pavlovic 1968). Results are compared with those of Tatsumoto and Okamoto (1959) and Gersdorf (1961).

2.3. Physical and mechanical properties

Several of the important physical and room temperature mechanical properties of fully annealed irons are given in tables 6 and 7. The resistivity behavior of dilute alloys of iron as a function of solute concentration is shown in fig. 4 (Bozorth 1951). High resistivities are desirable in order to decrease eddy current losses and of the solutes shown, silicon is the most potent – the factor responsible for the great utility of iron–silicon alloys. The resistivity behavior of concentrated iron–cobalt and iron–nickel alloys is shown in figs. 5 and 6.

The values of mechanical properties are sensitive to very small traces of impurities, particularly at low temperatures, and therefore one cannot unambiguously ascribe a definite value to each property. Yield and tensile strengths decrease with increasing purity, while ductility, especially uniform elongation, progressively increases. The values, particularly ductility, are also sensitive to grain size. Yield and tensile strengths increase, while ductility decreases, with increasing amounts of prior cold deformation.

2.4. Iron–carbon alloys

Low carbon iron–carbon alloys or steels that can be classified as soft magnetic alloys contain carbon contents under 0.10% by weight. Since the metallurgy of these alloys can be discussed primarily in terms of the Fe–C phase diagram, a portion of this diagram pertinent to carbon steels and their heat treatment is

TABLE 6
Some physical properties of high purity iron*

Melting point (°C)	1536
Boiling point (°C)	2860
Crystal structure at 20°C	bcc†
Lattice constant at 20°C (Å)	2.8664
Density at 20°C (g/cm³)	7.87
Temperature of $\alpha \rightarrow \gamma$ transformation on heating (°C)	910
Temperature of $\gamma \rightarrow \delta$ transformation on heating (°C)	1400
Specific heat** at 20°C (cal/g °C)	0.106
Thermal conductivity for 0–100°C (W/mK)	78.2
Resistivity at 20°C ($\mu\Omega$ cm)	10.1
Temperatute coeff. of resistivity** for 0–100°C (10^{-3} K)	6.5
Coefficient of expansion** for 0–100°C (10^{-6} K)	12.1

* Metals reference book (1976), Metals handbook (1961)
**This data as a function of temperature can be obtained from the Metals handbook (1961)
†bcc: body centered cubic

TABLE 7
Some room temperature mechanical properties of pure iron*

Compressibility at 30°C, ingot iron (cm²/kg × 10⁶)	0.566
Modulus of elasticity, average (lbs/in² × 10⁻⁶)	28.5
E[100] ingot iron (lbs/in² × 10⁻⁶) single crystal	19.0
E[111] ingot iron (lbs/in² × 10⁻⁶)	41.0
Shear modulus, G	
G[100] (lbs/in² × 10⁻⁶)	16.2
G[110] (lbs/in² × 10⁻⁶)	9.6
G[111] (lbs/in² × 10⁻⁶)	8.5
Tensile strength, fully annealed irons (lbs/in² × 10⁻³)	
vacuum-melted electrolytic irons	35–40
carbonyl irons	28–40
Yield strength, fully annealed irons (lbs/in² × 10⁻³)	
vacuum-melted electrolytic irons	10–20
carbonyl irons	15–24
Elongation, fully annealed irons (%)	
vacuum-melted electrolytic irons	40–60
carbonyl irons	30–40
Strain, fully annealed irons (A_i/A_f)**	
vacuum-melted electrolytic irons	1.2–2.3
carbonyl irons	1.2–1.6

*Metals handbook (1961)
**A_i = initial area, A_f = final area

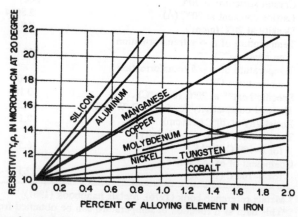

Fig. 4. Dependence of resistivity of iron on concentration of various alloying elements (Bozorth 1951).

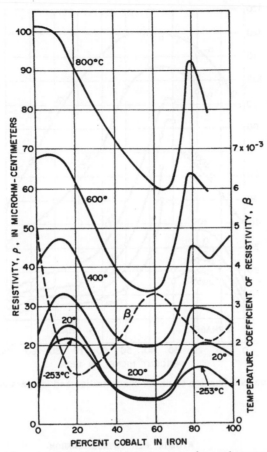

Fig. 5. Resistivity of iron as a function of cobalt concentration and temperature (Bozorth 1951).

shown in fig. 7. Figure 8 is an enlarged portion of this diagram relevant to iron–carbon soft magnetic alloys.

Swisher et al. (1969) and Swisher and Fuchs (1970) investigated by systematic "doping" experiments of laboratory melts and a commercial steel the role of nonmetallic impurities on the magnetic properties and susceptibility to aging of low-carbon steels. Carbon and nitrogen behaved similarly on magnetic properties before aging, but only nitrogen promoted aging at 100°C. Sulfur and oxygen, present as MnS and MnO inclusions, did not promote aging, although they increased the coercive force significantly.

The effect of nitrogen on the coercivity of a high oxygen, decarburized steel (carbon content <10 ppm) is shown in figs. 9 and 10. The coercive force rises markedly with nitrogen content in solid solution (fig. 9) while fig. 10 shows that aging (as indicated by the increase in coercivity ΔH_c) due to nitride precipitation for nitrogen concentration in excess of 30 ppm is significant.

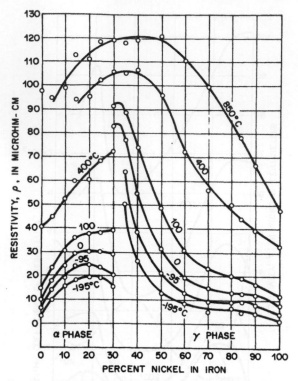

Fig. 6. Resistivity of iron as a function of nickel concentration and temperature (Bozorth 1951).

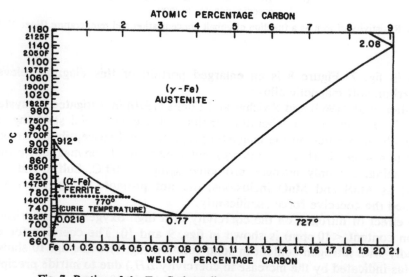

Fig. 7. Portion of the Fe–C phase diagram (Metals handbook 1973a).

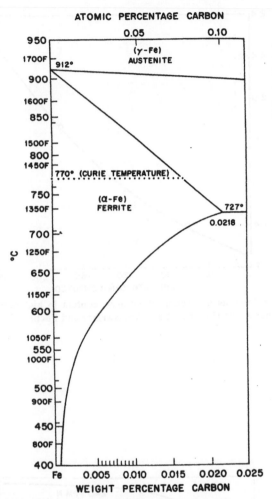

Fig. 8. Enlarged portion of the low carbon region of the Fe–C phase diagram (Metals handbook 1973b).

The effect of carbon on the coercivity of a high purity steel intentionally doped with carbon is shown in fig. 11. These data, when compared with similar data for nitrogen (fig. 9), show that carbon and nitrogen are comparable in their effects on magnetic behavior before aging. The limiting coercive forces at zero nitrogen or carbon content is due to the fact that the oxygen contents of the two steels employed in this investigation were different.

The role of carbon in nitrogen free steels regarding aging is illustrated in fig. 12. The results show that carbon-bearing, nitrogen-free steels do not age at 100°C at a measurable rate. While some carbon remains in supersaturated solid solution after cooling, this carbon will precipitate during aging at higher temperatures.

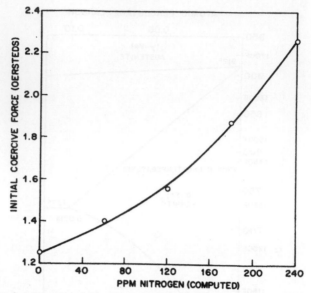

Fig. 9. Initial coercive force as a function of nitrogen content in high-oxygen steel after decarburization (Swisher et al. 1969, Swisher and Fuchs 1970).

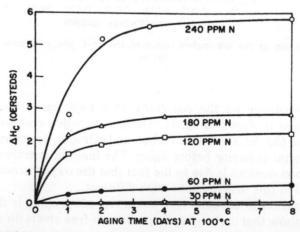

Fig. 10. Magnetic aging, as indicated by the increase in coercivity, ΔH_c, due to nitrogen in a high-oxygen steel after decarburization (Swisher et al. 1969, Swisher and Fuchs 1970).

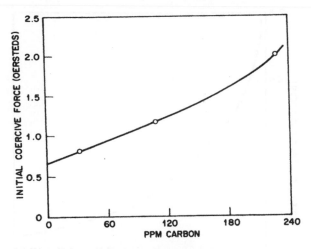

Fig. 11. Initial coercive force as a function of carbon content in high-purity steel (Swisher et al. 1969, Swisher and Fuchs 1970).

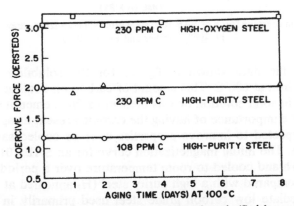

Fig. 12. Magnetic aging due to carbon in experimental steels (Swisher and Fuchs 1970).

Coercive forces of one oersted and lower have been obtained by Wiseman* (in Swisher et al. 1969) in low-carbon steels without complete decarburization. This is due to the fact that the cementite (Fe_3C) particles in the microstructure were larger and spherical and this did not affect magnetic properties. On the other hand, if the Fe_3C is present as fine lamellae or as a grain boundary network, magnetic properties are affected adversely. Swisher et al. (1969) and Swisher and Fuchs (1970) subsequently showed that a non-decarburizing anneal at 680°C (atmosphere 25% CH_4, 75%H_2) to spheroidize the insoluble carbon in an SAE 1010 steel markedly decreased the coercive force. These results are shown in fig. 13. The limiting value of 1.4 Oe (1.1 A/cm) shown in this figure was

* Wiseman, A.S., 1967, Hawthorne Station, Western Electric Co., Chicago Il. 60623.

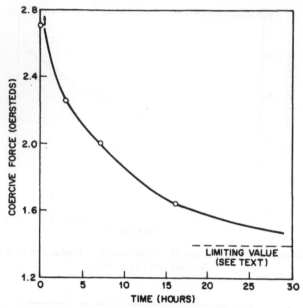

Fig. 13. Decrease in coercive force of SAE 1010 steel after spheriodizing anneal at 680°C (Swisher et al. 1969, Swisher and Fuchs 1970).

obtained from the data shown in fig. 11 for the solubility limit of carbon (150 ppm) in α-iron at 680°C. This amount of carbon on cooling from 680°C, was either retained in solid solution or was present as fine cementite particles.

Because of the importance of having the carbide present as spheroids, furnace cooling was expected to improve properties over air cooled samples. This was indeed the case. The initial magnetization curve for an SAE 1010 steel annealed at 680°C for 16 h and cooled to room temperature over a period of 4 h is shown in fig. 14 and compared with a high purity steel (H_2 annealed at 1300°C for 24 h) and an intermediate low carbon grade steel used primarily in pole pieces for relays. The coercive force and remanance of the spheroidized sample were 0.84 Oe (0.67 A/cm) and 15 000 G (1.5 T) and for the high-purity sample, 0.25 Oe and 6400 G. The results show that spheroidized SAE 1010 steel has a lower initial permeability and higher induction at intermediate field strengths compared to the high-purity steel. In addition, the spheroidized steel more than satisfies the requirements for the intermediate grade low-carbon steel and could very well be used in more critical applications.

Summarizing the results of Swisher et al. (1969) and Swisher and Fuchs (1970), carbon, nitrogen, and oxygen increase the coercivity of steels rapidly cooled from high temperatures, aging at 100°C in nitrogen-bearing samples is quite severe, while aging is not measurable at 100°C in low carbon bearing samples. Aging due to oxygen, sulfur, and phosphorus is insignificant, although moderate and high concentrations of MnO and MnS inclusions contribute to coercive force. Magnetic properties obtained without complete decarburization are

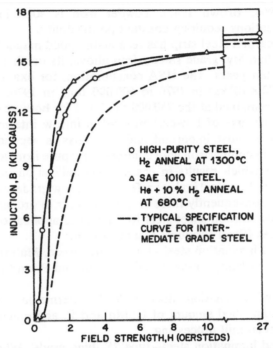

Fig. 14. Initial magnetization curves of high-purity steel after annealing at 1300°C for 24 h, and of SAE 1010 steel after annealing in helium plus 10% hydrogen at 680°C for 16 h (Swisher et al. 1969, Swisher and Fuchs 1970).

sufficient for many applications provided the carbide particles are spheriodized by annealing at temperatures below 727°C.

Rastogi (1972) investigated the influence of Mn and S up to 0.65 and 0.02% respectively on the texture and magnetic properties of non-oriented low carbon stesls. The results indicated that at a particular S level, the 15 kG core loss decreased linearly with increasing Mn concentration. The reduction is associated with increasing grain size, decreasing eddy current loss and changes in texture. The presence of higher Mn and lower S contents is conducive to the development of the {110} texture component at the expense of the {222} component in cold rolled and decarburized low C steels. A significant improvement in permeability at 15, 17 and 18 kG was generally observed as a function of Mn content and explained in terms of texture and microstructure. For a given Mn concentration, magnetic properties generally improved by lowering the S level from 0.02 to 0.005%.

Commercially pure irons are not suitable for ac applications because of their low resistivity. Nevertheless, in addition to being economical, pure iron exhibits very useful magnetic properties and enjoys wide use for dc applications (parts for relays, brakes, plungers, pole pieces, frames, etc.). Relays of various designs are extensively used (tens of millions) in the communications industry (Slauson

1964, Keller 1964, Brown 1967). Powder iron is sometimes used for high frequency applications requiring constant permeability.

Because low-carbon steel strip has reasonably good magnetic properties but is less expensive than low grade silicon–iron alloys, its use has risen considerably within the past ten years. The USA consumption, for example, has risen from 300 000 tons (270×10^6 kg) in 1970 to 600 000 tons in 1976, while the total for silicon–iron has remained at the 750 000 ton level for both years (Luborsky et al. 1977). The primary use of low-carbon steel is in low intermittent duty motors where first cost is most important and core loss is secondary. According to Littmann (1971), in comparison with commercially pure iron, low-carbon steel has high Mn content which results in easier recrystallization, and also has higher resistivity. Low-carbon steel is also preferred over low grade silicon irons due to its lower hardness. Consequently there is less die wear in preparing laminations.

Commercial low-carbon sheet steel for electrical applications are furnished in one of four grades (Steel Products Manual 1975b).

(i) Cold rolled lamination steel – type 1, low grade. Material is furnished to a controlled chemical composition, usually C, Mn and P, and may be full-hard or annealed.

(ii) Cold rolled lamination steel – type 2, intermediate grade. Material is furnished to a controlled amount of C, Mn and P, and magnetic properties are improved by special mill processing.

(iii) Cold rolled lamination steel – type 2S, high grade. Whereas no guarantee is made on types 1 and 2, core loss limits are set for type 2S. Table 8 shows such limits after annealing at 788°C for one hour in a decarburizing atmosphere to reduce the carbon level to about 0.005%.

(iv) Hot rolled sheet steel. Used primarily for dc applications, this steel shows great variability throughout coils.

2.5. Iron dust cores

It is appropriate to consider briefly low coercivity and relatively high permeability iron powders which have found use in the form of cores exhibiting relatively low loss at frequencies up to ~1 MHz. An early application of iron

TABLE 8
Maximum core losses type 2S cold rolled lamination steel*

| Normal thickness | | Max. core loss at 15 kG, 60 Hz | |
(in)	(mm)	(W/lb)	(W/kg)
0.016	0.41	3.3	7.3
0.018	0.46	3.6	7.9
0.021	0.53	4.1	9.0
0.025	0.64	4.9	10.8
0.028	0.71	5.6	12.3

*Steel Products Manual (1975b)

dust cores was for loading coils for telephony at kilohertz frequencies where high stability and very low eddy current losses were required.

The most important magnetic properties that affect core performance are the permeability μ, its temperature coefficient TC, and the Q-factor (quality factor). In a circuit design, a certain minimum Q-value is chosen for the particular frequency used. This then dictates the value of the permeability of the magnetic core.

Cores are manufactured by compressing carbonyl iron powder together with insulating material, a binder and a lubricant. The insulating material coats the iron particles and suppresses eddy current losses. The binder serves to add mechanical strength to the core and acts as diluent for control of permeability. In general, the permeability of a core is of the form $\mu = \mu_i^P$, where μ_i is the intrinsic permeability of the magnetic material and P is the volume fraction of the magnetic material. The intrinsic permeability is highly sensitive to the amount and state of dispersion of the C (and N) which exists up to about 0.8% each in non-decarburized carbonyl iron powder, and to the strain introduced by the method of preparation of the powder and by the molding operation. The volume fraction P is obviously affected by the insulator, binder and molding pressure. Hence the permeability and its temperature dependence are sensitive to the above factors.

Eddy current losses change rapidly with particle size. At low frequencies hysteresis loss predominates and larger particle sizes are most appropriate for use, while at higher frequencies, at which eddy current losses predominate, the finer powders have the lowest total loss in spite of their greater hysteresis (Tebble and Craik 1969).

The Q-factor of the core is governed by the permeability and the resistance. Both quantities are affected by the factors discussed above. Thus, the problem is rather complex and not well understood. Experimental data, as far as they are available, showing the effect of several materials and processing parameters on μ, TC, and Q, are summarized in table 9. It is apparent that an increase in the particle size of the iron powder increases μ and TC, but decreases Q, and an increase in the molding pressure will increase the value of all three magnetic properties. The effect of carbon and nitrogen content noted above was seen by

TABLE 9

The effect of several materials and processing parameters on μ, TC and Q of iron cores (I = increases, D = decreases)

Magnetic property of iron core	Trend on μ, TC and Q by increasing value of				
	Fe particle size	C, N content	binder	insulator	molding pressure
μ	I	D	D	D	I
TC	I	D	D	I	I
Q	D	D	I	I	I

workers who decarburized and denitrided the iron and found an increase in all three properties. On the other hand, if the C and N content varies as a result of the original preparation, then TC goes down with lower C and N content.

Thus, it is concluded that the composition, particle size and structure of the iron particles and the processing variables during core preparation – binder, insulator, molding pressure – have complex effects on the magnetic properties. Reproducibility of these properties call for close control of the powder and processing operations. The reader is referred to the books by Tebble and Craik (1969) and Heck (1974d) for extensive treatments of the subject of dust cores.

Finally, a comparison of dust core materials is presented in fig. 15. Here the loss factor $(\tan \delta)/\mu_i$ is plotted against frequency for Sendust (an Fe–Al–Si alloy), Mo-Permalloy (Fe–Ni–Mo), carbonyl iron-C (decarburized carbonyl iron), and two other carbonyl irons (Kornetzki 1953, Heck 1974d) (the Sendust family of alloys and Mo-Permalloy cores are treated in later sections.) It will be noted that at high frequencies, the loss is smaller the lower the initial permeability of the core.

3. Iron–silicon alloys

3.1. Introduction

While plain carbon steels were the earliest magnetic materials used in electrical machinery, silicon steels containing about 3% Si have catapulted into prommimence as a result of pioneering studies by Barrett et al. (1900, 1901) [as cited by Bozorth (1951, p. 67)] in England just before the turn of the century. Today the silicon steels are the most heavily used of soft magnetic materials.

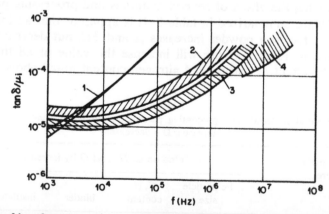

Fig. 15. Ranges of loss factor, $\tan \delta/\mu_i$, as a function of frequency for various dust cores (measured values extrapolated to zero field strength): 1. Sendust ($\mu_i = 80\mu_0$), Mo–Permalloy ($\mu_i = 125\mu_0$); 2. carbonyl iron C ($\mu_i = 50\mu_0$); 3. carbonyl iron HF ($\mu_i = 14\mu_0$); 4. carbonyl iron HFF ($\mu_i = 5-10\mu_0$) (Kornetzki 1953, Heck 1974d).

The US consumption of silicon steel in 1976 amounted to approximately 680 million kg (750 000 tons), split equally between the grain oriented and non-oriented varieties, while the figure is 550 million kg for low carbon steel Luborsky et al. (1978). Figure 16 shows the change in production rate of silicon steel, now commonly known as Flat Rolled Electrical Steel, cold rolled electrical steel or simply electrical steel, in the period 1959–1976 for a number of countries. (It does not include low carbon steel which, as noted above, is used in the US almost as extensively as silicon steel.) The production rate has increased twofold during this period, reflecting the growth in the consumption of electrical energy worldwide. The major producers today are Japan, USSR and the US.

As with the case of low carbon steel, the primary reasons for the prominent use of silicon steel are high induction, low loss and low cost. Compared with carbon steel, silicon steel is lower in loss but higher in cost and hence likely to be even more important in the years ahead in the total energy cost equation. Silicon steel is used almost exclusively in large rotating machinery and transformers while low carbon steel is primarily used in small intermittent duty motors. In Japan, low grade silicon steel (<0.5% Si) is generally used instead of low carbon steel.

As a result of intense research and development efforts over the years, a

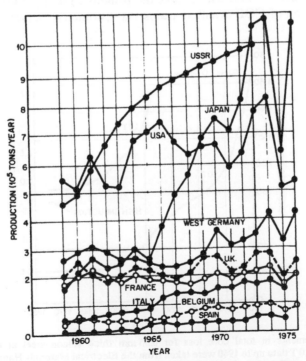

Fig. 16. Production rate of silicon steel in several countries (Wada 1975) (Data: Wada 1974, 1975, 1976, private communication).

number of significant improvements in the quality of silicon steel have emerged. These are reflected in the total core loss curves of fig. 17. The 1900–1930 period saw the improvement primarily coming from advances in the art of steel making and processing such as the control of carbon and other impurities and improvements in hot-rolling techniques. A new breakthrough came in 1934 with the development by Goss (1934) of {110}⟨001⟩, or cube-on-edge (COE), grain oriented material, where most grains have a {110} crystal plane nearly parallel to the sheet surface and an ⟨001⟩ direction nearly parallel to the rolling direction. Guaranteed core loss limits were initially quoted at 15 kG along with the conventional value of 10 kG at the time. Improvements to date have primarily come from modern developments in manufacturing technology. These include the use of the basic oxygen furnace and electric furnace instead of the open hearth furnace for melting and refining, and continuous hot and cold rolling mills instead of the hot-mill pack method for fabrication to sheet stock. Today both oriented and nonoriented grades are prepared by cold rolling, since this technique permits better gage control and improved flatness over hot rolling.

Along with commercial improvements, research in magnetism and in physical metallurgy has led to a greater understanding of the physical bases of core loss and permeability as well as the mechanisms controlling grain orientation. The paper by May and Turnbull (1957) is generally credited with the first recognition of the importance of MnS particles as the principal grain growth inhibitor in the Goss production of COE oriented grains via secondary recrystallization.

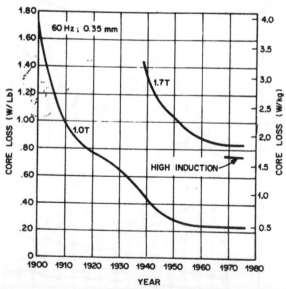

Fig. 17. Improvements in total core loss for 0.35 mm thick silicon steel at 60 Hz and various inductions. The 1.0 T data up to 1960 were taken from the Electrical Materials Handbook (1961). Some 1.7 T data were provided by Littmann (private communication), other data were compiled by the authors from literature.

The search for other such inhibitors finally culminated in the commercial introduction in 1968 of new grades of high-induction low-loss material of exceptionally sharp COE texture using AlN particles along with MnS (Taguchi et al. 1974, 1976). As a result core loss guarantees are now quoted at 17 kG for these new high induction grades (fig. 17).

Another notable development was that of cube-on-face, {100}⟨001⟩, oriented material (Assmus et al. 1957). This texture results in high permeability and low loss in both transverse and longitudinal directions. Due to the high cost of producing this material, however, utilization has been rather limited and it is only used for applications where good permeability and loss are required in all directions (Littmann 1971).

3.2. Iron–silicon phase diagram

Figure 18 shows the most recently drawn iron rich portion of the Fe–Si phase diagram. This diagram is based on the construction by Hultgren et al. (1973) using the evaluation of Köster and Gödecke (1968). The primary difference between fig. 18 and that given by Bozorth (1951, p. 71) is the appearance of the DO_3 ordered α_1 phase, based on Fe_3Si, and of the B2 ordered α_2 phase. Below 540°C the α_2 phase decomposes eutectoidally to the disordered bcc and α_1 phases. Thus changes in mechanical behavior, density and elastic constants in the vicinity of 5–6% Si (~9–11 at.%), as described later, are attributed to the bcc–DO_3 order–disorder transition.

In the solid solution area, the tip of the γ-loop extends to about 2.5% Si. One of the benefits of Si addition to Fe is to enable high temperature heat treatment for grain orientation control without encountering the deleterious effect of the α–γ transformation.

The size of the γ-loop, however, is rather sensitive to small additions of carbon. When 0.1% C is present, the tip could extend to 7 or 8% Si (Bozorth 1951, p. 71, for later references to the Fe–Si–C phase diagram, see Brewer et al. 1973). Thus in practice carbon is suppressed to well below 0.01% through a combination of melting control and decarburization during heat treatment. In addition to being a γ-phase former, carbon along with S, O and N decreases permeability and causes aging effects.

3.3. Physical and mechanical properties

Recent measurements of *density* in Fe–Si alloys have been reported by Machova and Kadeckova (1977) and are presented in fig. 19. These are in good agreement with the data of Bozorth (1951). The density decreases with silicon content and shows a kink in the vicinity of 5% (9.5 at.%) that is attributed to the bcc–DO_3 order–disorder phase transition. For commercial electrical steels which contain less than 4% Si and which may also contain aluminum, studies conducted by ASTM showed that the following equation describes the density quite well (ASTM standard A34)

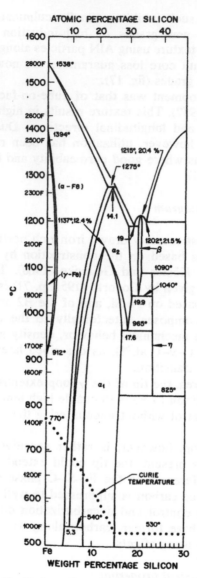

Fig. 18. Fe–Si phase diagram (Metals handbook 1973b).

$$\text{Density (g/cm}^3) = 7.865 - 0.065[\%Si + 1.7(\%Al)].$$

The electrical resistivity is shown in a later figure (fig. 22). Its value rises rapidly with silicon content. Thus besides suppressing γ-phase formation, silicon imparts a second benefit by decreasing eddy current losses. A fairly good ap-

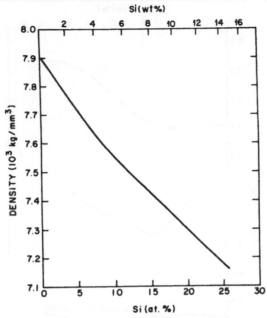

Fig. 19. Density of Fe–Si alloys (Machova and Kadeckova 1977).

proximation (Heck 1974a)* relating the silicon content and resistivity in com-mercial alloys is given by

$$\%\text{Si} = (\rho - 13.25)/11.3 \, (\mu\Omega \, \text{cm}).$$

Since the contribution of aluminum to resistivity is almost the same as silicon, the above approximate formula holds for Si + Al as well, for up to ~0.3% Al often found in non-oriented grades.

Figure 20 shows the variation of *elastic constants* with silicon content (Machova and Kadeckova 1977). Again the minimum in C_{11} near 5% Si is attributed to the order–disorder transition. It can be seen that the elastic constants are rather anisotropic. For 3.2% Si alloy, the value of Young's modulus of COE grain oriented 3.2% Si steel more than doubles in going from 216 GPa respectively (Machova and Kadeckova 1977). For this reason, Young's modulus of COE grain oriented 3.2% Si steel more than doubles in going from the longitudinal direction (~[100]) to 55 degrees away (~[111]) (Armco oriented electrical steels 1974).

Yield and *tensile strengths* generally increase with silicon content, while the ductility as indicated by the percent *elongation* decreases. These are presented in fig. 21. Because of excessive decrease in ductility at high silicon content, practically all commercial grades contain less than 3.5% Si today.

* Cited from Bull. Ass. Suisse Elect., 1953, **44**, 121.

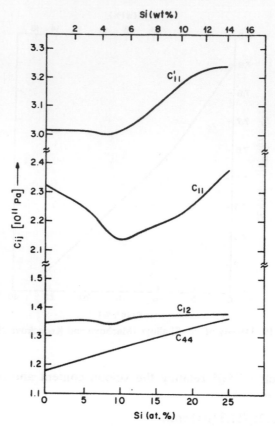

Fig. 20. Elastic constants of Fe–Si alloys (Machova and Kadeckova 1977) $[C'_{11} = \frac{1}{2}(C_{11} + C_{12} + C_{44})]$.

3.4. Intrinsic magnetic properties

Figure 22 shows the change with silicon content of the four important intrinsic magnetic properties: *saturation induction, Curie temperature, magnetocrystalline anisotropy* constant K_1 and *saturation magnetostriction* λ_{100}. A more detailed plot for λ_{111} and λ_{100}, including data for fast and slowly cooled materials reflecting the order–disorder transition in the 4–6% Si regime, is presented in fig. 23 based on measurements of Hall (1959).

It can be seen that the addition of silicon lowers the saturation induction and Curie temperature. These effects are undesirable from the magnetic properties standpoint. On the other hand, the decrease in K_1 and λ, particularly near 6% Si where both λ_{111} and λ_{100} have nearly zero values, is beneficial in terms of increased permeability and lower core loss. For the commercially important grain oriented 3.2% Si composition, the room temperature values of physical, mechanical and magnetic parameters are summarized in table 10.

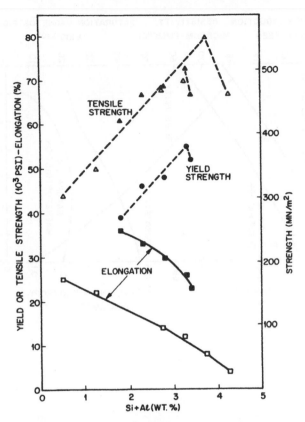

Fig. 21. Mechanical properties of commercial silicon steels. Open symbols refer to values given by Bozorth (1951, p. 78) while closed symbols refer to those typical of current practice (Armco non-oriented electrical steels 1973). Approximate %(Si + Al) is calculated from the resistivity curve, fig. 1, of Littmann (1971).

Fe–Si alloys also develop a uniaxial anisotropy when annealed in the presence of a magnetic field. The value of K_u, however, amounts to less than 10^3 ergs/cm^3 (Fahlenbrach 1955, Fielder and Pry 1959, Sixtus 1962, 1970) which is very small in comparison with the value $\sim 400 \times 10^3$ ergs/cm^3 for K_1. Hence magnetic annealing effects are inconsequential in commercial practice, in contrast to the case of Fe–Ni alloys discussed in section 5. The magnetic annealing induced anisotropy is explained by the theory of directional ordering of atom pairs and will be discussed in section 5.

Summarizing the various effects of silicon on iron, it may be said that on the plus side silicon suppresses the γ-loop and hence suppresses internal stresses, aids in grain growth and in the development of grain orientation, thereby enhancing permeability and reducing the coercivity. It increases the electrical resistivity, thus lowering the eddy current losses. It reduces the magnetocrystalline anisotropy and, in the case near 6% Si, the magnetostriction, thereby further

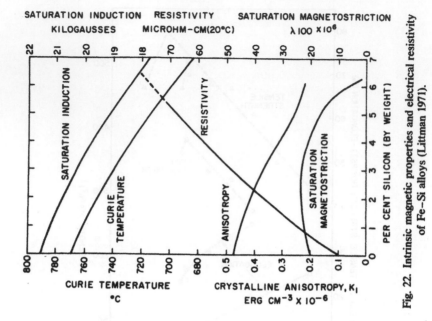

Fig. 22. Intrinsic magnetic properties and electrical resistivity of Fe–Si alloys (Littman 1971).

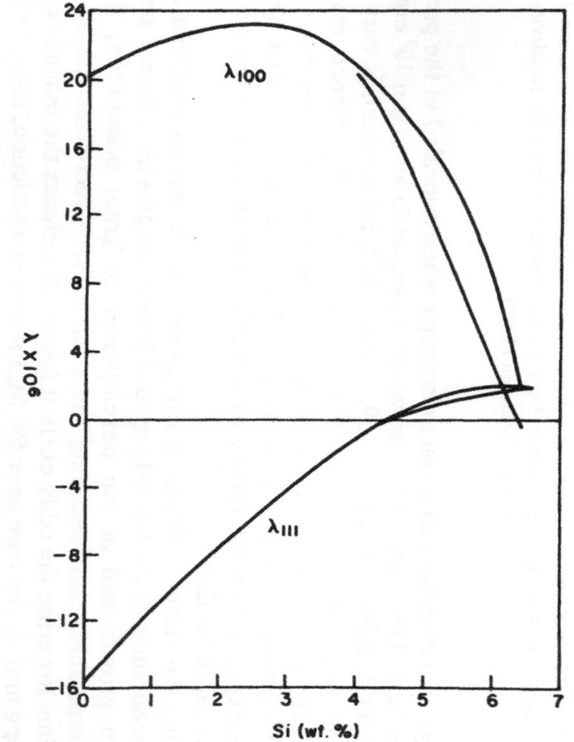

Fig. 23. Magnetostriction constants of Fe–Si alloys (Hall 1959). The lower curves in the 4–7% Si regime are for more ordered (slower cool) samples.

TABLE 10

Magnetic, physical and mechanical properties of 3.2% Si–Fe alloy

Saturation induction (kG)	20
Curie temperature (°C)	745
Magnetocrystalline anisotropy, K_1 (10^3 ergs/cm^3)	360
Saturation magnetostriction (10^{-6})	
λ_{100}	23
λ_{111}	−4
Density (g/cm^3)	7.65
Resistivity (10^{-6} Ω cm)	48
Young's modulus (GPa)	
Single crystals	
[100] direction	120
[110] direction	216
[111] direction	295
(110) [001] grain oriented	
Rolling direction	122
45° to rolling direction	236
transverse direction	200
Yield strength, rolling direction (MPa)	324
Tensile strength, rolling direction (MPa)	345
Percent elongation in 2 in	10
Rockwell B-scale hardness	76

lowering the losses. On the negative side, silicon reduces the saturation induction and Curie temperature, thus diluting the value of the flux density. It also increases the mechanical strength and decreases the ductility, rendering fabrication more difficult.

3.5. Magnetic properties control, core loss

The single most important technical magnetic property of silicon iron is the core loss. The major variables affecting core loss are: composition (silicon content), impurities, grain orientation (texture), applied stress, grain size, thickness and surface condition.

3.5.1. Composition (silicon content)

It has been well established that core loss decreases with silicon content. Figure 24 shows one recent example in the case of non-oriented grades. The primary reasons for the decreasing loss with silicon content are the decrease in magnetocrystalline anisotropy, the elimination of harmful impurities and defects through annealing at higher temperature (due to vanishing γ-loop), and the increase of electrical resistivity with Si content (fig. 22).

It will be seen later (Sakakura et al. 1975) that the influence of Si content on core loss is carried over to the new high induction material as well.

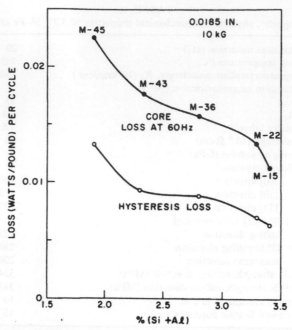

Fig. 24. Hysteresis and 10 kG/60 Hz core loss of Armco non-oriented grades of silicon iron (Armco non-oriented electrical steels 1973).

3.5.2. Impurities

Bozorth (Bozorth 1951, pp. 83–87) summarized much of the early work on the effect of impurities on core loss of silicon iron. In particular, the rather harmdul effect of carbon was recognized very early. Extensive experiments by Yensen (1914, 1924) and Yensen and Ziegler (1936), whose results are summarized in fig. 25, demonstrated the detrimental effects of carbon, sulfur and oxygen. These elements, plus nitrogen, are particularly bad in solid solution form since they tend to be dissolved interstitially, causing large elastic distortion in the lattice and thereby decreasing the permeability and increasing the loss. In addition, they can lead to aging (disaccommodation) effects due to low temperature diffusion. For these reasons, these elements are generally kept below 0.01% in today's silicon steels. In this context an important recent development has been the more widespread use of various liquid steel refining techniques such as AOD (argon oxygen), VAD (vacuum arc) and other vacuum degassing processes. These techniques permit the reduction of carbon levels to as low as 0.005% (Cooke 1975).

Insoluble inclusions and precipitates also increase the loss as they present barriers to domain wall motion and are also sites for spike domains. In section 3.6.2 an example will be given of improvement in core loss of the new high

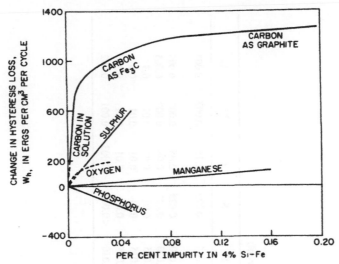

Fig. 25. Effect of various impurities on the hysteresis loss of 4%Si–Fe at B = 10 kG (Bozorth 1951).

induction material through removal of SiO_2 and Al_2O_3 inclusions formed during melting practice.

Of the substitutional solid solution elements, manganese and aluminum are the most commonly encountered in silicon steels. Manganese causes only a moderate loss increase and is generally added to about 0.25% to improve workability in non-oriented grades, although it is generally kept well below 0.1% in oriented grades. Aluminum also increases the ductility of silicon iron and is present to about 0.3% in non-oriented grades, but is again kept low in the oriented grades. A more recent summary of the effect of impurities as well as other factors is given by Littmann (Littmann 1971). Tables 11 and 12, taken from his paper, provide some typical chemical analyses and magnetic properties respectively.

3.5.3. Grain orientation (texture)

Grain orientation is another extremely important variable that affects core loss. It accounted for the breakthrough by Goss (1934) with the development of COE texture. The more recent breakthrough in the development of extra low loss high induction grades, to be discussed in a later section, is again partly a result of sharper COE texture. Commercial exploitation of texture is still well-guarded and much revolves around production conditions. An attempt at some basic known procedures will be described later in connection with the new high induction grades.

Data relating grain orientation and core loss are generally presented in two stages. First, core loss is plotted as a function of μ_{10}, the permeability at 10 Oe (8 A/cm), as in fig. 26 (Littman 1967). Secondly, permeability is plotted as a

TABLE 11

Typical chemical analyses* (in weight percent)

Material	Form	Si	C	Mn	P	S	O	N	Al
Magnetic ingot iron	Mill sheet	0.003	0.015	0.030	0.005	0.025	0.15	0.007	0.003
Low-carbon steel (SAE 1008)	Mill sheet	0.005	0.08	0.40	0.020	0.025	0.05	0.007	0.005
Low-carbon steel	Decarburized	0.005	<0.01	0.40	0.020	0.025	0.05	0.007	0.005
M36 cold rolled	Fully processed	2.2	<0.01	0.25	0.015	0.01	0.01	0.01	0.3
M22 cold rolled	Fully processed	2.8	<0.01	0.25	0.015	0.01	0.01	0.01	0.3
M15 hot rolled	Fully processed	3.3	<0.01	0.25	0.015	0.01	0.01	0.01	0.3
(110)[001] oriented	Fully processed	3.2	<0.002	0.07	0.010	<0.002	<0.003	0.001	0.002

*Littmann (1971)

TABLE 12
Typical magnetic properties* (fully annealed)

	B_s (G)	Density (g/cm³)	Resistivity (μΩ cm)	H_c $B_m = 10$ kG (Oe)	Permeability $H = 1$	$H = 10$	Core loss- 15 kG/60 Hz 60 W/lb at thickness shown 14 mil	18.5 mil	25 mil
Mild steel, 0.2%C normalized	21 400	7.85	–	4.0	–	–	–	–	–
Cast magnetic ingot iron	21 500	7.85	10.7	0.85	3500	1500	–	–	–
Magnetic ingot Fe, 0.08 in sheet	21 500	7.85	10.7	1.11	1800	1575	–	–	(6.0)
Electromagnet Fe, 0.08 in	21 500	7.85	12	1.02	2750	1575	–	–	–
Low-carbon steel, decarburized	21 400	7.85	12.5	0.9	2000	1530	3.7	4.2	5.2
Oriented Fe	21 500	7.85	10.7	0.4	–	1925	2.0	2.4	3.9
M36, cold rolled	20 400	7.75	41	0.45	7400	1485	–	1.75	2.15
M22, cold rolled	19 800	7.65	49	0.39	8100	1450	–	1.65	1.95
M15, hot rolled	19 700	7.65	53	0.35	7800	1390	1.25	–	–
(110)[001] 3.2%Si–Fe	20 300	7.65	48	0.07	16 000	1820	0.63	–	–
(100)[001] 3.2% Si–Fe	20 300	7.65	48	0.07	14 000	1600	1.0	–	–

*Littmann (1971)

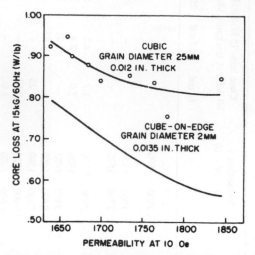

Fig. 26. Effect of permeability at $H = 10$ Oe on core loss for 3.15%Si–Fe with two types of texture
(Littmann 1967).

function of misorientation from the ideal COE texture (fig. 27). It is evident that
improved orientation results in greater permeability which in turn lowers the
core loss. Littmann (1971) estimated that for COE oriented 3.15% Si–Fe the
effect of only 1° less average misorientation is to improve total core loss at 15 kG
and 60 Hz by about 0.03 W/lb or about 5%.

In general the value of μ_{10} is a reasonably sensitive measure of grain
orientation and fairly insensitive to impurity content, residual stress level and
grain size (Littmann 1967, Shilling and Houze 1974). The review by Shilling and
Houze (1974) indicated that the correlation is explainable in terms of a model of
domain structure, proposed by Becker and Döring (1939) for the knee of the
magnetization curve. The model assumes that only the three easy directions
closest to the field can be occupied. Calculations based on the model appear to
fit well for the (100)[001] oriented material (see upper curve of fig. 27). However,
the agreement is not as good for the (110)[001] oriented material (Shilling and
Houze 1974, Swift et al. 1973). In addition, more recent experimental data by
Shilling et al. (1977, 1978) indicate that μ_{10} cannot be used to describe misorientation for angular deviations $\leqslant 2°$. In a 3% Si–Fe (110)[001] crystal, for example,
$\mu_{10} = 2030$ for both $\theta = 0$ and $\theta = 2°$, where θ is the tilt angle of [001] out of the
sheet plane. For such small misorientations, the magnetization is aligned in the
field direction against crystal anisotropy (away from easy direction).

Magnetic domain observations, summarized by Shilling and Houze (1974)
indicate two components of the domain structure in (110)[001] oriented silicon
iron: the *main* structure consisting of large flux carrying slab domains with
magnetizations along [001], and the *supplementary* structure which flux closes
the grain surface with magnetizations along [010] and [100]. Core loss associated
with rearrangement of the supplementary structure is considerably higher than

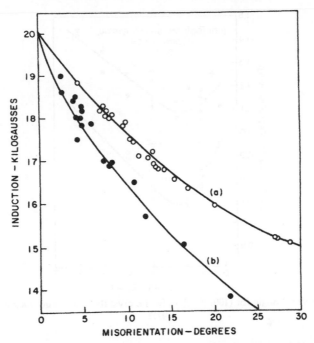

Fig. 27. Induction at $H = 10$ Oe as a function of misorientation (Littmann 1971). (a) Cube texture
data (Foster and Kramer 1960), (b) COE texture data (Littmann 1967).

that of the main structure. Supplementary structure is found to increase with
misorientation; hence an increase in core loss with misorientation. However,
recently both Shilling et al. (1977, 1978) and Nozawa et al. (1977, 1978) have
discovered that for a tilt angle of [001] out of the surface of less than 2°, core loss
increases with decreasing tilt. This result is illustrated in fig. 28. Domain
observations by both groups of workers indicate that below a tilt of 2°, the 180°
main domain wall spacing increases rapidly. It is well known (Shilling and Houze
1974) that losses increase with an increase in domain wall spacing. The change in
domain wall spacing and supplementary structure with misorientation is shown
in the domain patterns of Shilling et al. (1977, 1978) (fig. 29) and quantitatively
plotted in fig. 30 from the work of Nozawa et al. (1977, 1978). The latter data
refer to samples stretched to a tensile stress of 1.5 kg/mm².

3.5.4. Applied stress
The above mentioned studies (Shilling et al. 1977, 1978, Nozawa et al. 1977,
1978) were also aimed at a change in domain structure and core loss with applied
stress. An applied tensile stress in the rolling direction decreases the core loss
for all values of tilt angle as seen in figs. 28 and 31. Domain observations,
illustrated in fig. 32, indicate a decreasing domain wall spacing with applied
stress, thus accounting for the decreased core loss. It may be noted that the

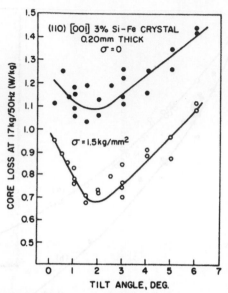

Fig. 28. Core loss vs tilt angle for (100) [001] 3%Si–Fe crystals of 0.20 mm thickness (Nozowa et al. 1977).

largest reduction in wall spacing occurs at very small angles ($\leqslant 2°$). In addition, while a tensile stress is effective in suppressing or even eliminating the supplementary structure when a small tilt (2°) sample is magnetized to 17 kG, such stress is far less effective for larger tilts (5°) Shilling et al. (1977, 1978). Hence the stress effect on core loss is larger for smaller tilt angles.

This dependence of core loss on tilt angle is most likely responsible for the larger improvement of higher permeability material over lower permeability material when both are stressed. In fig. 33 the decrease in core loss with stress is greater for $\mu_{10} = 1913$ than for $\mu_{10} = 1810$, both grain sizes being the same. As discussed in a previous section, higher permeability is associated with better orientation (smaller tilt). It will be seen later that one of the additional benefits of the newer high induction oriented grades is a further decrease in core loss resulting from a tensile stress exerted by the inorganic insulation coating, the effect of which is greater than that in conventional oriented grades.

Similar studies on the stress dependence of core loss for specimens cut from different angles to the rolling direction of (110)[001] oriented 3% Si–Fe sheet have also been conducted (Swift 1976, Yamamoto and Taguchi 1975). According to Swift (1976): (i) in the region $0 \leqslant \theta \leqslant 10°$ and $55° \leqslant \theta \leqslant 90°$, tensile stress produces lower losses than no stress or compressive stress (here θ is the angle between the rolling ([001]) direction and stress direction, (ii) losses are stress-independent for $\theta = 55°$ as predicted by magnetoelastic theory, and (iii) compressive stress produces lower losses than tensile stress of the same magnitude for $30° \leqslant \theta \leqslant 55°$; however, unstressed losses are lowest here. Yamamoto and

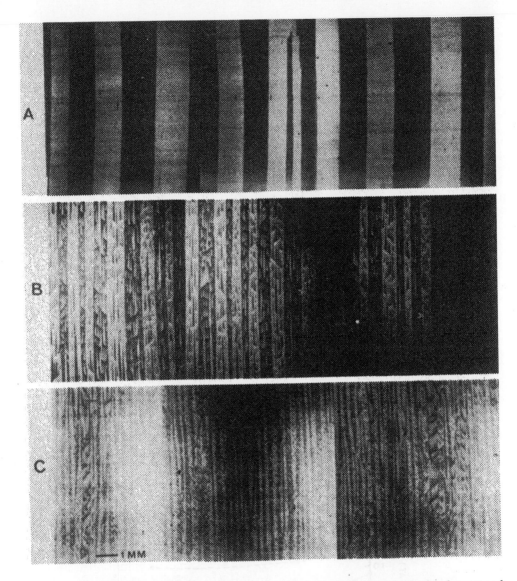

Fig. 29. Domain structure of (100) [001] 3%Si–Fe crystals in the 60 Hz demagnetized and unstressed state. (a) Tilt angle $\theta = 0$, (b) $\theta = 2°$, (c) $\theta = 5°$ (Shilling et al. 1978).

Taguchi (1975) reported similar findings, with these additional observations: (a) for $55° < \theta \leq 90°$, tensile stress beyond a certain point increases the loss again, (b) compared with uncoated specimens, coated samples show less variation with tension for $\theta < 55°$, and more variation for $\theta > 55°$ (these observations are attributed to an isotropic tensile stress imparted by the surface coating).

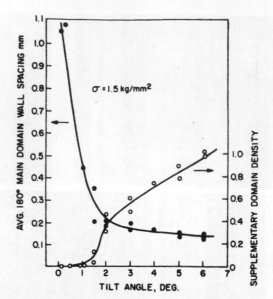

Fig. 30. Dependence of 180° main domain wall spacing and supplementary domain density (normalized at 1 for 6°) on tilt angle for crystals of fig. 28 (Nozowa et al. 1977).

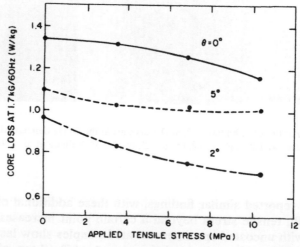

Fig. 31. Core loss vs tensile stress for (100) [001] 3%Si-Fe crystals of various tilt angles (Shilling et al. 1978).

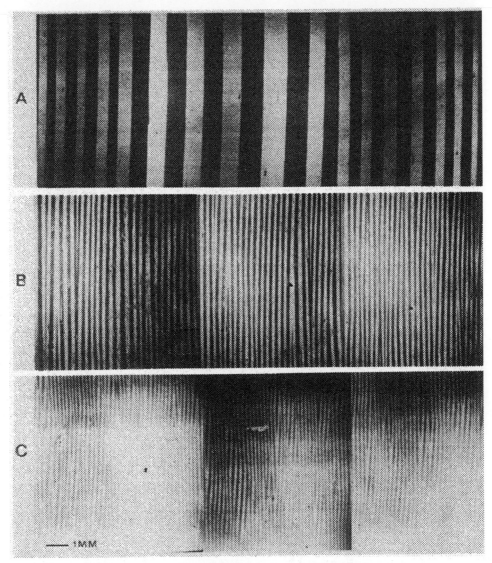

Fig. 32. Domain structure of crystals of fig. 29 stressed to 13 790 kPa, showing decreased wall spacing under stress (Shilling et al. 1978). (a) $\theta = 0$, (b) $\theta = 2°$, (c) $\theta = 5°$.

3.5.5. Grain size

Littmann (1967) has demonstrated that everything being equal, the core loss goes through a minimum with grain size at a grain diameter of about 0.5 mm (fig. 34). The results of fig. 34 are for a wide range of laboratory specimens of 3.15% Si–Fe with comparable purity and COE texture. The earlier controversies as reviewed by Bozorth (1951, p. 86) are due to masking effects such as purity and

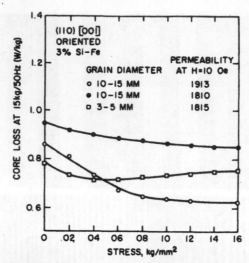

Fig. 33. Core loss vs tensile stress for (100) [001] grain-oriented 3%Si–Fe of different degrees of orientation (permeability) and grain size (Yamamoto and Nozawa 1970).

grain orientation which often accompany efforts at producing materials of varying grain size. As pointed out by Littmann (1971), for the non-oriented commercial grades, better permeability and core loss are associated with the larger grains as a result of annealing at high temperature to remove carbon and nitrogen. For the oriented grades, core loss is generally lower with the larger grains as well because of better orientation. Obviously the key to improvement is to achieve high purity, well-oriented material with a grain diameter of ~0.5 mm.

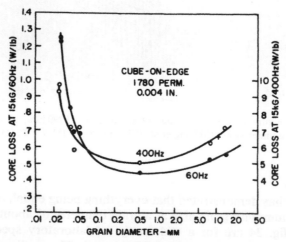

Fig. 34. Core loss vs grain size for 3.15%Si–Fe sheets of comparable texture and purity (Littmann 1967).

It has been observed (Shilling and Houze 1974) that in the larger grain size range (~5 mm), a decrease in grain size is accompanied by a decrease in the 180° main domain wall spacing. Thus a decrease in losses is expected on this basis. On the other hand, at the finer grain size end (~0.05 mm), large magnetostatic energy is built up at the grain boundaries (Shilling 1971). The result is an increase in losses due to domain wall pinning.

3.5.6. Thickness

In general, a decrease in sheet thickness decreases the core loss due to geometrical effects. At very small thicknesses, surface pinning effects become dominant and core loss rises again. The opposing effects have been demonstrated by Littmann (1967) and by Sharp and Overshott (1973). The results in fig. 35 show a core loss minimum in the vicinity of 0.005 in for grain oriented material and 0.010 in for unoriented material. Thus the reduced surface pinning associated with better oriented material drives the thickness minimum to smaller values. It will be seen later that this improvement continues with the new high induction oriented grades.

Domain observations have revealed different structures for samples of different thickness. Hubert (1965) for example, has observed a "chessboard" domain structure for thick samples where magnetoelastic energy dominates, and a "columnar" structure for thin samples where domain wall energy is important. However, a systematic evaluation of domain structure with thickness and its correlation with core loss has not been done.

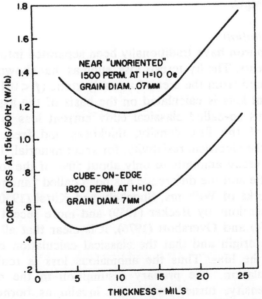

Fig. 35. Core loss vs sheet thickness for 3.15%Si–Fe of comparable purity (Littmann 1967).

3.5.7. Surface morphology

The increased losses for thin specimens mentioned in the previous section is primarily the result of domain pinning effects at the surface due to magnetostatic energy associated with grain misorientation. In addition, pinning can also occur at rough spots on the surface. Therefore it is expected, and generally observed, that achievement of a smooth surface will lead to lower losses. This has been demonstrated by several workers (Yamamoto and Nozawa 1970, Littmann 1975b, Swift et al. 1975) using chemical polishing techniques. One interesting synergistic effect of a smooth surface is the interaction with a tensile stress either applied externally or inherent from a surface coating. Yamamoto and Nozawa (1970) showed that beyond a certain stress level, the core loss of both smooth and rough surfaced samples becomes identical. However, this stress level is higher in the case of lower permeability material.

In commercial production of grain oriented material an inorganic coating is developed by applying MgO to the surface and annealing at the final high temperature ($\sim 1200°C$) hydrogen purification cycle. Therefore the formation of coating is also intimately related to the resulting surface morphology. Swift et al. (1975) conducted a detailed investigation in this regard. Using optical and electron microscopic techniques to characterize the surface morphology and relating the latter to core losses, they found that smooth surfaces were obtained by annealing with an Al_2O_3 separator and that rough surfaces were produced by annealing with an MgO separator, annealing for a time too short to reduce surface oxides, or annealing in an oxidizing atmosphere. The Al_2O_3 and MgO results confirmed an earlier finding by Jackson (1974) that lower losses were obtained when COE textured 3% Si–Fe sheet was annealed with an Al_2O_3 coating as compared with MgO.

3.5.8. Loss separation

Losses in silicon iron have traditionally been separated into hysteresis and eddy current components. The hysteresis loss, such as that shown in figs. 24 and 25, is generally measured from the area of a nearly static ($f < 0.1$ Hz) hysteresis loop. The eddy current loss is calculated on the basis of the assumption of uniform flux density. This so-called classical eddy current loss is proportional to the second power of the flux density, thickness and frequency and inversely proportional to the electrical resistivity, for sheet material. The sum of these two components generally amounts to only about 50% of the total measured loss in oriented 3% Si–Fe and the difference has been called "anomalous loss". Following the early works of Williams, Shockely and Kittel (1950) and Pry and Bean (1958), and discussions by Becker (1959) and more recent reviews by Shilling and Houze (1974) and Overshott (1976), it is clear that all losses are basically eddy current in origin and that the classical calculation of eddy current loss underestimates the loss. Thus the anomalous loss is really the result of an unrealistic calculation. The primary assumption in the classical calculation, uniform flux density, turned out to be invalid as borne out by numerous experiments, particularly those involving actual observation of domain wall motion during magnetization. Basically, these experiments indicate that during the magnetization cycle and with changing frequency, the number and nature of moving domain walls are altered, and the moving walls move with different velocities depending on such factors as pinning at local inhomogeneities and proximity of neighboring walls, etc. The situation is rather complicated and even semi-quantitative conclusions cannot be drawn at present. The most recent work by Hill and Overshott (1976) addressing the question of the effect of pinning points, for example, showed that artificially induced pinning points affect the behavior of single crystals and single grains within polycrystalline material differently. Thus one cannot calculate the losses based on physically realistic models at the present time.

3.5.9. Rotational power loss

While losses such as discussed in previous sections are usually referred to measurements in alternating magnetic fields, materials are often used in fields which rotate. Obvious examples are ac generators and induction motors, but rotational processes are also important in the T-joints of three phase transformers and behind the teeth of electrical machines.

Early measurements by Brailsford (1939) indicate that under nearly static conditions, the rotational hysteresis loss is generally larger than the alternating hysteresis loss at low flux density, but becomes smaller at high flux density. The changeover occurs at a value of $(B - H)/B_s$ in the 0.6–0.7 range for a number of materials including silicon iron. Corresponding measurements at 50 Hz were measured more recently by Boon and Thompson (1965) for hot-rolled ("unoriented"), cube-on-edge and cube textured 3% Si–Fe. Some of their results are shown in fig. 36. The trend is similar to that under static conditions. It is generally concluded that (i) the alternating power loss in cube-textured material is slightly less than in cube-on-edge textured material and that (ii) the rotational power loss of the former is considerably lower than that of the latter.

Detailed studies of rotational hysteresis loss have been correlated with domain observations in several instances (Boon and Thompson 1964, Narita 1974, 1975, Phillips and Overshott 1974), including the superposition of dc bias and applied stress (Phillips and Overshott 1974). Theoretical treatments have also been made (Kornetzki and Lucas 1955, Brailsford 1970). In addition, studies have also been conducted on localized power loss and flux distributions at corners and T-joints of three phase transformers to determine the extent of rotational flux (Moses et al. 1972, Nakata et al. 1975).

3.6. New high induction oriented materials

The major event in silicon steel during the past two decades has been the development of new high induction oriented grades from Japan. For some time now conventional grain oriented silicon steel has been employed almost exclusively as core material for power and distribution transformers. Because of a fairly sharp COE texture, this type of steel exhibits a high induction of about

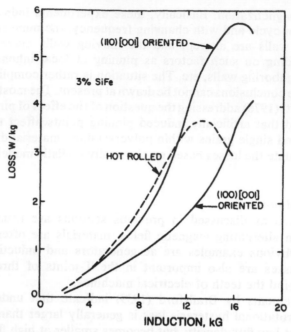

Fig. 36. Rotational loss at 50 Hz vs induction for three grades of 3%Si–Fe (Boon and Thompson 1965).

18.3 kG at $H = 10$ Oe (18.3 T at $H = 8$ A/cm) and a low core loss at high induction (~1.11 W/kg at 15 kG and 50 Hz for 0.35 mm strip). This core loss is nearly two and a half times lower than that of the best non-oriented fully processed grades.

In 1968, Nippon Steel introduced commercially a newly developed grain oriented steel Orientcore · HI-B, commonly called HI-B, which exhibits even greater induction (19.2 kG at $H = 10$ Oe) and lower loss than conventional oriented grades (Taguchi et al. 1974). These new grades are primarily distinguished by a much sharper COE texture, with average misorientations around 3° instead of 7° for conventional oriented grades. Nippon Steel has now been joined by Kawasaki Steel with the latter's introduction in 1974 of RG-H grades made by another new process (Goto et al. 1975). And in 1977 Allegheny Ludlum announced the commercial development of similar grades (Malagari 1977) made by yet another process developed at the General Electric Company (Grenoble 1977, Fiedler 1977a). The HI-B and RG-H grades are also produced in several countries through licensing agreements. Hence, it is fairly certain that new developments of new high-induction grades will continue for some time to come.

Table 13 shows a comparison of core loss at 17 kG and 50 Hz for several popular lamination thicknesses. It may be seen that the core losses of the new high induction grades are lower by about 20%. These lower losses can thus be

TABLE 13*
Core loss of grain oriented silicon steels at 17 kG (W/kg)

Type	Grade	0.27 mm		0.30 mm		0.35 mm	
		50 Hz	60 Hz**	50 Hzt	60 Hz	50 Hzt	60 Hz
High induction	M0H	–	–	1.05	(1.41)	–	–
High induction	M1H	(1.04)	1.40	1.11	(1.49)	1.16	(1.55)
High induction	M2H	(1.11)	1.49	1.17	(1.56)	1.22	(1.63)
High induction	M3H	(1.17)	1.57	1.23	(1.65)	1.28	(1.72)
High induction	M4H	–	–	–	–	1.37	(1.84)
Conventional	M4(27H076)††	1.27	1.67	–	–	–	–
Conventional	M5(30H083)††	–	–	1.39	1.83	–	–
Conventional	M6(35H094)††	–	–	–	–	1.57	2.07

* Data in parenthesis calculated on the basis of Loss(60 Hz)/Loss(50 Hz) = 1.34, typical of high induction material
** Malagari (1977), data for 0.28 mm thick samples
† Taguchi (1977)
†† ASTM designation A725-75

translated into smaller transformers for a given rating and/or same size transformers with higher design flux densities.

The new high induction grades differ from conventional oriented grades in the use of grain growth inhibitors, the processing procedure, and the type of glass coating. A comparison is made below.

3.6.1. Manufacturing process
The conventional commercial process for developing the COE texture, outlined in table 14, involves hot-rolling a cast ingot near 1370°C to a thickness of about 2 mm, annealing at 800–100°C, and then cold rolling to a final thickness of 0.27–0.35 mm in two steps, with a recrystallization anneal (800–1000°C) sandwiched in-between. The final cold reduction is about 50%. The cold-rolled material is first decarburized to ~0.003%C at about 800°C in wet $H_2 + N_2$, a step which also results in primary recrystallization. It is then box annealed in dry hydrogen at 1100–1200°C to form the COE texture by secondary recrystallization. During the box anneal, impurities are removed and inclusions of the grain growth inhibitors are dissolved and absorbed in the glass film formed on the surface. Manganese sulfide has been the most common grain growth inhibitor used.

As far as can be ascertained from the reports of the HI-B developers (Taguchi et al. 1974, Taguchi and Sakakura 1969, Yamamoto et al. 1972, Takashima et al. 1976, Taguchi et al. 1976, Taguchi 1977) the HI-B process, also outlined in table 14, differs from the conventional process in two major areas: (i) the use of AlN in addition to MnS as the grain growth inhibitor and (ii) a one-stage cold rolling with large thickness reduction (80–85%). A recent improvement was achieved by reducing the number of small inclusions by increasing the

TABLE 14

Comparison of manufacturing processes for grain oriented silicon steel

Conventional	HI-B
Steelmaking (MnS)	Steelmaking (AlN + MnS)
Hot rolling (1370°C)	Hot rolling
Annealing (800–1000°C)	Annealing (950–1200°C)
Cold rolling (70%)	Cold rolling (85%)
Annealing (800–1000°C)	
Cold rolling (50%)	
Decarburizing (800°C, wet $H_2 + N_2$)	Decarburizing
Box annealing (1200°C, dry H_2)	Box annealing
RG-H	GE-AL
Steelmaking (Sb + MnSe or MnS)	Steelmaking (B + N + S or Se)
Hot rolling	Hot rolling (1250°C)
Annealing	Annealing
Cold rolling	Cold rolling (>80%)
Annealing	
Cold rolling (65%)	
Decarburizing	Decarburizing
Box annealing (820–900°C, then 1200°C, dry H_2)	Box annealing

temperature of the molten steel when alloying elements are added (Sakakura et al. 1975).

Usually a large final cold reduction results in a sharper COE texture in the finished strip, but it also enhances the tendency for grain growth of the primary grain structure, which is undesirable. Strong inhibitors such as AlN permit the larger cold reduction by restraining the primary grain growth. For the AlN particles to be effective, however, the annealing step following hot-rolling is generally carried out at a higher temperature than conventionally, followed by rapid cooling (Taguchi et al. 1976). In addition, the glass film which contains the absorbed aluminum has a smaller coefficient of thermal expansion than the usual glass film. This imparts a large tensile stress on the strip and lowers the core loss.

In the Kawasaki RG-H process (Goto et al. 1975, Shimanaka et al. 1975) (table 14) these are the distinguishing features:

(i) Antimony was added along with MnSe or MnS as a grain growth inhibitor. It is thought that Sb segregates to boundaries of the primary recrystallized grains, further aiding the growth retardation mechanism of the MnSe or MnS inclusions.

(ii) In the cold rolling process, RG-H material is given a two-stage rolling as in the conventional process, but the second stage is closer to a 60–70% reduction rather than 50%.

(iii) In the final box annealing process, the coiled strip is first kept for 5–50 h at 820–900°C, followed by purification above 1100°C. The thought here is that when the secondary recrystallization is carried out at low temperature, the orientation becomes sharper.

(iv) Finally, in the RG-H process, a new low thermal expansivity inorganic phosphate coating was developed to impart a larger tensile stress to the finished strip (Shimanaka et al. 1975).

In the General Electric-Allegheny Ludlum process (Malagari 1977, Grenoble 1977, Fiedler 1977a, b, Maucione and Salsgiver 1977) (table 14) these features differ from conventional techniques: (i) use of boron and nitrogen together with sulfur or selenium as grain growth inhibitors, (ii) large final cold reduction in excess of 80%. It is thought that controlled amounts of B,N and S (or Se) in excess of MnS segregate to the grain boundaries to retard grain growth. Hence inclusions are not involved in the process. In addition, it has been pointed out (Malagari 1977) that since Mn and S are rather low in this new steel (Mn = 0.030–0.040%, S = 0.015–0.025%), MnS solubility temperature is lowered and hence hot working can be done below 1300°C. This is advantageous in terms of lower fuel costs, added mill life, etc.

3.6.2. Structure and magnetic properties

The various factors affecting core loss, as discussed in section 3.5 are seen to have important implications in the new high induction materials as well. Figure 37 shows the core loss for three ranges of *silicon content* (Sakakura et al. 1975).

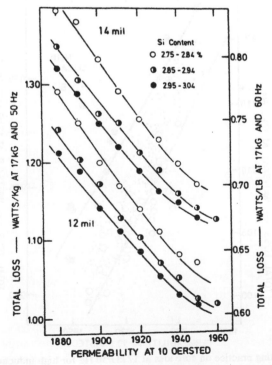

Fig. 37. Effect of Si content on loss at 17 kG/50 Hz for high induction material (Sakakura et al. 1975).

As expected, lower core loss goes with high Si content. The effect is somewhat greater for the thicker samples (14 mil) presumably because of the greater eddy current effect there. Figures 38 and 39 show the reduction of core loss as a result of minimizing oxide *inclusions* through melting practice. The temperatures of the molten steel when alloying elements are added are 1660°C, 1630°C and 1600°C for melts A, B and C respectively. Already mentioned was the fact that special processing techniques have resulted in a much sharper COE *texture*, with an average misorientation of 3° instead of 7° for conventional oriented material. In addition to giving rise to a higher permeability which leads to a lower core loss per se, the sharper texture results in several secondary features:

(i) In fig. 40, it is seen that the high induction material shows a more rapid increase in core loss with increasing angle from the rolling direction. This fact thus needs to be taken into account in practical transformer design.

(ii) The sharper texture leads to a greater loss improvement under tensile stress conditions (fig. 41). Hence in commercial practice the tensile stress imparted by the inorganic coating is more effective for the high induction material over the conventional material (Yamamoto et al. 1972). This effect is further exploited by the recent development of a new inorganic phosphate coating with reduced thermal expansivity, 4×10^{-6} vs 8×10^{-6} for conventional

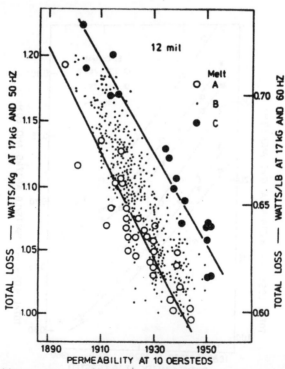

Fig. 38. Effect of melting practice on core loss at 17 kG/50 Hz for high induction material (Sakakura et al. 1975). See text for details.

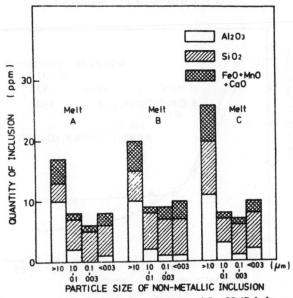

Fig. 39. Inclusion content and particle size of melts of fig. 38 (Sakakura et al. 1975).

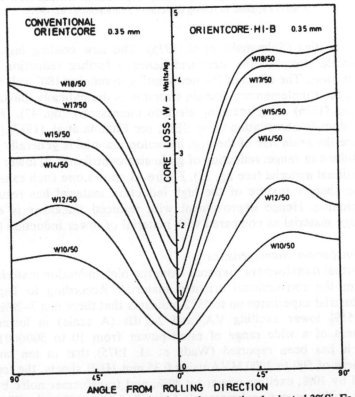

Fig. 40. Comparison of core loss of high induction and conventional oriented 3%Si–Fe as a function of angle from rolling direction. Curves for various induction levels are shown (Taguchi et al. 1974).

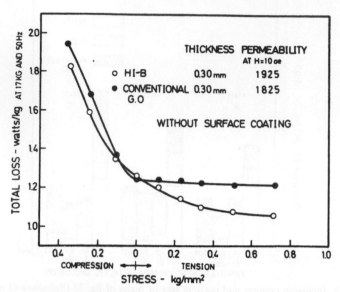

Fig. 41. Comparison of core loss of high induction and conventional oriented 3%Si–Fe as a function of stress applied in rolling direction (Yamamoto et al. 1972).

phosphate coating (Shimanaka et al. 1975). The new coating imparts a much greater tensile stress to the steel and hence a further reduction in loss and magnetostriction. The effect of the new coating in refining 180° wall spacings and supressing the supplementary domain structure is dramatically illustrated by Irie and Fukuda (1976) using scanning electron microscopy (fig. 42). [For a recent review of domain observation using SEM, see Fukuda et al. (1977).]

(iii) Since the *grain size* of the high induction material is generally large, in the several millimeter range, reduction of grain size contributes to lower core loss as in conventional material (see fig. 22). Figure 43 shows one such example.

(iv) The sharper texture of the high induction material has resulted in less surface pinning. Hence improvement with reduced *thickness* is extended to thinner gage material as compared with material of lower induction (fig. 44).

3.6.3. Transformer characteristics

To date actual transformers prepared from the high induction material appear to outperform the conventional oriented material. According to Taguchi et al. (1974) industrial experience up to 1972 indicates that there is a 7–20% lower core loss, 20–55% lower exciting VA, and 2–7 dB (A scale) in lower noise for transformers of a wide range of rated power from 10 to 500 000 kVA. More recently, it has been reported (Wada et al. 1975) that in ten large 3-phase transformers of 500 to 1000 MVA using 0.35 mm HI-B sheets, the total loss was decreased by 10%, exciting power by 40%, and transformer noise by 4–7 dB as compared with conventional material. A detailed comparison by Thompson and Thomas (1975) in a practical 3-phase transformer core showed that localized

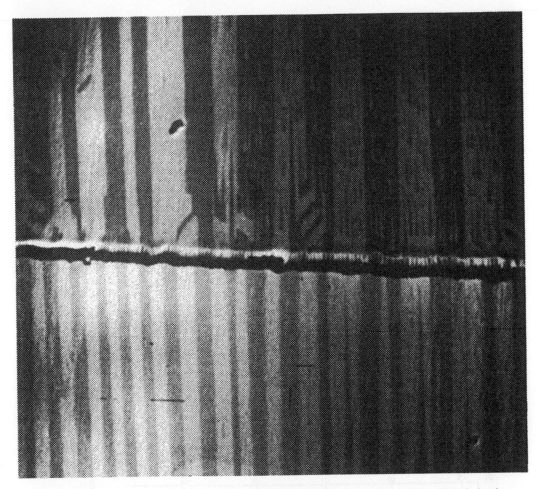

Fig. 42. Effect of inorganic glass coating on domain structure of RG-H, a new high induction 3%Si–Fe, using a SEM technique (Irie and Fukuda 1976, Fukuda et al. 1977). Lower half, with coating, fine domain spacing. Upper half, coating removed, coarse domain spacing plus supplementary structure.

power loss within the T-joint area is 8% higher for the high induction material, but 18% lower in the yokes and limbs. As a result, the high induction material reduced the overall power loss of the core by 13%. This figure is in line with previous findings (Taguchi et al. 1974, Wada et al. 1975).

3.6.4. Outlook
The recent trend towards increased texture perfection, smaller grain size, improved purity and controlled tensile stress should continue to drive down the core loss. This trend is summarized in table 15. State of the art in 1975 commercial production reached a value of 0.95 W/kg at 17 kG/50 Hz for the core

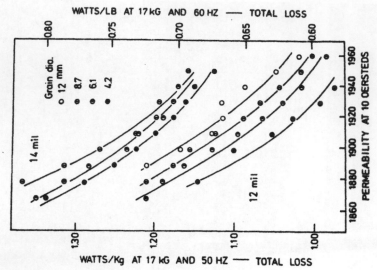

Fig. 43. Effect of grain size on core loss of high induction 3%Si-Fe (Sakakura et al. 1975).

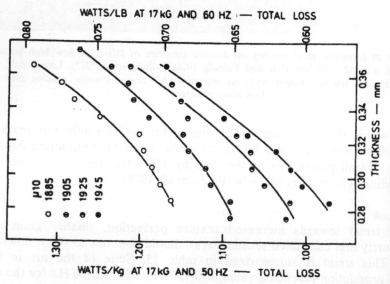

Fig. 44. Effect of thickness on core loss of high induction 3%Si-Fe (Sakakura et al. 1975).

TABLE 15

Recent trends in total loss (in W/kg at 17 kG/50 Hz) of oriented 3%Si–Fe, values of μ_{10} in parentheses

Note	0.25 mm	0.28 mm	0.30 mm	0.35 mm
[a]	–	0.95 (1960)	0.98 (1958)	1.13 (1965)
[b]	–	0.86	0.95	1.10
[c]	0.58 (1980)	–	–	–
[d]	0.53 (2030)	–	–	–
[e]	–	–	0.55	–

[a] State of the art commercially, 1975 (Taguchi et al. 1976)

[b] State of the art commercially, projected for 1980 (Taguchi et al. 1976)

[c] Chemical polish + 10 mm array scratches in transverse direction + 2.8 kg/mm² applied tension (Yamamoto and Taguchi 1975)

[d] Single crystal with [001] tilted 2° out of surface + 10 340 kPa tension (Shilling et al. 1977, 1978). Value calculated from 60 Hz data assuming Loss(60 Hz)/Loss(50 Hz) = 1.32 based on table 1 of Taguchi et al. (1976)

[e] Theoretical prediction (Taguchi et al. 1976)

loss of 0.28 mm material and it has been predicted to go down to 0.86 W/kg by 1980 (Taguchi et al. 1976). Laboratory experiments (Yamamoto and Taguchi 1975) involving a chemical polish of the surface followed by the application of scratches spaced at 10 mm intervals in the transverse direction and then stretching the sample in tension, resulted in a loss of 0.58 W/kg for a 0.24 mm sample of $\mu_{10} = 1980$. Another laboratory experiment (Shilling et al. 1977, 1978) involving a (110) [001] single crystal with [001] tilted out of the surface by 2° showed a loss of 0.7 W/kg at 17 kG/60 Hz, or about 0.53 W/kg at 17 kG/50 Hz using a conversion ratio of 1.32 based on 50 and 60 Hz values measured on very high induction samples (Taguchi et al. 1976) A theoretical value of 0.55 W/kg for 0.30 mm material has been given by Taguchi et al. (1976). Thus it appears that a further 50% decrease in loss from the present already low values might be achievable in the future.

In addition, since the early work of May and Turnbull (1957) there has been a steady stream of fundamental investigations into the nature of recrystallization and behavior of solutes and inclusions: Fiedler (1958, 1961, 1964, 1967), Hu (1961, 1964), Taguchi and Sakakura (1966), Flowers (1967), Grenoble and Fielder (1969), Littmann (1969), Sakakura (1969), Sakakura and Taguchi (1971), Matsumoto et al. (1975), Takashima and Suga (1976), Nichol and Shilling (1976), Stroble (1976), Salsgiver and McMahon (1976), Datta (1976). Improved understanding along these lines as well as new revelations from domain studies will undoubtedly contribute to future developments.

3.7. Other iron–silicon alloy developments

(i) High silicon alloys. It is well known that a silicon content near 6% is highly attractive because of near zero magnetostriction and hence reduced strain sensitivity and transformer noise (Getting 1967). However, because of the brittleness beyond about 3.5% Si as mentioned previously, there has been no wide scale commercial development of the high silicon products. Ames et al. (1969) reported laboratory studies of siliconizing commercial grain oriented 3.25% Si–Fe via decomposition of $SiCl_4$. In this way improvements were obtained at higher Si as expected, but the reduced saturation resulted in smaller improvements at higher induction (17 kG). In addition, the excitation fields had to be increased at these levels. Variations in hot rolling schedules have been attempted to improve the ductility of 6.5% Si steels with some improvement noted, but the resulting magnetic properties have so far been inferior (Ishizaka et al. 1966).

(ii) Cube-on-face texture. Cube textured material has the potential advantage of lower core loss in rotating machinery. However, as reviewed by Littmann (1971) and Taguchi (1977) due to difficulties in obtaining a sharp texture and the expense of fabricating this type of material in commercial practice, the cube textured material saw only limited production and use.

(iii) Thin-gage material. Grain oriented thin-gage material in thicknesses of 0.1, 0.05 and 0.025 mm are being used mostly in tape wound cores as high quality high frequency transformers. These generally have higher saturation but also higher losses and lower permeability as compared with Permalloy cores (see section 5).

(iv) Strand cast material. Since strand or continuous casting offers the advantages of more uniform composition and less scrap, studies had been conducted to see if low loss grain oriented material could be produced by this process (Littmann 1975a). The grain structure of the strand cast strip generally consists of small equiaxed grains at the surface, a large columnar grained region with $\langle 100 \rangle$ directions normal to the slab surface, and a centervane of small equiaxed grains of random orientation. Littmann (1975a) found that by a pre-rolling treatment at 1100°C the large elongated structure could be broken up and a fairly uniform grain structure with reasonably good texture could be developed in the final product. For a 0.27 mm thickness material, the average core loss at 17 kG/60 Hz is 1.59 W/kg and an average permeability of $\mu_{10} = 1840$.

3.8. Core loss limits of electrical steel

In the US flat-rolled electrical steel is available in the following classes according to AISI (Steel Products Manual 1975a):

(i) Non-oriented fully processed. The properties are developed by the producer and the core loss limits are provided by table 16.

(ii) Non-oriented semiprocessed. The properties must be developed by the consumer. The core loss limits when tested after a quality evaluation anneal are

TABLE 16

Maximum core losses, non-oriented fully processed types, flat rolled electrical steel (ASTM A677)

AISI type	ASTM designation	Thickness		Max. core loss at 15 kG, 1.5 T	
		in	mm	(W/lb) 60 Hz	(W/kg) 50 Hz
M-15	36F145	0.014	0.36	1.45	2.53
M-15	47F168	0.0185	0.47	1.68	2.93
M-19	36F158	0.014	0.36	1.58	2.75
M-19	47F174	0.0185	0.47	1.74	3.03
M-19	64F208	0.025	0.64	2.08	3.62
M-22	36F168	0.014	0.36	1.68	2.93
M-22	47F185	0.0185	0.47	1.85	3.22
M-22	64F218	0.025	0.64	2.18	3.80
M-27	36F180	0.014	0.36	1.80	3.13
M-27	47F190	0.0185	0.47	1.90	3.31
M-27	64F225	0.025	0.64	2.25	3.92
M-36	36F190	0.014	0.36	1.90	3.31
M-36	47F205	0.0185	0.47	2.05	3.57
M-36	64F240	0.025	0.64	2.40	4.18
M-43	47F230	0.0185	0.47	2.30	4.01
M-43	64F270	0.025	0.64	2.70	4.70
M-45	47F305	0.0185	0.47	3.05	5.31
M-45	64F360	0.025	0.64	3.60	6.27
M-47	47F460	0.0185	0.47	4.60	8.01
M-47	64F575	0.025	0.64	5.75	10.01

TABLE 17

Maximum core losses, non-oriented semiprocessed types, flat rolled electrical steel (ASTM A683)

AISI type	ASTM designation	Thickness		Max. core loss at 15 kG, 1.5 T	
		in	mm	(W/lb) 60 Hz	(W/kg) 50 Hz
M-27	47S178	0.0185	0.47	1.78	3.10
M-27	64S194	0.025	0.64	1.94	3.38
M-36	47S188	0.0185	0.47	1.88	3.27
M-36	64S213	0.025	0.64	2.13	3.71
M-43	47S200	0.0185	0.47	2.00	3.48
M-43	64S230	0.025	0.64	2.30	4.01
M-45	47S250	0.0185	0.47	2.50	4.35
M-45	64S280	0.025	0.64	2.80	4.88
M-47	47S350	0.0185	0.47	3.50	6.10
M-47	64S420	0.025	0.64	4.20	7.31

TABLE 18

Maximum core losses grain-oriented fully processed types flat rolled electrical steel (ASTM A665-15kG, ASTM A725-17kG)

AISI type	ASTM designation	Thickness		Maximum core loss			
		in	mm	at 15 kG, 1.5 T		at 17 kG, 1.7 T	
				(W/lb) 60 Hz	(W/kg) 50 Hz	(W/lb) 60 Hz	(W/kg) 50 Hz
M-4	27G053	0.0106	0.27	0.53	0.89	–	–
M-4	27H076	0.0106	0.27	–	–	0.76	1.27
M-5	30G058	0.0118	0.30	0.58	0.97	–	–
M-5	30H083	0.0118	0.30	–	–	0.83	1.39
M-6	35G066	0.0138	0.35	0.66	1.11	–	–
M-6	35G094	0.0138	0.35	–	–	0.94	1.57

shown in table 17. The anneal customarily involves a one hour soak at 844°C (except 788°C for M47) using a suitable atmosphere to reduce the carbon level to ≤0.005%.

(iii) Non-oriented full hard.   The material is produced to a hardness of $R_B \geqslant 84$. Magnetic properties, to be developed by the customer, are generally inferior to the semiprocessed grades.

(iv) Grain oriented fully processed.   The properties are fully developed by the producer and the core loss limits are provided by table 18.

A summary table is given in table 19 to provide equivalent qualities of several popular grades of fully processed steels according to major internationally-accepted standards. Table 20 summarizes the salient characteristics and typical applications of various grades according to AISI. As of 1978, the new high induction oriented grades have not been standardized. The properties can be found in table 13. Their applications are the same as for grades M4–M6.

TABLE 19

Major internationally-accepted standards (maximum core losses in W/kg at 15 kG/50 Hz, values inside parentheses at 17 kG/50 Hz)

Thickness	USA AISI (1975) ASTM (1975)	Japanese JIS C 2553 (1975)	German DIN 46400: blatt 1 (1973)	British BS 601: part 2 (1973)
		Grain oriented		
0.30 mm 0.012 in	M-5 0.97 (1.39)	G09 (1.33) G10 (1.47) G11 (1.62)	VM97-30 0.97 (1.50)	30M5 0.97 30M6 1.05
0.35 mm 0.014 in	M-6 1.11 (1.57)	G10 (1.51) G11 (1.66) G12 (1.83)	VM111-35 1.11 (1.65)	35M6 1.11 35M7 1.23
		Non-oriented		
0.35 mm 0.014 in	M-15 2.53 M-19 2.75 M-22 2.93 M-36 3.31 M-43 4.01	S12 3.10 S14 3.60 S18 4.40 S20 5.00 S23 5.50	V110-35A 2.70 V130-35A 3.30	250 2.50 265 2.65 315 3.15 335 3.35
0.50 mm 0.020 in	M-19 3.28 M-22 3.48 M-36 3.86 M-43 4.34 M-45 5.74 M-47 8.66	S12 3.60 S14 4.00 S18 4.70 S20 5.40 S23 6.20 S30 8.00 S40 10.50 S50 13.00 S60 15.50	V135-50A 3.30 V150-50A 3.50 V170-50A 4.00 V200-50A 4.70 V230-50A 5.30 V260-50A 6.00 V360-50A 8.10	355 3.55 400 4.00 450 4.50

TABLE 20
Some characteristics and typical applications for flat rolled electrical steel*

AISI Type	Some characteristics	Typical applications
	Oriented types	
M-4 M-5 M-6	Highly directional magnetic properties due to grain orientation. Very low core loss and high permeability in rolling direction.	Highest efficiency power and distribution transformers with lower weight per KVA. Large generators and power transformers.
	Non-oriented types	
M-15	Lowest core loss, conventional grades. Excellent permeability at low inductions.	Small power transformers and rotating machines of high efficiency.
M-19 M-22 M-27	Low core loss, good permeability at low and intermediate inductions.	High-reactance cores, generators, stators of high-efficiency rotating equipment.
M-36 M-43	Good core loss, good permeability at all inductions and low exciting current. Good stamping properties.	Small generators, high-efficiency, continuous duty rotating ac and dc machines.
M-45 M-47	Ductile, good stamping properties, good permeability at high inductions.	Small motors, ballasts and relays.

* (Steel products manual 1975a)

4. Iron–aluminum and iron–aluminum-silicon alloys

4.1. Introduction

Commercially important iron–aluminum and iron–aluminum–silicon soft magnetic alloys are characterized by high resistivities, high hardness, high permeability and low loss. The trade names for these alloys are Alfer (13% Al), Alfenol (16% Al), Alperm (17% Al) and the Sendust family of iron–aluminum–silicon alloys (4–7% Al, 7–13% Si). These alloys are primarily used as recording head material nowadays (see table 30 in section 5) although the binary alloys may also find limited application for magnetostrictive transducer applications.

The iron-rich portion of the iron–aluminum phase diagram is shown in fig. 45. Aluminum stabilizes the bcc form of iron (α) i.e., suppresses the formation of the fcc γ-phase, and is soluble in iron up to the ordered alloy Fe_3Al (β_1) at the lower temperatures. Ordered Fe_3Al (β_1) is also bcc and ordering can be suppressed by quenching from temperatures above 600°C. The iron-rich portion of the iron–silicon phase diagram is similar to the iron–aluminum diagram and is shown in fig. 18 in section 3. Silicon stabilizes the bcc form of iron and is soluble in α-Fe to the extent of ~5% at 500°C.

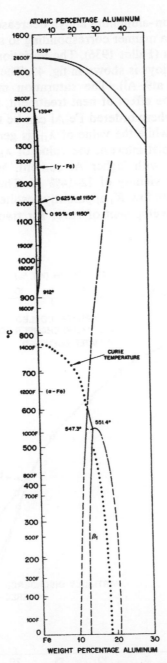

Fig. 45. The Fe-rich portion of the Fe–Al phase diagram (Metals handbook 1973a).

4.2. Iron–aluminum alloys

The magnetization of iron–aluminum alloys decreases with increasing aluminum content up to 15% Al in a manner corresponding to simple dilution, similar to the behavior of Fe–Si alloys (Fallot 1936). The variation of the anisotropy constant K_1 for iron–aluminum alloys is shown in fig. 46. The anisotropy constant goes to zero at ~13% Al (~25 at% Al). The saturation magnetostriction behavior is shown in fig. 47. Note the effect of heat treatment, i.e. state of atomic order, in the composition range where ordered Fe_3Al can be formed.

It may be noted that while the value of λ_{111} is generally small in the range of composition of commercial interest, the value of λ_{100} first rises sharply with Al content and then drops with higher Al content. Meanwhile, the value of K_1 decreases to zero in the vicinity of 12–14% Al. Thus, a compromise of a large value of λ_s consistent with low K_1 apparently resulted in the choice of 13% Al in the original Japanese development of Alfer for transducer use (Masumoto and Otomo 1950).

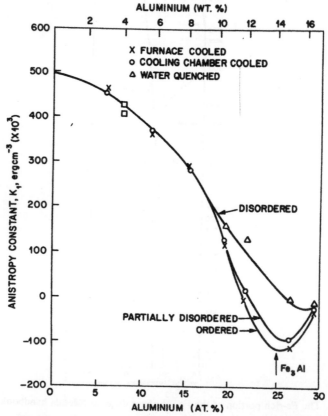

Fig. 46. Anisotropy constant K_1 of Fe–Al alloys (Hall 1959).

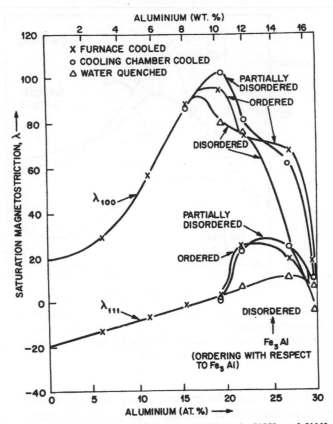

Fig. 47. Saturation magnetostriction ($\lambda \times 10^6$) of Fe–Al alloys in [100] and [111] crystallographic directions (Hall 1959).

The rather large value of λ_{100} led Borodkina et al. (1960) to investigating the possibility of improving λ_s via texture control. Their early efforts on a Fe–10% Al alloy are shown in fig. 48. Adopting the Goss technique of producing {110} ⟨001⟩ texture for silicon steel (Goss 1935), Borodkina et al. (1960) prepared sheet materials by rolling to thickness reductions of 50–60% sandwiched by intermediate annealing at ~900°C. X-ray examination showed that in the rolled condition the ideal sheet texture contained three prominent components: {111} ⟨112⟩, {100} ⟨011⟩ and {112} ⟨$\bar{1}$10⟩, the first being the strongest. The saturation magnetostriction is fairly low, between 10 and 20×10^{-6}, with the 45° direction having the highest value. After a two-hour anneal at 900°C, the sheet recrystallized to a dual texture of {110} ⟨001⟩ and {100} ⟨001⟩, with the value of λ_s rising to 42×10^{-6} in the rolling direction. The texture is still far from perfect, however, as λ_{100} is over 90×10^{-6} for this alloy (see fig. 47). If the rolled sheet is annealed at temperatures greater than 900°C, the texture becomes diffuse and λ_s is decreased. Subsequent work (Bulycheva et al. 1965) showed that oriented material can be made with $\lambda_s \sim 75 \times 10^{-6}$.

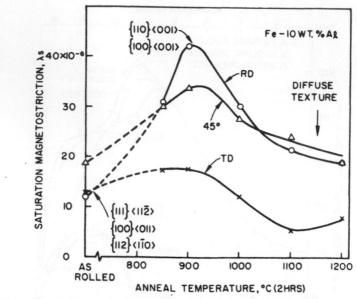

Fig. 48. Values of saturation magnetostriction measured in three directions of Fe–10%Al alloy sheet, as a function of annealing temperature. RD = rolling direction, TD = transverse direction. Ideal textures after particular annealing are indicated (Chin 1971a, data plotted from table 3 of Borodkina et al. 1960).

Aluminum and silicon have the greatest effect on the resistivity of Fe and both are nearly equal in their effect (see fig. 4 in section 2). Sugihara (1960a) and Helms [unpublished, cited by Adams (1962)] have shown that isotropic Fe–Al alloys possess magnetic properties comparable to equivalent composition Fe–Si alloys. This is illustrated in table 21. Examination of this data reveals that the properties of the dilute alloys favor the Fe–Al system while the converse is true for the more concentrated alloys.

The Fe–Al alloys are sensitive to magnetic annealing particularly in composition ranges where K_1 is small. The effect of heat treating Fe–Al alloys in a magnetic field has been studied by several researchers. Masumoto and Saito (1951, 1952a, b) studied alloys up to 17% Al and showed that initial and maximum permeabilities, coercive forces, and residual induction were sensitive to cooling rates, particularly in the 14–16% Al range. The highest permeabilities were obtained for alloys containing ~16% Al when quenched from 600°C. The 17% alloy was named Alperm and has found applications as recording head material on account of its high hardness and resistivity along with good permeability. The 13% Al alloy exhibited high magnetostriction and was named Alfer. It has been used as magnetostrictive material.

The effect of magnetic annealing is shown in figs. 49 and 50. Sugihara (1960b) studied vacuum melted alloys prepared from electrolytic iron (99.9%) and high purity aluminum (99.99%). The specimens were ring-shaped (0.8 mm thick)

TABLE 21
Comparison of magnetic properties of Fe–Al and Fe–Si alloys*

% Al or Si[a]	dc magnetic properties					Core loss at 10 kG (W/lb) 60 Hz
	$\mu_{20}^{d)}$	μ_{max}	H_c (Oe)	B_r (G)	$B_m^{b)}$ (G)	
1.0 Al	2041	20 214	0.329	14 101	16 348	0.70
0.94 Si	860	14 780	0.442	14 140	16 400	0.69
3.0 Al	1923	12 815	0.450	13 741	16 278	0.70
2.81 Si	1720	17 440	0.192	–	15 340	0.58
5.0 Al	2381	12 625	0.329	10 153	15 446	0.64
4.8 Si	1410	17 160	0.201	7410	15 230	0.43
8.0 Al	513	3461	0.550	3242	13 392	1.07
6.4 Si	1720	28 860	0.147	11 310	14 050	0.40
8.0 Al[c]	1140	50 900	0.269	10 900	12 800	0.45
6.4 Al[c]	1390	66 940	0.197	12 100	14 600	0.37

* Adams (1962)
[a] Annealed at 1000°C in hydrogen
[b] $H = 30$ Oe
[c] Magnetic anneal at 1.7 Oe
[d] $B = 20$ G

prepared from hot rolled strip and then annealed at 1000°C for one hour and furnace cooled. The rings were then cooled from 800°C at 150°C per hour in a field of 12.5 Oe. The results, shown in fig. 49, and Sugihara's hysteresis loops, indicate that magnetic annealing is most effective for alloys of ~10% Al. Sugihara also showed that non-magnetically annealed 15–16% Al alloys exhibited larger initial and maximum permeabilities compared to the magnetically annealed Al alloys. The response of Fe–Al alloys to annealing in a magnetic field of 1.7 Oe is shown in fig. 50. The data shown in this figure were obtained on 0.014 in laminated cores.

The cold ductility of Fe–Al alloys is good. Six percent Al alloys can be cold reduced by as much as 90%, vacuum melted sixteen percent alloys can be reduced from 0.007 in to 0.003 in at room temperature (Nachman and Buehler 1954). Nachman and Buehler determined 1000°C as the optimum hot working temperature for the 16% Al alloy (Alfenol). They have also shown that "cold rolling" of 0.125 in strip can be successfully carried out at temperatures near 575°C. This temperature is slightly above the Fe_3Al order–disorder temperature (Nachman and Buehler 1954).

4.3. Textured iron–aluminum alloys

A cube texture, (100) [001], can be produced in Fe–Al alloys (Bozorth 1942, Foster and Pavlovic 1963). Foster and Pavlovic produced predominantly cube textured strip (0.025–0.008 in) from a cold rolled nominally Fe–3% Al alloy 0.125 in thick. The 0.125 in strip was obtained from a vacuum melted ingot by a

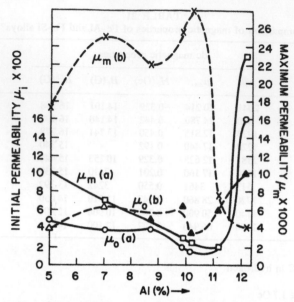

Fig. 49. Initial and maximum permeabilities of Fe–Al alloys heat treated without (a) an applied field and (b) with an applied field (Sugihara 1960).

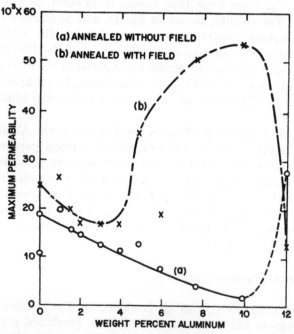

Fig. 50. Response of Fe–Al alloys to annealing in a magnetic field of 1.7 Oe. Samples were 0.014 in thick laminated cores (Adams 1962).

combination of hot forging, cold rolling and annealing. The texturing anneal consisted of 5 h at 1000°C in dry hydrogen. The magnetic properties of the predominantly cube textured material are shown in table 22. The 0.025 in strip was 70% cube textured; the inductions at 0° and 90° were not quite equal, indicating some deviation out of the {100} plane towards the {110} plane. The degree of texturing decreased to some degree with decreasing strip thickness, while the 60 Hz power losses decreased considerably.

Pavlovic and Foster (1965) produced (111) textured Fe–Al alloy (8–9% Al) which exhibited low remanence and coercive force. The strip was produced by a combination of hot and cold working. Texture was developed by heat treating at either 1000°C or 1200°C in dry hydrogen. Some of their results are shown in table 23. The low remanence is consistent with the (111) texture, i.e., lack of an easy direction in the plane. Other typical properties exhibited by the (111) textured alloy were $\mu_{max} = 2700$, $B_{20} = 12.9\,\text{kG}$, $B_{100} = 16.1\,\text{kG}$, resistivity $= 90\mu\Omega$ cm, and Rockwell B hardness $= 87$. The low remanence and hardness suggested that this material would be suitable for relay cores and could be substituted for the Fe–3.25% Si alloy for this application.

TABLE 22

Properties of cube-textured Fe–3%Al alloy at different angles from the rolling direction*

Sheet thickness (in)	Angle to rolling direction	Coercive force (Oe)	Induction at 10 Oe (kG)**	60 Hz, 15 kG Core loss (W/lb)
0.025	0°	0.25	16.1	1.99
	45°	0.32	13.9	–
	90°	0.29	15.3	2.32
0.018	0°	0.24	15.8	1.50
	45°	0.32	14.1	–
	90°	0.27	15.2	1.61
0.014	0°	0.22	15.8	1.25
	45°	0.28	14.2	–
	90°	0.27	15.4	1.43
0.012	0°	0.27	15.5	–
	45°	0.26	14.2	–
	90°	0.28	15.2	–
0.008	0°	0.22	14.7	–
	45°	0.21	13.6	–
	90°	0.21	14.3	–

* Foster and Pavlovic (1963)
** The table in the original paper (Foster and Pavlovic 1963) printed "gausses"; it should be "kilogausses"

TABLE 23
Direct current magnetic properties of 8% and 9% Fe–Al sheets*

Alloy	Rolled	Heat treatment	Coercive force (Oe)	Residual induction (kG)
8%Al–Fe	Hot[a]	2 h at 1000°C	0.768	7.66
8%Al–Fe	Hot[a]	2 h at 1000°C		
		and 2 h at 1200°C	0.246	2.68
9%Al–Fe	Hot[a]	2 h at 1000°C	1.150	7.04
9%Al–Fe	Hot[a]	2 h at 1000°C		
		and 2 h at 1200°C	0.285	1.94
9%Al–Fe	Cold[a]	2 h at 1000°C	0.716	8.50
9%Al–Fe	Cold[a]	2 h at 1200°C	0.278	2.01
9%Al–Fe	Cold[b]	2 h at 1000°C	0.709	4.33
9%Al–Fe	Cold[b]	2 h at 1200°C	0.30	1.66

* Pavlovic and Foster (1965)
[a] Epstein strips cut in the rolling direction
[b] Punched rings

4.4. Iron–aluminum–silicon alloys

Within the Fe–Al–Si ternary system, initial permeabilities as high as 35 000, maximum permeabilities as high as 110 000, resistivities of ~100 $\mu\Omega$ cm, and hysteresis losses about 15% of that exhibited by grain oriented 4% silicon–iron have been achieved. Because of their high resistivities, strain insensitivity and hardness, alloys in the ternary system are used chiefly for synchronous motors and magnetic recording heads.

The excellent dc magnetic properties of Fe–Si–Al alloys (Sendust family of alloys) were first reported by Masumoto (1936). Certain compositions within this alloy system possess very low magnetostriction and anisotropy, high initial permeability and high resistivity (fig. 51). Optimum magnetic properties are obtained over a narrow composition centered at 9.6% Si and 5.4% Al, the alloy is generally referred to as *Sendust*. These alloys are hard and brittle and thus limited in their use, in cast form, to dc applications and as insulated powder cores for high frequency inductors. However, Adams (1962) and Helms and Adams (1964) produced sheet by various powder metallurgy techniques, such as slip casting, liquid-phase sintering, and sintering of loose powder. The magnetic properties of a Sendust alloy (~9.5% Si and ~5.6% Al) produced by various processes are shown in table 24.

The sheet preparation and processing technique described by Helms and Adams (1964) consisted of (i) preparing a cast ingot of the Sendust alloy deliberately deficient in iron by 25%, (ii) pulvering the ingot to −120 mesh powder, (iii) adding to this an amount of iron powder or iron flakes to restore the iron deficiency, (iv) thoroughly mixing the mixture and (v) rolling the powder to

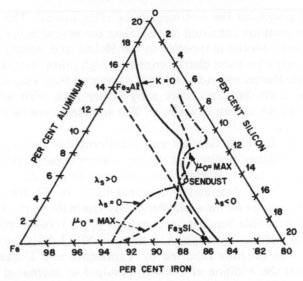

Fig. 51. Lines of highest initial permeability μ_0 and lowest magnetostriction λ and crystal anisotropy K, in Fe–Si–Al alloys (Bozorth 1951, p. 100, Zaimovsky and Selissky 1941).

produce 0.018–0.035 in strip. The size of the ductile iron powder determines the strip thickness and serves as an excellent bonding agent for the hard, brittle, iron-deficient Sendust particles. Sintering this strip in argon for a short time (10–30 min at 1200°C) improves the strength of the strip without embrittling it for further cold rolling. Green strip containing iron flakes can be cold rolled directly

TABLE 24
Magnetic properties of Sendust alloys produced by various fabricating processes[††]

Process	Thickness (in)	Density (g/cm³)	dc permeability 20	dc permeability max.	B_r (G)	B_m H = 30 Oe	H_c (Oe)
Cast ring**	0.250	6.96	15 000	110 000	7000	14 000	0.04
Sintered*							
Liquid phase[a]	0.350	6.50	6200	25 000	5000	8200	0.10
Liquid mix[b]	0.260	6.51	10 000	33 700	4500	9500	0.07
Slip case[†c]	0.022	5.50	2000	15 000	4400	7900	0.23
Loose powder[d]	0.400	6.30	11 500	46 000	5200	8800	0.08
Hot pressed[e]	0.250	6.86	–	–	–	–	–

* Sintered alloys:
  [a] (FeSi–AlSi–Fe) mix, 30 ksi 1175°C, He atm
  [b] 20%(FeSi–AlSi–Fe) + 80% Sendust powder mix, 30 ksi
  [c] (FeSi–AlSi–Fe) mix, water-alginate slurry
  [d] Sendust powder, no pressure
  [e] Sendust powder, 2 ksi 1260°C
** Cast alloy: 1000°C H_2 anneal
† Losses = 0.2 W/lb, 60 Hz
†† Adams (1962)

to final strip size without the intermediate bonding anneal. The final annealing treatment on laminations consisted of annealing for seveal hours at 1200–1300°C in dry argon. Cast material is superior to the Helms and Adams sintered sheet.

Because Sendust alloys are characterized by high initial permeability and low saturation values, Helms and Adams (1964) compared their magnetic results with similar data for a 4% Mo–Permalloy alloy rather than with higher saturation alloys. These data are shown in table 25. The samples were in the form of ring core configurations.

The core losses for both Sendust and Mo–Permalloy (at $B_m = 6000$ G) were comparable but at higher frequencies the loss values became more favorable for Sendust laminated cores (Helms and Adams 1964). At a frequency of 1.25 MHz, Haas (1953) determined a permeability of 71 for a 0.01 in thick Sendust laminate versus 46 for a Mo-Permalloy laminate of the same thickness.

Commercially available Sendust cores exhibit $\mu_i = 10\,000$–$30\,000$, $\mu_m = 80\,000$–$160\,000$ and losses of 20–100 erg/cm^3. A new Fe–Si–Al–Ni alloy, referred to as *Super Sendust*, was recently reported by Yamamoto and Utsushikawa (1976). They showed that the addition of 2–4% Ni resulted in decreased coercivity and increased permeability. A 5.33% Si, 4.22% Al and 3.15% Ni alloy, annealed at 750°C in H_2 for one hour and cooled to room temperature at a rate of 1.7°C/s, exhibited $\mu_m = 56\,800$, $H_c = 0.096$ Oe, $B_r = 8100$ G and $B_m = 9800$ G. However, by cooling in a magnetic field of 20 Oe from 700°C after annealing at 1250°C in H_2 for one hour, this alloy exhibited $\mu_m = 165\,000$, $H_c = 0.030$ Oe, $B_r = 9540$ G and $B_m = 10\,900$ G.

A "squeeze casting" technique, which consists of forcing molten Sendust alloys into a die of predetermined shape, was recently reported for producing recording heads (Senno et al. 1977). The squeeze-cast alloys, after annealing at 900°C, exhibited $\mu_i \geqslant 10\,000$ at 1 kHz (thickness of 150 μm) before and after molding cores, $H_c \leqslant 0.03$ Oe and $B_{10} \geqslant 8000$ G. Head wear was reduced to $\frac{1}{3}$ of that of conventional Fe–Si–Al alloy heads and comparable to that of ferrite heads.

TABLE 25

Direct current magnetic properties of Sendust and 4%Mo–Permalloy alloys[*][a]

Material	Core Type	Thickness (in)	μ_{20}	μ_{max}	H_c (Oe)	B_m (G)
Sendust	Cast	0.250	15 000	110 000	0.040	11 000
Sendust	Compact	0.250	18 180	39 125	0.034	8715
Sendust[b]	Laminated	0.014	16 130	36 625	0.062	8964
4%Mo–Permalloy[c]	Laminated	0.014	35 700	580 300	0.006	8920

[*] Helms Jr. and Adams (1964)
[a] Nominal composition of Sendust strip: 9.5% Si, 5.7% Al
[b] Resistivity = 106 $\mu\Omega$ cm
[c] Resistivity = 55 $\mu\Omega$ cm

A proprietary precision casting process for the Sendust alloy has been developed by the Furukawa Electric Co., Ltd. for the production of sound castings, which, if necessary, can withstand cutting and grinding procedures (Negishi 1977). The magnetic properties were said to be superior to hard Permalloy and ferrite for recording head applications.

Sputtered thick (12μm) Sendust films suitable for recording heads have recently been prepared (Shibaya and Fukuda 1977). The films exhibit saturation flux densities of 11 000 G and effective permeabilities of 240 at 20 MHz.

5. Nickel–iron alloys

5.1. Introduction

The nickel–iron alloys in the Permalloy range, from about 35 to 90% nickel, are probably the most versatile soft magnetic alloys in use today. With suitable alloying additions and proper processing, the magnetic properties can be controlled within wide limits. For high quality transformers and for magnetic shielding, values of initial permeability up to 100 000 have been achieved at cryogenic temperatures as well as at room temperature. Values of coercive force from 0.002 to 10 Oe can now be adjusted with reasonable precision. Thus applications have been found in magnetic memories where the concern is no longer on striving for the lowest possible coercivity, but on controlling its value to 1 or 2 or 5 Oe as may be dictated by device design. For magnetic amplifiers, inverters, converters and other saturable reactors, square loop material with remanence practically equal to saturation can be provided. And for unbiased unipolar pulse transformers where low remanence and constant permeability are called for, such properties can be attained as well.

Aiding in the attainment of wide ranging properties and applications is the ease of fabrication of the Permalloys. Material can be precision cold-rolled to foils of 2.5μm thick or drawn to wires of 10μm diameter. Thin films can easily be made by vapor deposition, sputtering and electroplating. In fact, as an example of one of newest applications, Permalloy films have been used to define the circuit patterns in magnetic bubble device technology, and have also been successfully utilized to detect bubble movement based on the magnetoresistance effect.

These ever-widening applications could not have been achieved, however, without the knowledge gained from the extensive studies of the fundamental magnetic parameters and their dependence on composition and structure, as well as the application of innovative processing techniques such as vacuum induction melting.

A detailed account of nickel–iron alloys up to 1950, including historical developments, is given in the classic work of Bozorth (1951). The present chapter principally emphasizes the more recent developments. In order to present a coherent picture, however, selected data from Bozorth's account are presented, particularly in the light of present understanding. Only bulk properties and applications are discussed.

5.2. Nickel–iron phase diagram

The most recent equilibrium phase diagram for binary nickel–iron alloys was compiled by Goldstein (1973) and is shown in fig. 52. This diagram differs little from the one from Hansen (1958a) with the exception of some minor details near the ordered Ni₃Fe composition.

The equilibrium solvus lines are difficult to establish due to slow diffusion kinetics in the solid state. For alloys that are usually cooled from high temperature to room temperature, the face-centered cubic γ phase exists from about 30% to 100% nickel. In the low nickel regime the γ phase undergoes a martensitic transformation to the body-centered cubic α phase. There is considerable hysteresis in the transformation, as noted in fig. 53.

Of prime importance to the magnetic behavior of nickel–iron alloys is the appearance of the long-range ordered Ll_2 structure at the Ni₃Fe composition below $500 \pm 5°C$. The ordering kinetics is very slow, e.g., annealing at 450°C for one week is required to reach the near maximum value of order. Above the order–disorder temperature, the alloys in the vicinity of the Ni₃Fe composition exhibit short range order, the kinetics of which are rather fast. As will be seen later, the most pronounced effect of ordering is a large change in the value of the

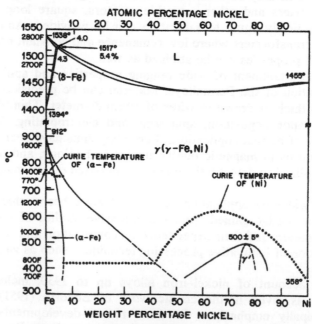

Fig. 52. Iron–nickel phase diagram (Goldstein 1973).

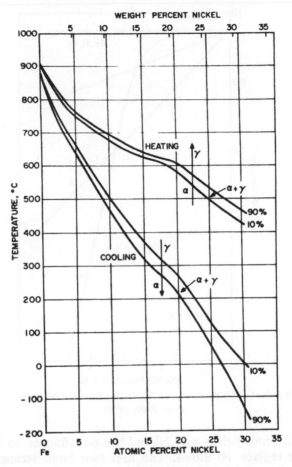

Fig. 53. Hysteresis in the $\alpha \rightarrow \gamma$ and $\gamma \rightarrow \alpha$ phase transformations in the low Ni regime (Hansen 1958b).

magnetocrystalline anisotropy energy, making K_1 less positive or more negative as the case may be. Hence variations in isothermal treatment temperature and/or cooling rates within the ordered alloy phase field can have a drastic effect on the technical magnetic properties such as permeability. Addition of alloying elements such as Mo, Cr, Cu and V tend to slow down the ordering kinetics and lowers the degree of long range order, while additives such as Mn, Si and Ge tend to stabilize it (Goman'kov et al. 1969, 1970). Figure 54 shows the variation of the long range order parameter S with alloy content for alloys of approximately Ni/Fe = 3/1 composition, compiled from the data of Gomankov et al. (1969, 1970). The general rules drawn by Goman'kov et al. regarding the influence of alloying elements on the Ni_3Fe structure are as follows: V, Cr, Mo and W replace Fe atoms and strongly bond with Ni, causing destruction of the Ni_3Fe

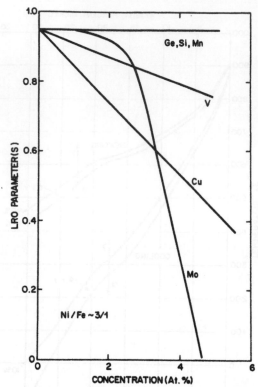

Fig. 54. Long range order (LRO) parameter of Ni₃Fe as a function of various additives (Goman'kov et al. 1969, 1970).

superlattice. Mn bonds lightly with Ni and has no influence on Ni₃Fe order. Cu, Co, Si and Ge replace Ni atoms. The first two bond strongly with Fe and perturbs Ni₃Fe order, while the latter two bond lightly, causing no perturbation. It is thought (Goman'kov et al. 1969) that the creation of short-range order with predominant MoNi and CrNi bonds in these ternaries is mainly responsible for the rise in electrical resistivity associated with the so-called K-state (Thomas 1951, Pfeifer and Pfeiffer 1964). The creation of long range order in Ni₃Fe ordinarily results in a decrease in resistivity.

Another finding, not shown in fig. 52, is that alloys near the Ni₅₀Fe₅₀ composition orders to the Ll_0 structure below 320°C when exposed to neutron irradiation to increase the kinetics (Neel et al. 1964, Marchand 1967). This point will be amplified in section 5.4.6.

5.3. Physical and mechanical properties

Data on density, lattice parameter, thermal expansion and electrical resistivity of nickel–iron alloys at various temperatures, as provided by Bozorth (1951) are still appropriate. Figure 55 summarizes the room temperature values of these

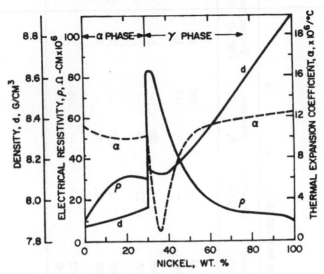

Fig. 55. Density, electrical resistivity and thermal expansion coefficient of Ni–Fe alloys (Bozorth 1951).

physical constants except lattice parameters. It is noted that considerable variation occurs near the α–γ phase change region of 30% Ni, and the values are subject to the work history of the alloy. In the vicinity of the ordering composition Ni_3Fe, the electrical resistivity is quite sensitive to cooling rate; the slower the cooling rate, the lower the value of the resistivity as the alloy orders.

Mechanical properties such as yield strength, tensile strength, elongation and hardness are highly sensitive to purity and work history. Typical values for commercial alloys in various conditions are tabulated in table 26. Even so-called structure-insensitive properties such as Young's modulus can be subject to variation, particularly as a result of different degrees of preferred crystallographic orientation (texture) brought about by working and annealing procedures. The elastic constants of nickel–iron alloys are highly anisotropic. The Young's modulus of nickel, for example, ranges from a low value of 130 GPa in the $\langle 100 \rangle$ direction to a high value of 300 GPa in the $\langle 111 \rangle$ direction, a difference of a factor of nearly three.

5.4. Intrinsic magnetic properties

5.4.1. Curie temperature
The Curie temperature of nickel–iron alloys is given in fig. 56 as a function of composition. In the Fe-rich end (<30%Ni), the Curie temperature decreases with nickel content. In the Ni-rich end (>30% Ni), the Curie temperature rises with iron additions, reaching a maximum value of 612°C at 68% Ni and then decreases thereafter.

TABLE 26
Mechanical properties of typical commercial nickel–iron magnetic alloys

Type	Trade name	Yield strength (ksi)	Yield strength (MPa)	Tensile strength (ksi)	Tensile strength (MPa)	Elongation (2 in)	Hardness R_b	Hardness H_v	Youngs modulus (10^6 psi)	Youngs modulus (GPa)
36Ni	Radiometal 36									
	cold rolled	–	–	140	990	–	–	290	19	130
	annealed	–	–	77	540	–	–	115	19	130
50Ni	High permeability 49									
	cold rolled	–	–	130	910	5	100	–	24	170
	annealed	40	280	80	560	32	68	–	24	170
4Mo–80Ni	HyMu 80									
	cold rolled	–	–	135	950	4	100	–	30	210
	annealed	21	150	77	540	38	58	–	30	210
4Mo–5Cu–77Ni	Mumetal									
	cold rolled	–	–	130	910	–	–	290	26	185
	annealed	–	–	77	540	–	–	110	26	185

5.4.2. Saturation induction
Figure 56 gives the value of saturation induction at both room temperature and near 0 K. The drop in $4\pi M_s$ near 30% Ni at room temperature corresponds to the steep decrease in Curie temperature of the γ phase in this region. Atomic ordering at the Ni_3Fe composition increases the value of $4\pi M_s$ by about 6% as compared to the disordered state.

5.4.3. Magnetocrystalline anisotropy
The value of the magnetocrystalline anisotropy constant K_1 is plotted in fig. 57 as a function of nickel content in the γ phase regime, for both quenched (disordered) and slowly-cooled (ordered) alloys. As mentioned previously, the effect of ordering is to make the value of K_1 less positive or more negative displacing the $K_1 = 0$ composition from about 75% Ni to 63% Ni. The influence of ordering is particularly severe at the Ni_3Fe composition, changing the value of K_1 from zero to -2.5×10^4 erg/cm^3 if the alloy is slowly cooled instead of quenched from above 600°C. Figure 58 shows this relationship rather clearly, where both the long range order parameter S and the anisotropy constant K_1 are plotted as a function of long term annealing at indicated temperatures, for a 75.5 at% Ni alloy (Liashchenko et al. 1957).

5.4.4. Magnetostriction
The magnetostriction constants λ_{111} and λ_{100} for the binary alloys are shown in fig. 59. Here the effect of ordering is less dramatic as compared with the case of magnetocrystalline anisotropy, the two λ's being brought closer together as a result.

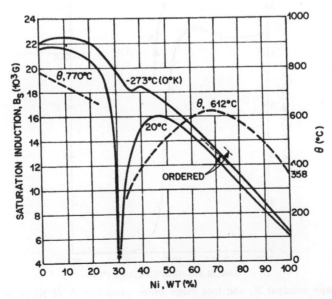

Fig. 56. Saturation induction and Curie temperature of Ni–Fe alloys (Bozorth 1951).

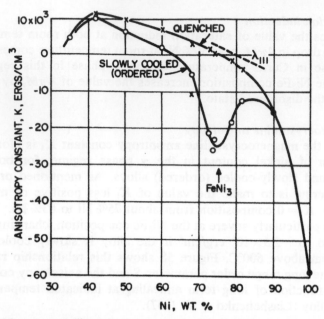

Fig. 57. Magnetocrystalline anisotropy constant K_1 of Ni–Fe alloys (Bozorth 1953).

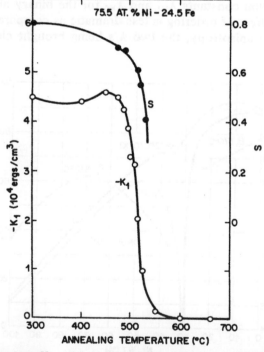

Fig. 58. Anisotropy constant K_1 and long range order parameter S of Ni$_3$Fe as a function of annealing temperature (Liashchenko et al. 1957).

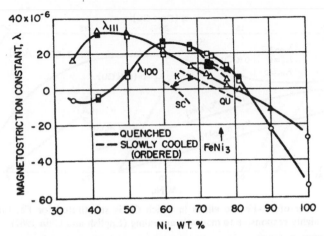

Fig. 59. Magnetostriction constants of Ni–Fe alloys (Bozorth 1953).

5.4.5. Ternary and complex alloys

Since the most widely used Ni–Fe alloys contain Mo and/or Cu, the composition dependence of the magnetostriction constant λ and magnetocrystalline anisotropy constant K_1 are summarized in figs. 60 and 61 respectively. Figure 60 shows the line of zero magnetostriction $\lambda_s = 0$ for polycrystalline material in

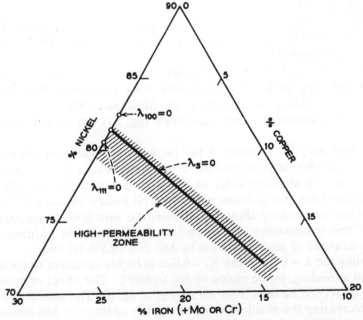

Fig. 60. Zero magnetostriction composition in the Ni–Cu–Fe ternary. English and Chin (1967) adopted from Pfeifer (1966b).

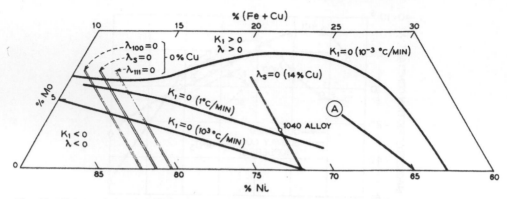

Fig. 61. Nickel corner of Ni–Fe–Mo alloys in which Cu is substituted for Fe. Line A alloys are highly responsive to magnetic annealing (English and Chin 1967).

the Fe–Ni–Mo–Cu system (Pfeifer 1966b). Since zero magnetostriction can be obtained by proper choice of the Ni and Cu contents without regard to Mo (or Cr), Mo (or Cr) can be considered substituted for Fe. It may be noted that:

(i) The $\lambda_s = 0$ line is almost a function of Cu content alone and essentially occurs at a constant Ni composition. In fact, for the special case of 0% Cu (pure Ni–Fe–Mo ternary), Puzei and Mototilov (1958) observed that λ_{111}, λ_{100} and λ_s are each zero at a constant Ni (79.5, 81 and 82.5% respectively), independent of any Mo present (fig. 61).

(ii) The $\lambda = 0$ lines are not appreciably affected by the degree of order (i.e. cooling rate), shifting toward each other by no more than $\sim\frac{1}{2}\%$ Ni similar to the case of binary alloys (fig. 59).

(iii) The $\lambda = 0$ lines appear to be independent of measurement temperature. Generally the magnitude of λ decreases with increasing measurement temperature without change in sign (Puzei and Molotilov 1958).

As for the magnetocrystalline anisotropy (fig. 61) the salient features are as follows:

(i) Copper may be substituted for Fe without significant alteration of the $K_1 = 0$ line, as inferred from labeling of the axes (Puzei 1961). Away from this line, however, the absolute value of K_1 decreases with Cu and Mo additions. Figure 62 illustrates this difference in a 76% Ni binary versus a 77%Ni–14%Fe–5%Cu–4%Mo quarternary alloy. In addition, the saturation magnetization along with the Curie temperature are reduced by the Cu and Mo additions. Figure 63 shows an example of such reduction by Mo in a 78.5% Ni alloy.

(ii) Unlike the $\lambda = 0$ lines, the $K_1 = 0$ line is highly sensitive to cooling rate, or isothermal annealing, in the region of 300 to 600°C. The effect of cooling rate on the value of K_1 can be seen from the curves of fig. 62. The salient feature is that, besides decreasing the absolute value of K_1, the addition of Mo and/or Cu slows down the cooling rate required to achieve $K_1 = 0$. Similar data have been reported by Aoyagi (1965).

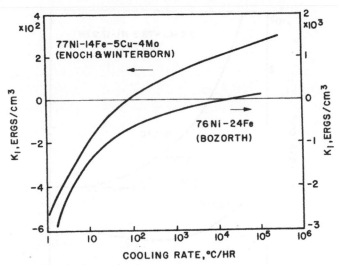

Fig. 62. Magnetocrystalline anisotropy constant K_1 as a function of cooling rate for Ni–Fe binary (Bozorth 1953) and Ni–Fe–Cu–Mo quarternary (Enoch and Winterborn 1967).

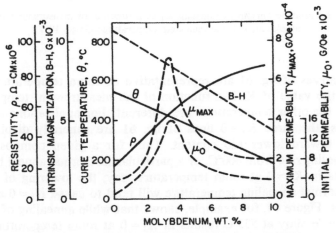

Fig. 63. Changes in magnetic parameters with Mo additions to a 78.5%Ni–Fe alloy (Bozorth 1951, pp. 136, 137, Bradley 1971a).

Figure 64 shows the variations of K_1 as a function of measurement temperature for a 3.8% Mo–Ni–Fe alloy after isothermal annealing at different temperatures (Puzei 1957). It can be seen that lowering the isothermal annealing temperature tends to make the value of K_1 less positive or more negative, as in the case of the binary alloys (fig. 58) and is related to the degree of atomic ordering.

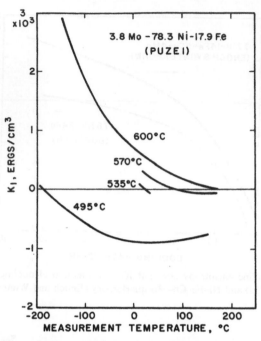

Fig. 64. Magnetocrystalline anisotropy constant K_1 as a function of measurement temperature for samples annealed at indicated temperatures (Puzei 1957).

(iii) The curves of fig. 61 are representative of room temperature condition only. Since the value of K_1 is a function of the measurement temperature, the $K_1 = 0$ lines for other temperatures are different. For applications at cryogenic temperatures, the three $K_1 = 0$ lines of fig. 61 are generally displaced downwards, i.e., toward lower Mo content. Thus for a given composition and a particular cooling rate from 600°C, or a particular isothermal anneal below 600°C which results in $K_1 = 0$ at room temperature, say, the lowering of the cooling rate or isothermal annealing temperature will tend to cause $K_1 = 0$ at cryogenic temperatures. Figure 64, for example, shows that while annealing of a 3.8%Mo–78.3Ni–17.9%Fe alloy at 535°C results in $K_1 = 0$ at room temperature, lowering of the anneal temperature to 495°C causes $K_1 = 0$ at −195°C (Puzei 1959). The conditions for achieving $K_1 = 0$ at cryogenic temperatures in the Ni–Fe–Mo–Cu system are given by Pfeifer (1969).

The position of the $K_1 = 0$ lines for disordered alloys with Mo, Cr, W, Cu and Si additions are shown in fig. 65.

5.4.6. Thermomagnetic anisotropy

Along with a number of magnetic alloys, the Ni–Fe alloys acquire a magnetic anisotropy when annealed below the Curie temperature (Graham 1959, Slonzewski 1963). If the annealing is done in the presence of an applied magnetic

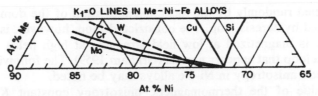

Fig. 65. The $K_1 = 0$ lines for disordered ternaries. The solid curves are from Puzei (1961), the dashed curve is from Hoffmann (1972).

field, the hysteresis loop squares up; if the applied field is absent, the hysteresis loop may become constricted. Figure 66 shows such examples for the case of a 65% Ni-alloy (Bozorth 1951, p. 121).

The origin of the thermomagnetic anisotropy has been explained by Neel (1954), Taniguchi (1955) and Chikazumi and Oomura (1955) as due to a short-range directional ordering of atom pairs. Since the magnetic (pseudo-dipolar) coupling energy of a pair of atoms generally depends on the identity of the atoms, e.g., Fe–Fe, Fe–Ni, and Ni–Ni, annealing below the Curie temperature in the presence of an applied magnetic field tends to align in the field direction those pairs with the minimum coupling energy. Fast cooling to a low temperature then freezes in the directional order structure. This gives rise to a uniaxial magnetic anisotropy with the easy axis of magnetization in the field direction. Subsequent measurement in this direction results in a square hysteresis loop, while measurement perpendicular to this direction results in a skewed hysteresis loop of extremely low remanence. If an applied field is absent, the anisotropy

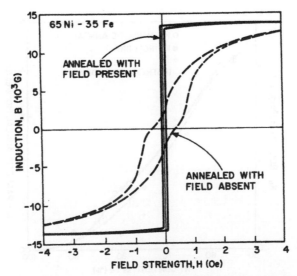

Fig. 66. Hysteresis loops of 65% Ni alloy annealed with and without the presence of a magnetic field (Bozorth 1951, p. 121).

becomes aligned randomly by the local magnetization of the domains, resulting in a constricted hysteresis loop. The constricted loop, however, is retained only if the sample is magnetized at low field strength; at high fields it irreversibly gives way to a loop that appears normal (Graham 1959). The following features of thermomagnetic anisotropy in Ni–Fe alloys may be noted.

(i) The value of the thermomagnetic anisotropy constant K_u is $1-4 \times 10^3$ erg/cm^3 and increases with Fe content (fig. 67) as a result of an increased number of aligned atom pairs. In addition, as fig. 68 shows, the value of K_u is higher the lower the annealing temperature.

(ii) Since the thermomagnetic anisotropy is established by annealing below the Curie temperature, diffusion becomes sluggish and attainment of an appreciable value of K_u becomes impractical if the Curie temperature is too low. For this reason, large values of K_u can be obtained easily in binary alloys near 65% Ni where the Curie temperature is maximum (fig. 68). As will be discussed later, there are commercial alloys in this composition range where hysteresis loops are squared up by magnetic annealing. Conversely, thermomagnetic treatment is less effective in ternary and quarternary alloys where the Curie temperature is depressed. The kinetics of the diffusion process, however, may be enhanced by neutron (Schindler and Salkovitz 1960, Schindler et al. 1964, Nesbitt et al. 1966) and electron (Sery and Gordon 1964, 1966, Gordon and Sery 1965) irradiation. Significant changes in hysteresis loop shape by irradiation treatment at temperatures as low as 100°C have been observed.

In this connection, large dosages of neutron irradiation ($\sim 10^{19}$ n/cm^2, $E >$ 1 MeV) below 320°C have been observed to produce a previously unreported Ll_0 ordered structure in a 50 at% Ni alloy (Neel et al. 1964, Marchand 1967). The

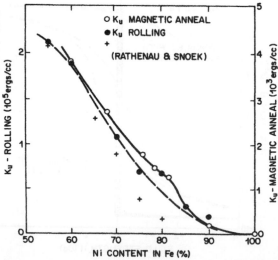

Fig. 67. Uniaxial anisotropy constant K_u obtained by rolling and by magnetic anneal in high Ni alloys (Chikazumi 1964).

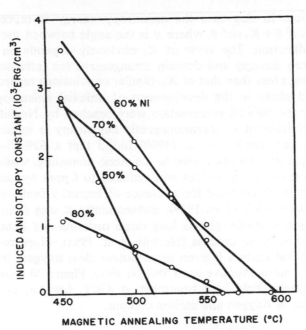

Fig. 68. Thermomagnetic anisotropy constant as a function of magnetic annealing temperature for various Ni compositions (Ferguson 1958).

ordered crystallites have dimensions of about 300 Å and exhibit a strong uniaxial anisotropy in the ⟨100⟩ tetragonal axis direction, with values of $K_u \sim 10^7$ erg/cm³. Figure 69 shows the value of anisotropy constant K_1 as a function of neutron dosage for a single crystal having an applied magnetic field along the [100] axis

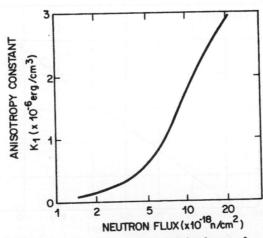

Fig. 69. Anisotropy constant K_1 induced by neutron bombardment of a 50 at% Ni single crystal (Néel et al. 1964). Magnetic field was directed in [100] direction during bombardment.

during irradiation. In this case the anisotropy energy is represented by the expression $K_1 \sin^2 \theta + K_2 \sin^4 \theta$, where θ is the angle between the magnetization and the [100] direction. The value of K_1 obviously depends on the degree of ordering (neutron dosage) and domain arrangement (as influenced by applied field) and is always less than that of K_u. Similar conclusions regarding the role of the Ll_0 ordered phase in the development of uniaxial anisotropy by neutron irradiation near the 50% Ni composition were reached by Nesbitt et al. (1966).

(iii) The establishment of thermomagnetic anisotropy is influenced by small amounts of oxygen. Nesbitt et al. (1960) found that a 63%Ni–35%Fe–2%Mo alloy with 57 ppm oxygen responded to magnetic annealing, while there was no response when the oxygen content was reduced to 6 ppm by careful hydrogen annealing. Early work revealed the presence of impurity faults associated with the oxygen (Heidenreich et al. 1959). Subsequently it was found that a small amount of oxygen strongly inhibits long-range ordering and hence may aid the local directional ordering process (Nesbitt et al. 1965). More recently, Ste'fan (1974) reported that carbon is even more potent than oxygen, its effect on K_u being ~3.2 times more effective for a 58% Ni alloy. Figure 70 shows the value of K_u after a magnetic field heat treatment at 430°C for 4 h, as a function of weighted average of oxygen and carbon content.

5.4.7. Slip-induced anisotropy

The phenomenon of slip- or deformation-induced magnetic anisotropy was discovered in Europe in the early 1930's during a search for a ferromagnetic core material for loading coils. The chief property requirements for loading coils are a constant permeability with applied field, and a low hysteresis loss. The search culminated in the development of Isoperm (Dahl et al. 1933), an alloy near the

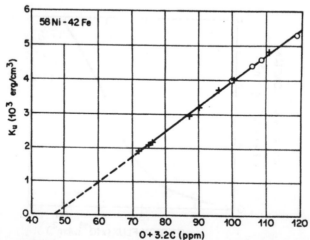

Fig. 70. Uniaxial anisotropy constant K_u induced by magnetic annealing of a 58% Ni alloy as a function of oxygen and carbon contents (Stéfan 1974).

50%Ni–50%Fe composition, often containing small amounts of copper or aluminum. The material is first subjected to a large cold reduction, usually exceeding 90%, and then recrystallized at a high temperature around 1000°C. When the recrystallized material is again cold-rolled about 50%, the magnetic hysteresis loop in the rolling direction exhibits low remanence, constant permeability, and low hysteresis loss. Figure 71 illustrates such a curve.

The early developments and more recent investigations have led to the conclusion that, like the case of magnetic annealing, the induced anisotropy here is also due to a directional order of atom pairs (Chikazumi and Graham 1969, Chin 1971b) In this case the directional order is established mechanically, by plastic deformation. The presently accepted theory was first proposed by Bunge and Muller (1957) and Chikazumi et al. (1957) and later extended by Chin (1965, 1966) and Chin et al. (1967). The theory is generally known as the "slip-induced directional order" theory and the anisotropy as "slip-induced" anisotropy because the directional order is produced by the crystallographic slip process accompanying the deformation.

A simple case of slip-induced directional order is illustrated in fig. 72. In fig. 72a, the AB alloy is thermally ordered so that all nearest neighbor atom pairs are of the AB type and the pair distribution is symmetric. In fig. 72b, a shear stress is applied horizontally, causing the atoms to slip over one another to form an antiphase domain boundary. A number of AA and BB atom pairs have now been created across the domain boundary in the vertical direction but not in the horizontal direction. If the pseudodipolar magnetic coupling energy is lowest for an AB pair (which is the case of Ni–Fe alloys), the vertical direction becomes a hard (or high-energy) magnetic axis as a result of the slip-induced directional order. Clearly, the number of induced AA and BB pairs, and hence the strength of the induced anisotropy, depends on the degree of order and the slip character. For example, in fig. 72b a second or even-numbered slip step on the same plane would recover the original arrangement, wiping out the atom pairs created by the first step. Both the number of atom pairs and the direction of their alignment also depend on which of several equivalent slip systems will actually operate to

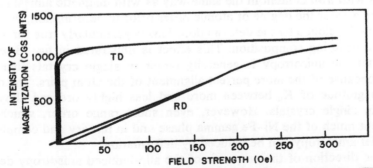

Fig. 71. Portions of hysteresis loop of Isoperm sheet in rolling (RD) and transverse (TD) directions (Snoek 1935).

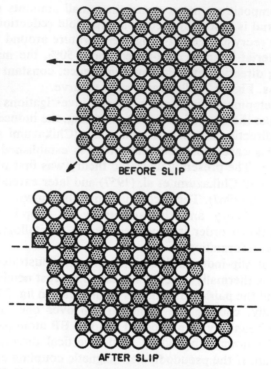

BEFORE SLIP

AFTER SLIP

Fig. 72. Schematic illustration of slip-induced directional order in simple AB alloy. (a) before slip, (b) after slip on horizontal planes (Chin 1971b).

accommodate a given deformation. The active slip systems in turn depend on crystal orientation and on the imposed shape change, e.g. rolling vs wire drawing. For a recent review, see Chin (1971b).

One can summarize the salient features of slip-induced anisotropy as follows:

(i) The magnitude of the slip-induced anisotropy is of the order of 10^5 erg/cm^3, about 50 times that obtained by magnetic annealing. It generally increases with iron content in the same way as with magnetic annealing (fig. 67).

(ii) The greater the degree of atomic order prior to deformation, the larger the anisotropy developed by the deformation. This is particularly true for alloys near the Ni$_3$Fe ordering composition. This effect is illustrated in fig. 73 which also shows that the anisotropy is generally larger in single crystals than in poly-crystals because of the more perfect alignment of the atom pairs. The difference in the magnitude of K_u between more and less highly ordered alloys is also greater in single crystals. However, even short range order, which prevails throughout much of the Ni–Fe gamma phase and in ternary and complex alloys, substantial anisotropy can be induced by deformation.

(iii) The direction of the easy axis of the slip-induced anisotropy depends on (a) the type of order (long or short range), (b) the crystal orientation (or texture if polycrystalline), and the geometry of the imposed deformation. Table 27

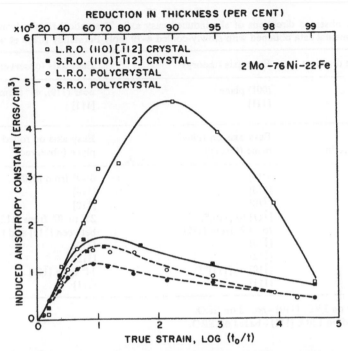

Fig. 73. Slip-induced anisotropy constant K_u as a function of thickness strain for single crystals compressed in $(110)[\bar{1}12]$ orientation and for rolled polycrystalline material (Chin et al. 1967). Short range order (SRO): air quenched from 700°C; long range order (LRO): air quench plus 190 h at 475°C.

summarizes the theoretical and experimental results for Ni–Fe alloys. With few exceptions, the agreement is good.

For polycrystalline material, starting with randomly oriented material, it has been found experimentally that rolling or wire drawing causes the rolling direction or wire axis to become an easy axis of magnetization. While detailed calculations have not been made, table 27 suggests that with the exception of those orientations aligned with the $\langle 100 \rangle$ direction, the easy axis generally coincides with the rolling or wire drawing direction. Thus it is not unreasonable to expect that in a randomly oriented polycrystal, a majority of the orientations behave as above.

On the other hand, in the Isoperm process, the recrystallized strip exhibits a $\{100\}\langle 001 \rangle$ sheet texture. Thus when this material is rolled, the rolling direction becomes a hard axis of magnetization (table 27) resulting in the skewed loop shape with low remanence (fig. 71).

(iv) With increasing deformation, the magnitude of the induced anisotropy first rises, reaches a maximum, and then decreases. This is evident in fig. 73. For polycrystalline material deformed by cold rolling, the peak occurs near 50% thickness reduction. The peak generally occurs at a higher reduction for single

TABLE 27

Predicted and observed directions of slip-induced easy axis. Theoretical results from Chin (1965). Experimental data obtained with slowly-cooled 4-79 moly-Permalloy (Chin et al. 1967)

Swaging (wire drawing)	Easy axis (theory)	Easy axis (observed)
[001]	(001) plane	near ⟨110⟩, 45° from [001]
[111]	[111]	[111]

Plane strain compression (rolling)	Easy axis on rolled plane (theory)	Easy axis on rolled plane (observed)
(001)[100]	[010]	0°–7° from [010]
(001)[110]	[110]	[110]
(110)[001]	[1̄10]	[1̄10]
(110)[1̄12]	[1̄11] to [1̄10]*, 19.5° to 54.7° from [1̄12]	25° to 40° from [1̄12] between [1̄11] and [1̄10]
(110)[1̄10]	[1̄10]	[1̄10]
(111)[1̄12]	[1̄12]	[1̄12]
(112)[1̄10]	[1̄10] or [111̄]**	14° from [1̄10]
(112)[1̄1̄1]	[1̄11]	[1̄11]

* [1̄11] – based on LRO, [1̄10] – based on SRO.
** [1̄10] – based on LRO, [1̄11] – based on SRO.

crystals vs polycrystalline material, and for more highly ordered vs less ordered condition. The occurrence of the peak was explained by Chin (1971b) as due to opposing tendencies on the number of induced atom pairs. This number increases with strain as the number of slipped planes increases, but then it also decreases with strain as the structure becomes generally randomized by intersecting slip systems.

5.5. Magnetic properties control

The technical hysteresis properties of a magnetic material such as permeability, coercive force, and remanence are governed mainly by the magnetization processes of domain nucleation, domain-wall motion and domain rotation. These processes in turn are governed by the intrinsic properties such as saturation magnetization and magnetic anisotropy energy as well as microstructural details such as inclusion content and grain size. In this section the attainment of several important magnetic properties of Ni–Fe alloys is illustrated.

Table 28 lists some commercial Ni–Fe alloys and their magnetic and physical properties. For convenience of comparison, properties of a 3%Si–Fe and a 2%V–49%Co–Fe alloy are also included. It may be noted that the alloys are grouped into three types of magnetic characteristics: high initial permeability, square loop and skewed (or flat) loop. The alloys are generally used in sheet form and devices operate to about 50 kHz. Beyond that, eddy current limitations dictate that (metal) powder or ferrite cores be used.

TABLE 28
Materials for laminations, cut cores and tape wound cores

Alloy type	μ_i ×10⁻³	μ_m ×10⁻³	H_c (Oe)	(A/cm)	B_s (kG)	B_r (kG)	$\frac{B_t}{B_m}$	T_c (°C)	Resistivity (μΩ cm)	Density (g/cm³)	Notes
				High initial μ							
36Ni	3	20	0.2	0.16	13			250	75	8.15	a)
48Ni	11	80	0.03	0.024	15.5			480	48	8.25	b)
56Ni	30	125	0.02	0.016	15			500	45	8.25	c)
4Mo–80Ni	40	200	0.015	0.012	8.0			460	58	8.74	d)
4Mo–5Cu–77Ni	40	200	0.015	0.012	8.0			400	58	8.74	e)
5Mo–80Ni	70	300	0.005	0.004	7.8			400	65	8.77	f)
4Mo–5Cu–77Ni	70	300	0.005	0.004	8.0			400	60	8.74	g)
				Square loop							
4Mo–80Ni			0.03	0.024	8.0	6.6	0.80	460	58	8.74	h)
4Mo–5Cu–77Ni			0.03	0.024	8.0	6.6	0.80	400	58	8.74	i)
3Mo–65Ni			0.025	0.02	12.5	10.5	0.94	520	60	8.50	j)
50Ni			0.10	0.08	16.0	15.0	0.95	500	45	8.25	k)
3Si			0.40	0.32	20.3	16.3	0.85	730	50	7.65	l)
2V–49Co			0.20	0.16	23.0	20.1	0.90	940	26	8.15	m)
				Skewed (flat) loop							
4Mo–5Cu–77Ni			0.016	0.012	8.0	1.2		400	58	8.74	n)
3Mo–65Ni			0.12	0.10	12.5	1.5		520	60	8.50	o)

a) Permenorm 3601 K2, Hyperm 36M, Radiometal 36. 0.3 mm/50 Hz
b) 4750 Alloy, High Permeability 49, Hyperm 52, Alloy 48, Superperm 49, PB-2, Super Radiometal, Permenorm 5000 H2. 0.15 mm/60 Hz
c) Permax M. 0.15 mm/60 Hz
d) 4–79Mo–Permalloy, HyMu 80, Round Permalloy 80, Superperm 80. 0.1 mm/60 Hz
e) Mumetal Plus, Hyperm 900, Vacoperm 100. 0.1 mm/60 Hz
f) Supermalloy, HyMu 800, Hyperm Maximum. 0.1 mm/60 Hz
g) Supermumetal, Ultraperm 10. 0.1 mm/60 Hz
h) Square Permalloy, HyRa 80, Square Permalloy 80, Square 80. 0.05 mm/0.5 Oe dc
i) Ultraperm Z, Othomumetal. 0.015 mm/0.5 Oe dc
j) Permax Z. 0.05 mm/0.5 Oe dc
k) Deltamax, HyRa 49, Hyperm 50T, Othonol, Square 50, PE-2, HCR Alloy, Permenorm Z. 0.05 mm/0.5 Oe dc
l) Silectron, Microsil, Oriented T–S, Hyperm 5T/7T, Thin-Gage Orientcore. 0.1 mm/3 Oe dc
m) Supermendur, Vacoflux Z, Hyperm Co50. 0.1 mm/3 Oe dc
n) Ultraperm F
o) Permax F

There are three popular commercial design configurations for utilizing sheet material: laminations, cut-cores and tape would (toroidal) cores. Laminations are usually 0.014 and 0.006 in thick for operations from 60 Hz to 400 Hz. They are stamped to shapes such as EI, EE, F and DU. The DU lamination shape is most nearly like a toroid and preferred for saturating devices utilizing square loop materials. The other lamination shapes have the advantage of best space utilization and are preferred for devices requiring high permeability. Cut cores have C, O and E shapes and are generally wound from thin gage material ranging from 0.001 in to 0.004 in to enable operating frequencies to go up to about 25 kHz for the 0.001 in material of the 4–79 Mo-Permalloy type. Tape wound cores were originally developed for use in magnetic amplifiers. Now they are mainly used in inverters, converters, instrumentation transformers and power transformers. They are generally made from strips 0.001, 0.002 and 0.004 in thick, although 0.006 and 0.014 in materials are not uncommon. For miniature high frequency operation to 500 kHz, bobbin cores made of material from 0.000125 to 0.001 in thick, with case dimensions as small as 0.095 in ID, 0.225 in OD and 0.105 in high are available.

5.5.1. Initial permeability

High initial permeability was the original label attached to the Ni–Fe alloys; it still is today. From domain theory, rotational processes lead to the general expression

$$\mu_i \propto M_s^2/K_{eff}$$

for the initial permeability μ_i, where M_s is the saturation magnetization and K_{eff} is an effective anisotropy constant covering all sources of anisotropy energy (English and Chin 1967). A domain-wall motion model taking grain size into effect (Hoekstra et al. 1978) has the expression

$$\mu_i \propto M_s^2 d/(AK)^{1/2}$$

where A is the exchange constant and d the grain size. Hence the maximization of initial permeability requires a maximization of M_s and d and minimization of K. The latter has been exploited the most, as epitomized by the development of two broad classes of alloys. The low Ni alloys, near 50% Ni, are characterized by moderate permeability ($\mu_{40} \sim 10\,000$)* and high saturation ($B_s \sim 15\,000$ G). This region corresponds to near zero values of λ_{100} coupled to low positive K_1 (figs. 57 and 59). The high Ni alloys, near 79% Ni, are characterized by high permeability ($\mu_{40} \sim 50\,000$) but lower saturation ($B_s \sim 8000$ G). This region corresponds to near zero values of K_1. Some suppliers also offer alloys near 36% Ni (table 28). These have lower permeability than the 50% alloys, but a higher resistivity and lower cost.

5.5.2. 50% Ni alloys

Alloys in the 50% Ni vicinity generally range from 45 to 50% nickel, balance iron, and may contain up to 0.5% Mn and 0.35% Si. The carbon level is generally kept below 0.03%. Most current alloys contain 48% Ni, and some are available in two grades: rotor grade of $\mu_{40} \sim 6000$ and transformer grade of $\mu_{40} \sim 10\,000$.

---

*In the USA, μ_i values are generally given as μ_{40}, measured at $B = 40\,G$ and 60 Hz for 0.014 in thick material, whereas in Europe, μ_4, measured at $H = 4\,mA/cm$ and 50 Hz for 0.1 mm thick material is adopted. The μ_4 value is equivalent to one used in the older notation called μ_5, measured at $H = 5\,mOe$. In general, $\mu_4 > \mu_{40}$ as defined above.

They are generally used in audio and instrument transformers, instrument relays, for rotor and stator laminations and for magnetic shielding.

The rotor grade material is unoriented and hence useful for rotors and stators and applications in which magnetic properties must be non-directional. To achieve the essentially random texture, the final cold reduction is held low (~50–60%). Annealing is done in dry hydrogen for several hours at ~1200°C followed by furnace cooling (~150°C/h). The transformer grade, on the other hand, is oriented such that high permeability is achieved in directions parallel and perpendicular to the rolling direction. Such a material is achieved by a large final cold reduction (~95%) so that a secondary recrystallization texture is obtained after an annealing treatment similar to that for the rotor grade material. Figure 74 shows some typical permeability–flux density curves for an 0.014 in (0.35 mm) transformer grade lamination at various frequencies. Typical core loss curves are shown in fig. 75. Figures 76 and 77 illustrate the magnetic behavior of thin gage (0.004 in) materials suitable for cut cores and tape wound cores. Comparison among various alloys is indicated.

A three to four fold increase in initial and maximum permeability can be achieved by annealing alloys in the 56–58% Ni regime in the presence of a magnetic field after the usual high temperature anneal (Pfeifer 1966a, Kang et al. 1967, Rassmann and Wich 1960, Fahlenbrach 1968). Figure 78 shows the permeability improvement of a 58% Ni alloy after such heat treatment as compared with one prior to such treatment and a typical 50% Ni alloy. It may be noted

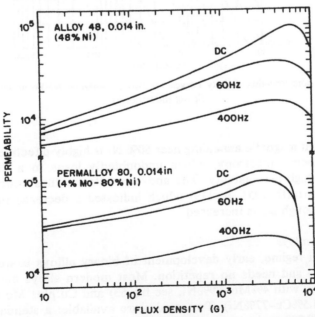

Fig. 74. Typical permeability–flux density curves for two commercial Ni–Fe alloys (Magnetics Div. Spang Industries a).

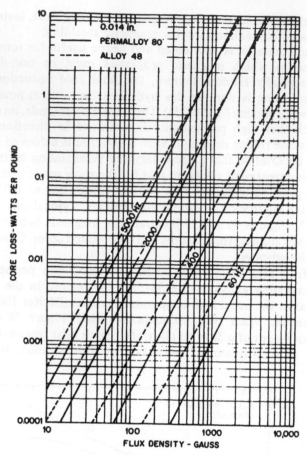

Fig. 75. Typical core loss–flux density curves for two commercial Ni–Fe alloys (Magnetics Div. Spang Industries a).

from fig. 68 that magnetic annealing near 60% Ni is highly effective in developing a thermomagnetic anisotropy, which undoubtedly leads to a domain pattern conducive to high permeability. The above results are thus contrary to earlier experience (Bozorth 1951, p. 119) which indicated a decrease in μ_i with field treatment, although μ_m is increased.

5.5.3. 79% Ni alloys

In the 79% Ni regime, early development of binary allloys is well detailed by Bozorth (1951) and needs no repetition. Most modern alloys contain Mo (e.g. 4–79 Permalloy with 4%Mo–79%Ni, see fig. 63) and Cu plus Mo (e.g. Mumetal with 4%Mo–4.5%Cu–77%Ni). Two grades are available: a standard grade with $\mu_{40} \sim 30\,000$–$40\,000$ and a very high permeability grade with $\mu_{40} \sim 60\,000$. The latter is based on the development of Supermalloy (Boothby and Bozorth 1947)

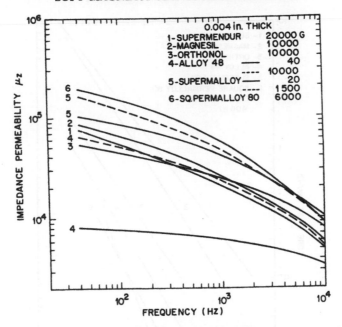

Fig. 76. Impedance permeability vs frequency of thin gage (0.004 in) tapes of various commercial alloys (Magnetics Div. Spang Industries b). The operating flux density is shown in the legend. See table 28 for alloy designation.

with $\mu_{40} > 100\,000$ where impurities such as carbon (~0.01%) and Si (~0.15%) are minimized and careful attention is paid to melting and fabrication practice. The Mo content is also generally higher (5%) as compared with the standard grade (4%). The general applications of the 79% Ni alloys are the same as for the 50% Ni alloys but where superior qualities are sought, particularly where compactness and weight factors are important. Figures 74 and 75 show some typical permeability and loss curves for a standard grade 0.014 in molybdenum Permalloy lamination and figs. 76 and 77 indicate generally superior properties of the 79% Ni alloys over others in the low flux density regime.

The ternary and quaternary alloys achieve superior permeability over the binary alloys primarily as a result of achieving $K_1 = 0$ and $\lambda_s = 0$ simultaneously. The principles governing the attainment of $K_1 = 0$ and $\lambda_s = 0$ have been discussed in section 5.4.5. Since the $K_1 = 0$ line is much more dependent on heat treatment than the $\lambda_s = 0$ line, the first selection is for a composition near $\lambda_s = 0$ and then the heat treatment is varied to achieve $K_1 = 0$. For the Mo–Ni–Fe ternary, $\lambda_s = 0$ is near 80% Ni; and for the 5% Cu–Mo–Ni–Fe quaternary, $\lambda_s = 0$ is near 78% Ni. For the usual practical ranges of cooling rates, optimum Mo content is within the 4–6% range.

Within the above ranges, it can be seen from fig. 61 that increasing the nickel content requires a faster cooling rate to achieve $K_1 = 0$ and hence optimum permeability. This is demonstrated in figs. 79 and 80(a). Likewise, since increas-

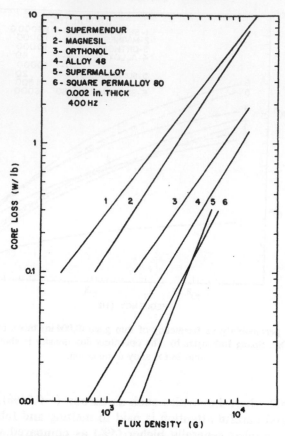

Fig. 77. Core loss characteristics of several commercial alloys (Magnetics Div. Spang Industries b). (0.002 in thick, 400 Hz).

ing cooling rate is equivalent to an increasing baking temperature as far as ordering is concerned, the increase in the optimum baking temperature with increasing nickel content, as shown in fig. 81, is also explained. Figure 81 also shows that the absolute value of μ_i goes down at either the high or low Ni end as a result of deviation from the $\lambda_s = 0$ composition.

Other variations can also be considered. Figure 61 shows, for example, that for a fixed cooling rate, an increase in the Ni content requires an increase in Mo to stay on the $K_1 = 0$ line. The permeability results of fig. 80(b) can thus be rationalized this way.

More extensive discussions and experimental results have been provided by Pfeifer (1966b). Results of other studies of the effect of composition and/or cooling rate (Walters 1956, Lykens 1962, 1966, Puzei 1963, Jackson and Lee 1966, Ames and Stroble 1967, Snee 1967, Enoch and Winterborn 1967) on μ_i of the high Ni alloys are generally consistent with the above rationale.

The above discussions pertain chiefly to the attainment of high permeability at

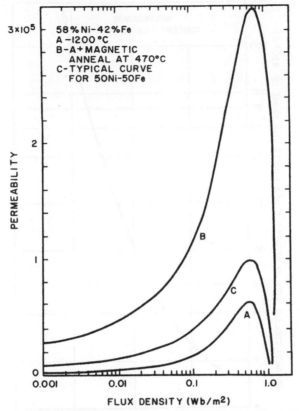

Fig. 78. Permeability–flux density curves for a 58% Ni alloy (Kang et al. 1967).

room temperature. For some applications, such as magnetic shielding inside liquid nitrogen or helium chambers, the materials must be heat treated for high μ_i at the cryogenic temperature. Based on the composition and temperature dependence of K_1 and λ_s, Pfeifer (1969) developed the necessary heat treatments for high μ_i at 4.2 and 77 K, for alloys of Mumetal composition. Since for these alloys the low temperature branch of the K_1–T curve has a negative slope (fig. 64), and this curve is shifted lower (to more negative K_1) the more ordered is the structure; the general rule is to make K_1 negative at room temperature. Thus starting with a known heat treatment which gives high μ_i ($K_1 = 0$) at room temperature, high μ_i at cryogenic temperature could be obtained by lowering (i) the Mo content, (ii) the isothermal annealing temperature and/or (iii) the cooling rate. It may be re-emphasized that since λ_s does not change sign with temperature (Puzei and Mototilov 1958) the $\lambda_s = 0$ line is mainly a function of composition alone and not of heat treatment.

In connection with temperature effects, Pfeifer and Boll (1969) have reported the processing of Mumetal to achieve temperature stability of the initial per-

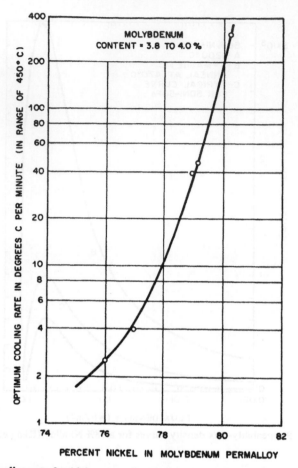

Fig. 79. Optimum cooling rate for high μ_i as a function of Ni content in 4%Mo–Permalloy (Bozorth 1951, p. 138).

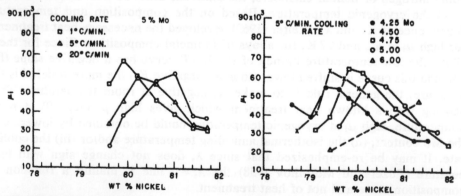

Fig. 80. Effect of composition and cooling rate on initial permeability of Mo–Permalloy (Cohen 1967). Left–varying cooling rate for 5% Mo, right–varying Mo for 5°C/min cooling rate.

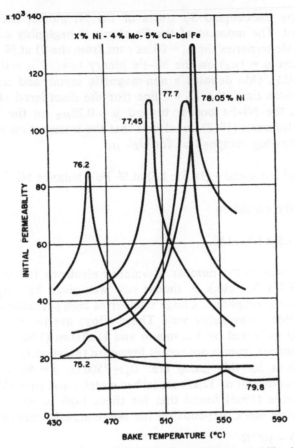

Fig. 81. Influence of annealing ("bake") temperature on initial permeability for alloys of Mumetal type (Scholefield et al. 1967).

meability near room temperature. Such stability is needed, for example, in earth leak transformers or telephone equipment which are exposed to the elements. In the Pfeifer and Boll work, temperature stability was achieved by making K_1 slightly negative with a lower temperature anneal or slower cooling rate. A similar development in molybdenum–Permalloy was recently reported by Finke (1974). Puzei (1957) also discussed the attainment of temperature stability of μ_i by heat treatment such that $K_1 \sim 0$ throughout a wide temperature range.

In addition to Mo and Cu, high permeability alloys have been developed in complex alloys. Useful rules for attaining $\lambda_s = 0$ and $K_1 = 0$ compositions have been given by Enoch and Fudge (1966) and Rassmann and Hofmann (1968). Both groups of workers based their ideas on the empirical finding that non-magnetic elements such as Cu, Mo, Cr, etc. decrease K_1 and λ_s of Ni–Fe alloys. The decrease is rationalized by assuming that the valence electrons of the added element enter the unoccupied 3d states of the Ni atoms, decreasing the Ni magnetic moment. The moment of the Fe atoms is negligibly affected. On this basis, in Ni–Fe–Me ternaries the $\lambda_s = 0$ line runs from the 81 at.% Ni composition (magnetic moment $\mu = 1\mu_B$) on the Ni–Fe binary toward $\mu = 0$ on the Ni–Me border (Went 1951). (Me denotes a non-magnetic metal, and μ_B is Bohr magneton). At the same time, the $K_1 = 0$ line (for the disordered state) runs from ~75 at.% Ni on the Ni–Fe border toward $\mu = 0.28\mu_B$ on the Ni–Me border. Rassmann and Hofmann (1968) concluded that the intersection of the two lines occurs at the following composition for high μ_i:

$$\sum_i C_i \text{ at.\% added metal} + (14.5 \pm 1.5) \text{ at.\% Fe + balance Ni}$$

where C_i obeys the condition

$$\sum_i C_i Z_i = (19.5 \pm 1.5) \text{ at.\%.}$$

The symbol Z_i refers to the number of valence electrons for the added metals ($Z_i = 1$ for Cu, 6 for Mo etc.). A similar rule was given by Enoch and Fudge (1966). Although oversimplified, a large number of high permeability alloys fulfill the above conditions reasonably well. These alloys are listed in table 29. The data were mostly collected by Rassmann and Hoffman (1968); we have added the bottom two entries from more recent literature (Masumoto et al. 1972, 1974). The average values for this group are: $\bar{B}_s = 7500\,\text{G}$, $\bar{\theta} = 400°\text{C}$, $\bar{\rho} = 65\mu\Omega\,\text{cm}$ (transition metal addition) or $32\mu\Omega\,\text{cm}$ (other metals) and $\mu_i = 100\,000$.

Enoch and Fudge (1966) found that for these high μ_i alloys, the optimum cooling rate R from 600°C depends on the non-magnetic content Z as follows.

$$\exp(-0.32Z) \sim 10^{-8}R$$

where R is in °C/h and $Z = \Sigma_i C_i Z_i / \mu_{Ni}$ with $\mu_{Ni} = 0.6\mu_B$. Thus the optimum cooling rate decreases with increasing non-magnetic content due to the suppression of atomic order. In practice it was found (Enoch and Murrell 1969b) necessary to limit the non-magnetic content to below about 33 at.% to maintain high permeability. Although several reasons were cited, it appears that excessive decrease in saturation magnetization is a large factor, since $\mu_i \propto M_s^2$.

5.5.4. Effect of alloy preparation and fabrication on initial permeability

Most of the alloys in use today are produced pyrometallurgically. Air melting is still practiced to some extent because of low cost. The resulting product is generally inferior, but then for some applications high quality grades are not necessary. The overwhelming major trend today, however, is towards vacuum melting, which minimizes harmful ingredients such as oxygen and sulfur. Figure 82, for example, shows the progress in improvement of initial permeability of commercial grade nickel–iron alloys for the past 35 years (Jacobs 1969). The sudden jump in μ_i in 1960 for 4–79 Mo–Permalloy is the result of introducing vacuum melting to commercial practice. Recent examples of quantitative studies

TABLE 29

Properties of high-permeability alloys

Alloy	Chemical composition (wt.%)	$\Sigma_i C_i Z_i$ of Me (at.%)	θ (°C)	B_s (G)	K_1 (erg cm⁻³)	$\lambda_{100} \times 10^6$	$\lambda_{111} \times 10^6$	$\mu_i \times 10^{-3}$	ρ ($\mu\Omega$ cm)
Supermalloy	79 Ni, 15.5 Fe, 5.0 Mo, 0.5 Mn	18.6	400	7900	-	-	-	100	60
Supermalloy	79.7 Ni, 14.5 Fe, 5.1 Mo, 0.7 Mn	19.0	394	7270	0	1.79	0.17	163	68
Mumetal	76 Ni, 16.9 Fe, 4.8 Cu, 1.9 Cr, 0.4 Mn	17.2	400	8000	-	-	-	100	55
Mumetal	77.0 Ni, 16.6 Fe, 4.7 Cu, 1.7 Cr	15.8	461	8310	1000	5.5	0	90	60
1040	72 Ni, 11 Fe, 14 Cu, 3 Mo	24.4	290	6000	-	-	-	40	56
Mo–Cr–Permalloy	80.5 Ni, 15.1 Fe, 2.6 Mo, 1.8 Cr	21.8	411	7500	-	-	-	108	64
Mo–Cu–Cr–V Permalloy	78.7 Ni, 15.3 Fe, 1.4 Mo, 2.4 Cu, 1.0 Cr, 1.2 V	20.8	405	7640	0	3.01	0.44	150	72
V–Permalloy	82.3 Ni, 13.3 Fe, 3.9 V	22.3	371	6720	0	0.77	-0.52	128	65
W–Permalloy	75.8 Ni, 15.1 Fe, 9.1 W	18.4	414	7200	300	3.13	0.63	82	66
Cr–Permalloy	81.4 Ni, 15.0 Fe, 3.6 Cr	24.0	421	7800	-	-	-	34	61
Cu–Permalloy	69.5 Ni, 13.9 Fe, 15.9 Cu	14.9	413	7500	-	-	-	74	30
Ta–Permalloy	73.0 Ni, 12.0 Fe, 15.0 Ta	27.0	-	7000	-	-	-	57	64
Nb–Permalloy	79.2 Ni, 11.8 Fe, 9.0 Nb	29.2	-	6000	-	-	$0.35\lambda_s$	125	75

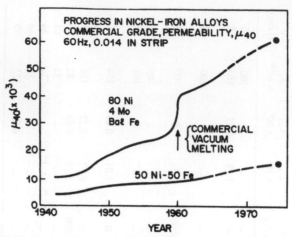

Fig. 82. Progress in initial permeability of commercial grade Ni–Fe alloys. Solid curves given by Jacobs (1969).

showing the effect of sulfur and oxygen on the initial permeability of the two classes of Ni–Fe alloys are shown in fig. 83.

In this connection, it should be mentioned that attempts at deoxidation by adding strong deoxidizers such as Ca, Si and Al have generally resulted in

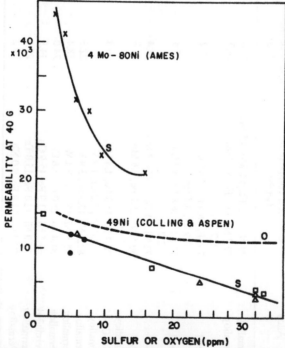

Fig. 83. Influence of sulfur and oxygen on initial permeability (Colling and Aspen 1969, 1970, Ames 1970).

reduced permeability rather than increasing it (Bozorth 1951, p. 138, Hoffman 1970, 1971). The most likely explanation is that fine oxide precipitates are left in the alloy matrix, which directly contributes to lower permeability by blocking domain wall motion. In addition, the precipitates restrict grain growth during subsequent annealing operations, which contributes further to decreased permeability. The complex interplays among deoxidizer actions, S and O contents and grain size have recently been studied in detail by Hoffmann (1970, 1979). Figure 84 shows, for example, the initial permeability μ_5, i.e. at $H = 5$ mOe [4 mA/cm], as a function of grain diameter for Al-containing and Al-free samples of a 50% Ni alloy. Annealing was done in very dry hydrogen (dew point $-45-60°C$); the insert gives the annealing temperature. It may be noted that (i) for both types of samples, annealing at 1250°C results in a larger grain size and generally higher μ_5, although μ_5 tends to level off at larger values of grain size; and (ii) for the same grain size, the Al-containing samples show a lower value of μ_5. The latter observation can be attributed to oxide precipitates, as is the phenomenon of levelling off of μ_5 at large grain sizes.

More recently, Adler and Pfeiffer (1974) separated the effects of grain size and inclusion content and arrived at the following expression for the coercivity of 47.5% Ni alloys:

$$H_c = H_{co} + H_{ci} + H_{cK} = 8\,\text{mA/cm} + (28\,\text{mA/cm})N_F \times 10^{-6}\,\text{cm}^2 + 0.29\,\text{mA}/d_K$$

where H_{co} is the residual coercivity of the pure alloy, H_{ci} represents the effect of inclusions with N_F being the number of particles per cm^2 in the range 0.02 to 0.5m, and H_{cK} represents the effect of grain size with d_K as the grain diameter in cm. These effects are illustrated in figs. 85 and 86. A similar study was done on a Mumetal composition (Kunz and Pfeifer 1976).

To achieve the desired deoxidation during melting and yet not retaining oxide inclusions afterwards, Ichinose (1967) proposed that: (i) additives have a stronger affinity for oxygen than Fe or Ni, (ii) oxides of additives have a very high vapor pressure (gaseous oxides) and (iii) good solubility of additive in the solid. Ichinose selected Ge as the most appropriate additive for this purpose and

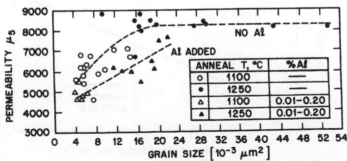

Fig. 84. Dependence of initial permeability on grain size for a commercial 50% Ni alloy with and without Al added as deoxidizer (Hoffmann 1971).

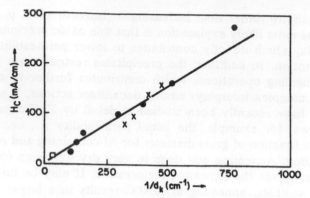

Fig. 85. Coercivity as a function of inverse grain diameter for 47.5% Ni sintered alloys (Adler and Pfeiffer 1974). Circles – normal microstructure, crosses – cube texture, square – secondary recrystallization.

found that it lowers the oxygen content and inclusion count while improving the magnetic properties of 50% Ni alloys. The effect of Si was also investigated (Ichinose 1968) but is less potent compared with Ge due to the higher vapor pressure of SiO over GeO.

Since the magnetic properties are highly sensitive to composition, powder metallurgy offers an excellent means of composition control. To this end, several studies have been conducted (Cohen 1967, Walker and Walters 1957, Enoch and Murrell 1969a, 1971, Murrell and Enoch 1970). In general, properties comparable to, but no better than, those obtained by vacuum melting have been achieved. However, in this case it is very important to control the purity and particle size of the initial powders, blending procedure, and sintering condition. Sufficient time must be allowed for blending and sintering to achieve homogeneous, complete alloying. Blending in of ternary and quaternary alloys involving

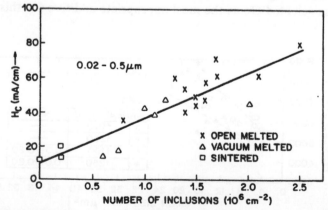

Fig. 86. Coercivity as function of inclusion content of 47.5% Ni alloys prepared by different techniques (Adler and Pfeiffer 1974).

elements such as Cu and Mo is particularly tricky. Enoch and Murrell (1969a) showed, for example, that in a 77Ni–14Fe–5Cu–4Mo pressed powder compact, sintering for 100 h at 1300°C is necessary to achieve uniform copper dispersion. In addition, cost of powders, particularly carbonyl iron, is fairly high. Thus the powder metallurgy market remains small. One salient advantage of the powder metallurgy approach is in the production of square-looped, cube-on-face oriented 50% Ni alloy to be discussed in section 5.5.6. The texture is achieved by large final cold reduction (>95%) followed by primary recrystallization and purification at temperatures in the 1100–1200°C range. Annealing in this range, however, could cause secondary recrystallization and consequent degradation of properties. It has been found (Walker and Walters 1957) that powder metallurgically prepared material tends to be more resistant than cast material to secondary recrystallization. Presumably, the small pores in the powder metallurgy product resist the motion of the secondary recrystallized grain boundaries.

5.5.5. Square or skewed hysteresis loop

While the primary objective of attaining high initial permeability is to minimize all sources of anisotropy, the major prerequisite for achieving a square or skewed hysteresis loop is an aligned nonzero anisotropy. If the magnetization is carried out in the easy-axis direction of aligned anisotropy, the remanence is high and the hysteresis loop square. If magnetized perpendicular to the easy-axis direction, the material exhibits low remanence and a skewed hysteresis loop. Square loop materials are used in magnetic amplifiers, memory devices, inverters and converters, while skewed loop materials are primarily developed for unipolar pulse transformers.

Since the magnitude and directional dependence of the various types of anisotropy depend on composition and heat treatment as discussed, a general rule is to aim for the dominance of one particular type to the exclusion of others unless the easy directions from two or more types can be aligned parallel. In this respect, magnetocrystalline anisotropy has been the most exploited, although thermomagnetic anisotropy has shown recent prominence. Slip induced anisotropy has also been utilized.

5.5.6. Square loops via magnetocrystalline anisotropy

The most celebrated square loop material is the cube-on-face, {100}(001), textured 50% Ni alloy. Early studies by Dutch (Snoek 1935, Burgers and Snoek 1935) and German (Dahl and Pfaffenberger 1931, Dahl and Pawlek 1935) researchers were responsible for this development. The texture is obtained by extensive prior rolling (>95%) followed by primary recrystallization in the vicinity of 1100–1200°C. Among factors favoring texture formation are small penultimate grain size, large cold reduction, and a lowering of impurity elements such as Si and C (Benford and Linsary 1963, Casani et al. 1966). Figure 87 shows a typical 60 Hz hysteresis loop of a commercial oriented 50% Ni alloy (Orthonol) along with those of high permeability alloys. Permeability and loss characteristics of thin gage materials including Orthonol have previously been illustrated in fig. 76

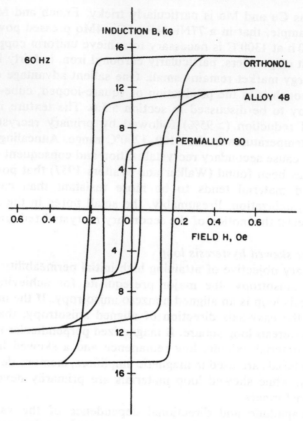

Fig. 87. 60 Hz hysteresis loops of three commercial Ni–Fe alloys (Magnetics Div. Spang Industries a, b).

and 77. As discussed in section 5.5.4 the developed texture appears to be more stable in powder metallurgically prepared material versus cast material. The Orthonol alloy is prepared by powder metallurgy.

Squareness control in the high Ni alloys is done in two ways, with λ_{111} set to zero for $K_1 < 0$ and $\lambda_{100} = 0$ for $K_1 > 0$. The usual way applies to weakly textured material normally encountered in commercial processing. Here compositions of alloys such as 4–79 Mo–Permalloy and Mumetal lie close to the $\lambda_{111} = 0$ line. Thus according to theory a heat treatment to achieve $K_1 < 0$ is desired for "quasi-squareness" (Pfeifer 1966b). This is achieved by lowering the isothermal annealing temperature or decreasing the cooling rate, as compared to the heat treatment for high μ_i where $K_1 = 0$ is desired. The results are illustrated in fig. 88 for a series of alloys studied by Pfeifer (1966b) in which Ni is held at 80%. A representative hysteresis loop is shown in fig. 89 labeled as Square Permalloy.

The second way of squaring up high Ni alloys is via the cube-on-face texture, just as in the case of the 50% Ni alloys. Here Odani (1964) found that the best

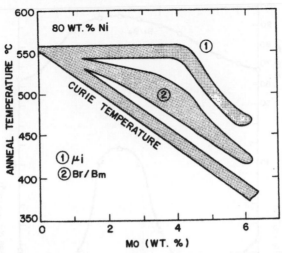

Fig. 88. Optimum annealing temperatures for high μ_i and B_r/B_m in Mo–Permalloys (Pfeifer 1966b).

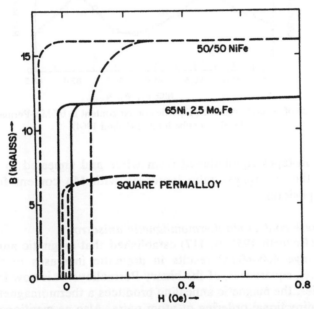

Fig. 89. Upper half of hysteresis loops of three square loop grades of Ni–Fe alloys (Pfeifer and Boll 1969). The 50Ni alloy is standard cube textured; the other two are unoriented but annealed in magnetic field.

squareness is obtained in a 6%Mo–81.3%Ni alloy (fig. 90). For the high Mo Odani used, $K_1 > 0$. In addition, the optimum B_r/B_s at 81.3% Ni corresponds to $\lambda_{100} = 0$. It may be pointed out that with 4% Mo, K_1 is most likely negative and a cube texture would then be detrimental to squareness. Thus Chin et al. (1971)

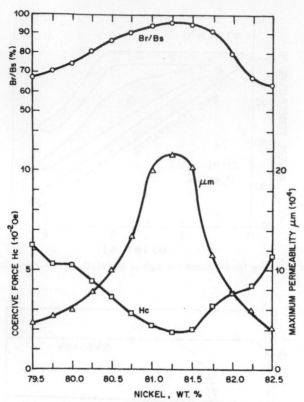

Fig. 90. Dependence of several magnetic properties on Ni content in 6%Mo–Permalloy processed to develop a cube texture (Odani 1964).

observed that in tapes roll-flattened from wires and annealed to achieve ⟨100⟩ axial texture, the squareness ratio increases with Mo content for a constant 80.5% Ni composition.

5.5.7. Squareness control via thermomagnetic anisotropy

Early studies (Bozorth 1951, p. 117) established that magnetic annealing in the temperature range 400–600°C results in dramatic increases in the maximum permeability, i.e., squareness, of the binary Permalloys. It is now known that, as discussed earlier, the magnetic annealing produces a thermomagnetic anisotropy as a result of directional ordering of atom pairs. Also as mentioned previously, although early work indicated that the initial permeability was lowered by the magnetic annealing, recent studies have shown that both μ_i and μ_m are increased (see fig. 78).

By means of columnar casting of ingots, special hot and cold reduction and final annealing, Howe (1956) obtained a {210}⟨001⟩ textured sheet of 2%Mo–65%Ni alloy (Dynamax). With additional magnetic annealing the thermomagnetic anisotropy is superposed onto the magnetocrystalline anisotropy, resulting in an extremely square hysteresis loop with very low dc coercivity. The importance of texture in the response of the material to magnetic annealing is demonstrated by Kang et al. (1967). They found that of the two types of secondary recrystallization of Ni–Fe alloys, the {210}⟨001⟩ type is highly responsive to magnetic annealing, while the {112} type showed poor response. There is evidence that the {210} type is favored by a sharp {100}⟨001⟩ primary texture and by the presence of strong deoxidizers (Aspen and Colling 1969).

Pfeifer, on the other hand, has found that magnetically annealed unoriented material possessing a fine grained structure shows lower dynamic coercivity than oriented material (Boothby and Bozorth 1947, Pfeifer and Boll 1969). The fine-grained unoriented material was obtained by repeated intermediate anneal as a strip is rolled to its final thickness, with a final anneal at 1150–1200°C (Pfeifer 1966a). To produce the maximum squareness in unoriented material, it is important the magnetic annealing heat treatment be done such that $K_u \gg K_1$, i.e. aligning the thermomagnetic anisotropy at the expense of crystal anisotropy. Since a low annealing temperature (~430°C) results in small K_1 (Puzei 1957) but high K_u (Ferguson 1958), this anneal is expected to give large squareness. Such reasoning by Pfeifer (1966a) was confirmed by his own findings as well as those of Kang et al. (1967). A typical hysteresis loop is given in fig. 89. Note that the remanent induction is nearly twice that of 4%Mo–80%Ni (Square Permalloy) but with considerably lower coercivity than the 50% Ni alloy. Figure 91 shows the constant current flux reset (CCFR) characteristics of the three alloys; such characteristics are important in amplifier type design considerations.

5.5.8. Squareness control via slip-induced anisotropy

As mentioned in section 5.4.7, plastic deformation of Ni–Fe alloys leads to the development of a uniaxial anisotropy energy $\sim 10^5$ erg/cm^3 (see figs. 67 and 73), far exceeding the value of 10^3 or less for magnetocrystalline anisotropy. Thus the direction of the magnetic easy axis and consequently the squareness behavior of cold-worked Ni–Fe alloys are controlled by this slip-induced anisotropy. The hysteresis loops of 4–79 Mo–Permalloy deformed by several techniques are shown in fig. 92. It may be noted that a cold-drawn wire exhibits a very square loop. When this wire is flattened to tape by rolling, the squareness drops. The drop is small, however, if the same wire is flattened by drawing through rectangular dies. A fairly square loop is obtained in a rolled strip. Such characteristics have been interpreted in terms of slip-induced anisotropy coupled to the texture in the material and the manner of cold working (Chin 1966, Chin et al. 1967).

Other studies relating slip-induced anisotropy and loop shape of flattened wires have been made (Chin et al. 1964). Such wires have actually been utilized in Twistor memories (Chin et al. 1971, Bobeck 1957) which have found successful applications in telephone electronic switching systems (Ault et al. 1964, Lounquist et al. 1969). An important consideration in processing is the goal of achieving stress insensitivity and a controlled value of coercivity along with a square hysteresis loop.

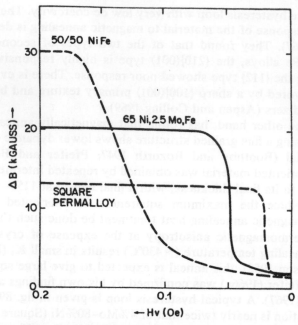

Fig. 91. Constant-current flux reset characteristics of three square loop grades of Permalloy whose hysteresis loops are shown in fig. 89 (Pfeifer and Boll 1969).

5.5.9. Skewed hysteresis loop

While the alignment of a magnetic anisotropy results in a square hysteresis loop in the direction of alignment, it also results in a skewed loop in the perpendicular direction. Such a low remanence, constant permeability characteristic is highly desirable in loading coils and unipolar pulse transformers. Isoperm is one material developed to exhibit this characteristic for loading coil use (see fig. 71). For unipolar transformers, a lower coercivity and higher (but still constant)

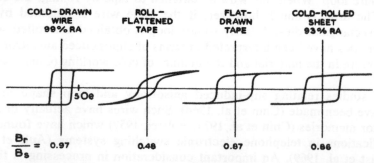

Fig. 92. 60 Hz hysteresis loops of 4–79 Mo–Permalloy deformed under various condition (Chin et al. 1971).

permeability than those obtained in Isoperm are needed (Pfeifer and Boll 1969). Pfeifer and Deller (1968) found that coercivity can be lowered by high temperature anneal, which also reduces K_1. A small K_u was then put in by a transverse field treatment below the Curie temperature to produce a high but constant permeability. The authors also developed a 2.5Mo–65Ni Permalloy (Permax Z) with lower permeability but larger excursion range of induction. The pulse characteristics of the two new materials are shown in fig. 93 along with that of a standard 5–79 Permalloy which had not been field treated.

5.6. Other recent developments

5.6.1. Wear-resistant alloys for recording heads

For applications such as recording heads for tape recorders and magnetic tape information storage devices, a combination of high mechanical hardness and magnetic softness is clearly desirable. Standard Permalloys exhibit high μ_i, but poor wear-resistance. The 16% Al–Fe and Fe–Si–Al (Sendust) alloys and soft ferrite are mechanically hard, but the permeability is poor. It has been found that alloys near the 79% Ni composition with Nb and Ti additions (Miyazaki et al. 1972) or with Nb alone (Masumoto et al. 1974) can greatly enhance the hardness while retaining the high permeability. Figure 94 shows how the mechanical and magnetic parameters change as a function of aging time at 630°C, for a 3%Ti–2.8%Nb–3.8%Mo–10.5%Fe–bal Ni alloy (Miyazaki et al. 1972). It may be noted that after aging for 4 h, the permeability and coercivity went through a maximum and minimum respectively, while the Vickers hardness increases. This phenomenon is explainable in terms of precipitation where the particles are of optimum size to resist dislocation motion but too small to pin magnetic domain walls. Table 30 shows some comparisons of performance among various recording head alloys (Hitachi Metals E-172). A similar material based on Mumetal composition has also been developed (Vacuumschmelze 1976).

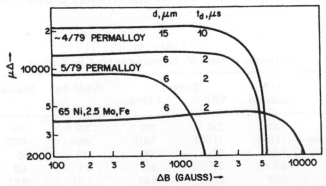

Fig. 93. Pulse permeability vs excursion range of induction for transverse field treated 4–79 Permalloy (upper 2 curves) and 65% Ni–2.5% Mo alloy in comparison with nonfield treated 5–79 Permalloy (Pfeifer and Boll 1969) (d = tape thickness, t_d = pulse duration).

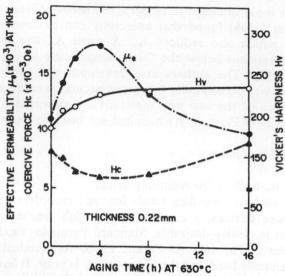

Fig. 94. Magnetic and mechanical properties of a 3%Ti–2.8%Nb–3.8%Mo–10.5%Fe–bal Ni alloy as function of aging time at 630°C (Miyazaki et al. 1972).

5.6.2. High permeability low nickel alloys

Alloys in the 36% Ni range have been available for some time. Those alloys generally exhibit values of $\mu_i \sim 2000$–3000 which are fairly constant at low field intensities, and have high resistivity (0.75 $\Omega\text{mm}^2/\text{m}$ vs 0.45 for 50% Ni). Various grades have been developed for such applications as low-distortion transformers, broadband transformers and relays. The lower Ni content means lower cost, but the market in the transformer area is being eroded by soft ferrites.

Recently Pfeifer and Cremer (1973) restudied this low Ni regime and found that the addition of 2% Mo improves the initial permeability substantially. The material was rolled to over 90% reduction and annealed at over 1100°C in H_2, the

TABLE 30
Properties of materials for recording head applications

Variable	4–79 Permalloy	Tufperm* YEP-H	YEP-S	16%Al–Fe	Sendust	Ferrite
Hardness H_V	120	230	290	290	480	580
μ at 1 kHz, 0.2 mm	11 000	11 000	7000	4000	8000	45 000
ρ ($\mu\Omega$ cm)	100	60	100	150	85	10^6–10^7
B_s (T)	0.8	0.5	0.5	0.8	1.0	0.4
H_c (A/cm)	0.02	0.01	0.02	0.03	0.02	0.06
T_c (°C)	460	280	280	350	500	150

* Ti–Nb–Mo–Permalloy (Hitachi Metals Ltd.)

deformation and annealing adjusted so as to cause secondary recrystallization. As a result of this study, an alloy containing 2% Mo and 34.5% Ni exhibited a remarkable permeability $\mu_4 = 55\,000$ at 25°C, with a large value of electrical resistivity (0.9 Ωmm^2/m) and a Curie temperature of 160°C. It was found that the permeability peaks at various temperatures, this temperature increasing with Ni content, corresponding to the position of $K_1 = 0$. The Mo addition decreased the magnetostriction constant, leading to a further increase in permeability. Such properties could be cost effective competing with ferrites in the medium frequency area.

5.7. Other nickel–iron alloys

5.7.1. Molypermalloy powder cores
Molybdenum Permalloy powder (MPP), of composition near 2%Mo–81%Ni, is used to prepare cores for applications such as high Q filters, loading coils, resonant circuits, and RFI filters. Generally a range of permeabilities ranging from about 14 to 550 are commercially available, with operating frequencies ranging up to about 300 kHz for the low permeability cores. Compared with ferrites, MPP cores have higher saturation and hence superior inductance stability after high dc magnetization or under high dc bias conditions. Due to lower resistivity, MPP cores are replaced by ferrite cores at high frequencies. Carbonyl iron powder cores are also used in frequencies up to about 1 MHz. An early account of Molypermalloy powder was given by Bozorth (Bozorth 1951, p. 144) and an excellent summary of various dust cores was recently provided by Heck (1974b). Table 31 shows a summary of some representative core material properties as provided by Bradley (1971b).

5.7.2. Nickel and nickel alloys for magnetostrictive applications
Nickel and its alloys are being used as magnetostrictive transducers in ultrasonic devices such as soldering irons and ultrasonic cleaners. The pertinent properties of nickel and several of its alloys of interest as magnetostrictive devices are shown in table 32 along with a Co–Fe alloy, an Fe–Al alloy, and a Ni–Cu–Co ferrite (Heck 1974e, Bradley 1971c). Compared with Ni, the 50Ni–50Fe alloy has higher resistivity (hence better at high frequency) and the 95Ni–5Co alloy has greater magnetostriction and a larger electromechanical coupling coefficient. The Co–Fe alloy has higher magnetostriction, but is more expensive. The Fe–Al alloy has higher resistivity, but is considerably more brittle. The ferrite, being an oxide ceramic, is also brittle. As an insulator, however, it has outstanding properties at high frequencies. Chin (1971a) has a review in this area.

Today piezoelectric devices using materials such as PZT (lead zirconate–titanate) have taken over much of the ultrasonic device market. Because the electrostrictive strain of $\sim 1000 \times 10^{-6}$ is much greater than the magnetostrictive strain ($\sim 30 \times 10^{-6}$), most high power acoustic radiators such as sonar and heavy duty ultrasonic cleaners and welders make use of PZT. Because of good fatigue strength and corrosion resistance, however, nickel is still used appreciably,

TABLE 31
Representative core materials**

Material	Initial permeability	Resistivity (Ω cm)	Hysteresis coeff.* $a \times 10^6$	Eddy current coeff.* $e \times 10^9$	Residual loss coeff.* $c \times 10^6$
96% Fe, 4% Si (14 mil)	400	60×10^{-6}	120	870	75
54% Fe, 45% Ni:					
14 mil	4000	44×10^{-6}	0.4	1550	14
6 mil	4000	44×10^{-6}	0.4	280	14
17% Fe, 79% Ni, 4% Mo:					
14 mil	20 000	55×10^{-6}	0.05	950	0.05
6 mil	20 000	55×10^{-6}	0.05	175	0.05
17% Fe, 81% Ni,	550	1	1.3	83	96
2% Mo insulated	200	1	1.3	40	40
powder	125	1	1.6	19	30
	60	1	2.5	10	50
Carbonyl iron	55	1	9	7	80
insulated powder	16	1	2.5	8	80
$Mn_{0.8}Zn_{0.2}Fe_2O_4$	750	80	15	0.42	10
$Mn_{0.6}Zn_{0.4}Fe_2O_4$	2000	80	1.5	0.04	3
$Ni_{0.6}Zn_{0.4}Fe_2O_4$	100	106	40	1	50

* $R/\mu fl = aB_m + ef + c$
** Bradley (1971b)

TABLE 32
Magnetostrictive materials*

Material	Density (g/cm^3)	Youngs modulus (GPa)	Resistivity ($\mu\Omega$ cm)	SMS** $\times 10^6$	EMCC† (km)
Ni	8.9	200	7	-28	0.30
50Ni–50Fe	8.2	200	40	$+28$	0.32
95Ni–5Co	8.8	200	10	-35	0.50
49Co–49Fe–2V	8.2	220	30	-65	0.30
87Fe–13Al	6.7	130	90	$+30$	0.22
$Ni_{0.42}Cu_{0.49}Co_{0.01}Fe_2O_4$	5.3	160	10^9	-28	0.27

* Heck (1974e), Bradley (1971c)
** SMS = Saturation magnetostriction
† EMCC = Electromechanical coupling coefficient

primarily in units under 100 W. Newer magnetostrictive alloys of the rare earth–iron family such as $SmFe_2$ and $TbFe_2$, with magnetostrictive strains in the 2000×10^{-6} regime, are being explored for a number of applications (Clark 1977).

5.7.3. Temperature compensator alloys

Alloys in the 30–32%Ni regime have Curie temperatures slightly above room temperature. Hence the magnetic permeability decreases steeply in this region so that these alloys can be used in electrical circuits as a shunt to maintain constant magnet strength against ambient temperature changes. Electric meters, voltage regulators and speedometers are examples of such applications. Figure 95 shows the permeability (at 46 Oe) vs temperature for three types of commercial temperature compensator alloys (Carpenter Technology 1970). The control of nickel content is very important for these materials.

5.7.4. Expansion alloys

The 52%Ni–Fe alloy has a mean coefficient of thermal expansion about $10 \times 10^{-6°}C^{-1}$ that closely matches that of soft glasses and is known as the glass sealing alloy. Today a variety of glasses and ceramics have been developed for electron tubes, lamps and bushings. Therefore a number of alloy compositions have become commercially available to match the thermal expansivities of these glasses and ceramics. In general, the alloys are prepared by vacuum melting using high purity ingredients to provide high quality seals. In other applications,

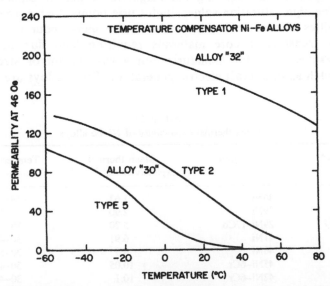

Fig. 95. Permeability at 46 Oe as a function of temperature for three types of commercial temperature compensator alloys (Carpenter Technology 1970). Type 1 contains ~32% Ni and types 2 and 5 contain ~30% Ni.

the thermal expansion characteristic is utilized in temperature controls, thermostats, measuring instruments, condensers, and others. Table 33 provides values of thermal expansivity for several commercial expansion and sealing alloys (Magnetics Div. Spang Industries c).

6. Iron–cobalt alloys

6.1. Introduction

Iron–cobalt alloys near the equiatomic composition are technologically important because of their high saturation magnetization, low magnetocrystalline anisotropy and associated high permeability. Commercially important alloys based on this system are 2V-Permendur (White and Wahl 1932) and Supermendur (Gould and Wenny 1957), containing 49% Fe, 49% Co and 2% V.

Equiatomic alloys of Fe and Co containing 2–5% V (ex: 48%Co–48%Fe–3.5%V) are referred to as Remendurs (Gould and Wenny 1957) and bridge the gap between the high coercive force Vicalloys (Wenny 1965, Nesbitt and Kelsall 1940, Nesbittt 1946) (8–15% V) and high permeability 2V-Permendur. The coercive force can be controlled by varying the V content. Remendur is a semi-hard magnet ($H_c \sim 30$ Oe) and Vicalloy is a permanent magnet ($H_c \sim 200$ Oe). Both are ductile, as are 2V Permendur and Supermendur.

These particular alloys are relatively costly and therefore are used for specialized applications requiring the magnetic properties exhibited by these materials. The high saturation values, high Curie points, high permeabilities, and low coercivities obtainable in Permendur and Supermendur have made them suitable for transformer core materials requiring high flux densities. High saturation values allow for miniaturization where weight or size savings are important, such as in aircraft electric generators. These alloys are also used for

TABLE 33
Mean thermal expansivity of Ni–Fe alloys

Alloy*	Composition	Mean thermal expansivity ($°C^{-1} \times 10^6$)	Temperature range (°C)
36	36Ni	1.63	−18–175
39	39Ni	2.90	25–200
29–17	29Ni–17Co	5.20	30–450
42	41Ni	6.91	30–450
46	46Ni	8.5	30–500
45–6	45Ni–6Cr	10.05	30–425
42–6	42Ni–6Cr	10.1	30–425
52	51Ni	10.4	30–550
22–3	22Ni–3Cr	19.9	30–450

* Magnetics Div. Spang Industries (C)

receiver coils, switching and storage cores, high temperature components, as diaphragms (Permendur) in telephone handsets because of the high incremental permeability over a wide induction range, and to a limited extent for magnetostrictive transducers for sonar. Several physical properties of Permendur and Supermendur are given in table 34.

6.2. The iron–cobalt and iron–cobalt–vanadium phase diagrams

The iron–cobalt phase diagram is shown in fig. 96. Iron and cobalt form a complete series of disordered fcc solid solutions (γ) at elevated temperatures. At low temperatures the bcc solid solution (α) exists above 730°C up to ~75% Co. The α-bcc solid solutions near the equiatomic composition undergo below 730°C an atomic ordering to the CsCl structure-type (α_1) and this order–disorder transformation plays an important role in determining the magnetic and mechanical properties of these materials (see section 6.3).

TABLE 34
Several physical properties of Permendur and Supermendur

Material	Density (g/cm³)	Room temperature resistivity ($\mu\Omega$ cm)	Melting point (°C)
2V Permendur	8.3	6-7	1470
Supermendur	8.2	26	1480

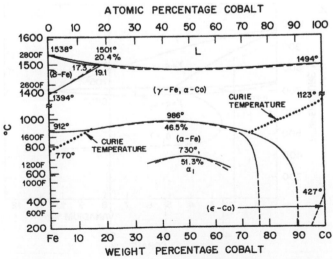

Fig. 96. Iron–cobalt phase diagram (Metals handbook 1973a).

The effect of V in the vicinity of 50Fe–50Co is shown in fig. 97. The effect of a 2% V addition is to lower the $\alpha + \gamma$ transformation to about 880°C on heating as well as expanding the $\alpha + \gamma$ region. The critical temperature for the order–disorder reaction ($\alpha \leftrightarrows \alpha_1$) is lowered somewhat and the $\alpha + \alpha_1$ region expanded (Martin and Geisler 1952, Stoloff and Davies 1964).

6.3. Intrinsic magnetic properties

Alloys of iron and cobalt have the highest known values of saturation magnetization ($\sim 24\,500$ G at room temperature for a 35% Co alloy). Although cobalt exhibits a lower moment/atom compared to iron, the addition of Co to Fe increases its magnetization. This is shown in fig. 98, where the variation of the room temperature saturation moment, σ_s, with composition is given, peaking at about 35% Co. The Curie temperature, shown as dotted lines in fig. 96, rises as Co is added to Fe up to $\sim 15\%$ Co, then follows (and is nearly coincident) the boundaries of the $(\alpha + \gamma)$ region to $\sim 73\%$ Co. A maximum in permeability is obtained at 50%Co–50%Fe.

The variation of the anisotropy constant, K_1, with composition is shown in fig. 99. K_1 goes to zero at $\sim 40\%$ Co when in the disordered state, and near 50% when in the ordered state. The easy direction of magnetization changes from [100] to [111] near 41% Co.

The saturation magnetostriction (λ) in the [100] and [111] crystallographic directions is shown in fig. 100 for Fe–Co alloys in the disordered and ordered states. Although the data for the ordered state are scattered, it appears that ordering lowers λ_{100} and raises λ_{111}.

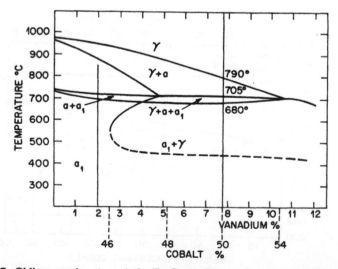

Fig. 97. Oblique section through the Fe–Co–V diagram (Martin and Geisler 1952).

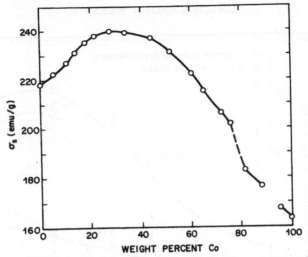

Fig. 98. Variation of the saturation magnetization σ, of Fe–Co alloys as a function of composition after Weiss and Forrer (1929) as given by Hoselitz (1952).

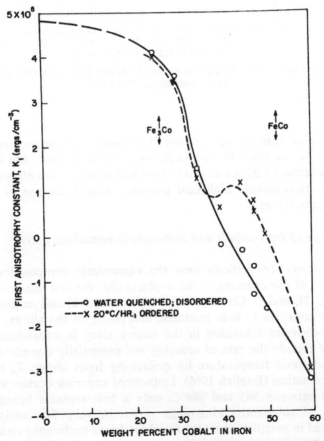

Fig. 99. The first anisotropy constant, K_1, for iron–cobalt alloys (Hall 1960).

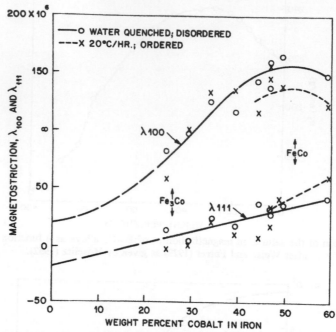

Fig. 100. The saturation magnetostriction (λ) in the [100] and [111] crystallographic directions for iron–cobalt alloys (Hall 1960).

Further examination of figs. 99 and 100 shows that the ordered 50Fe–50Co alloy exhibits λ_{100} of 150×10^{-6} while $K_1 \simeq 0$. For this reason, the 2V-Permendur (discussed in section 6.5) is one of the best magnetostrictive alloys available. As noted above, its relatively high cost prevents large scale use of this alloy for transducer applications.

6.4. Metallurgy of iron–cobalt and iron–cobalt–vanadium alloys

The Fe–Co alloys, particularly near the equiatomic composition, tend to be brittle because of the ordering of the α-phase (fig. 96) and the presence of trace amounts of C, H, and O. Consequently, the mechanical and magnetic properties attainable are sensitive to heat treatment and purity of the alloys.

The order–disorder transition in the binary alloy is exceedingly rapid. The addition of V retards the rate of ordering and essentially complete disorder can be retained at room temperature by quenching from above T_c but below the $\alpha \rightarrow \gamma$ transformation (English 1966). Isothermal ordering occurs very rapidly at temperatures between 565 and 680°C, only a few seconds being required for development of substantial long-range order. A very fine antiphase domain structure found in samples annealed for short times undergoes coalescence with further annealing (English 1966). The antiphase domain structure was described as being "Swiss Cheese"-like, consisting of two multiply connected and inter-twined domain volumes.

Because of the retardation of the ordering of the α-phase by quenching, the ternary alloy can be produced in the form of thin sheets by heavy cold reduction. The heavy cold working generates crystallographic texture (Josso 1974). The texture is duplex and the components are (112) [$\bar{1}$10] and (001) [$\bar{1}$10] (Josso 1974). Because of texture the mechanical properties become anisotropic. Table 35 shows tensile strength and Young's modulus data for 0.25 mm thick strip samples cut from annealed cold-rolled sheet (91.7% reduction). The diagonal direction (45° from the rolling direction) shows poorer strength, ductility and modulus (Josso 1974).

Stoloff and Davies (1964) observed that ordering markedly affected the deformation mode of the 2V alloy. Wavy glide was shown to be the predominant slip mode in the disordered alloy and planar glide the slip mode in the ordered material. The dependence of the room temperature flow stress for 0.1% strain on quenching temperature is shown in fig. 101 (upper curve). A moderate increase in strength was observed with decreasing quench temperature until the degree of long range order reached 0.2, followed by a rapid decrease in strength. The ductility, as indicated by the percent uniform elongation (lower curve) also decreased rapidly with decreasing quench temperature below the order–disorder temperature (i.e. with increasing degree of long range order). More recent work (see for example Thornburg 1969) by many researchers have established that a peak in the ductility exists at ~690°C (see also fig. 102). The data of Stoloff and Davies (1964) shown in fig. 101 also suggests a peaking of the ductility.

6.5. Magnetic properties of Permendur and Supermendur

Elmen (1929, 1935), the discoverer of Permendur, showed that the high permeability of this alloy persists ("endures", thus the name Permendur) to high flux densities. In general, the alloy is usually hot-rolled above the $\alpha \rightarrow \gamma$ tem-

TABLE 35

Several mechanical properties of cold rolled and annealed strip of Co–Fe–2V as a function of direction in the plane of the strip*

Direction	Tensile properties anneal: 2 h at 670°C		Young's modulus	
	Tensile strength (ksi)	% Elongation	Cold worked (10^6 psi)	Anneal: 4 h at 850°C (10^6 psi)
Longitudinal (Rolling direction)	212	19	32.9	25.3
Diagonal	154	7	26.3	22.2
Transverse	200	12.5	36.2	31.3

* Josso (1974)

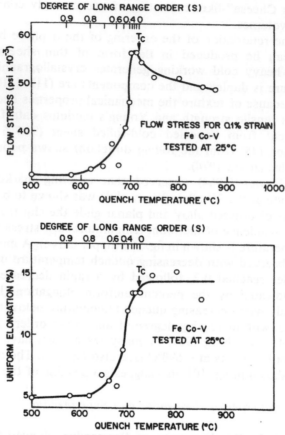

Fig. 101. The dependence of room temperature flow stress of Fe–Co–V on quench temperature (upper curve) and the effect of long-range order on the ductility (Stoloff and Davies 1964).

perature. This is then followed by a magnetic anneal at 850°C to develop the magnetic structure and then rapidly cooled to room temperature should it be necessary to further work the alloy. White and Wahl (1932) found that the addition of 2.0% V improved the ductility of the alloy so that it could be rolled into sheets as thin as a few mils ($= 10^{-6}$ in). This amount of V imparts a minimal effect on magnetic properties but has a pronounced (and beneficial) effect on the resistivity (table 34). The resistivity increase is nearly fourfold, going from $7\mu\Omega$ cm with zero V to $26\mu\Omega$ cm for the 2% V alloy.

Now Gould and Wenney (1957) showed that if the 2% V alloy is made rather pure by using the highest purity commercial V, Fe, and Co and melting in a controlled atmosphere furnace using wet and dry hydrogen treatments (hydrogen also purified to prevent contamination) so as to minimize the incorporation of interstitial impurities, and by annealing in a magnetic field, the magnetic properties of the 2% V alloy can be improved considerably. The 2% V alloy made in this fashion was named Supermendur.

Gould and Wenny's (1957) working and annealing procedure of their laboratory ingots was as follows. The ingot was hot-rolled in two stages; the first from $1-\frac{1}{2}$ in to $\frac{1}{2}$ in at 1200°C followed by reduction to 0.090 in at 1000°C. The resulting strip was then quenched in ice H_2O from 910°C for cold-rolling to the desired thickness. The 0.090 in strip could be cold-reduced without intermediate anneals as thinly as 0.0003 in. Even at this thickness, the material is still ductile. To develop the magnetic properties, a heat treatment below the $\alpha \rightarrow \gamma$ temperature is carried out followed by cooling at a rate of 1°C/min. For optimum magnetic properties, Gould and Wenny performed the above final heat treatment in a magnetic field of 1 Oe or greater. A comparison of the properties of "regular" and high purity 2 V alloy is shown in table 36. Note the large maximum permeability and low coercivity of Supermendur.

Thus, to develop the best soft magnetic behavior, the final heat treatment consists of annealing below the $\alpha \rightarrow \gamma$ transformation but above the ordering temperature (~850°C). Because of the ordering reaction, cooling rate is expected to affect magnetic properties. Some information regarding the effect of cooling rate is shown in table 37. The best values of μ_{max} and H_c were obtained at

TABLE 36
Properties of regular and high-purity Co–Fe–2V

Variable	Regular	High purity (Supermendur)
Electrical resistivity	$25 \times 10^{-6} \Omega$ cm	$25 \times 10^{-6} \Omega$ cm
Saturation induction	24 000 G	24 000 G
Remanent induction	15 000 G	22 150 G
μ_m	4000–8000	92 500
H_c	5 Oe	0.20 Oe
Magnetostriction coeff.	$+60 \times 10^{-6}$	$+60 \times 10^{-6}$
	(at 100 Oe)	(at 100 Oe)

TABLE 37
Effect of cooling rate on maximum permeability (μ_{max}) and coercive force (H_c). Heat treatment was an 8 h anneal at 870°C and cooled to room temperature as shown*

Melt	Slow cooling (25°C/h)		Furnace cooling (250°C/h)		Rapid cooling (quenched in H_2, 10 000°C/h)	
	μ_{max}	H_c (Oe)	μ_{max}	H_c (Oe)	μ_{max}	H_c (Oe)
A	4850	1.20	19 900	0.67	5700	1.10
B	3300	1.20	20 800	0.60	6300	1.10

* Josso (1974)

"intermediate" cooling rates, faster than that employed by Gould and Wenny (1957). They are not as good as those obtained by Gould and Wenny but the purity of the Co–Fe–2V alloy used for this study was not given. Ordering lowers the resistivity (Rossmann and Wick 1962); the minimum resistivity being obtained by a heat treatment at 640°C (Karmanova et al. 1965).

Carefully controlled heat treatments have been developed for the regular 2V Permendur in order to increase the room temperature strength and ductility at a small sacrifice in maximum magnetic properties. This work was motivated by the need for developing a soft magnetic material with superior strength for high speed rotor applications for aerospace use. Thornburg (1969) showed that annealing cold-rolled commercial sheet for 2 h in dry H_2 at 695°C (complete recrystallization occurs at 710°C) and cooling to room temperature at a rate of ~ 900°C/h yielded material exhibiting a yield strength of 100×10^3 psi and 13% elongation. B_{tip} (at $H = 100$ Oe) was 22.5 kG and $H_c = 6.3$ Oe. The microstructure of these samples indicated only partial recrystallization. In contrast to the work of Stoloff and Davies (1964) (fig. 101) the ductility maximum occurred at 690°C and the 0.02% offset yield strength did not drop off at this temperature but continued to rise with decreasing annealing temperature, reaching $\sim 180 \times 10^3$ psi at 650°C.

Thomas (1975) developed two separate heat treatments for high speed rotor and stator applications. For the rotor material, heat treatment at 650°C produces sufficient strength and ductility in the rolling direction and acceptable dc magnetic properties. For stator material, heat treatment in vacuum or H_2 for 1–2 hours at 875°C is desirable followed by cooling between 100°C/h and 250°C/h, to about 400°C below which the cooling rate is unimportant.

Moses and Thomas (1976) have shown that when 2V Permendur is annealed above 710°C to develop good magnetic properties at the expense of mechanical properties, compressive stress is very detrimental to power loss but tension has a marginal effect, although it reduces the magnetostriction considerably. The magnetic properties are most influenced by the initial domain distribution which depends on the residual internal stress and changes in anisotropy with annealing temperature. The effect of roll anisotropy is removed by annealing above 710°C, the recrystallization temperature.

6.6. Iron–cobalt–vanadium–nickel alloys

As indicated in section 6.5, the poor ductility of the basic regular 2V Permendur alloy has restricted its use. For example, it is commonly used in stator assemblies of aircraft electrical generators but not as a rotor material, although special heat treatments, as noted above, have been devised to utilize the basic 49/49/2 composition in high speed aircraft generators. In order to extend the usefulness of 2V Permendur so that it could be used in higher speed generators, a new alloy containing Ni has recently been developed (Major et al. 1975).

Major et al. (1975) have shown that the addition of 4.5% Ni to the 49/49/2 Fe–Co–V alloy results in enhanced ductility and strength over a wider heat

treating temperature range, thus eliminating the need for relatively close temperature control. As shown in fig. 102, the maximum in ductility which occurs in
the vicinity of 700°C on heat treating cold rolled Permendur is associated with
the occurrence of loss of long range order and recrystallization (see also fig. 101).
The addition of Ni (denoted as New Alloy in fig. 102) results in the ductility
remaining high after heat treating over a wider temperature range (650–740°C).
For example, the ductility as given by the percent elongation, is ~20% after heat
treating at 650°C. Similarly, the yield strength of the Ni-bearing alloy remains
relatively high beyond 690°C (fig. 103). In addition, the ultimate tensile strength
only drops off slightly after heat treating between 670 and 690°C whereas it falls
off drastically for the ternary alloy in this range. It appears that the magnetic
performance is not seriously degraded by the addition of Ni and heat treatment
in the vicinity of 700°C. X-ray examination of this alloy showed the presence of
some γ-phase and ordered α-phase. It is suggested that the enhanced ductility
and strength stems from the restriction of grain growth due to the presence of
these phases. Major et al. (1975) suggest that the Fe Co V Ni alloy will find
application as a rotor material for high speed generators. Another claimed
advantage for this alloy is that it can be manufactured in forged and heavy
sections exhibiting good mechanical properties in recrystallized or partially
recrystallized microstructures.

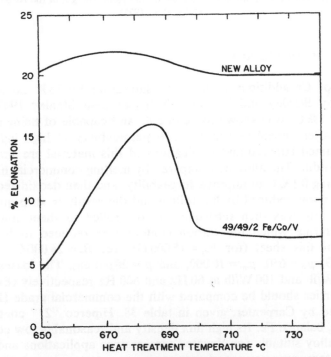

Fig. 102. The effect of heat treatment temperature on the ductility of the Ni-bearing Permendur (new
alloy) (Major et al. 1975).

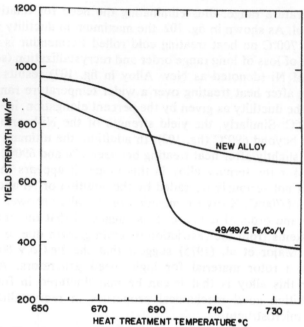

Fig. 103. The effect of heat treatment temperature on the yield strength of the Ni-bearing Permendur (Major et al. 1975).

6.7. Other iron–cobalt alloys

The effect of Cr additions to the high saturation Fe–35% Co alloy was investigated by Stanley and Yensen (1947) (see also Stanley 1947). The alloy containing 0.5% Cr was shown to be ductile and capable of being rolled to thin sheets and after annealing possess high permeability at high inductions. The alloy was named Hiperco and applications of this material are similar to those for V Permendur. The alloy was prepared by melting commercial high-purity Fe and Co, adding 0.1% C to improve forgeability, and then deoxidized with Si and Ti. The ingot was reduced by hot-rolling and the resulting strip quenched from 910°C. The strip was then subsequently cold rolled to sheet and annealed at 875–925°C during which the carbon content was reduced to below 0.005%. Properties of this sheet (for $B_m = 15\,000\,G$) are: $B_s = 24\,000\,G$, $H_c = 0.63\,Oe$, $B_r = 11\,500\,G$, $\mu_o = 650$, $\mu_m = 10\,000$, and $p = 20\,\mu\Omega$ cm. The losses for 0.017 in sheet are 3 W/lb and 100 W/lb at 60 Hz and 600 Hz respectively (Bozorth 1951). These properties should be compared with the commercial grade Hiperco "27", manufactured by Carpenter, given in table 38. Hiperco "27" contains 27% Co and 0.6% Cr, balance Fe. Its high flux density and reasonably low coercive force makes this alloy suitable for elevated temperature applications and for laminations for motors, generators, and cores.

The binary alloy Hyperm Co 35, containing 35% Co, is manufactured by

TABLE 38

Some mechanical and dc magnetic properties of Hiperco "27" and Hyperm Co 35

Variable	Hiperco "27"	Hyperm Co 35
Tensile strength (10^3 psi)	75	–
Yield strength (10^3 psi)	38	–
Elongation (%)	10	–
Reduction of area (%)	50	–
Rockwell hardness	B85/92	–
Magnetic force, H_{max} for B = 10 000 G	–	–
Coercive force (Oe)	2.5	2.5
B_r(kG), from $B = 10\,000$	–	–
Permeability (initial)	–	–
Permeability (max.)	2800	–
Flux density at max. permeability (kG)	16	–
Saturation induction (kG)	23.6	24.0
Resistivity ($\mu\Omega$ cm)	19	40

Krupp and is supplied in the form of castings or forgings or as parts machined from them. Because of cost considerations, this alloy enjoys limited applications confined to parts for aircraft electronic systems and magnet pole pieces.

Appendix: Some commercial soft magnetic alloys:

Type	Trade name	Company
Iron	AME	Edelstahlwerke Buderus AG, Röchlingsche Eisen-u. Stahlwerke
	CDW	C.D. Wälzholz
	Carpenter Core Iron	Carpenter Technology
	Hyperm O	F. Krupp Widiafabrik
	Motork	Magnetic Metals
	Sumitomo (F) iron	Sumitomo Special Metals
	Vacofer	Vacuumschmelze
Low carbon steel	Forgelec	Gueugnon
	Ly-Core	JLA (Australia)
	Newcore	British Steel Corp.
	Nomatil	Cockerill
	Norcor	Norsk Jernverk

Appendix, continued

Type	Trade name	Company
Non-oriented silicon steel	Armco DI-MAX	Armco Steel
	Dyba	Hoesch
	Hilitecore, Homecore	Nippon Steel
	Hyperm 1, 4, 7	F. Krupp Widiafabrik
	Kawasaki RM	Kawasaki Steel
	Lohy, Special Lohy	Tisco
	Losil	British Steel Corp.
	RSP	Hindustan Steel
	SWB Electrosheet	Elektrobleck-Gesellschaft
	Traba	Hoesch
	Trancore	Elektroblech-Gesellschaft
	Transil	British Steel Corp.
Oriented 3.2% silicon steel	Alphasil	British Steel Corp.
	Armco Oriented, Oriented LS	Armco Steel
	Hyperm 5T	F. Krupp Widiafabrik
	Kawasaki RG	Kawasaki Steel
	Imphisil	Societé Metallurg. d'Imphy
	Orientcore	Nippon Steel
	ORSI, ORSI-H	Grillo Funke
	Sempersil	Tohoku Metal Ind.
	Silectron	Allegheny Ludlum
	Unisil	British Steel Corp.
High-induction oriented 3% silicon steel	Silectperm	Allegheny Ludlum
	Kawasaki RG-H	Kawasaki Steel
	Orientcore H1-B	Nippon Steel
	Tran-Cor H	Armco Steel
Thin gage oriented 3.2% silicon steel	Hyperm 5T, 7T	F. Krupp Widiafabrik
	Magnesil	Magnetics-Div. Spang Ind.
	Microsil	Magnetic Metals
	Silectron Z	Arnold Engineering
	Thin Gage Orientcore	Nippon Steel
	Trafoperm	Vacuumschmelze
30–32% Ni (temperature compensator)	Carpenter "30", "32"	Carpenter Technology
	Thermoflux	Vacuumschmelze
	Thermoperm	F. Krupp Widiafabrik
	Sumitomo MS-1, MS-2, MS-3 (Fe–Ni–Cr)	Sumitomo Special Metals
36% Ni	Anhyster B	Societé Metallurg. d'Imphy
	Hyperm 36, 36M	F. Krupp Widiafabrik
	Permalloy D	ITT CG-E Mag. Mat'ls Div.
	Permenorm 3601	Vacuumschmelze
	Radiometal 36	Telcon Metals
	Suprahnyster 36	Societé Metallurg. d'Imphy
45–50% Ni round loop	Allegheny 4750	Allegheny Ludlum
	Anhyster D	Societe Metallurg. d'Imphy
	Carpenter "49"	Carpenter Technology

Appendix, continued

Type	Trade name	Company
	45 Permalloy	Western Electric
	48 Alloy	Magnetics-Div. Spang Ind.
	Hyperm 50, 52,	F. Krupp Widiafabrik
	Hyrho Radiometal	Telcon Metals
	Monimax	Allegheny Ludlum
	Niron 52	Wilbur B. Driver
	Permalloy B	ITT CG-E Mag. Mat'ls Div.
	Permenorm 5000 H2	Vacuumschmelze
	Sandvik Safeni 48	Sandvik
	Sumitomo PB, PD	Sumitomo Special Metals
	SuperPerm 49	Magnetic Metals
	Superradiometal	Telcon Metals
	Supranhyster 50	Societe Metallurg. d'Imphy
	Viperm 5	Vimetal
50% Ni	Deltamax	Arnold Engineering
square loop	HCR Alloy	Telcon Metals
	Hy-Ra 49	Carpenter Technology
	Hyperm 50T	F. Krupp Widiafabrik
	Hyrem Radiometal	Telcon Metals
	Orthonol	Magnetics-Div. Spang Ind.
	Permalloy F, G	ITT CG-E Mag. Mat'ls Div.
	Permenorm 5000 Z	Vacuumschmelze
	Rectimphy	Soc. Metallurg. d'Imphy
	Sendelta	Tohoku Metals
	Square 50	Magnetic Metals
	Sumitomo PE	Sumitomo Special Steels
	Hyperm 53, 54	F. Krupp Widiafabrik
54–68% Ni	Permax M (56 Ni, high μ)	Vacuumschmelze
	Permax Z, F (65Ni–3Mo)	Vacuumschmelze
4Mo–80Ni	4-79 Mo-Permalloy	Western Electric; Arnold Engineering
round	HyMu 80, 800	Carpenter Technology
loop	Hyperm Maximum	F. Krupp Widiafabrik
	Hypernorm (shielding)	Carpenter Technology
	Permalloy 80	Magnetics-Div. Spang Ind.
	Sandvik Safeni 80 Mo	Sandvik
	SuperPerm 80	Magnetic Metals
	Supermalloy	Arnold Eng.; Magnetics; Mag. Metals
4Mo–80Ni	Hy-Ra 80	Carpenter Technology
square	Square 80	Magnetic Metals
loop	Square Permalloy	Arnold Engineering
	Square Permalloy 80	Magnetics-Div. Spang Ind.
	Super Square 80	Magnetic Metals
4Mo–5Cu–77Ni	Hyperm 900	F. Krupp Widiafabrik
round	Hyperm 766, 767	F. Krupp Widiafabrik
loop	(2Cr–5Cu–77Ni)	
	Mumetal, Mumetal Plus	Telcon Metals

Appendix, continued

Type	Trade name	Company
	Mumetal	Vacuumschmelze
	Permalloy C	ITT CG-E Mag. Mat'ls Div.
	Perimphy 1, 2, 3	Soc. Metallurg. d'Imphy
	Sumitomo PA, PC	Sumitomo Special Metals
	Supermumetal	Telcon Metals
	Vacoperm 100	Vacuumschmelze
	Ultraperm 10	Vacuumschmelze
4Mo–5Cu–77Ni square loop	Ultraperm Z	Vacuumschmelze
Fe–Al and Fe–Al–Si	Alfernol (16 Al)	–
	Alfer (13 Al)	Tohoku Metal Ind.
	Alperm (17 Al)	–
	Sendust (7–13Si, 4–7 Al)	Tohoku Metal Ind.
	Vacodur 16 (16 Al)	Vacuumschmelze
27–35% Co	AFK 1	Soc. Metallurg. d'Imphy
	Hiperco 27, 35	Carpenter Technology
	Hyperm Co 35	F. Krupp Widiafabrik
2V–49Co	AFK 502	Soc. Metallurg. d'Imphy
	Hiperco 50	Carpenter Technology
	Hyperm Co 50	F. Krupp Widiafabrik
	Supermendur	Arnold Eng.; Magnetics; Mag. Metals
	2V–Permendur	Western Electric; Arnold Eng.
	Vacoflux 48, 50, Z	Vacuumschmelze

References

Adams, E., 1962, J. Appl. Phys. 33, 1214.

Adler, E. and H. Pfeiffer, 1974, IEEE Trans. Mag. MAG-10, 172.

Ames, S.L., 1970, J. Appl. Phys. 41, 1032.

Ames, S.L. and C.P. Stroble, 1967, J. Appl. Phys. 38, 1168.

Ames, S.L., G.L. Houze Jr. and W.R. Bitler, 1969, J. Appl. Phys. 40, 1577.

Aoyagi, K., 1965, Japan J. Appl. Phys. 4, 551.

Armco non-oriented electrical steels, 1973, p. 7.

Armco oriented electrical steels, 1974, p. 7.

Aspen, R.G. and D.A. Colling, 1969, Trans. Met. Soc. AIME 245, 2598.

Assmus, F., R. Boll, D. Ganz and F. Pfeifer, 1957, Z. Metallk. 48, 341.

ASTM standard A34, Annual book of ASTM standards, part 44.

Ault, C.F., L.E. Gallaher, T.S. Greenwood and D.C. Koehler, 1964, Bell Syst. Tech. J. 43, 2097.

Barrett, W.F., W. Brown and R.A. Hadfield, Sci. Trans. Roy. Dublin Soc. 7, 67 (1900).

Barrett, W.F., W. Brown and R.A. Hadfield, 1901, J. Inst. Elec. Engrs. 31, 674 [cited by Bozorth (1951, p. 67)].

Becker, J.J., 1959, Magnetic properties of metals and alloys (Am. Soc. Met., Metals Park, OH) p. 68.

Becker, R. and W. Döring, 1939, Ferromagnetismus (Edwards, Ann Arbor, MI) p. 139.

Benford, J.G. and R.W. Linsay, 1963, J. Appl. Phys. 34, 1307.

Bobeck, A.H., 1957, Bell Syst. Tech. J. 36, 1319.

Boon, C.R. and J.E. Thompson, 1964, Proc. IEE 111, 605.

Boon, C.R. and J.E. Thompson, 1965, Proc. IEE 112, 2147.

Boothby, O.L. and R.M. Bozorth, 1947, J. Appl. Phys. 18, 173.

Borodkina, M.M., Z.N. Bulycheva and Ya. P. Sellinski, 1960, Phys. Met. Metallogr. 9, 61.

Bozorth, R.M., 1937, J. Appl. Phys. 8, 575.

Bozorth, R.M., 1942, US Patent 2 300 336.

Bozorth, R.M., 1951, Ferromagnetism (Van Nostrand, New York).

Bozorth, R.M., 1953, Rev. Mod. Phys. 25, 42.

Bradley, F.N., 1971a, Materials for magnetic functions (Hayden Book Co., New York) p. 127.

Bradley, F.N., 1971b, Materials for magnetic functions (Hayden Book Co., New York) p. 230.

Bradley, F.N., 1971c, Materials for magnetic functions (Hayden Book Co., New York) p. 238.

Brailsford, F., 1939, J. IEE 84, 399.

Brailsford, F., 1970, Proc. IEE 117, 1052.

Brewer, L., J. Chipman and S.G. Chang, 1973, Metals handbook, Vol. 8 (Am. Soc. Met., Metals Park, OH) p. 413.

Brown, C.B., 1967, Bell Syst. Tech. J. 46, 117.

Bulycheva, Z.N., M.M. Borodkina and V.L. Sandomirskaya, 1965, Phys. Met. Metallogr. 19, 147.

Bunge, H.J. and H.G. Muller, 1957, Z. Metallk. 48, 26.

Burgers, W.G. and J.L. Snoek, 1935, Z. Metallk. 27, 158.

Carpenter Technology, 1970, Technical data sheet: Carpenter temperature compensator "30", Carpenter temperature compensator "32" (July).

Casani, R.T., W.A. Klawitter, A.A. Lykens and F.W. Ackermann, 1966, J. Appl. Phys. 37, 1202.

Chikazumi, S., 1964, Physics of magnetism (Wiley, New York) p. 363.

Chikazumi, S. and C.D. Graham Jr., 1969, Magnetism and metallurgy, Vol. 2, eds. Berkowitz and Kneller (Academic, New York) Ch. 12.

Chikazumi, S. and T. Oomura, 1955, J. Phys. Soc. Japan 10, 842.

Chikazumi, S., K. Suzuki and H. Iwata, 1957, J. Phys. Soc. Japan 12, 1259.

Chin, G.Y., 1965, J. Appl. Phys. 36, 2915.

Chin, G.Y., 1966, Mater. Sci. Eng. 1, 77.

Chin, G.Y., 1971a, J. Metals 23, 42.

Chin, G.Y., 1971b, Adv. Mater. Res. 5, 217.

Chin, G.Y., L.L. Vanskike and H.L. Andrews, 1964, J. Appl. Phys. 35, 867.

Chin, G.Y., E.A. Nesbitt, J.H. Wernick and L.L. Vanskike, 1967, J. Appl. Phys. 38, 2623.

Chin, G.Y., T.C. Tisone and W.B. Grupen, 1971, J. Appl. Phys. 42, 1502.

Cioffi, P.P., H.J. Williams and R.M. Bozorth, 1937, Phys. Rev. 51, 1009.

Clark, A.E., 1977, AIP Conf. Proc. 34, 13.

Cohen, P., 1967, J. Appl. Phys. 38, 1174.

Colling, D.A. and R.G. Aspen, 1969, J. Appl. Phys. 40, 1571.

Colling, D.A. and R.G. Aspen, 1970, J. Appl. Phys. 41, 1040.

Cooke, H., 1975, Metal Bull. Monthly (Dec.) p. 1.

Cullity, B.D., 1972, Introduction to magnetic materials (Addison-Wesley, Reading, MA).

Dahl, O. and F. Pawlek, 1935, Z. Phys. 94, 504.

Dahl, O. and J. Pfaffenberger, 1931, Z. Phys. 71, 93.

Dahl, O., J. Pfaffenberger and H. Sprung, 1933, Elektrische Machrichten Tech. 10, 317.

Da Silva, J.R.G and R.B. McLellan, 1976, J. Less Common Met. 49, 407.

Datta, A., 1976, IEEE Trans. Mag. MAG-12, 867.

Dunn, C.G. and J.L. Walker, 1961, Trans. Met. Soc. AIME 221, 413.

Electrical materials handbook, 1961, (Ludlum Steel Co., Allegheney) p. IV/I.

Elmen, G.W., 1929, US Patent 1 734 752.

Elmen, G.W., 1935, Elec. Eng. 54, 1292.

English, A.T., 1966, Trans. Met. Soc. AIME 236, 4.

English, A.T. and G.Y. Chin, 1967, J. Appl. Phys. 38, 1183.

Enoch, R.D. and A.D. Fudge, 1966, Br. J. Appl. Phys. 17, 623.

Enoch, R.D. and D.L. Murrell, 1969a, J. Phys. D 2, 357.

Enoch, R.D. and D.L. Murrell, IEEE Trans. Mag. MAG-5, 370 (1969b).

Enoch, R.D. and D.L. Murrell, 1971, J. Phys. D 4, 133.

Enoch, R.D. and A. Winterborn, 1967, Br. J. Appl. Phys. 18, 1407.

Fahlenbrach, H., 1955, Tech. Mitt. Krupp 13, 84.

Fahlenbrach, H., 1968, Tech. Mitt. Krupp 26, 17.

Fallot, M., 1936, Ann. de Phys. 6, 305.

Ferguson, E.T., 1958, J. Appl. Phys. 29, 252.

Fiedler, H.C., 1958, J. Appl. Phys. 29, 361.

Fiedler, H.C., 1961, Trans. AIME 221, 1201.

Fiedler, H.C., 1964, Trans. AIME 230, 95.

Fiedler, H.C., 1967, J. Appl. Phys. 38, 1098.

Fiedler, H.C., 1977a, IEEE Trans. MAG. MAG-13, 1433.

Fiedler, H.C., 1977b, Met. Trans. 8A, 1307.

Fiedler, H.C. and R.H. Pry, 1959, J. Appl. Phys. 30, 109S.

Finke, G.B., 1974, IEEE Trans. Mag. MAG-10, 144.

Flowers, J.W., 1967, J. Appl. Phys. 38, 1085.

Foster, K. and J. Kramer, 1960, J. Appl. Phys. 31, 2335.

Foster, K. and D. Pavlovic, 1963, J. Appl. Phys. 34, 132.

Fukuda, B., T. Irie, H. Shimanaka and T. Yamamoto, 1977, IEEE Trans. Mag. MAG-13, 1499.

Gersdorf, R., 1961, Thesis (University of Amsterdam, Holland).

Getting, M.P., Jr., 1967, J. Appl. Phys. 38, 1074.

Goldstein, J.I., 1973, Metals handbook, Vol. 8 (Am. Soc. Met., Metals Park, OH) p. 304.

Goman'kov, V.I., I.M. Puzei, A.A. Loshmanov and Ye. I. Mal'tzev, 1969, Phys. Met. Metallogr. 28, 77.

Goman'kov, V.I., I.M. Puzei and Ye. I. Mal'tzev, 1970, Phys. Met. Metallogr. 30, 237.

Gordon, D.I. and R.S. Sery, 1965, J. Appl. Phys. 36, 1221.

Goss, N.P., 1934, US Patent 1 965 559.

Goss, N.P., 1935, Trans. Am. Soc. Met. 23, 511.

Goto, I., I. Matoba, T. Imanaka, T. Gotoh and T. Kan, 1975, Proc. 2nd EPS Conf. on Soft Magnetic Materials (Wolfson Centre for Magnetic Technology, Cardiff, Wales) p. 262.

Gould, H.L.B. and D.H. Wenny, 1957, AIEE Spec. Publ. T-97, 675.

Graham, C.D., Jr., 1958, Phys. Rev. 112, 1117.

Graham, C.D., Jr., 1959, Magnetic properties of metals and alloys (Am. Soc. Met., Metals Park, OH) Ch. 13.

Grenoble, H.E., 1977, IEEE Trans. Mag. MAG-13, 1427.

Grenoble, H.E. and H.C. Fiedler, 1969, J. Appl. Phys. 40, 1575.

Haas, P.H., 1953, J. Res. Nat. Bur. Stand. 51, 221.

Hall, R.C., 1959, J. Appl. Phys. 30, 816

Hall, R.C., 1960, Trans. Met. Soc. AIME 218, 268 (1960).

Hansen, M., 1958a, Constitution of binary alloys (McGraw-Hill, New York) p. 678.

Hansen, M., 1958b, Constitution of binary alloys (McGraw-Hill, New York) p. 681.

Heck, C., 1974a, Magnetic materials and their applications (Crane and Russak, New York) p. 317.

Heck, C., 1974b, Magnetic materials and their applications (Crane and Russak, New York) p. 430.

Heck, C., 1974c, Magnetic materials and their applications (Crane and Russak, New York) p. 442.

Heck, C., 1974d, Magnetic materials and their applications (Crane and Russak, New York) p. 448.

Heck, C., 1974e, Magnetic materials and their applications (Crane and Russak, New York) p. 663.

Heidenreich, R.D., E.A. Nesbitt and R.D. Burbank, 1959, J. Appl. Phys. 30, 995.

Helms Jr., H.H., unpublished [cited by Adams (1962)].

Helms Jr., H.H. and E. Adams, 1964, J. Appl. Phys. 35, 871.

Hill, Mrs. S., and K.J. Overshott, 1976, AIP Conf. Proc. 34, 129.

Hitachi Metals Ltd., Tokyo, Japan. Report E-172 (TUFPERM).

Hoekstra, B., E.M. Gyorgy, P.K. Gallagher, D.W. Johnson Jr., G. Zydzik and L.G. Van Uitert, 1978, J. Appl. Phys. 49, 4902.

Hoffmann, A., 1970, Z. Angew. Phys. 30, 72.

Hoffmann, A., 1971, Z. Angew. Phys. 32, 236.

Hoffmann, U., 1972, Phys. Stat. Sol. (a) 11, 145.

Honda, K. and S. Kaya, 1926, Sci. Rep. RITU 15, 721.

Honda, K., H. Masumoto and S. Kaya, 1928, Sci. Rep. RITU 17, 111.

Hoselitz, K., 1952, Ferromagnetic properties of metals and alloys (Clarendon, Oxford) p. 317.

Howe, G.H., 1956, Elec. Eng. 75, 702.

Hu, H., 1961, Trans. AIME 221, 130.

Hu, H., 1964, AIME 230, 572.

Hubert, A., 1965, Z. Angew. Phys. **18**, 474.

Hultgren, R., P.D. Desai, D.T. Hawkins, M. Gleiser and K.K. Kelley, 1973, Selected values of the thermodynamic properties of binary alloys (Am. Soc. Met., Metals Park, OH) p. 871.

Ichinose, Y., 1967, Trans. Japan. Inst. Met. **8**, 81–89.

Ichinose, Y., 1968, Trans. Japan. Inst. Met. **9**, 166.

Irie, T. and B. Fukuda, 1976, AIP Conf. Proc. **29**, 574.

Ishizaka, T., K. Yamabe and T. Takahashi, 1966, J. Japan Inst. Met. **30**, 552.

Jackson, J.M., 1974, US Patent 3 785 882.

Jackson, R.C. and E.W. Lee, 1966, J. Mater. Sci. **1**, 362.

Jacobs, I.S., 1969, J. Appl. Phys. **40**, 917.

Josso, E., 1974, IEEE Trans. Mag. **MAG-10**, 161.

Kang, I.K., H.H. Scholefield and A.P. Martin, 1967, J. Appl. Phys. **38**, 1178.

Karmanova, Ye. G., V.D. Kuleshova, A.A. Roytman and M.M. Knoroz, 1965, Fiz. Metall. Metallov. **20**, 785.

Keller, A.C., 1964, Bell Syst. Tech. J. **43**, 15.

Kohler, D.M., 1967, J. Appl. Phys. **38**, 1176.

Kornetzki, M., 1953, Ferrit-Kerne für die Nachrichtentechnik, Siemens-Z 27, 212.

Kornetski, M. and I. Lucas, 1955, Z. Phys. **142**, 70.

Köster, W. and T. Godecke, 1968, Z. Metallk. **59**, 602.

Kunz, W. and F. Pfeifer, 1976, AIP Conf. Proc. **34**, 63.

Liashchenko, B.G., D.F. Litvin, I.M. Puzei and Iu. G. Abov, 1957, Sov. Phys. Crystallogr. **2**, 59.

Littmann, M.F., 1967, J. Appl. Phys. **38**, 1104.

Littmann, M.F., 1969, Trans. AIME **245**, 2217.

Littmann, M.F., 1971, IEEE Trans. Mag. **MAG-7**, 48.

Littmann, M.F., 1975a, Met. Trans. **6A**, 1041.

Littmann, M.F., 1975b, AIP Conf. Proc. **24**, 721.

Lounquist, C.W., J.C. Manganello, R.S. Skinner, M.J. Skubiak and D.J. Wadsworth, 1969, Bell Syst. Tech. J. **48**, 2817.

Luborsky, F.E., J.J. Becker, P.G. Frischmann and L.A. Johnson, 1978, J. Appl. Phys. **49**, 1769.

Lykens, A.A., 1962, J. Appl. Phys. **33**, 1232S.

Lyhens, A.A., 1966, J. Appl. Phys. **37**, 1204.

Machova, A. and S. Kadeckova, 1977, J. Phys. **B27**, 555.

Magnetics Div. Spang Industries, a, Catalog MG-3037: Magnetic lamination.

Magnetics Div. Spang Industries, b, Catalog TWC-300R: Tape wound cores.

Magnetics Div. Spang Industries, c, Bull. CEA-01 and G5A-03.

Major, R.V., M.C. Martin and M.W. Branson, 1975, Proc. 2nd EPS Conf. on Soft Magnetic Materials (Wolfson Centre for Magnetics Technology, Cardiff, Wales) p. 103.

Malagari, F.A., 1977, IEEE Trans. Mag. **MAG-13**, 1437.

Marchand, A., 1967, Mem. Sci. Rev. Met. **114**, 141.

Martin, D.L. and A.H. Geisler, 1960, Trans. Am. Soc. Met. **44**, 461.

Masumoto, H., 1936, Sci. Rep. RITU, **25**, 388.

Masumoto, H. and G. Otomo, 1950, Sci. Rep. RITU **A2**, 413.

Masumoto, H. and H. Saito, 1951, Sci. Rep. RITU **A3**, 523.

Masumoto, H. and H. Saito, 1952a, Sci. Rep. RITU, **A4**, 321.

Masumoto, H. and H. Saito, 1952b, Sci. Rep. RITU, **A4**, 338.

Masumoto, H., Y. Murakami and M. Hinai, 1972, J. Japan. Inst. Met. **36**, 63.

Masumoto, H., Y. Murakami and M. Hinai, 1974, J. Japan, Inst. Met. **38**, 238.

Matsumoto, F., K. Kuroki and A. Sakakura, 1975, AIP Conf. Proc. **24**, 716.

Maucione, C.M. and J.A. Salsgiver, 1977, IEEE Trans. Mag. **MAG-13**, 1442.

May, J.E. and D. Turnbull, 1957, Trans. AIME, **212**, 7.

Metals handbook, 1961, Vol. 1 (Am. Soc. Met., Metals Park OH).

Metals handbook, 1973a, Vol. 8 (Am. Soc. Met., Metals Park OH).

Metals handbook, 1973b, Vol. 8 (Am. Soc. Met., Metals Park OH) p. 306.

Metals reference book, 1976, eds. Smithel and Brandes (Butterworths, London).

Miyazaki, T., R. Sawada and Y. Ishijima, 1972, IEEE Trans. Mag. **MAG-8**, 501.

Moses, A.J. and B. Thomas, 1976, IEEE Trans. Mag. **MAG-12**, 103.

Moses, A.J., B. Thomas and J.E. Thompson, 1972, IEEE Trans. Mag. **MAG-8**, 785.

Murrell, D.L. and R.D. Enoch, 1970, J. Mater. Sci. **5**, 478.

Nachman, J.F. and W.J. Buehler, 1954, J. Appl. Phys. 25, 306.

Nakata, T., Y. Ishihara, K. Yamada and A. Sasano, 1975, Proc. 2nd EPS Conf. on Soft Magnetic Materials (Wolfson Centre for Magnetic Technology, Cardiff, Wales) p. 57.

Narita, K., 1974, IEEE Trans. Mag. MAG-10, 165.

Narita, K., 1975, IEEE Trans. Mag. MAG-11, 1661.

Neel, L., 1954, J. Phys. Radium 15, 225.

Neel, L., J. Pauleve, R. Pauthenet, J. Langier and D. Dautreppe, 1964, J. Appl. Phys. 35, 873.

Negishi, A., 1977, private communication (Furukawa Electric Co., Tokyo).

Nesbitt, E.A., 1946, Trans. AIME 166, 415.

Nesbitt, E.A. and G.A. Kelsall, 1940, Phys. Rev. 58, 203.

Nesbitt, E.A., R.D. Heidenreich and A.J. Williams, 1960, J. Appl. Phys. 31, 228S.

Nesbitt, E.A., B.W. Batterman, L.D. Fullerton and A.J. Williams, 1965, J. Appl. Phys. 36, 1235.

Nesbitt, E.A., G.Y. Chin, A.J. Williams, R.C. Sherwood and J. Moeller, 1966, J. Appl. Phys. 37, 1218.

Nichol, T.J. and J.W. Shilling, 1976, IEEE Trans. Mag. MAG-12, 858.

Nozawa, T., T. Yamamoto, Y. Matsuo and Y. Ohya, 1977, Proc. 3rd EPS Conf. on Soft Magnetic Materials, Bratislava, Paper 21-1.

Nozawa, T., T. Yamamoto, Y. Matsuo and Y. Ohya, 1978, Trans. Mag. MAG-14, 252.

Odani, Y., 1964, J. Appl. Phys. 35, 865.

Overshott, K.J., 1976, IEEE Trans. Mag. MAG-12, 840.

Pavlovic, D. and K. Foster, 1965, J. Appl. Phys. 36, 1237.

Pfeifer, F., 1966a, Z. Metallk. 57, 240.

Pfeifer, F., 1966b, Z. Metallk. 57, 295.

Pfeifer, F., 1969, Z. Angew. Phys. 28, 20.

Pfeifer, F. and R. Boll, 1969, IEEE Trans. Mag. MAG-5, 365.

Pfeifer, F. and R. Cremer, 1973, Z. Metallk. 64, 362.

Pfeifer, F. and R. Deller, 1968, Electrotech. Z. A 89, 601.

Pfeifer, F. and I. Pfeiffer, 1964, Z. Metallk. 55, 398.

Phillips, R. and K. Overshott, 1974, IEEE

Trans. Mag. MAG-10, 168.

Pry, R.H. and C. Bean, 1958, J. Appl. Phys. 29, 532.

Puzei, I.M., 1957, Bull. Acad. Sci. USSR Phys. Ser. 21, 1083.

Puzei, I.M., 1961, Phys. Met. Metallogr. 12, 136.

Puzei, I.M., 1963, Phys. Met. Metallogr. 14, 47.

Puzei, I.M. and B.V. Mototilov, 1958, Bull. Acad. Sci. USSR Phys. Ser. 22, 1236.

Rassmann, G. and U. Hofmann, 1968, J. Appl. Phys. 39, 603.

Rassmann, G. and H. Wich, 1960, Ber. Arbeitsgemeinschaft Ferromagnetismus (Stahleisen, Dusseldorf) p. 18.

Rassmann, G. and H. Wich, 1962, Metallische Spezialwerkstoffen (Dresden) p. 142.

Rastogi, P.K., IEEE Trans. Mag. MAG-13, 1448 (1972).

Sakakura, A., 1969, J. Appl. Phys. 40, 1534.

Sakakura, A. and S. Taguchi, 1971, Met. Trans. 2, 205.

Sakakura, A., T. Wada, F. Matsumoto, K. Ueno, K. Takashima and M. Kawashima, 1975, AIP Conf. Proc.24, 714.

Salsgiver, J.A. and D.J. McMahon, 1976, IEEE Trans. Mag. MAG-12, 864.

Sato, H. and B.S. Chandrasekhar, 1957, J. Phys. Chem. Solids 1, 228.

Schindler, A.I. and E.I. Salkovitz, 1960, J. Appl. Phys. 31, 245S.

Schindler, A.I., R.H. Kernoham and J. Weertmann, 1964, J. Appl. Phys. 35, 2640.

Scholefield, H.H., R.V. Major, B. Gibson and A.P. Martin, 1967, Br. J. Appl. Phys. 18, 41.

Senno, H., Y. Yanagiuchi, M. Satomi, E. Hirota and S. Hayakawa, 1977, IEEE Trans. Mag. MAG-13, 1475.

Sery, R.S. and D.I. Gordon, 1964, J. Appl. Phys. 35, 879.

Sery, R.S. and D.I. Gordon, 1966, J. Appl. Phys. 37, 1216.

Sharp, M.R.G. and K.J. Overshott, 1973, Proc. IEEE 120, 1451.

Shibaya, H. and I. Fukuda, 1977, IEEE Trans. Mag. MAG-13, 1029.

Shilling, J.W., 1971, J. Appl. Phys. 42, 1787.

Shilling, J.W. and G.L. Houze Jr., 1974, IEEE Trans. Mag. MAG-10, 195.

Shilling, J.W., W.G. Morris, M.L. Osborn and P. Rao, 1977, Proc. 3rd EPS Conf. on Soft

Magnetic Materials, Bratislava.

Shilling, J.W., W.G. Morris, M.L. Osborn and P. Rao, 1978, IEEE Trans. Mag. MAG-14, 104.

Shimanaka, H., I. Matoba, T. Ichida, S. Kobayashi and T. Funahashi, 1975, Proc. 2nd EPS Conf. on Soft Magnetic Materials (Wolfson Centre for Magnetic Technology, Cardiff, Wales) p. 269.

Sixtus, K., 1962, Z. Angew. Phys. 14, 241.

Sixtus, K., 1970, Z. Angew. Phys. 28, 270.

Slauson, W.C., 1964, Bell Syst. Tech. J. 43, 2905.

Slonzewski, J.C., 1963, Magnetism, Vol. 1, eds. Rado and Suhl (Academic, New York) Ch. 5.

Snee, D.J., 1967, J. App. Phys. 38, 1172.

Snoek, J.L., 1935, Physica 2, 403.

Stanley, J.K., 1947, US Patent 2 442 219.

Stanley, J.K. and T.D. Yensen, 1947, Trans. AIEE 66, 714.

Steel products manual, 1975a, Flat rolled electrical steel (AISI, Washington).

Steel products manual, 1975b, Flat rolled electrical steel (AISI, Washington) p. 29.

M. Stéfan, 1974, IEEE Trans. Mag. MAG-10, 136.

Stoloff, N.S. and R.G. Davies, 1964, Acta. Met. 12, 473.

Stroble, C.P., 1976, IEEE Trans. Mag. MAG-12, 861.

Sugihara, M., 1960a, J. Phys. Soc. Japan 15, 8.

Sugihara, M., 1960b, J. Phys. Soc. Japan 15, 1456.

Swift, W.M., 1976, AIP Conf. Proc. 29, 568.

Swift, W.M., W.R. Reynolds and J.W. Shilling, 1973, AIP Conf. Proc. 10, 976.

Swift, W.M., W.H. Daniels and J.W. Shilling, 1975, IEEE Trans. Mag. MAG-11, 1655.

Swisher, J.H. and E.O. Fuchs, 1970, J. Iron and Steel Inst., August, p. 777.

Swisher, J.H., A.T. English and R.C. Stoffers, 1969, Trans. Am. Soc. Met. 62, 257.

Taguchi, S., 1977, Trans. ISIJ 17, 604.

Taguchi, S. and A. Sakakura, 1966, Acta Met. 14, 405.

Taguchi, S. and A. Sakakura, 1969, J. Appl. Phys. 40, 1539.

Taguchi, S., A. Sakakura and H. Takashima, 1966, US Patent 3 287 183.

Taguchi, S., T. Yamamoto and A. Sakakura, 1974, IEEE Trans. Mag. MAG-10, 123.

Taguchi, S., A. Sakakura, F. Matsumoto, K.

Takashima and K. Kuroki, 1976, J. Mag. and Mag. Mater. 2, 121.

Takashima, K. and Y. Suga, 1976, IEEE Trans. Mag. MAG-12, 852.

Takashima, K., T. Sato and F. Matsumoto, 1976, AIP Conf. Proc. 29, 566.

Taniguchi, S., 1955, Sci. Rep. RITU 1, 269.

Tatsumoto, E. and T. Okamoto, 1959, J. Phys. Soc. Japan 14, 1588.

Tebble, R.S. and D.J. Craik, 1969, Magnetic materials (Wiley-Interscience, New York).

Thomas, H., 1951, Z. Phys. 129, 219.

Thomas, B., 1975, Proc. Conf. 2nd EPS on Soft Magnetic Materials (Wolfson Centre for Magnetics Technology, Cardiff, Wales) p. 109.

Thompson, T.J. and B. Thomas, 1975, Proc. 2nd EPS Conf. on Soft Magnetic Materials (Wolfson Centre for Magnetics Technology, Cardiff, Wales) p. 40.

Thornburg, D.R., 1969, J. Appl. Phys. 40, 1579.

Vacuumschmelze GmbH, Hanau, W. Germany, 1976, RECOVAC Data sheet MO42 (Feb.).

Wada, S., 1975, Proc. 2nd EPS Conf. on Soft Magnetic Materials (Wolfson Centre for Magnetics Technology, Cardiff, Wales) p. 1.

Wada, S., S. Taguchi and S. Shimizu, 1975, Proc. 2nd EPS Conf. on Soft Magnetic Materials (Wolfson Centre for Magnetics Technology, Cardiff, Wales) p. 37.

Walker, E.V. and R.E.S. Walters, 1957, Powder Metallurgy 4, 23.

Walters, R.E.S., 1957, Proc. 2nd EPS Conf. on Magnetism and Magnetic Materials (Wolfson Centre for Magnetics Technology, Cardiff, Wales) p. 258.

Weiss, P. and R. Forrer, 1929, Ann. Phys. 12, 279.

Wenny, D.H., 1965, Bell Labs. Record 43, 257.

Went, J.J., 1951, 17, 98.

White, J.H. and C.V. Wahl, 1932, US Patent 1 862 559.

Williams, G.M. and A.S. Pavlovic, 1968, J. Appl. Phys. 39, 571.

Williams, H.J., W. Shockley and C. Kittel, 1950, Phys. Rev. 80, 1090.

Yamamoto, T. and T. Nozawa, 1970, J. Appl. Phys. 41, 2981.

Yamamoto, T. and S. Taguchi, 1975, Proc. 2nd EPS Conf. on Soft Magnetic Materials (Wolfson Centre for Magnetics Technology, Cardiff, Wales) p. 15.

Yamamoto, T. and Y. Utsushikawa, 1976, J.
  Jap. Inst. Met. **40**, 975.
Yamamoto, T., S. Taguchi, A. Sakakura and T.
  Nozawa, 1972, IEEE Trans. Mag. MAG-8,
  677.
Yensen, T.D., 1914, Trans. AIEE 33, 451.

Yensen, T.D., 1924, Trans. AIEE, **43**, 145.
Yensen, T.D.. and N.A. Ziegler, 1936, Trans.
  Am. Soc. Met. **24**, 337.
Zaimovsky, A.S. and I.P. Selissky, 1941, J.
  Phys. (USSR) **4**, 563.

chapter 3

FERRITES FOR NON-MICROWAVE APPLICATIONS

P.I. SLICK

Bell Telephone Laboratories
Allentown, PA 18103
USA

Ferromagnetic Materials, Vol. 2
Edited by E.P. Wohlfarth
© North-Holland Publishing Company, 1980

CONTENTS

1. Introduction

The unique properties of non-microwave ferrites are high magnetic permeability and high electrical resistivity. The resultant lower eddy current losses allows their use to higher frequencies than possible with metals. Non-microwave ferrites which will henceforth be referred to as ferrites are used at frequencies from audio to about 500 MHz. Since ferrites are ceramic, they are easily moldable by low cost ceramic processing and are machinable. On the other hand, their ceramic structure poses a major problem in a way because the magnetic properties are very sensitive to microstructure. In addition, the attainment of high densities with uniform and controlled microstructure has been the major challenge in advancing the technology.

These magnetic ceramics have been vital to the development of the telecommunication, electronic entertainment and digital computer industries. Since the appearance of the first commercial ferrite products in about 1945, the thrust of the development has been for smaller components of greater reliability and hence the continued need for higher quality and stability.

In this chapter we briefly review the definitions and applications of soft non-metallic materials and the important magnetic properties required for these applications. After a brief description of the nature of ferrites we present the scientific understanding of these properties and their technological importance. Numerous reviews dealing with this subject exist in the literature that the reader can refer to such as Craik (1975), Sibelle (1974), Owens (1956), Snelling (1962, 1964, 1969, 1972), Roess (1977), Heck (1974), Tebble and Craik (1969), Van Groenou et al. (1969), Greifer (1969) and Peloschek (1963).

2. Definitions

The prime important properties in non-microwave non-rectangular ferrite applications are the permeability and its frequency spectrum. The initial permeability, μ_i, called simply the permeability of a material, μ, is defined as the limit of the change of the induction B with respect to the applied field H in the demagnetized state as H approaches zero and is represented as

$$\mu_i = (1/\mu_0) \lim_{H \to 0}(B/H) \tag{1}$$

where μ_0 = magnetic constant = $4\pi \times 10^7$ and B and H are in W/m^2[10^4 G] and A/m[0.0126 Oe] respectively.

The permeability concept can be extended to include the losses. For time harmonic fields, $H = H_0 \exp(j\omega t)$ where ω is the angular frequency and t time, the dissipation can be described by a phase difference, δ, between H and B. In the complex notation, the frequency dependence of the permeability becomes

$$\mu(\omega) = B \exp\{j(\omega t + \delta)\}/H \exp(j\omega t) = \mu' - j\mu''. \qquad (2)$$

The real part, μ', describes the stored energy expressing the component of B in phase with H and the imaginary part, μ'', describes the energy dissipated expressing the component of B out of phase with H. The loss factor or loss tangent then becomes $\tan\delta = \mu''/\mu'$. The Q factor or quality factor is defined as the reciprocal of $\tan\delta$. The loss factor divided by μ is a material figure of merit referred to as the normalized loss factor and written as:

$$\frac{\tan\delta}{\mu} = \frac{1}{\mu Q} = \frac{\mu''}{(\mu')^2}. \qquad (3)$$

The loss factor of a component is obtained by multiplying $(\tan\delta)/\mu$ by the effective permeability, μ_e, of the structure. Here $\mu_e = \mu/(1 + (g/a)\mu)$ where g is the ratio of air gap length to total length and a is the ratio of effective cross section of gap to core cross section.

The stabilities of permeability with temperature and time are also very important. The temperature coefficient of permeability, TC, is defined as the change in permeability per degree centigrade for a specific temperature range (TC = $\Delta\mu/\mu\Delta T$). The TC of materials is generally tailored so that it compensates for

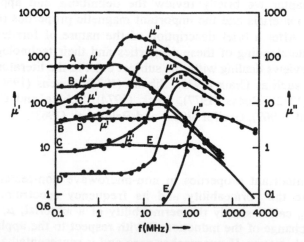

Fig. 1. Permeability spectra for NiZn ferrites with ratios of NiO:ZnO of (A) 17.5:33.2, (B) 24.9:24.9, (C) 31.7:16.5, (D) 39.0:9.4, (E) 48.2:0.7 the balance being Fe$_2$O$_3$ (after Smit and Wijn 1954).

the temperature coefficient of the capacitor in the component circuit. The decrease of permeability with log time is termed disaccommodation, $D = \Delta\mu/\mu\,\Delta\log t$. Disaccommodation is also of importance since it can pose reliability problems in component circuits. The component stabilities with temperature and time are obtained from the expressions $TC \cdot \mu_e/\mu$ and $D \cdot \mu_e/\mu$, respectively.

The magnetic spectra of μ and μ'' contain the important properties for the material scientist and the design engineer. Figure 1 shows typical permeability spectra at room temperatures of several NiZn ferrites (Smit and Wijn 1954). The frequency at which μ'' maximizes (loss resonance) is nearly inversely proportional to the low frequency permeability according to the equation (Snoek 1948)

$$f_r(\mu - 1) = \tfrac{4}{3}\gamma M_s \tag{4}$$

where f_r is the loss resonance frequency, γ is the gyromagnetic ratio and M_s is the saturation magnetization. This is a rather general feature of ferrites which show only one maximum of the μ'' curve. This maximum occurs at about twice the frequency at which μ' begins to decrease with frequency and is a loss resonance due to wall permeability. Verweel (1964) has shown that with high density $Ni_{0.36}Zn_{0.64}Fe_2O_4$ resolution of the two dispersion regions is apparent (fig. 2). The lower frequency dispersion at about 15 MHz is associated with domain wall processes and the higher frequency dispersion (70 MHz), with rotational processes. From the component design engineer's standpoint, the normalized loss tangent, $(\tan\delta)/\mu$ $[= \mu''/\mu'^2]$ is a more convenient manner to represent component losses (see fig. 3). Note that the separation between the MnZn ferrites and NiZn ferrite is about 1 MHz.

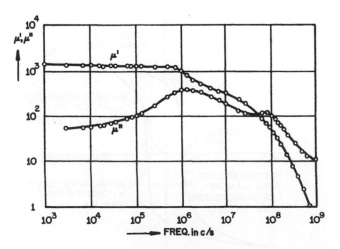

Fig. 2. Permeability spectra of a dense NiZn ferrite showing loss resonances for domain wall and rotational processes (after Verweel 1964).

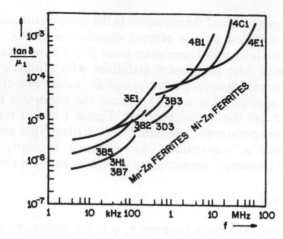

Fig. 3. Loss factor, tan δ/μ, as a function of frequency for different types of Ferroxcube ferrites.

3. Applications

The non-microwave applications fall roughly into categories determined by the magnetic properties that can be described with reference to the B/H behavior in fig. 4. The categories are:

(1) Linear B/H, low flux density
(2) Non-linear B/H, medium to high flux density
(3) Highly non-linear B/H, square loop or rectangular.

In the linear region, high stability of μ_i along with low losses are important for

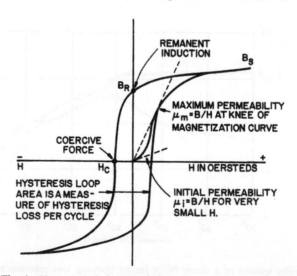

Fig. 4. Characteristic hysteresis loop for a magnetic material.

high quality inductors, antenna rods, tuning slugs for resonant circuits, recording heads, etc., whereas high permeability is required for transformers. In the non-linear region the maximum permeability μ_m at the knee of the magnetization curve along with B_s is important in TV flyback transformers, large inductors, and chokes, etc. For square loop ferrites, a high ratio of B_r/B_s and controlled coercive force are important in cores for magnetic amplifiers, memories and switches.

Table 1 summarizes the design applications of ferrites, their functions and the properties of ferrites on which they depend. Note that the MnZn and NiZn ferrites are the only materials used in the first two categories and that the most common properties are high permeability and low losses.

An additional application not describable by the B/H loop is for magnetomechanical resonators and transducers. High saturation magnetostriction $(\lambda_s \geqslant 20 \times 10^{-6})$, high magnetomechanical coupling factor and low TC of modulus of elasticity are required. The main compositions are NiZn, NiCuCo, NiCo and NiCu ferrites. These types of materials will not be covered here. The reader can refer to Heck (1974) for more information.

The largest usage of ferrite measured in terms of material weight is the non-linear medium to high flux category such as for deflecting yokes and flyback transformers in the entertainment industry. However, the greatest technical challenge has been with the linear low flux density ferrites. In this case the losses and the stability requirements of permeability are very stringent. For non-rectangular loop ferrites, discussions will be concentrated on the linear $B-H$ low flux density materials but the principles will apply for the other applications.

4. Evolution of magnetic properties

Shown in figs. 5 and 6 are the values of maximum permeability and μQ product as a function of the year that the values were attained. These improvements were obtained by compositional selections, high purity raw materials and refined processing techniques. In both cases the commercial materials have improved about an order of magnitude in about 30 years. The properties of time and temperature stability of permeability are considered acceptable for their practical use, namely that the normalized TC [TC$/\mu$] is about 0.5 ppm/°C and the normalized D [D/μ] is less than about 5 ppm/decade. Because of their relative insensitivity to sintering atmosphere, NiZn ferrites were the early high permeability ferrites using simple air firing. For permeability of about 5000 and above, MnZn ferrites became the prevalent material (fig. 5). The greatest advances for the MnZn ferrites were made in the mid-sixties from selection of low anisotropy composition, use of highly pure raw materials and refinement of the process in atmospheric control and sintering to reduce ZnO loss and Fe_2O_3 precipitation. Figure 5 shows a large divergence in the permeability values obtained in the laboratory compared with those commercially available especially after 1964. This divergence is due mostly to the difficulty in making a per-

TABLE 1

Summary of non-microwave ferrite applications

Device	Ferrite chemistry	Device function	Frequencies	Desired ferrite properties
		Linear B/H, low flux density		
Inductor	MnZn, NiZn	Frequency selection network Filtering and resonant circuits	≤ 1 MH (MnZn) ~ 1–100 MHz (NiZn)	High μ, high μQ, high stability of μ with temperature and time
Transformer (pulse and wide band)	MnZn, NiZn	V and I transformation Impedance matching	Up to 500 MHz	High μ, low hysteresis losses
Antenna rod	NiZn	Electromagnetic wave receival	Up to 15 MHz	High μQ, high resistivity
Loading coil	MnZn	Impedance loading	Audio	High μ, High B_s, high stability of μ with temp., time and dc bias
		Non-linear B/H, medium to high flux density		
Flyback transformer	MnZn, NiZn	Power converter	<100 kHz	High μ, high B_s, low hysteresis losses

Composition	Device	Application	Frequency	Properties
MnZn	Deflection yoke	Electron beam deflection	<100 kHz	High μ, high B_s
MnZn, NiZn	Suppression bead	Block unwanted ac signals	Up to 250 MHz	Mod. high μ, high B_s, high hysteresis losses
MnZn, NiZn	Choke coil	Separate ac from dc signals	Up to 250 MHz	Mod. high μ, high B_s, high hysteresis losses
MnZn, NiZn	Recording head	Information recording	Up to 10 MHz	High μ, high density, high μQ, high wear resistant
MnZn	Power transformer	Power converter	<60 kHz	High B_s, low hysteresis losses
		Non-linear B/H, rectangular loop		
MnMg, MnMgZn, MnCu, MnLi, etc.	Memory cores	Information storage	Pulse	High squareness, low switching coefficient and controlled coercive force
MnMgZn, MnMgCd	Switch cores	Memory access transformer	Pulse	High squareness, controlled coercive force
MnZn	Magnetic amplifiers			

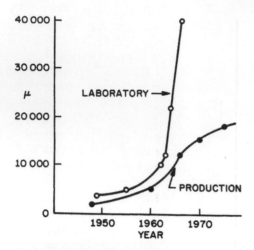

Fig. 5. Maximum permeability as a function of
the year attained.

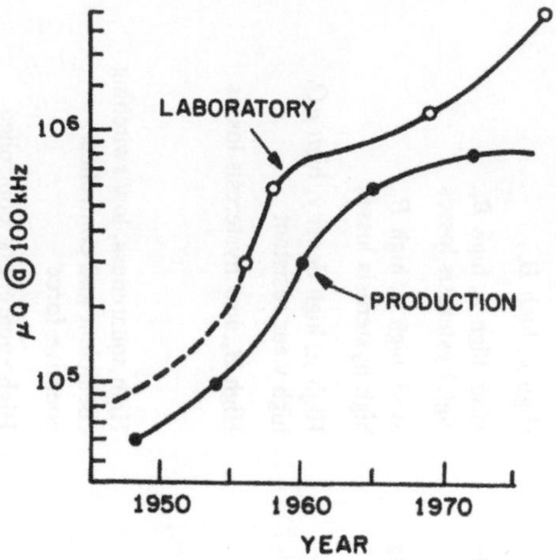

Fig. 6. Maximum μQ product as a function of
the year attained.

meability greater than about 12 000 in a manufacturing environment. This difficulty
stems from the incompatible combination of the need for increased continuous
grain size without the loss of zinc from the ferrite during sintering causing
stresses.

The early improvements in low loss MnZn ferrites (fig. 6) were due mainly to
improvements in eddy current losses with Ca and Si additions and more recently
(since about 1965) to hysteresis loss improvements by way of Co additions and
annealing treatments. A ferrite with μQ product of 600 K at 100 kHz was
manufactured in 1960 and a ferrite with a μQ product of 1.2 million was
manufactured in about 1972. However, these ferrites were only available for
internal Japanese usage.

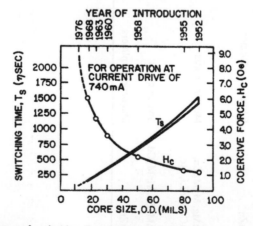

Fig. 7. Coercive force and switching time versus core size and year of commercial appearance.

The criterion for improved performance of square loop ferrites in practical applications is related to the reduction in the time required for flux reversal, τ_s, at the lowest drive fields. For coincident current applications $\tau_s \sim S_w/0.3 H_c$ (Quartly 1962) where S_w is the switching coefficient for flux reversal described in section 6.3.4 below and H_c is the coercive force. Since materials for practical use have about the same S_w, H_c must be increased. Figure 7 shows the progression cores have made since 1952 to faster speeds by reducing core size and increasing coercive force. The smallest commercial core settled at around 18 mil outside diameter (OD) after 1968 with a memory cycle time of about 500 ns for $2\frac{1}{2}$D organization. This compares to a cycle time of about 9 μs for 3D organization in 1955. A 3D memory system is one in which reading and writing are accomplished by the coincidence of 2 and 3 current pulses, respectively, on 2 and 3 orthogonal axes, respectively. In a $2\frac{1}{2}$D memory system, both reading and writing are accomplished by the coincidence of 2 current pulses on the *same* 2 orthogonal axes. Experimental cores as small as 12 mil OD have been made producing cycle times of 300 ns for $2\frac{1}{2}$D–3 wire organization (Ohta et al. 1971). Cores from 12–16 mil OD are now beginning to become the standard.

5. Nature of ferrites

5.1. Chemistry

A ferrite can be considered a material containing mostly iron which is derived from magnetite ($Fe^{2+}O \cdot Fe_2^{3+}O_3$) by substituting divalent metal ions in place of Fe^{2+}. Trivalent metal ions substitute for Fe^{3+} and other valencies (+1, +4, +5, and +6) can be incorporated into the lattice by charge compensation by the appropriate change in the Fe^{2+}/Fe^{3+} ratio. In all cases the ionic radii of the substituting ion should be between about 0.5 to 1.0 Å (Gorter 1954). The more common divalent metals are from the first transition elements: Mn, Fe, Co and Ni and metals such as Cu, Zn, Mg, Cd and Ge. Trivalent Al, Cr, Ga and Mn can replace Fe^{3+} while monovalent Li and tetravalent Ti and Sn are incorporated into the lattice by the respective decrease or increase in the Fe^{2+}/Fe^{3+} ratio. Chemistry is insufficient to define the properties of the ferrites; the distribution of the cations and point defects among the crystal lattice sites is of utmost importance.

5.2. Structure

Ferrite materials have a crystalline structure similar to the mineral spinel, $MgAl_2O_4$ where the divalent ions replace Mg and the trivalent ions replace Al. Spinel is a close-packed cubic structure of oxygen atoms with eight formula units per unit cell. A unit cell of the spinel lattice is represented in fig. 8. Two kinds of sites are available for the cation: tetrahedral sites surrounded by four oxygen ions situated at corners of a tetrahedron (A site) and octahedral sites surrounded by six oxygen ions situated at corners of an octahedron (B site). In a

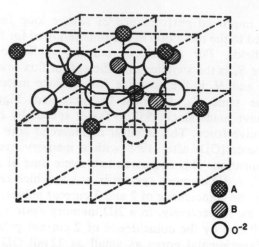

Fig. 8. Two octants of the spinel unit cell. A ions are on tetrahedral sites and B ions on octahedral
sites of the O^{2-} anion packaging.

unit cell there are 64 tetrahedral sites of which 8 are filled and 32 octahedral sites
of which 16 are filled. A compound is called "normal" where M occupies the
tetrahedral site $(M[Fe_2]O_4)$ and "inverse" where M is in the octahedral site
$(Fe[MFe]O_4)$. The distribution of the ions between the two types of sites is
determined by a delicate balance of contributions, such as the magnitude of the
ionic radii, their electronic configuration and the electrostatic energy of the
lattice.

Based on the Coulombic energy of charged ions (Verwey and Heilman 1947,
Verwey et al. 1948) and their influence in the polarization of anions (Smit et al.
1962) large divalent ions favour tetrahedral occupancy and large trivalent ions
favour octahedral occupancy. From these considerations, it follows that 2–3
spinels favour normal configuration. However, the only consistency of experiment with predictions are most likely for Mn^{2+} and Zn^{2+} (table 2) since these ions
have solely symmetrical electron orbitals, half filled for Mn^{2+} and full for Zn^{2+}.
Applying crystal field theory and optical data Dunitz and Orgel (1957) generated
semiquantitative data relating the site preference for the various transition
metals. The highest octahedral site preferences are the d^3 and d^8 ions and the
highest tetrahedral site preferences are d^0, d^5 and d^{10} ions. These theoretical
results agree well with experiments.

Ferrites used in practical applications are mixed ferrites of the type $(M_A +
M_B)_{1-x}Fe_{2+x}O_4$ where there are two or more metals besides iron and where x
varies between -0.3 to 0.3. The ferrite is considered to be stoichiometric from a
formula standpoint when $x = 0$ (or $Fe_2O_3 = 50$ mole %). In contrast to this *formula stoichiometry*, there can exist an *atomic stoichiometry* where there are
exactly 3 metal atoms for every 4 oxygen atoms. The spinel lattice can tolerate a
high concentration of cation vacancies designated as □. This tendency to form
cation vacancies increases as x increases from 0 to a more positive value by the

TABLE 2
Properties of spinel ferrites (after Smit and Wijn 1959 and Von Aulock 1965)

Ferrite and ionic distribution	η_B at 0 K calc.	exp.	M_s	$T_c(K)$	ρ (ohm cm)	K_1 (10^3 ergs/cm³)	$\lambda_s \times 10^6$	$M_s^2 \times 10^{-7}/|K_1|$
$Mn^{2+}_{0.8}Fe^{3+}_{0.2}[Mn^{2+}_{0.2}Fe^{3+}_{1.8}]O_4$	5	4.6	400	573	10^4	-40	-5	4.0
$Fe^{3+}[Fe^{2+}Fe^{3+}]O_4$	4	4.1	480	858	10^{-3}	-130	$+40$	1.8
$Fe^{3+}[Ni^{2+}Fe^{3+}]O_4$	2	2.3	270	858	10^4	$-54^{e)}$	-26	1.4
$Fe^{3+}[Li^+_{0.5}Fe^{3+}_{1.5}]O_4$	2.5	2.6	310	943	10^2	-83	-8	1.2
$Fe^{3+}_{0.5}Mg^{2+}_{0.5}[Mg^{2+}_{0.5}Fe^{3+}_{1.5}]O_4$	1.0	$1.1^{a)}$	110	713	10^7	-40	-6	0.3
$Fe^{3+}[Cu^{2+}Fe^{3+}]O_4$	1.0	$1.3^{b)}$	$135^{b)}$	728	10^5	-63	-10	0.3
$Fe^{3+}[Co^{2+}Fe^{3+}]O_4$	3	3.7	425	793	10^7	$+2000$	-110	0.09
$Fe^{3+}[\square_{1/3}Fe^{3+}_{5/3}]O_4$	3.3	$2.3^{c)}$	$417^{c)}$	1020				
$Zn^{2+}[Fe^{3+}_2]O_4$	10	$0^{d)}$	0	$(T_n = 9.5)$	10^2			

a) Varies with heat treatment
b) After slow cooling: $\eta_B = 2.3$ after quenching
c) Varies with heat treatment and atmosphere
d) $\eta_B = 5$ below Néel point
e) After Van Groenou and Schulkes (1957)

combination of excess α-Fe_2O_3 with rhombohedral structure dissolving into a ferrite of spinel structure

$$4\,Fe_2^{3+}O_3(rh) \rightarrow 3\,Fe_{8/3}^{3+}\square_{1/3}O_4(sp) \tag{5}$$

and the equilibration of the dissolved species with temperature and atmosphere

$$3Fe_{8/3}^{3+}\square_{1/3}O_4 \rightleftarrows \tfrac{8}{3}Fe^{2+}Fe_2^{3+}O_4 + \tfrac{4}{3}O_2\uparrow, \qquad \Delta H = 23\,\text{kcal/mole}. \tag{6}$$

Increased temperature and decreased atmospheric oxygen tend to reduce the cation vacancy content. From this analysis, *atomic stoichiometry* can exist without having *formula stoichiometry*. Cation vacancies play an extremely important role in the ferrite's sintering kinetics and magnetic properties.

5.3. Magnetism

The origin of magnetism in ferrites is due to: (i) unpaired 3d electrons, (ii) superexchange between adjacent metal ions and (iii) nonequivalence in number of A and B sites. In the free state the total magnetic moment of an atom containing 3d electrons is the sum of the electron spin and orbital moments. In an oxide compound such as ferrite, the orbital magnetic moment is mostly "quenched" by the electronic fields caused by the surrounding oxygens about the metal ion (crystalline field). The atomic magnetic moment (m) then becomes the moment of the electron spin and is equivalent to ($m = \mu_B n$) where μ_B is a Bohr magneton unit and n is the number of unpaired electrons.

Indirect exchange interaction (superexchange) takes place between adjacent metal ions separated by oxygen ions (Anderson 1950). The strength of this interaction depends on the degree of orbital overlap of oxygen p orbits and the transition metal d orbitals. The interaction will decrease as the distance between the metals increases and as the angle of Me–O–Me decreases from 180° to 90°. The only important interaction in ferrites is the AB interaction since the angle is about 125° while the BB is negligible because the angle is 90°. This interaction is such that antiparallel alignment of the moments results between the two metal ions when both ions have 5 or more 3d electrons or 4 or fewer 3d electrons. Parallel alignment occurs when one ion has $\leqslant 4$ electrons and the second has $\geqslant 5$ electrons. Since the common ferrite ions (Mn^{2+}, Fe^{2+}, Ni^{2+}, Co^{2+}) have more than 5 d electrons, the magnetic moments are aligned antiparallel between A and B sites. Since there are twice as many B sites occupied as A sites, the B site will dominate over the A site resulting in ferrimagnetism (Néel 1948).

Most of the ferrites of formula MFe_2O_4 are inverse ($Fe^{3+}\uparrow[M^{2+}\downarrow Fe^{3+}\downarrow]O_4$) and the moment of the compound is the moment of M^{2+}. The difference at 0 K in the theoretical and experimental values in the magnetic moments for some of the more common ferrites is given in table 2. The agreement between theory and experiment is fairly close except for $CoFe_2O_4$ where orbital contributions are at play, and Zn ferrite. In this case the exchange interaction between the trivalent ions on the octahedral sites is too weak so the material is paramagnetic at room temperature.

5.4. *Magnetic anisotropy*

Magnetic anisotropy refers to the dependence of the internal energy on the direction of magnetization. That is, the spins of the magnetic ions are not free to rotate but are bound to a specific crystallographic direction. The source of this magnetic anisotropy is the interaction of magnetic ions with electrostatic fields generated mainly by the first nearest neighbours of oxygen. These anisotropies of common origin are called magnetocrystalline, magnetostriction and induced. The magnetocrystalline anisotropy possesses the crystal symmetry of the material. Magnetostriction anisotropy is the phenomenon wherein the shape of the material changes during the process of magnetization. Also with the application of mechanical stress, the magnetization will change direction relative to crystallographic directions. The magnetic anisotropy where moments assume specific crystallographic directions by thermomagnetic treatment or cold working is termed induced anisotropy.

For spinels which have cubic symmetry, the anisotropy energy, E_a, is written

$$E_a = K_1(\alpha_1^2\alpha_2^2 + \alpha_2^2\alpha_3^2 + \alpha_3^2\alpha_1^2) + K_2\alpha_1^2\alpha_2^2\alpha_3^2 \tag{7}$$

where K_1 and K_2 are the first- and second-order magnetocrystalline anisotropy constants respectively and the α_1, α_2 and α_3 are the direction cosines of the moment of the ion relative to the three crystal directions. In the case of MnZn and NiZn ferrites at environmental temperatures (-40 to $80°C$), K_2 can be considered negligible compared to K_1.

For polycrystalline materials of cubic symmetry, the average magnetostriction or deformation (λ_s) is:

$$\lambda_s = \tfrac{2}{5}\lambda_{100} + \tfrac{3}{5}\lambda_{111} \tag{8}$$

where λ_{100} and λ_{111} are the maximum linear deformation with field in the [100] and [111] crystallographic directions, respectively. An extra deformation can be brought about by means of an average external stress, σ, which for polycrystalline materials increases the energy E by the expression, $E = \tfrac{3}{2}\lambda_s\sigma \sin\theta$, where θ is the angle between the directions of stress and magnetization.

By subjecting a magnetic material of cubic structure to an annealing treatment in the presence of a magnetic field, the induced magnetic anisotropy becomes uniaxial. Conditions for the occurrence of this anisotropy are that the ions in the crystal should exhibit no complete ordering and that the Curie temperature should be sufficiently high to allow ion diffusion below this temperature to take place rapidly. Upon slow cooling to room temperature the high temperature state is "frozen in" manifesting itself in a preferred direction of magnetization. For polycrystalline specimens, the induced anisotropy energy is

$$E = K_u \sin^2(\theta - \theta_u) \tag{9}$$

where K_u is the induced anisotropy constant and $(\theta - \theta_u)$ is the angle of the measuring field (θ) relative to the annealing field (θ_u).

The total anisotropy of a magnetic material can be determined using the single ion model (Slonczewski 1961), namely the weighted sums of the anisotropies of the individual ions. The ions that play a *major role* in all magnetic properties of technical ferrites are Co^{2+} and Fe^{2+}. The effects of those ions are great because the orbital angular momentum in octahedral sites is not fully quenched by the crystal field. Their moments are hence coupled strongly to the trigonal [111] axis of the crystal field producing a uniaxial anisotropy superposed on the cubic anisotropy. The couple for Co^{2+} is much stronger than for Fe^{2+}. (These ions are most influential in the compensation of K_1 and the induced anisotropy in magnetic after-effects and thermomagnetic treatments.)

These magnetic anisotropies (K_1, λ_s and K_u) play an important role in a variety of magnetic properties as, for example, in domain structure, magnetization processes, shape of the hysteresis loop, magnitudes of permeability, hysteresis losses and coercive force.

Shape anisotropy which is of different origin than the magnetic anisotropies above also affects the magnetic properties. Shape anisotropy is mainly due to demagnetization fields, which create domains of reverse magnetization and closure domains. These demagnetization fields originate on the surface of the sample and from internal surfaces such as inclusions, pores, grain boundaries, second phase etc. Their major effect is in reducing the remanent induction and increasing the coercive force which results in a reduced squareness ratio.

6. Science and technology of ferrites

6.1. Ferrite preparation

The preparation of polycrystalline ferrites with optimum properties is considered to be difficult and complex. The main problems involved are due to the fact that most of the properties needed for ferrite applications are not intrinsic but extrinsic. That is, the ferrite is not completely defined by its chemistry and crystal structure but also requires knowledge and control of the parameters of its microstructures, such as density, grain size, and porosity and their intra- and inter-granular distribution.

The preparation falls into four categories: (i) powder preparation, (ii) compact formation, (iii) sintering and (iv) machining to final shape. Each of these interrelated steps is necessary but not sufficient for the successful formation of the final ferrite part. Also, the sintering process is irreversible in terms of the microstructure so that constant care must be maintained to keep conditions constant prior to and during sintering.

6.1.1. Powder preparation

The purpose of this operation is to provide a powder that has physical and chemical homogeneity for uniform compaction and subsequent uniform sintering. There are two techniques used to prepare powders: solid state reaction of mixed oxides (used typically for production) and by decomposition of a solution of mixed salts. The wet chemical techniques consist of coprecipitation (Takada and Kiyama 1971), freeze drying of mixed sulphates by water sublimation (Schnettler and Johnson 1971), spray drying of mixed sulphates (DeLau 1970) and spray roasting of mixed chloride solutions (Ruthner et al. 1971). The use of coprecipitated ferrite has been beneficial for reducing TC variability (Akashi et al. 1971) and the use of spray roasting has been used for preparing large volume entertainment quality ferrites (Ruthner et al. 1971).

The production method begins with the mixing of oxides, or compounds that readily decompose into oxides, in a ball mill as a water or alcohol slurry. This mixture is dried, granulated, prefired at 900–1200°C in air, ground in a ball or vibratory mill, to which binders, plasticizers and pressing lubricants are added, and finally spray dried. A recently developed solid state reaction is the fast reaction sintering (Ruthner 1977) where slurries of metal oxides, hydroxides or carbonates are spray roasted for very short times (~5 s) at temperatures up to 1050°C. This method may be considered an advanced alternative for producing ferrite powders that exhibit precisely controlled chemical and physical properties. The raw material particle size distribution and the milling time along with allowance for iron pickup from the iron balls and mill are very important in influencing the presintering reactions and the microstructure of the sintered ferrite (Chol 1971). The particle size bandwidth of the presintered powder rather than the average particle size has the greatest effect on the microstructure and electrical properties (Chol 1971).

During the prefiring stage, the reaction of Fe_2O_3 with a metal oxide (MO or $M_2'O_3$) takes place in the solid state to form spinel according to the reactions:

$$MO + Fe_2O_3 \rightarrow MFe_2O_4 \text{ (spinel)} \tag{10}$$

$$2M_2'O_3 + 4Fe_2O_3 \rightarrow 4M'Fe_2O_4 \text{ (spinel)} + O_2 \uparrow. \tag{11}$$

The spinel phase is formed at the interface between the reactants. The reaction proceeds by the counter diffusion of cations through the reaction layer, which is a rigid oxygen lattice (Wagner diffusion mechanism (1935)). The rate of diffusion of the cations (and hence rate of reaction) is dependent on the point defect concentration (vacancies and interstitials) which in turn are influenced by the temperature and oxygen partial pressure (Ogawa and Nakagawa 1967).

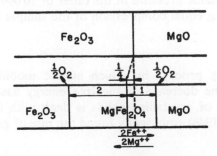

Fig. 9. The reaction $MgO + Fe_2O_3 \rightarrow MgFe_2O_4$ according to the Wagner mechanism.

Reaction (10) is well covered in the literature. Blackman (1959) and Reijnen (1967) have determined that Fe^{2+} ions exist in the reaction layer by the observed loss of oxygen for the initial reaction of $MgO + Fe_2O_3$. Reijnen proposed this reaction to take place with transfer of oxygen in the gaseous phase (Kirkendall Effect). Using compacts of divalent oxides and Fe_2O_3, Paulus (1971) proved that the diffusing ion is indeed Fe^{2+} and not Fe^{3+}. As shown schematically in fig. 9, oxygen is therefore evolved where the Fe_2O_3 goes into solution at the spinel/Fe_2O_3 interface and absorbed at the $M''O$/spinel boundary. The solution of the Fe_2O_3 into the spinel and hence spinel formation is enhanced with reducing conditions for reaction 10.

6.1.2. Compact formation

The pressing operation is a simple process in concept where a previously prepared powder is compressed by a 2–2.5:1 ratio to give a body that is 50–60% theoretically dense.

There are two methods in common use for compacting the powder. The most common is compaction using a punch-die assembly and the second is hydrostatic or isostatic compaction of a preshaped sample immersed in a liquid. Pressing a uniformly dense body by the first technique is difficult owing to the friction gradient of the powder at the walls of the die and between the particles themselves. This problem is somewhat overcome by the addition of external and internal lubricants to the powder such as stearic acid. For simple toroidal shapes the density gradient problem is minimized. But for pot core and other complex shapes, it is necessary to employ multiaction tooling and special pressing features to minimize density variations. For most ferrites produced on a large commercial scale the die-punch compaction is the most common procedure because pressing can be very rapid and economical by employing automatic presses.

Hydrostatic pressing for compacting powder is the preferred procedure in some small scale production and in most laboratories. It provides for higher and more uniform density with the possibility of achieving more complex parts. The technique involves filling a rubber or plastic mold with the powder, inserting the assembly in a pressure vessel filled with a hydraulic fluid such as water, glycerin or hydraulic oil. Hydraulic pressure of the order of 10 000 to 30 000 psi is applied to the liquid, imparting equal compression of the sample in all directions.

6.1.3. Sintering

Sintering is a heating process by which atomic mobility of the compact is sufficient to permit the decrease of the free energy associated with the grain boundaries. Sintering of crystalline solids is dealt with in detail by Coble and Burke (1964). Zener (1948) empirically found the rate of grain growth to be given by:

$$\bar{d} = Kt^n \tag{12}$$

where $\bar{d}$ is the mean grain diameter, n is about $\frac{1}{3}$ and K is temperature dependent. For MnZn ferrites K is also dependent on atmosphere varying from $0.55~\mu\text{m s}^{-1/3}$ for nitrogen (Roess 1971) to $1.8~\mu\text{m s}^{-1/3}$ for vacuum (Schichijo et al. 1964). Zener predicted that grain growth would be impeded by pores and/or inclusions at the grain boundaries to such an extent that the growth would stop at a diameter $d_c \leqslant 4D/3f$, where D is the average diameter of the pores and f is their volume fraction.

Ideally, the highest density with minimum uniform grain growth is desired for low loss ferrites (about $6~\mu\text{m}$, Knowles 1977) and with maximum uniform grain growth for high μ ferrites. In reality, however, some grains grow rapidly at the expense of other grains entrapping pores. This is denoted as duplex of discontinuous grain growth. It is generally felt that this duplex grain growth is a result of chemical or physical inhomogeneities leading to a porous structure. The rapid growth entraps pores within the grains where the annihilation of the pore is scarcely possible due to their distance from the grain boundaries. Duplex structure is deleterious to permeability and losses.

Discontinuous grain growth is minimized by keeping the grain boundary mobility low and the pore mobility high. Carpay (1977), extending the principles of pore and grain boundary mobility developed by Brook (1969), investigated the conditions for discontinuous grain growth in MnZn ferrites. Examples are shown where discontinuous grain growth occurs when the velocity of the grain boundary is much higher than the loaded boundary (boundary with pore) and the pore is left behind. He shows further that discontinuous grain growth increases with an increase in pore size and a decrease in the interpore separation. High oxygen firing decreases discontinuous grain growth by the increase in pore mobility through high flux of vacancies and Fe^{3+} (Reijnen 1968).

During conventional sintering it is difficult to control density and grain size independently, especially for high density and small grain size. With the use of external mechanical pressure during sintering (hot pressing) this can be achieved. NiZn ferrites with densities of 99.7% theoretical and an average grain size of $0.5~\mu\text{m}$ were obtained (DeLau 1968) using a continuous hot press developed by Oudemans (1968). Hot press sintering is a commercial technique for making high density and highly wear resistant material for recording-head applications. NiZn ferrites are considerably more wear resistant than MnZn ferrites (Foniok and Makolagawa 1977).

Control of the oxidation–reduction reactions by controlling the atmosphere during sintering and especially during cooling is very important. An equilibrium phase diagram of a MnZnFe ferrite (fig. 10) shows the importance of atmosphere on the oxidation state in the spinel phase and the hematite phase boundary. This figure agrees well with the universal equilibrium atmosphere diagram first shown by Blank (1961) and more recently worked out in more detail by Morineau and Paulus (1975). Care must be exercised so that minimal oxidation occurs by cooling along an isocomposition line and then quickly cooling through the phase boundary at the lowest temperatures when the kinetics are sluggish to minimize precipitation of $\alpha\text{-Fe}_2\text{O}_3$.

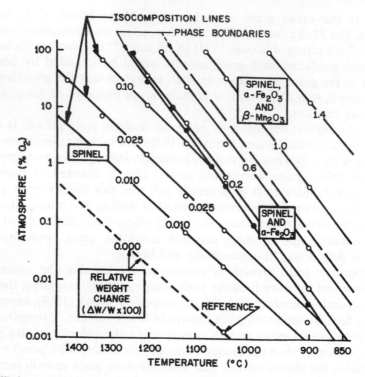

Fig. 10. Equilibrium weight changes as oxygen as a function of atmospheric oxygen content and temperature for the system: $(MnO)_{0.268}(ZnO)_{0.183}(Fe_2O_3)_{0.549}$ (after Slick 1971).

6.1.4. Machining

Machining of sintered ferrites is necessary to meet the final tolerances of the components since shrinkage control during sintering is inadequate. In telephony inductors require a controlled air gap whereas the transformers require a flat and smooth surface for minimum air gap. Machining is typically accomplished by precision grinding with an abrasive (diamond) wheel or tool. For transformer applications where very flat and smooth surfaces are necessary for mating surfaces, lapping is sometimes used.

According to Knowles (1970), machining stretches the ferrite surface (putting the surface in compression and the bulk in tension) affecting the properties. Stress levels of about $\frac{1}{10}$ the tensile strength were shown (Snelling 1974) to increase the losses and affect the permeability–temperature curve.

6.2. Linear ferrites

6.2.1. Permeability

There are two mechanisms in the phenomenon of permeability; spin rotation in the magnetic domains and wall displacement. The uncertainty of the contribution from each of the mechanisms, however, makes the theoretical interpretation of the experimental results difficult. In addition, a quantitative discussion of wall motion requires a detailed knowledge of domain wall configuration and wall energy but unfortunately few observations of domain walls have been made in polycrystalline materials. For a spherical grain with diametral and spherically bulging walls, Globus (1975) shows that the intrinsic rotational permeability (μ^R) and 180° wall permeability (μ^w) may be written:

$$\mu^R = 1 + 2\pi M_s^2/K, \qquad \mu^w = 1 + \tfrac{3}{4}\pi M_s^2 D/\gamma \tag{13}$$

where M_s is the saturation magnetization, K is the total anisotropy, D is the grain diameter and γ is the wall energy. The total anistropy K equals ($K_1 + \lambda_s\sigma$) where σ is the internal stress. Since the wall energy $\gamma \simeq K\delta_u$ (Smit and Wijn 1959) where δ_u is the wall thickness, μ^w becomes $1 + (3\pi/4\delta_u)(M_s^2/K)D$. The rotational permeability is dependent only on the intrinsic properties such as M_s, K_1 and λ_s which are controlled by chemistry. In addition the wall permeability is microstructurally sensitive to grain size and intragranular defects such as porosity, second phase, inclusions and dislocations that will effect the wall energy.

Of the intrinsic properties (M_s, K_1, and λ_s) M_s and K_1 are more important, λ_s only becoming significant when K_1 is very small. Examples of this are given in figs. 11 and 12. The exceptional case is when K_1 is large for NiZn ferrite, the permeability attains a maximum when the bulk magnetostriction becomes most negative (fig. 11). When K_1 is small the correlation between

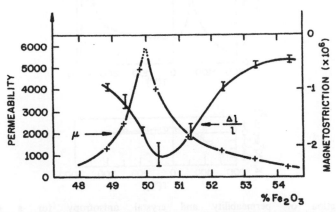

Fig. 11. Variation of permeability and magnetostriction with Fe_2O_3 content for NiZn ferrites with NiO:ZnO ratio equal to 15:35. (after Guillaud et al. 1960).

permeability and magnetostriction is much closer (fig. 12). However the permeability maximum does not occur at the point where $\lambda_s = 0$. Further evidence that the major contribution to permeability is K_1 is demonstrated in fig. 13. The upper half shows the permeability and the bottom half shows K_1 of a single crystal as a function of temperature. Along with the usual permeability maximum at the Curie temperature, a second maximum is noticed at about −55°C,

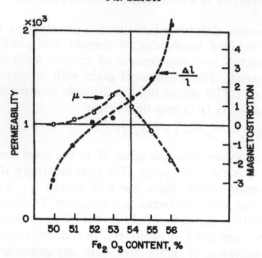

Fig. 12. Variation of permeability and magnetostriction with Fe_2O_3 content for MnZn ferrites whose MnO content ranged from 32 to 39% (after Guillaud 1957).

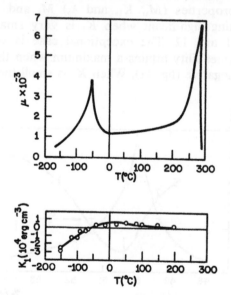

Fig. 13. Variation of permeability and crystal anisotropy for a single crystal $(MnO)_{0.31}(ZnO)_{0.11}(Fe_2O_3)_{0.58}$ (after Ohta 1963).

which is in excellent agreement with the compensation of the anisotropy $(K_1 = 0)$.

Table 2 shows properties of different spinel ferrites where the ferrites are listed according to the parameter M_s^2/K_1. In accord with discussions above, this parameter, neglecting λ_s, is considered a figure of merit for comparing materials for their intrinsic permeability capability. The best candidates for high permeability are Mn and Ni ferrite and the poorest is Co ferrite, Zn ferrite being paramagnetic. The resistivity of magnetite is so low as to be impractical and lithium ferrite possesses too low a resistivity along with high volatility of Li_2O to render it extremely process sensitive. The effect of zinc and ferrous ion substitution in Mn and Ni ferrites has considerable practical importance having a beneficial effect on all intrinsic properties except the Curie temperature for Zn additions. The non-magnetic Zn^{2+} ions enter the tetrahedral site preferentially and thus increases M_s at 0 K as originally shown by Guillaud (1951) by affecting an increase in the magnetic moment difference between the B and A site (fig. 14).

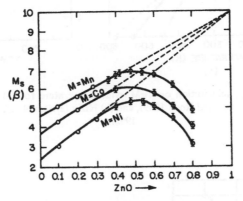

Fig. 14. Variation of spontaneous magnetization of spinel ferrites MFe_2O_4 with zinc substitution for M (after Guillaud 1951).

If the superexchange were unaffected, 100% Zn substitution would give 10 μ_B per formula unit from the two parallel Fe^{3+} moments, as indicated by the broken lines in fig. 14. But owing to the weakening AB interaction, M_s decreases with increasing Zn substitution above about 50% zinc.

Zn substitution also decreases the Curie temperature by this weakening of the AB interaction. Figure 15 shows the effect of Zn substitution on M_s with temperature for Mn and Ni ferrites. Since M_s decreases to zero at the Curie temperature, Zn substitution is more beneficial to M_s at 20°C for materials with the higher Curie temperature. Note that the M_s (20°C) for Ni ferrite is slightly higher than that for Mn ferrite for 40% Zn substitution even though the M_s (0 K) for $MnFe_2O_4$ is about twice that of $NiFe_2O_4$. An important effect of Zn substitution at 20°C is to contribute a positive K_1 and λ_s (Smit and Wijn 1954, Van Groenou and Schulkes 1967 and Arai and Tsuya 1976 for Ni ferrites and Ohta 1963 for Mn ferrites) increasing the normally negative values closer to zero.

Ferrous ions contribute to the magnetic anisotropy by an amount that varies in a complicated manner both with composition of the host ferrite and with temperature. This is demonstrated in figs. 16 (Miyata 1961)* and 17 where anisotropies are reduced further than that obtained for the simple Mn and Ni

* Similar to results obtained earlier by Penoyer and Shafer (1959) on MnFe spinels.

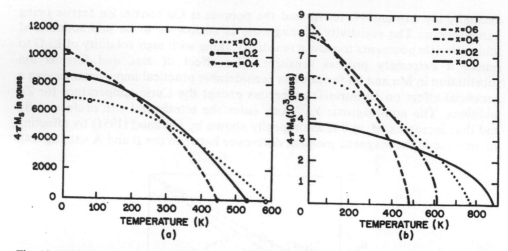

Fig. 15. Influence of zinc substitution on $4\pi M_s$ with temperature for (a) manganese zinc ferrites $Mn_{1-x}Zn_xFe_2O_4$ (after Guillaud and Roux 1949), and (b) NiZn ferrites $Ni_{1-x}Zn_xFe_2O_4$ (after Pauthenet 1952).

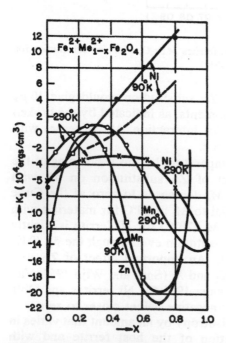

Fig. 16. Variation of crystal anisotropy, K_1, for $Me^{2+}_{1-x}Fe^{2+}_xFe^{3+}_2O_4$ as a function of x at 90 and 290 K (after Miyata 1961 and Usami et al. 1961).

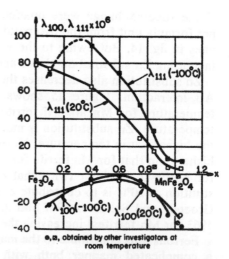

Fig. 17. Dependence of magnetostriction constants at -100 and $20°C$ for $Mn_xFe_{3-x}O_4$ as a function of x (after Miyata and Funatogawa 1962).

ferrites. Figure 16 shows the variation of K_1 as a function of x, the amount of Fe^{2+} per unit formula $Me^{2+}_{1-x}Fe^{2+}_x Fe^{3+}_2 O_4$ for two temperatures where Me is Mn or Ni. For small x which is the region of practical importance, the negative K_1 increases with increasing Fe^{2+} content resulting in a zero K_1 at 290 K at about $x = 0.18$ in Mn ferrite. Referring to fig. 17, magnetostriction ($\lambda_s = 0.4\lambda_{100} + 0.6\lambda_{111}$) is also near zero for the same composition. For $Ni^{2+}_{1-x}Fe^{2+}_x Fe^{3+}_2 O_4$, K_1 remains negative at 290 K for all values of x from 0 to 1. Another important finding here is the temperature insensitivity of K_1 at a value of $x = 0.25$ for MnFe ferrite.

Shown in fig. 18 are regions of small anisotropy K_1, λ_{111} and λ_{100} along with lines where $K_1 = 0$ and $\lambda_s = 0$ for the case where both Zn^{2+} and Fe^{2+} substitutions are incorporated into a Mn ferrite host. The successful use of these findings culminated in obtaining very high permeability by using pure raw materials and carefully controlled sintering conditions. Figure 19 shows contours of constant

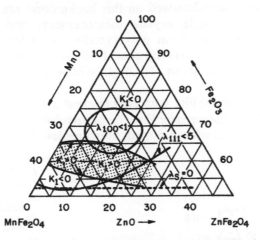

Fig. 18. Compositional dependence of crystal anisotropy and magnetostriction constants in the mixed oxide system (MnZnFe)–Fe$_2$O$_4$ at 20°C (after Ohta 1963).

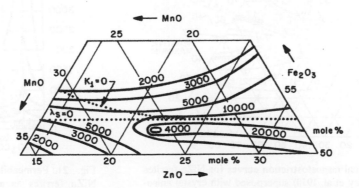

Fig. 19. Constant permeability contours relative to zero crystal anisotropy K_1, and saturated magnetostriction for the (MnZnFe)–Fe$_2$O$_4$ system (after Roess 1971).

permeability lines superimposed with zero lines of K_1 and λ_s from the previous figure. The maximum permeabilities obtained are very close to the composition where both K_1 and $\lambda_s = 0$ but are not exactly coincident. This slight difference perhaps is due to the contribution of K_2 and/or the fact that the processing of the polycrystalline samples was different from that used in preparing the single crystals from which the intrinsic data were determined. Further support for the latter interpretation is given for samples sintered in vacuum (Shichijo et al. 1964) where the maximum μ of 40k is obtained for a composition of about $(MnO)_{0.22}(ZnO)_{0.29}(Fe_2O_3)_{0.49}$.

Equivalent anisotropy and magnetostriction data do not exist for the NiZnFe ferrite system as with MnZnFe ferrites. Consolidated in fig. 20 are available magnetic anisotropies found in the literature. Loci of equal magnetostriction values are plotted showing negative and positive maxima along with a null magnetostriction line (Harvey et al. 1950). These data are qualitative but the trends are still valid. Superimposed on this background are magnetocrystalline anisotropy data from single crystal measurements and quantitative magnetostriction data measured from polycrystalline samples or calculated from single crystal measurements. Harvey et al. 1950 aslso showed that the best value of permeability occurs at about $(NiO)_{0.15}(ZnO)_{0.35}(Fe_2O_3)_{0.5}$ near the null magnetostriction line.

Similar trends in permeability with NiO content are given in fig. 21. Stuijts et al. (1964) clarified the discrepancy that the permeability maximum at 50 mole % Fe_2O_3 does not correlate with minimum anisotropies. The results are shown in

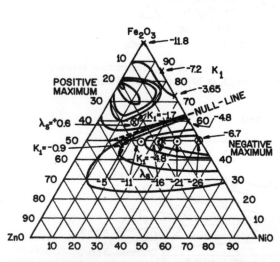

Fig. 20. Equal magnetostriction curves for NiZn ferrites (after Harvey et al. 1950) superposed with crystal anisotropy and magnetostriction values (after Smit and Wijn 1954, Elbinger 1961 and Ohta 1960).

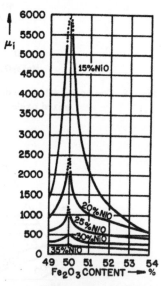

Fig. 21. Permeability of some NiZn ferrites as a function of Fe_2O_3 content (after Guillaud et al. 1960).

fig. 22 where sintering was performed in O_2 using very pure raw materials for control of grain growth. The decrease in permeability with increasing excess iron oxide $(x > 0)$ is attributed to the decreasing density caused by pore growth. The sharp decrease in permeability with increasingly deficient oxide $(x < 0)$ is due to increase in a wüstite-like second phase as observed by microscopic investigations. As a result of the presence of the second phase, stresses are set up and μ decreases since magnetostriction is still appreciable.

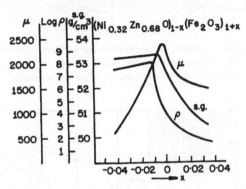

Fig. 22. Initial permeability, μ, specific gravity, SG, and resistivity as a function of composition for NiZn ferrites (after Stuijts et al. 1964).

For the use of MnZn and NiZn ferrites in filter inductor applications, the stability of permeability with temperature and time is extremely important. The temperature stability of permeability is controlled by the manner in which M_s and K_1 vary with temperature, K_1 being more important. The time stability of permeability is controlled by anisotropies induced in the ferrite. Figure 13 shows that permeability maxima occur where $K_1 = 0$, namely just before the Curie temperature despite M_s decreasing rapidly and at $-55°C$ denoted as the secondary peak maximum (SPM). It is the judicious choice of composition and careful process control to control the secondary permeability maximum that produces the desired change in permeability with temperature for a specific temperature range. The prime examples for controlling K_1 and hence TC are Co^{2+} and Fe^{2+} where their effect on the total anisotropies are shown schematically in fig. 23. Co^{2+} contributes a very large positive anisotropy changing rapidly with temperature. This results in an anisotropy compensation point that will increase with temperature with increasing Co^{2+} content. In the case of Fe^{2+} whose positive anisotropy contribution is considerably less in value and in temperature dependence, the anisotropy compensation point (fig. 23b) will decrease with increasing Fe^{2+} content as shown in fig. 24. In addition, Roess (1966) has shown that two compensation points exist with the presence of Fe^{2+} in MnZn ferrites. Figure 25 exemplifies the practical importance of this compensation phenomenon on the control of temperature coefficient of permeability for a MnZn ferrite. In practice the most interesting TC is nearly zero and this will be obtained in MnZn ferrites containing 53 to 55 mole % Fe_2O_3.

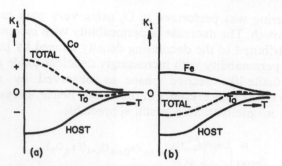

Fig. 23. Crystal anisotropy as a function of temperature for additives contributing a positive value to a negatively valued host. The compensation point, T_0, increases with an increase in Co^{2+}, a strong positive contribution and decreases with Fe^{2+}, a weak positive contributor.

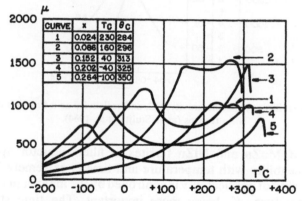

Fig. 24. Variation of permeability with temperature for different Mn ferrites $MnO_{(1-x)}FeO_xFe_2O_3$ as a function of x (after Lescroel and Pierrot 1960).

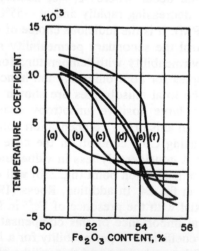

Fig. 25. Variation of temperature coefficient of permeability for MnZn ferrites with compositions (a) 27% MnO, (b) 29% MnO, (c) 31% MnO, (d) 32% MnO, (e) 33% MnO, and (f) 38% MnO (after Guillaud 1957).

(a) Effect of additives

The most important additives to MnZnFe and NiZnFe ferrites that are beneficial for the development of practical material are Ca, Si, Co, Ti, Sn and Li. Their major benefits, however, are in magnetic properties other than high permeability, namely improved losses, improved temperature and time stability of permeability and frequency extension of the properties.

Except for Li and Mg, the addition of alkali and alkaline earth metals decreases the permeability with increasing metal content (Guillaud 1957, Guillaud et al. 1960). The relationship between the ionic radii of the alkali metal and alkaline earth impurities in MnZn and NiZn ferrites is given in fig. 26. The

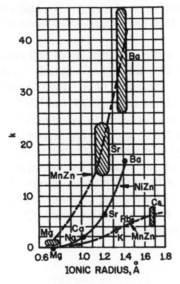

Fig. 26. Effect of ionic radius of impurities on permeability of MnZn and NiZn ferrites (after Guillaud 1957 and Guillaud et al. 1960).

ordinate $k = [(\mu_t/\mu) - 1]/\tau$ is a measure of the initial slope of μ_t/μ vs τ curve where μ_t and μ refer to the permeability of the undoped and doped ferrite, respectively and τ is the metal content in %. It is seen that the permeability decreases most with the impurity with the largest ionic radii. Since no important changes in magnetostriction and magnetocrystalline anisotropy have been observed with these impurity additions (Guillaud 1957), the lowering of permeability with impurity content is assumed to be due to changes in internal stress caused by the larger sized ions than are acceptable by the lattice. The solubility limit in percent for impurity ions is given by $1.8 \exp(-br)$ where r is the radius (Å) and b is 0.8 for alkali metals and 0.7 for alkaline earths (Guillaud 1957, Guillaud et al. 1960).

Figure 27 shows the change in μ/μ_t with small additions of MoO_3, CuO, CdO, C, Al_2O_3 and La_2O_3 to a MnZn ferrite whose permeability without additives is 2600 (Lescroel and Pierrot 1960). In all cases the permeability increases by about

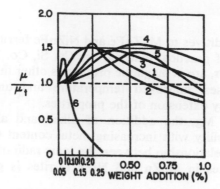

Fig. 27. Influence of small additions on the permeability of a MnZn ferrite where (1) MoO_3, (2) CuO, (3) CdO, (4) C, (5) Al_2O_3, and (6) La_2O_3 (after Lescroel and Pierrot 1960).

50% with initial increase in the impurity level followed by a permeability drop with continued impurity addition. Small additions of V_2O_3 to $(MnO)_{0.27}(ZnO)_{0.20}(Fe_2O_3)_{0.53}$ (Shichijo et al. 1961) show a 20% increase in permeability whereas small additions of Li_2O to $(NiO)_{0.15}(ZnO)_{0.35}(Fe_2O_3)_{0.50}$ (Guillaud et al. 1960) show a 50% increase in permeability. The substitution of Cr^{3+} and Al^{3+} for Fe^{3+} in both NiZn and MnZn ferrites decreases the permeability slightly but has a beneficial effect on the temperature coefficient of permeability (Guillaud 1957, Guillaud et al. 1960). The substitution of Cd for Zn in the composition $(MnO)_{0.322}$ $(ZnO)_{0.161}$ $(Fe_2O_3)_{0.517}$ shows that similar or higher permeabilities can be obtained by sintering at lower temperatures than that used for the undoped ferrite (Natansohn and Baird 1968). Cd is felt to promote sintering because its melting point is lower than Zn.

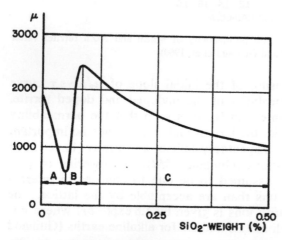

Fig. 28. Influence of silica content on the permeability of a MnZn ferrite (after Lescroel and Pierrot 1960).

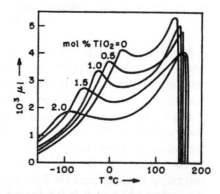

Fig. 29. Influence of TiO_2 additions on permeability versus temperature of $Mn_{0.567}Zn_{0.369}Fe_{2.062}$ (after Roess and Hanke 1970).

The effect of SiO_2 on the permeability of a MnZn ferrite is quite unusual (fig. 28). The grain size is very small in part A of fig. 28 where the permeability decreases with increasing SiO_2 content but increases along with increasing permeability with increasing SiO_2 in part B. The permeability decreases with further addition of SiO_2 in part C where discontinuous grain growth occurs. This trend is consistent with findings by Paulus (1962) where below 0.05%, SiO_2 forms inclusions which impede grain growth and above 0.05% it redissolves causing enhanced and discontinuous grain growth.

The substitution of (i) monovalent and divalent ions and (ii) trivalent and quadravalent ions to a ferrite have an indirect effect on properties in that the Fe^{2+} content is decreased for (i) and increased for (ii). Since Fe^{2+} on octahedral sites contributes a positive anisotropy, the addition of Ti^{4+}, Sn^{4+}, Ge^{4+}, V^{4+} and Al^{3+} will effect a decrease in the SPM or the temperature at which $K_1 = 0$ such as that given in fig. 29 and the opposite for Li^{1+} and Ni^{2+} (König 1974).

(b) Effect of grain size and porosity

Equation (13) predicts that the wall permeability varies linearly with the average grain size. Guillaud and Paulus (1956) were the first to relate μ with grain size (see fig. 30) where uncontrolled porosity within the grains and grain boundaries existed. In this case the permeability is not linearly dependent on grain size and the rate of permeability change with grain size becomes negligible above about

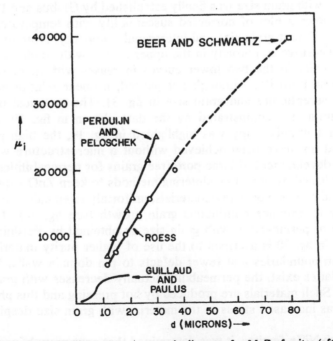

Fig. 30. Variation of permeability with average grain diameter for MnZn ferrites (after Guillaud and Paulus 1956, Roess 1966, Perduijn and Peloschek 1968 and Beer and Schwartz 1966).

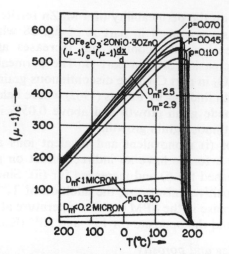

Fig. 31. The temperature dependence of the permeability of a NiZn ferrite for different levels of porosity and grain size. The permeability is corrected to correspond to single crystal density (after Globus and Duplex 1966).

$10 \, \mu$m. Guillaud (1957) further demonstrated that the permeability of a NiZn ferrite drastically decreased with increasing percentage of grains possessing pores despite increasing grain size above about $14 \, \mu$m. The correlation of permeability with grain size was finally established by Globus and Duplex (1966). Figure 31 shows a plot of corrected susceptiblity with temperature of a NiZn ferrite with different grain size and intergranular porosity. Note that the susceptibility is insensitive to porosity in the upper curves with similar grain sizes and the susceptibility in the two lower curves increases with increased grain size with the same porosity. Although not plotted, a linear relationship exists between the susceptibility and grain size in fig. 31. The technical importance of pore free grains is demonstrated by the data shown in fig. 19. Although the necessity of low anisotropy was highlighted in fig. 19, the high permeabilities shown could not have been achieved without a microstructure with pore free grains. The development of large pore free grains for permeabilities up to 40 000 was accomplished by adjusting sintering methods to keep ZnO vaporization to a minimum and by using pure raw materials, uniformly sized calcined particles and O_2 sintering to enhance continuous grain growth (see fig. 30). The increased dependence of permeability with grain size as obtained by Perduijn and Peloschek (1968) in fig. 30 is ascribed to the use of higher purity material and hence cleaner grain boundaries and fewer defects to pin domain walls. Where snaky grain boundaries exist, the permeability actually decreases with grain size (Löbl et al. 1977). Such materials are produced by hot pressing and this phenomenon is interpreted as increased stress in the material with grain size despite the low λ_s composition.

In iron rich NiZn ferrites, it was determined that pore growth increased (hence intragranular porosity decreased) with increasing iron content when sintering in

O_2 (see fig. 22). The explanation of this pore growth phenomenon was developed by the work of Reijnen (1968).

According to Reijnen (1968) the requirements for pore growth are large concentration of cation vacancies and the presence of cations with different valency states. Ferrites with excess Fe_2O_3 provide both requirements by the reaction given in section 5.2 when Fe_2O_3 dissolves into the spinel lattice.

The Reijnen model for pore growth is explained with reference to fig. 32. The

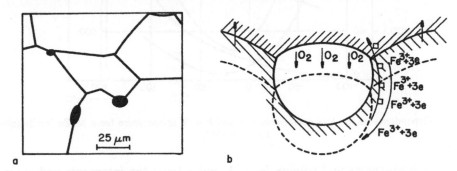

Fig. 32. Model for pore growth when sintering in an oxidizing atmosphere with excess Fe_2O_3. Different radii of curvature are adapted by a pore intersecting with a grain boundary (a) and moving with a grain boundary (b) (after Reijnen 1968).

progress of a migrating grain boundary is hampered when intersecting with a pore. This causes a pull on the pore which then acquires two radii of curvature (fig. 32a). As a result of this curvature difference, material transfer will tend toward the direction of least curvature. The pore will move with the migrating boundary by the scheme in fig. 32b. Note that the cation vacancy flux moves by surface or volume diffusion in the direction of the pore and boundary movement. The flux of $Fe^{3+} + 3$ electrons moves through the bulk of the material and O_2 through the gaseous phase in the opposite direction. Although the sintering rate is low, pores can move along with moving grain boundaries, coalesce and thereby grow.

6.2.2. Magnetic losses

Losses in ferrites can be separated into eddy-current losses, hysteresis losses and residual losses. For the ferrite core losses (R) per unit frequency (f) and inductance (L) we write, after Jordan (1924)

$$R/fL = ef + hH + r \qquad (14)*$$

where H is the field strength and e, h and r are the eddy current, hysteresis and residual coefficients, respectively. Separation of the losses is simply accomplished by plotting R/fL vs f for different field strengths or R/fL vs H at

---

* The Legg (1936) equation is similar in concept except the core losses are also normalized with respect to μ, namely $R/\mu fL = bf + aB_m + c$

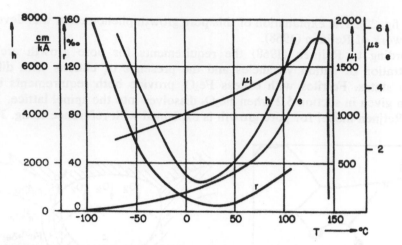

Fig. 33. Dependence of the loss coefficients e, r, and h with temperature for a MnZn ferrite (after Lescroel 1953).

different frequencies and solving for e, h and r from the intercepts and slopes. Figure 33 shows the contributions of the different loss coefficients and their dependence with temperature for a MnZn ferrite. The residual loss coefficient is higher for NiZn ferrites than for MnZn ferrites for the same permeability and generally rises more rapidly with temperature.

(a) Eddy current losses

The eddy current loss coefficient increases with increasing temperature whereas the hysteresis and residual loss coefficients reach a minimum at about room temperature increasing with both rising and falling temperatures. The eddy current loss factor $(\tan \delta_e)/\mu$ is directly related to the reciprocal of the resistivity, ρ. Since $\rho = \rho_\infty \exp(E/kT)$ (where $\rho = \rho_\infty$ at $T = \infty$, E = activation energy, k = Boltzmann constant and T the absolute temperature) for semiconductor ferrites, a plot of $\log[(\tan \delta_e)/\mu]$ vs $1/T$ should yield a straight line. This is shown in fig. 34 for a MnZnFe ferrite where the activation energy $E = 0.11\,\text{eV}$ and is consistent with electron conductivity (Roess 1971).

The resistivities of MnZn and NiZn ferrites with 50 mole % Fe_2O_3 are considerably higher than metals and hence have expected lower eddy current losses. But since increased Fe^{2+} is needed to reduce the crystal anisotropy to improve permeability and its stability with temperature the resistivity is reduced from electron mobility by the reaction $Fe^{2+} \rightleftarrows Fe^{3+} + e$. Figure 22 typifies the variation of resistivity with Fe_2O_3 content near the formula stoichiometry and is consistent with extensive findings by Van Uitert (1955). The bulk resistivity of ferrites containing Fe^{2+} can be increased and the eddy current losses decreased by the use of a high resistivity film to insulate the low resistivity grains. Guillaud (1957) first demonstrated this with the addition of Ca to a ferrite of composition $(MnO)_{0.28}(ZnO)_{0.19}(Fe_2O_3)_{0.53}$. The calcium segregated at the grain boundary (as

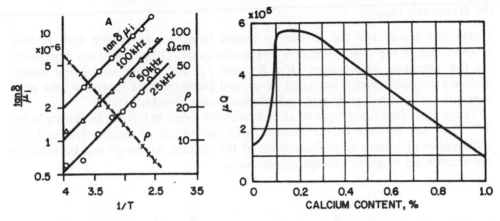

Fig. 34. Loss factor and resistivity as a function of temperature for $Mn_{0.65}Zn_{0.23}Fe_{2.12}O_4$ (after Roess 1971).

Fig. 35. Effect of calcium on μQ product of $Mn_{0.56}Zn_{0.38}Fe_{2.06}O_4$ at 40 kHz (after Guillaud 1957).

determined from radiography of labeled Ca) and the μQ improved as shown in fig. 35. The eddy current loss contribution was reduced by a factor of ten without changes in the hysteresis and residual losses and the initial permeability. Further improvement in resistivity and losses was shown by Akashi (1961a) with simultaneous additions of CaO and SiO_2 to a MnZn ferrite (fig. 36). The dramatic increase in resistivity across a grain boundary of a CaO and SiO_2 doped MnZn ferrite further substantiates this mechanism (Akashi 1961b).

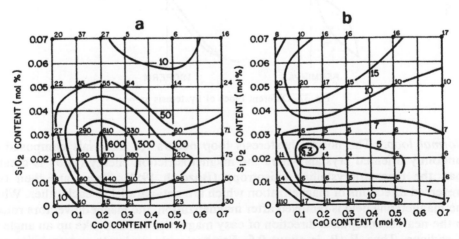

Fig. 36. Influence of SiO_2 and CaO additions on the resistivity (a) and loss factor, (b) of $Mn_{0.68}Zn_{0.21}Fe_{2.11}O_4$ at 100 kHz (after Akashi 1961a).

(b) Hysteresis losses

Hysteresis losses are due to energy losses from non-reversible wall motion proportional to the area of the B/H hysteresis cycle. These losses are dependent on the magnetic anisotropies (K_1, K_u and λ_s) and shape anisotropy which includes extrinsic properties such as crystal imperfections (porosity and non-magnetic inclusions), grain size and internal strain distributions.

There exist four main types of B/H hysteresis loops in ferrites as shown in fig. 37. They are (i) normal, (ii) isoperm, (iii) perminvar or constricted and (iv) rectangular or square loop. Discussion of the rectangular loops will be made in the section on magnetic storage applications.

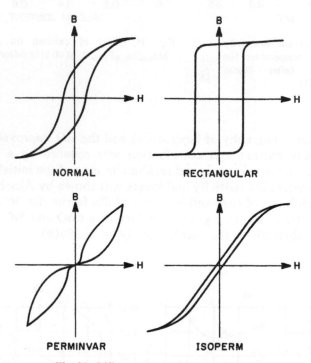

Fig. 37. Different types of hysteresis loops.

Normal loop. The normal hysteresis loop occurs for materials composed of randomly oriented crystals where the magnetostriction energy ($\frac{3}{2}\lambda_s\sigma$) is greater than the magnetocrystalline anisotropy (Bozorth 1951). There are then two preferred directions of magnetization which are anti-parallel to each other. When the magnetizing field is removed after magnetization, the magnetic vectors return to the nearest acceptable direction of easy magnetization and take up an angle of π radians. Thus B_R/B_s is about 0.5. Ferrites with moderate permeability and higher losses possess this type of hysteresis loop.

Isoperm loop. A prerequisite for an isoperm loop is a structure of homogeneous pore-free crystallites with a composition having small magnetocrystalline anisotropy and magnetostriction. In such a structure, domain wall movement is largely reversible with high mobility. Isoperm loops occur spontaneously in MnZn ferrites containing about 51 to 55 mole% iron oxide (Roess 1971) which is the compositional area where both K_1 and λ_s are very small and approach zero. Ferrites with high to very high permeability possess the isoperm loop with moderately low hysteresis losses. Induced isoperm loops can be produced in cobalt doped and iron excess ferrites by magnetically annealing normal to the direction of the measuring field during cooling (Kornetzki and Brockmann 1958).

Perminvar loop. The perminvar loop is characterized by low and constant permeability and very low hysteresis losses at low field strengths. At a particular field strength the permeability no longer remains constant but begins to increase with further increasing field strength. The loop opens on both ends at this particular field strength while the mid-region remains constricted. With rising field strengths, this constriction becomes less and finally disappears at high field intensities.

The perminvar loop is observed in MnZn and NiZn ferrite systems but only for iron contents around 60 mole % and at low and limited temperature ranges (Roess 1971). The perminvar loop occurs only just above the secondary peak maximum which is when K_1 is slightly positive. The perminvar loop can also occur just below the Curie temperature but the loop is quite unstable. The perminvar loop occurs more particularly in cobalt doped iron-rich ferrite compositions (any ferrite composition containing two divalent oxides of Mn, Ni, Cu, Mg, Pd, Zn and Cd) (Eckert 1957) and iron deficient NiZn ferrite (DeLau and Stuijts 1966). For the iron rich ferrites slow cooling from the Curie temperature to room temperature is needed (Kornetzki et al. 1955) so an induced anisotropy of Co^{2+} can be established to stabilize the domain walls into a narrow and deep energy well. The walls, therefore, move very little in a quite reversible mode giving rise to a low and constant permeability with extremely low hysteresis losses. Excess iron ferrites are required since cation vacancies are needed to enhance Co^{2+} diffusion (Iida 1960). Slowly cooling excess iron NiZn Co ferrites from above the Curie temperature has further improved the μQ product to that obtained from cooling slowly from the Curie temperature (Slick 1971). Dixon et al. (1977) interpret this improvement as due to reduced domain wall damping.

The domain walls of Co doped deficient iron NiZn ferrites are also stabilized by Co^{2+} but quite rapidly at room temperature through the reaction: $Co^{2+} \rightleftarrows Co^{3+} + e$ (DeLau and Stuijts 1966). This feature is very advantageous in practical application for the iron deficient ferrites in that the high quality properties recover quickly after a large disturbance. In the case of iron excess ferrites, after a disturbance the walls must be restabilized at a high temperature.

Domain wall stabilization with Co^{2+} has an additional benefit on the permeability spectrum. Figure 38 shows that effect of stabilization on the μ' and μ'' for a cobalt doped iron excess NiZn ferrite. μ' is decreased with stabilization

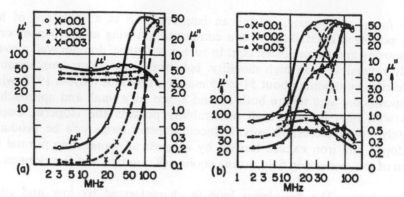

Fig. 38. Magnetic spectra of iron excess NiZn ferrites with compositions $Ni_{0.45-x}Zn_{0.25}$ $Co_xMn_{0.06}Fe_{2.24}O_{4+y}$. The domain walls are stabilized by annealing at 280°C (a) and after demagnetizing in an ac field are destabilized (b) (after Mizushima 1964).

but is overcompensated by a significant decrease in μ'' leading to an improved $(\tan\delta)/\mu$ $[\mu''/\mu'^2]$ and making the materials useful to higher frequencies. With increased cobalt content up to about 1 mole % ($x = 0.03$ in fig. 38a) the $(\tan\delta)/\mu$ improves and the materials become useful at higher frequencies. Small grain size can also stabilize domain walls by their pinning from the presence of many grain boundaries (DeLau 1975). DeLau showed that for $Ni_{0.8}Zn_{0.2}Mn_{0.01}Fe_{1.99}O_4$, the permeability decreased from about 90 with a grain size of 12 μm to about 15 for 0.3 μm grain size with a concomitant increase in the loss peak from about 20 to 400 MHz.

The use of cobalt addition and annealing to stabilize domain walls has recently been applied to excess iron MnZn ferrites (Akashi et al. 1971, 1972, Roess 1977). Shown in table 3 are comparisons of these ferrites with perminvar loops with commercial materials having isoperm loops.

TABLE 3

Comparison of magnetic properties at 100 kHz of excess iron MnZn ferrites with different B/H loops

Variable	Isoperm loops		Perminvar loops	
	Siemens N48	TDK H6H3	NEC superneferrite	Siemens stabilized perminvar
μ	2000	1300	1000	500
$(\tan\delta)/\mu$ ($\times 10^{-6}$)	<2.5	<1.2	0.8	0.3
h/μ^2 (cm/MA)	<0.4	<.13	0.05	0.03
TC/μ ($\times 10^{-6}$/°C) (−25 to 55°C)	0.7–0.9	1.2±0.8	0.3±0.05	0.3
D/μ ($\times 10^{-6}$/decade)	2	<5	2	2.5

6.2.3. Disaccommodation

Disaccommodation is a phenomenon by which the permeability of a ferrite decreases with time after demagnetization (either by use of a decaying ac field from above magnetization saturation or by cooling the material through the Curie temperature). Figure 39 shows typical time decrease of permeability at 0 and $-60°C$ for a MnZn ferrite of composition $(MnO)_{0.235}(ZnO)_{0.225}(Fe_2O_3)_{0.54}$ (Snoek 1947). Note that the rate of change depends on the temperature and that

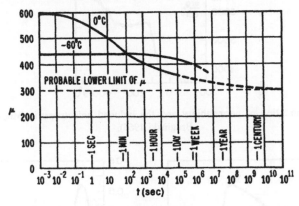

Fig. 39. Disaccommodation observed for a MnZn ferrite of composition $Mn_{0.46}Zn_{0.45}Fe_{2.08}O_4$ (after Snoek 1947).

the rate of change of permeability from a day to a century is small at $0°C$. This decrease in permeability with time is associated with the stabilization of the newly formed domain walls by a form of induced anisotropy. This stabilization is manifested from ionic or electronic diffusional processes.

A temperature spectrum of disaccommodation as measured 1 s and 30 min after ac demagnetization is shown in fig. 40 for a Mn ferrite. Similar spectra exist for other iron rich ferrites such as magnetite, Ni ferrite, Mg ferrite, etc. but the peaks are modified. The disaccommodation peaks I, II and III are clearly visible at $\sim 380°$, $100°$ and $0°C$, respectively with an indication of another process IV at about $-70°C$. The process IV has only been found in Mn containing iron rich spinels (Krupićka and Vilim 1961). The activation energies for these processes I, II, III and IV are about 2, 1, 0.8 and $\sim 0.6 eV$ respectively consistent with ionic diffusion. Measurements of the induced anisotropy (fig. 40) show that the D_{max} are definitely associated with the temperatures where abrupt changes in the anisotropy occur. D_{max} refers to the maxima in disaccommodation in the temperature spectrum.

The dependence of D_{max} with Mn in $Mn_xFe_{3-x}O_{4.01}$ and vacancy content in $Mn_{0.6}Fe_{2.4}O_4$ is shown in fig. 41a. Ni shows a similar but not exact dependence on D_{max}. The active elements are the Fe^{2+} and vacancy contents since the D_{max} for process I, II and III approach zero at zero $Fe^{2+}(x = 1)$ and zero ρ (fig. 41b). In addition, the intensity of the D_{max} of peak III for Li and Ti substituted MnZn

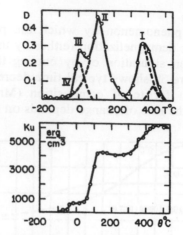

Fig. 40. Disaccommodation of permeability and the induced anisotropy in $Mn_{0.2}Fe_{2.8}O_4$, (b) (after Braginski 1965).

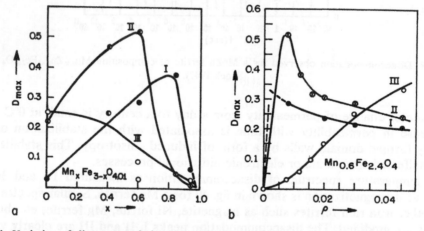

Fig. 41. Variation of disaccommodation peak intensities as a function of x in $Mn_xFe_{3-x}O_4$ (a) and cation vacancy content in $Mn_{0.6}Fe_{2.4}O_4$ (b) (after Braginski 1965).

ferrites is shown to increase linearly with the product of the Fe^{2+} content and the vacancy content (Marais and Merceron 1963a,b).

According to Braginski (1965) the mechanism for the disaccommodation processes I, II and III are the directional ordering of vacancies and Fe^{2+} ions on the non-equivalent octahedral sites. The model assumes a certain probability of Fe^{2+}-vacancy neighbourhood which reveals anisotropic properties of the Fe^{2+} (Slonczewski 1961).

There are three fundamental units to the anisotropy, namely a single ion with [111] symmetry, adjacent ion pair with [110] symmetry or non-adjacent ion pair with [100] symmetry. Process III seems to be due to the preferential occupation of Fe^{2+} ions of [111] local interaction symmetry axis which have the chance to interchange places with adjacent vacancies.

Processes I and II are ascribed to the ordering of Fe^{2+}–Fe^{2+} pairs through single vacancy jumps. Process II is caused by [110]-pairs of adjacent ferrous ions and process I to [100]-pairs of non-adjacent ferrous ions.

Anomalous aging effects (accommodation) have been observed in Co doped NiFe ferrites (Braginski and Kulikowski 1964) at temperatures above the K_1 compensation point. This phenomenon disappears after an induced anisotropy anneal and does not occur for technical ferrites.

(a) Effect of additives

Additives to MnZn and NiZn ferrites affect the disaccommodation in the manner in which they influence (i) the induced anisotropy, (ii) the Fe^{2+} content, and (iii) the cation vacancy content. Because of its high uniaxial anisotropy in octahedral sites, Co^{2+} additive plays a prevalent role in the induced anisotropy.

In an iron rich Mn ferrite, the intensity of the disaccommodation processes II and III increases with small increase in Co^{2+} while the process I intensity decreases (Petrescu and Maxim 1968). These changes were ascribed to the increased induced anisotropy from the Co^{2+} ion.

Additives which decrease the Fe^{2+} content such as Li_2O, Na_2O and Cu_2O have indeed shown a decrease with the D_{max} of process III and hence the decreased disaccommodation at room temperature (Marais and Merceron 1963a, Matsubara et al. 1971). Additives that increase the Fe^{2+} content such as TiO_2, SnO_2 and Ta_2O_5 have also shown an increase in the disaccommodation (Marais and Merceron 1963b, Matsubara et al. 1971). However, Ti^{4+} has an added feature that makes it useful for technical ferrites especially for long term stability at high operating temperatures. It localizes the Fe^{2+} so that the process III D_{max} is reduced by about 50°C from that in the unsubstituted ferrite and the intensity of the D_{max} of process II at about 100°C is reduced because the ability of the Fe^{2+} to participate in the Fe^{2+}–Fe^{2+} pair formation is diminished (Knowles 1974).

Other additives that show a decrease in disaccommodation are CaO, SiO_2, BaO and SrO (Matsubara et al. 1971). These are additives that tend to segregate at grain boundaries and increase the resistivity. The mechanism for this D decrease is not known. Perhaps the changed microstructure produced by these ions causes the material to respond differently to the sintering and cooling atmosphere affecting the Fe^{2+} and vacancy contents.

Figure 42 shows how the D can be decreased by controlling the vacancy content through the adjustment of the oxygen balance. Curve 1 shows the disaccommodation of a MnZn sample prepared by adequate industrial atmosphere control. The increased disaccommodation of curve 2 corresponds to the distribution of the excess oxygen from the surface layer of sample 1 throughout the specimen by a homogenizing heat treatment. Curve 3 shows the effect of the reduced vacancy content on a homogenized sample reduced by 0.1% of the oxygen. In technical linear ferrites there exists a certain amount of Fe^{2+} content and therefore it is necessary to minimize the vacancy content for low disac-

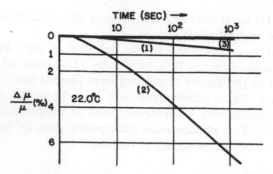

Fig. 42. Disaccommodation of the permeability after ac demagnetization for a MnZn ferrite with slight oxygen excess (1) with the excess oxygen from the surface distributed throughout the sample (2) and after reduction of 0.1% of the oxygen (3) (after Iida and Inoue 1962).

commodation. This has been mostly accomplished with proper atmospheric control during sintering and equilibrium cooling according to the schedules established by Blank (1961) and Slick (1971b) (fig. 10).

6.3. Square loop ferrites

The important characteristics of rectangular looped ferrites are high B/H squareness ratio, controlled coercive force, H_c, and a low switching time, τ_s. A sketch of saturation and minor rectangular B/H loops is shown in fig. 43. The minor loop best represents practical applications such as coincident currents. The remanence ratio ($R = B_r/B_s$) is a material constant dependent on K_1, λ_{111} and the shape anisotropy. A good square loop material will have a value

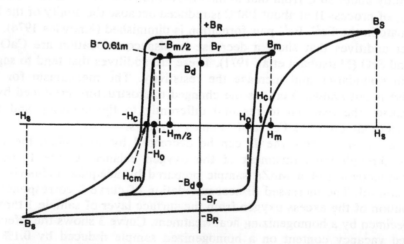

Fig. 43. Explanatory sketch of a hysteresis loop for the terms concerning the definition of squareness.

near unity. The squareness ratio ($R_s = B_{-m/2}/B_m$) can have an infinite number of squareness values for an infinite number of loops since H_m can vary continuously from 0 to H_s, the saturating field. There will always be one loop for which the squareness ratio is maximum denoted as $R_{s,max}$ which is also a material constant. For proper memory operation, the threshold field, H_0, must be larger than $H_{-m/2}$ (fig. 43). H_0 is approximately equal to the coercive field, H_{cm}, and hence the flank of the loop is steep. Finally, since the practical field ratio in coincident current memories is typically 0.61 to satisfy a 10% margin on H_m, the practical squareness ratio becomes $\alpha_{0.61} = B_{0.61m}/B_m$.

6.3.1. Natural squareness

Two types of rectangular loops exist in ferrites – natural and induced. Natural rectangularity is the type where no texture of the crystallites is present and is unaffected by low temperature heat treatments. Induced rectangular loop is the type that is made square by heat treatment in a magnetic field or by suitable directed external stresses. According to Wijn et al. (1954) a square loop ferrite can be expected when the material has a high negative magnetocrystalline anisotropy, K_1, along with a magnetostriction coefficient, λ_{111}, approaching zero. This can be understood in the following way. The magnetization vectors are bound to the [111] direction by the crystalline anisotropy K_1 with a certain force equivalent to the value of K_1. If λ_{111} becomes active by internal stresses distributed at random over all directions, the binding to the various (111) direction is changed and therefore the chance for assuming a position other than the nearest (111) position is increased and the remanence decreases and consequently the squareness will also decrease. Baltzer (1955) proposed an alternate but less acceptable model for high squareness where a zero effective domain anisotropy ($K_d = K_1 +$ magnetostrictive effects) was considered necessary.

In polycrystalline ferrites with $K_1 < 0$ and without apparent texture, the remanence ratio, B_r/B_s, has a theoretical maximum of 0.867 (Gans 1932). Innumerable compositions without apparent textures are reported in the literature with remanence ratios above 0.9. The understanding of this is not fully known but it is felt that either saturation is not fully reached or the toroidal shape itself induces a stress anisotropy such that a lower energy exists in the flux direction tangential to the toroidal diameter.

6.3.2. Induced squareness

Improvements in squareness ratio are possible by the use of stresses in ferrites having small K_1 and large λ_s or by the use of magnetic annealing in iron rich ferrites since Fe^{2+} can produce a uniazial anisotropy, K_μ (Weisz and Brown 1961). This latter effect is accentuated with the presence of a small quantity of Co^{2+} having considerably larger K_μ than Fe^{2+} (Gorter and Elsveldt 1957). The disadvantage in these ferrites with induced squareness is that the switching coefficient described in section 6.3.4 below and the coercive force are increased and hence these ferrites are not practical for commercial applications. The application of compressive stresses was demonstrated (Wijn and Van Der Heide

1957) on NiZn ferrites having large negative λ_s where the squareness ratio continually increased from 0.5 to 0.95 with increasing stress while S_w increased from 0.7 to 2.20 Oe μs.

The mechanism of magnetic annealing is an ordering phenomenon of Co^{2+}, predominantly in the presence of an ac magnetic field during cooling, that is parallel to the measuring field. Iida et al. (1958) found that cobalt containing ferrites are sensitive to the oxygen pressure during cooling and they consider that Co^{2+} and vacancy ordering are controlled by the oxygen pressure.

6.3.3. Coercive force

The most accepted theory for flux reversal from "1" to a "0" state is by domain wall motion. Domains of reverse magnetization are nucleated followed by irreversible expansion and annihilation of the original magnetization (Goodenough 1954). For high squareness and a sharp knee in the B/H loop, the distribution of the nucleating fields, H_n, must be narrow and $H_n \geq H_c$. Nucleation of domains on grain boundaries and lamellar precipitates are most effective here while intragranular inclusions and porosity are of secondary importance (Goodenough 1954).

Coercive force can be defined as the largest field required to cause the irreversible wall motion of reverse domains already nucleated. There are many independent factors that give rise to these field strengths, H_w, such as granular inclusions and porosity, grain boundaries, lamellar precipitates and wall surface tension. Goodenough (1954, 1957) made a detailed description of these factors concluding that the wall surface tension is the main source of the coercive force with the reverse domains adhering to defects at grain boundaries. In summary, a prerequisite for a low-coercivity ferrite is a dense, homogeneous, single phase material. The coercive force is given by Goodenough (1957) as

$$H_c \leq k[A(K_1 + \lambda_s\sigma)^{1/2}]/2M_sD + \tfrac{1}{6}\pi M_s\langle(\cos\theta_1 - \cos\theta_2)^2\rangle . \tag{15}$$

The term $\langle(\cos\theta_1 - \cos\theta_2)^2\rangle$ represents the mean difference of the cosines of the easy directions on either side of a grain boundary. The first term dominates since the value of M_s for ferrites is generally sufficiently small. Figure 44 shows support of this equation where the threshold field (approximately equal to H_c) decreases with increasing grain size (Schwabe and Cambell 1963).

The important functional parameters for the magnetic storage and retrieval of information are the voltage outputs dV_1 and dV_z. These voltage outputs are observed on a sense wire produced by nearly square current pulses on a second wire where both wires pass through a ferrite toroid. The voltage output produced by a change in magnetic flux from the disturbed remanent state (B^d in fig. 43) to the maximum negative induction ($-B_m$) is a measurement of dV_1. μV_1 is similar to dV_1 except the undisturbed remanent state, B_r, is the starting point. The voltage output due to a flux change from the disturbed remanent state ($-B_d$) and ($-B_m$) is a measurement of dV_z. The signal to noise ratio, dV_1/dV_z is a figure of merit for high quality memories and computers. High squareness and a narrow

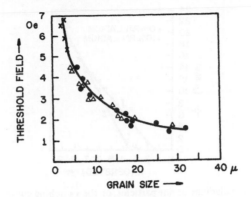

Fig. 44. Relation of threshold field with grain size of lithium ferrite cores of dimensions 0.050″ OD × 0.030″ ID × 0.015″ H (after Schwabe and Cambell 1963).

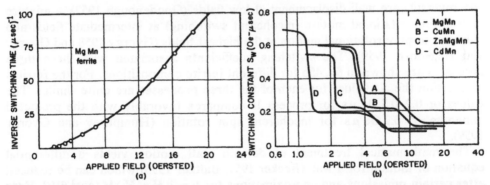

Fig. 45. Variation of switching time with applied field for a MnMg ferrite (a) and variation of switching coefficients with applied field for a range of representative ferrites as indicated (after Shevel 1959).

distribution of coercive forces (grain size uniformity) in the material are necessary for high dV_1/dV_z ratios.

6.3.4. Flux reversal

An important parameter to the design engineer is the switching time (τ_s) or the time necessary to reverse the magnetization. By convention, the switching time is defined as the time necessary to switch 90% of the irreversible flux. The switching time is not a constant but depends on the pulse shape and amplitude of the drive field. A plot of $1/\tau_s$ versus $(H - H_0)$, the drive and threshold fields, respectively yields three straight lines for low, intermediate and high values of H. Figure 45a shows this for a MgMn ferrite. The equation for each slope can be written as:

$$\tau_s = S_w/(H - H_0) \tag{16}$$

where S_w is the switching coefficient for each process. The mechanisms for these

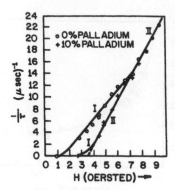

Fig. 46. Influence of 10% of palladium, as fine particles, on the switching curves of a MnMg ferrite (after Baba et al. 1963).

flux reversals are wall displacement at low fields (Goodenough 1957), a mixture of wall and rotational motion (incoherent switching) at intermediate fields and coherent or pure rotational motion at the highest fields (Gyorgy 1958 and Gyorgy and Hagedorn 1959). The switching coefficients associated with these three processes are shown in fig. 45b for different ferrite compositions. For the ferrites in question the S_w values for each of the three processes are quite similar. The values of 0.2–0.3 Oe μs for process II compares favorably with the predicted minimum of 0.2 Oe μs for incoherent spin rotation (Humphrey and Gyorgy 1959).

The mathematical treatment of domain wall motion yields a differential equation of motion for a wall (Becker 1951, Baltzer 1953) which can be reduced after certain omissions and approximations to: $v = 2(H - H_0)M_s(\cos^2 \theta)/\beta$. Here v is the domain wall velocity, β the viscous damping factor and θ the angle between the applied field and M_s. Since $v = d/\tau_s$, d being the distance moved by a wall, one derives when combining with $\tau_s = S_w/(H - H_0)$:

$$S_w = \beta d/2M_s \cos^2 \theta. \tag{17}$$

Wall relaxation is the major contributor to β since the eddy current contribution in ferrites is negligible due to their high resistivity. Satisfactory experimental support for this equation has not been forthcoming. The use of finely divided garnet particles with ferrite powder with the aim to produce more nucleating centers (decrease d) was insignificant (Peloschek 1963). Also, control of β through wall relaxation has not been fruitful since the exact mechanism is not known.

The switching time for coincident current mode can be improved by impeding wall motion (a slow process) and using incoherent switching (a fast process). This can be accomplished by increasing the porosity which will increase the coercive force without changing the threshold field for incoherent switching. Control of the properties are difficult here since firing must be stopped at a time the magnetic properties are changing rapidly. As an alternate to underfiring, a second phase of Pd or ThO_2 particles (Baba 1965, Baba et al. 1963) in a high density MnMgZn ferrite has successfully been used. With 10% Pd 1500 Å in diameter, H_c was raised from 0.9 Oe to 2.2 Oe while H_0 remained at 3.5 Oe as illustrated in fig. 46.

6.3.5. Chemistry

The patent literature is full of ferrite compositions that possess square loop properties. The reader can refer to Peloschek (1963) for a summary of the compositions and patents. The commercial materials are mostly Mn containing ferrites with spontaneous square loop such as: MnMg, MnCu, MnMgZn, MnCuZn, MnCuNi and MnLi with small quantities of Ca, Cr and other elements. The most famous of these is MnMg ferrite discovered by Albers-Schoenberg (1954). Figure 47 shows the influence of composition of this system upon the

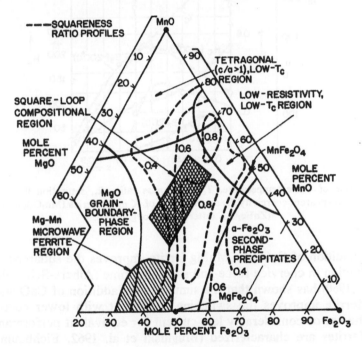

Fig. 47. Influence of composition upon squareness ratio and ferrite microstructure in the system MgO–MnO–Fe$_2$O$_3$ (after Goodenough 1957).

squareness ratio and microstructure (Goodenough 1957). According to Albers-Schoenberg (1956), the optimum squareness ratios are in the compositional band between 40 and 45 mole % Fe$_2$O$_3$ as compared to 26 to 38 mole % Fe$_2$O$_3$ (fig. 47) for the depicted square loop compositional region. Economos (1955) in an independent study determined that a composition of Mg$_{0.77}$Mn$_{0.58}$Fe$_{1.65}$O$_4$ (~38.9 mole % Fe$_2$O$_3$) had the maximum squareness. Goodenough (1957) established the fact that precipitation of MgO and α-Fe$_2$O$_3$ along with large lattice distortion associated with excess Mn are to be avoided if square loop ferrites are sought (fig. 47).

Figure 48 shows that the initial substitution of Zn in the MgMn ferrite system (Goodenough 1957) increases the squareness ratio but decreases the Curie temperature and coercive force. The latter lengthens the switching time since lower drive currents are required. By increasing the coercive force of these materials through control of smaller grain sizes and retaining the squareness, these materials are very useful for coincident current applications.

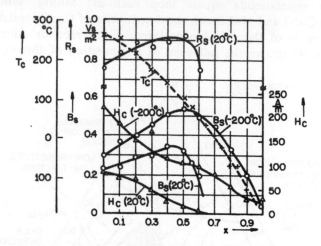

Fig. 48. Variation of Curie temperature, T_c, squareness ratio, R_s, saturation flux density B_s at $H = 2.4$ kA/m and apparent coercivity H_c as a function of X in the series $(Mn_3O_4)_{0.15}$ $(MgFe_2O_4)_{0.55-x}$ $(ZnFe_2O_4)_x$ (after Goodenough 1957).

A small addition of CaO to a MnMg ferrite improves the squareness ratio but with a reduction in coervice force and switching time (Albers-Schoenberg 1956). Eichbaum (1960) has shown that the simultaneous addition of CaO and Cr_2O_3 to a MnMg ferrite improves the switching coefficient with lower coercive force enabling the use of considerably less current for equivalent performance.

MnCd ferrites are characterized (Braginski et al. 1962, Eichbaum 1959) by moderate remanence ratios (0.8–0.9), low coercive forces and very low switching coefficients ($S_w \sim 0.2$ Oe μs). These rapid switching ferrites are not very suitable for ordinary coincident current memories but are useful in multipath operation. The addition of Mg to MnCd ferrite (Braginski et al. 1962) improves the squareness ratio while still retaining a low switching coefficient. When compared with a commercial MnMgZn ferrite (table 4), the equimolar MnMgCd ferrite offers some advantage especially when low coercive force is an essential feature. The low coercive force of Cd ferrites is due to enhanced grain growth by Cd addition (Braginski et al. 1962). However, Cd ferrites are more difficult to reproduce than those containing zinc since Cd evaporates more easily during heat treatment.

MnCu ferrites having spontaneous square loops were first discovered by Wijn

TABLE 4

Magnetic properties of square loop MnMgZn and MnMgCd ferrites

Type	B_m (G)	B_r (G)	H_c (Oe)	B_r/B_m	H_m (Oe)	T_c (°C)
MnMgZn ferrite	2500	2050	0.3	0.95	0.55	160
MnMgCd ferrite	2400	2000	0.2	0.92	0.35	150

et al. (1954) but a more detailed compositional study was performed by Weisz and Brown (1960). High squareness ratios with $S_w \sim 0.6$ Oe μs were obtained for a composition near $Mn_{1.7}Cu_{0.1}Fe_{0.12}O_4$. In table 5 a MnCu ferrite produces characteristics comparable with a conventional MgMn ferrite but using considerably less current.

TABLE 5

Comparison of pulse response of 80 mil OD memory cores of MnCu and MnMg ferrites

Type	I_f/I_d (mA)	μV_i (mV)	dV_z (mV)	τ_p (μs)	τ_s (μs)
MnCu ferrite	650/325	80	20	0.75	1.42
MnMg ferrite	800/400	100	18	0.70	1.40

Pure lithium ferrite possesses poor squareness and, because of its high Curie temperature, has high coercivity and a low TC of coercivity. With the addition of Ni, Zn or Mn (Wijn et al. 1954, West 1963 and Ohta et al. 1971) to Li ferrite the squareness ratio is improved considerably. According to Wijn et al. (1954) the squareness is improved by the compensation of negative λ_{111} of LiNi ferrite with ferrous ferrite produced by sintering at high temperature. The ferrous ferrite is undoubtedly produced by the loss of Li_2O. West (1963) produced a squareness ratio of about 0.96 for $Li_{0.455}Zn_{0.05}Ni_{0.04}Fe_{2.455}O_4$ but the switching constants H_0 and S_w were 2.15 Oe and 1.50 Oe μs respectively at 0°C. It has recently been shown (Ohta et al. 1971) that zero magnetostriction is most important for attaining squareness in LiMn ferrite and that the induced magnetic anisotropy via order–disorder of Li^{1+} and Fe^{3+} on the octahedral site is of secondary importance. Despite the fact that Li ferrites have the largest S_w, they are used exclusively for core sizes which are less than about 0.020″ OD since lithium ferrite possesses: (i) better temperature stability; (ii) higher coercive force providing for higher strength since for a given H_c lithium ferrite is more dense; and (iii) higher M_s to give a higher output.

High squareness has been found in the copper bearing ferrites of Mg, Li and Ni as a result of an investigation based on Baltzer's (1955a,b) criterion that a near zero value of the effective magnetocrystalline domain anisotropy is a necessary condition for loop squareness. Intensive investigation on the MgCu system (Greifer and Croft 1959) shows in fig. 49 that the squareness ratio is a function of the Mg/Cu ratio and the sintering condition. For the composition $Mg_{0.3}Cu_{0.7}Fe_2O_4$, a squareness ratio of 0.94 has been obtained but the switching coefficient of 3 Oe μs is much too high for practical high speed memories. For this composition, the saturation magnetostriction is minimized (-2×10^6) and the calculated λ_{111} goes through a minimum of (-3×10^6).

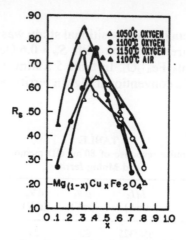

Fig. 49. Squareness ratio (R_s) as a function of x for the system $Mg_{(1-x)}Cu_xFe_2O_4$ (after Greifer and Croft 1959).

6.3.6. Processing

All the process steps from raw material through sintering outlined in section 6.1 must be closely controlled in order to develop the desired properties. Since a single spinel phase with homogeneous chemical and ceramic structure is desired (Goodenough 1957) the firing conditions (time, temperature and atmosphere) are of particular importance. The valency of Mn and Fe (Eschenfelder 1958) and their distribution over different lattice sites (and hence the properties) are determined by the firing and atmosphere cycle (McClure 1957, Goodenough and Loeb 1955). The distortion of the crystallographic structure due to the Jahn–Teller effect of Mn^{3+} and Cu^{2+} is also important because the crystal anisotropy can be disturbed (Irami et al. 1960).

The ceramic structure and the sintering behavior can be influenced mechanically by controlling the particle size of the raw material and pressing powder (Chol 1971) or chemically by adding certain ions. Elements known to enhance sintering are: Mn (Hegyi 1954), Cu and V (Blackman 1957), Zn (Palmer et al. 1957), and Cd (Eichbaum 1959). Usually these additions lower the Curie temperature and the coercive force. There are also additives such as Cr that keep the grain size small (Eichbaum 1960).

The trend toward smaller and faster cores has presented a great challenge to the ferrite core industry. Using conventional core manufacturing, particularly for smaller cores (0.022" OD and smaller), the uniform pressed density throughout the core is difficult to obtain in high volume production. A different approach (Wiechec 1968) incorporates ferrite powder into sheets with high binder content from which cores are stamped ready for firing. Excellent magnetic characteristics have resulted from this technique (Weber 1970) attributed to the improvements in uniformity of density.

References

Albers-Schoenberg, E., 1954, J. Appl. Phys. 25, 152.

Albers-Schoenberg, E., 1956, Ceramic Bull. 35, 276.

Anderson, P.W., 1950, Phys. Rev. 79, 350.

Akashi, T., 1961a, Trans. Jap. Inst. Metals 2, 171.

Akashi, T., 1961b, J. Jap. Appl. Phys. 30, 708.

Akashi, T., I. Sugano, Y. Kennoku, Y. Shima and T. Tsuiji, 1971, Proc. Int. Conf. on Ferrites, Kyoto, eds. Y. Hoshino et al. (University of Tokyo Press) pp. 183–186.

Akashi, T., I. Sugano, T. Okuda, Y. Onoda and T. Tsuji, 1972, US Patent 3 655 841.

Arai, K.I. and N. Tsuya, 1976, J. Phys. Chem. 36, 463.

Baba, P.D., 1965, J. Am. Cer. Soc. 28, 305.

Baba, P.D., E.M. Gyorgy and F.J. Schnettler, 1963, J. Appl. Phys. 34, 1125.

Baltzer, P.K., 1953, MIT Dig. Comp. Lab. Memo M-2275.

Baltzer, P.K., 1955a, MIT Lincoln Lab. Report 6R-236.

Baltzer, P.K., 1955b, Proc. Conf. Magnetism and Magnetic Materials, Pittsburgh (AIEE, USA) AIEE Spec. Publ. T-78.

Becker, R., 1951, J. Phys. Radium 12, 332.

Beer, A. and T. Schwartz, 1966, IEEE Trans. Mag. 2, 470.

Blackman, L.C.F., 1957, J. Appl. Phys. 28, 2511.

Blackman, L.C.F., 1959, J. Am. Cer. Soc. 42, 143.

Blank, J.M., 1961, J. Appl. Phys. 32, 378S.

Bozorth, R.M., 1951, Ferromagnetism (Van Nostrand Princeton NJ).

Braginski, A., 1965, Phys. Stat. Sol. 11, 603.

Braginski, A. and J. Kulikowski, 1964, Phys. Stat. Sol. 4, K9.

Braginski, A., W. Ciaston, J. Kulikowski, and S. Makolagawa, 1962, Proc. IEE 109B, 380.

Brook, R.J., 1969, J. Am. Cer. Soc. 52, 56.

Carpay, F.M.S., 1977, J. Am. Cer. Soc. 60, 82.

Chol, G., 1971, J. Am. Cer. Soc. 54, 34.

Coble, R.L. and J.E. Burke, 1964, 4th Int. Symp. on the Reactivity of Solids, Amsterdam, 1960 (Elsevier, Amsterdam) pp. 38–51.

Craik, D.J., 1975, Magnetic oxides (Wiley, London) Chs. 1, 5 and 6.

DeLau, J.G.M., 1968, Proc. Brit. Cer. Soc. 10, 275.

DeLau, J.G.M., 1970, Bull. Am. Cer. Soc. 49, 572.

DeLau, J.G.M., 1975, Philips Res. Rep. Suppl. No. 6.

DeLau, J.G.M., and A.L. Stuijts, 1966, Philips Res. Rep. 21, 104.

Dixon, M., T.S. Stakelon and R.C. Sundahl, 1977, IEEE Trans. Mag. MAG-13, 1351.

Dunitz, J.D. and L.E. Orgel, 1957, J. Phys. Chem. Solids 3, 318.

Eckert, O., 1957, Proc. IEE 104B, 428.

Economos, G., 1955, J. Am. Cer. Soc. 38, 408.

Eichbaum, B.R., 1959, J. Appl. Phys. 30, 49S.

Eichbaum, B.R., 1960, J. Appl. Phys. 31, 1175.

Elbinger, G., 1961, Naturwissenschaften 48, 498.

Eschenfelder, A.H., 1958, J. Appl. Phys. 29, 378.

Foniok, F. and S. Makolagawa, 1977, J. Mag. and Mag. Mat. 4, 95.

Gans, R., 1932, Ann. der Phys. 15, 28.

Globus, A., 1975, Proc. 2nd EPS Conf. on Soft Magnetic Materials (Wolfson Centre for Magnetics Technology, Cardiff, Wales).

Globus, A. and Duplex, P., 1966, IEEE Trans. Mag. 2, 441.

Goodenough, J.B., 1954, Phys. Rev. 95, 917.

Goodenough, J.B., 1957, Proc. IEE 104B, 400.

Goodenough, J.B. and A.L. Loeb, 1955, Phys. Rev. 98, 391.

Gorter, E.W., 1954, Philips Res. Rep. 9, 295.

Gorter, E.W. and C.J. Elsveldt, 1957, Proc. IEE **B104**, 418.

Greifer, A.P., 1969, IEEE Trans. Mag., **5**, 774.

Greifer, A.P. and W.J. Croft, 1959, J. Appl. Phys. **30**, 34S.

Guillaud, C., 1951, J. Phys. Rad. **12**, 239.

Guillaud, C., 1956, Compt. Rend. **242**, 2712.

Guillaud, C., 1957, Proc. IEE **104B**, 165.

Guillaud, C. and H. Crevaux, 1950, Compt. Rend. **230**, 1458.

Guillaud, C. and M. Roux, 1949, Compt. Rend. **229**, 1133.

Guillaud, C. and Paulus, 1956, Compt. Rend., **242**, 2525.

Gillaud, C., C. Villers, A. Maraid, and M. Paulus, 1960, Solid State Physics, Vol. 3, eds. M. Desirant and J.L. Michiels (Academic Pre, New York) pp. 71–90.

Gyorgy, E.M., 1958, J. Appl. Phys. **29**, 283.

Gyorgy, E.M. and F.B. Hagedorn, 1959, J. Appl. Phys., **30**, 1368.

Harvey, R.L., J.I. Hegyi and H.W. Leverenz, 1950, RCA Rev. **11**, 321.

Heck, C., 1974, Magnetic materials and their applications, (Crane and Russak, New York).

Hegyi, I.J., 1954, J. Appl. Phys. **25**, 176.

Humphrey, F.B. and E.M. Gyorgy, 1959, J. Appl. Phys. **36**, 935.

Iida, S., 1960, J. Appl. Phys., **31**, 215S.

Iida, S. and T. Inoue, 1962, J. Phys. Soc. Japan **17**, 281.

Iida, S., Y. Aizama and H. Sekigawa, 1958, J. Phys. Soc. Japan **13**, 58.

Ikeda, A., M. Satomi, H. Chiba and E. Hirota, 1971, Proc. Int. Conf. on Ferrites, Kyoto, 1970, eds. Y. Hoshino et al. (University of Tokyo Press) pp. 337–339.

Irami, K.S., A.P.B. Sinha and B. Biswassa, 1960, J. Phys. Chem. Solids **17**, 101.

Jordan, H., 1924, El. Nach. T., **1**, 7.

Knowles, J.E., 1970, J. Phys. **3**, 1346.

Knowles, J.E., 1974, Philips Res. Rep. **29**, 93.

Knowles, J.E., 1977, J. de Phys. **38**, C1–27.

König, U., 1974, Appl. Phys. **4**, 237.

Kornetzki, M. and J. Brockmann, 1958, Z. Siemens **32**, 412.

Kornetzki, M., J. Brockmann and J. Frey, 1955, Z. Siemens **29**, 434.

Krupićka, S. and F. Vilim, 1961, Czech. J. Phys. **B11**, 10.

Legg, V.E., 1936, Bell Syst. Tech. J. **16**, 39.

Lescroel, Y., 1953, Cables and Transm. **7**, 273.

Lescroel, Y. and A. Pierrot, 1960, Cables and Transm. **14**, 220.

Löbl, H., P. Neusser, M. Zenger and J. Frey, 1977, J. de Phys. **38**, C1–345.

Marais, A., 1965, Compt. Rend. **261**, 2188.

Marais, A. and T. Merceron, 1963a, Compt. Rend. **256**, 2560.

Marais, A. and T. Merceron, 1963b, Compt. Rend. **257**, 1760.

Matsubara, T., J. Kuwai, I. Sugano and T. Akashi, 1971, Proc. Int. Conf. on Ferrites, Kyoto, 1970, eds. Y. Hoshino et al. (University of Tokyo Press) pp. 214–216.

McClure, D.S., 1957, J. Phys. Chem. Solids **3**, 311.

Miyata, N., 1961, J. Phys. Soc. Japan **16**, 206.

Miyata, N. and Z. Funatogawa, 1962, J. Phys. Soc. Japan **17B**, 279.

Mizushima, M., 1964, Japan J. Appl. Phys. **3**, 82.

Morineau, R. and M. Paulus, 1975, IEEE Trans. Mag. **11**, 1312.

Natansohn, S. and P.H. Baird, 1968, J. Am. Cer. Soc. **51**, 533.

Néel, L., 1948, Ann. de Phys. **3**, 137.

Ogawa, S. and Y. Nakagawa, 1967, J. Phys. Soc. Japan **23**, 179.

Ohta, K., 1960, Bull. Kobayasi Inst. Phys. Res. **10**, 149.

Ohta, K., 1963, J. Phys. Soc. Japan **18**, 685.

Ohta, K. and Y. Nagata, 1971, Proc. Int. Conf. on Ferrites, Kyoto, 1970, eds. Y. Hoshino et al. (University of Tokyo Press) pp. 27–29.

Ohta, K., K. Shinozaki and J. Ezaki, 1971, Proc. Int. Conf. on Ferrites, Kyoto, 1970, eds. Y. Hoshino et al. (University of Tokyo Press) pp. 450–453.

Oudemans, G.J., 1968, Philips Tech. Rev. **29**, 45.

Owens, C.D., 1956, Proc. IRE **44**, 1234.

Palmer, C.G., R.W. Johnston and R.E. Schultz, 1957, J. Am. Cer. Soc. **40**, 256.

Paulus, M., 1962, Phys. Stat. Sol. **2**, 1325.

Paulus, M., 1971, Proc. Int. Conf. on Ferrites, Kyoto, 1970, eds. Y. Hoshino et al. (University of Tokyo Press) pp. 114–120.

Pauthenet, R., 1952, Ann. Phys. **7**, 710.

Peloschek, H.P., 1963, Progress in dielectrics, Vol. 5, eds. J.B. Birks and J. Hart (Academic Press, New York) pp. 38–93.

Penoyer, R.F. and M.W. Shafer, 1959, J. Appl. Phys. **30**, 3115.

Perduijn, D.J. and H.P. Peloschek, 1968, Proc.

Brit. Cer. Soc. 10, 263.

Petrescu, V. and G. Maxim, 1968, Rev. Roum. Phys. 14, 17.

Quartly, C.J., 1962, Square loop ferrite circuitry (Iliffe Books/Prentice Hall, London) p. 14.

Reijnen, P., 1967, Science of ceramics, Vol. 3, ed. G.H. Stewart (Academic Press, London) pp. 245–261.

Reijnen, P.J.L., 1968, Science of Ceramics, Vol. 4, ed. G.H. Stewart (Academic Press, London) pp. 169–188.

Roess, E., 1966, Z. Angew. Phys. 21, 391.

Roess, E., 1971, Proc. Int. Conf. on Ferrites, Kyoto, 1970, eds. Y. Hoshino et al. (University of Tokyo Press) pp. 187–190.

Roess, E., 1977, J. Mag. and Mag. Mat. 4, 86.

Roess, E. and I. Hanke, 1970, Phys. Stat. Sol. 2, 185.

Roess, E., I. Hanke and E. Moser, 1964, Z. Angew. Phys. 7, 504.

Ruthner, M.J., 1977, J. de Phys. 38, C1–311.

Ruthner, M.J., H.G. Richter and I.L. Steiner, 1971, Proc. Int. Conf. on Ferrites, Kyoto, 1970, eds. Y. Hoshino et al. (University of Tokyo Press) pp. 75–80.

Schnettler, F.J. and D.W. Johnson, 1971, Proc. Int. Conf. on Ferrites, Kyoto, 1970, eds. Y. Hoshino et al. (University of Tokyo Press) pp. 121–124.

Schwabe, E.A. and D.A. Cambell, 1963, J. Appl. Phys. 34, 1251.

Sibelle, R., 1974, Rev. Phys. Appl. 9, 837.

Shevel, W.L., 1959, J. Appl. Phys. 30, 475.

Shichijo, Y. and E. Takama, 1971, Proc. Int. Conf. on Ferrites, Kyoto, 1970, eds. Y. Hoshino et al. (University of Tokyo Press) pp. 210–213.

Shichijo, Y., N. Tsuga and K. Suzuki, 1961, J. Appl. Phys. 32, 3865.

Shichijo, Y., G. Asano and E. Takama, 1964, J. Appl. Phys. 35, 1946.

Slick, P.I., 1971a, US Patent 3 609 083.

Slick, P.I., 1971b, Proc. Int. Conf. on Ferrites, Kyoto, 1970, eds. Y. Hoshino et al. (University of Tokyo Press) pp. 81–83.

Slonczewski, J.C., 1961, J. Appl. Phys. 32, 253S.

Smit, J. and H.P.J. Wijn, 1954, Adv. Electr. and electr. phys. 6, 69.

Smit, J. and H.P.J. Wijn, 1959, Ferrites (Wiley, New York).

Smit, J., F.K. Lotgering, and R.P. Van Staple, 1962, J. Phys. Soc. Japan 17B, 268.

Snelling, E.C., 1962, Proc. IEE 109B, 234.

Snelling, E.C., 1964, Proc. Brit., Cer. Soc. 2, 151.

Snelling, E.C., 1969, Soft ferrites (Iliffe Books, London).

Snelling, E.C., 1972, IEEE Spectrum, 26 and 42.

Snelling, E.C., 1974, IEEE Trans. Mag. 10, 616.

Snoek, J.L., 1947, New developments in ferromagnetic materials (Elsevier, New York).

Snoek, J.L., 1948, Physica 14, 207.

Stuijts, A.L., J. Verweel and H.P. Peloschek, 1964, IEEE Trans. Commun. Electron. 83, 726.

Takada, T. and M. Kiyama, 1971, Proc. Int. Conf. on Ferrites, Kyoto, 1970, eds. Y. Hoshino et al. (University of Tokyo Press) pp. 69–71.

Tebble, R.S. and D.J. Craik, 1969, Magnetic materials (Wiley-Interscience, London) Ch. 7, 14 and 15.

US Patent 2 877 183.

Usami, S., N. Miyata and Z. Funatogawa, 1961, J. Phys. Soc. Japan 16, 2064.

Van Uitert, L.G., 1955, J. Chem. Phys. 23, 1883.

Van Groenou, A.B. and J.A. Schulkes, 1967, J. Appl. Phys. 38, 1133.

Van Groenou, A.B., F.P. Bonger and A.L. Stuijts, 1969, Mat. Sci. Eng. 3, 317.

Verweel, J., 1964, Philips Res. Rep. 19, 29.

Verwey, E.J.W. and E.L. Heilman, 1947, J. Chem. Phys. 15, 174.

Verwey, E.J.W., F. Du Boer and J.N. Van Santon, 1948, J. Chem. Phys. 16, 1091.

Von Aulock, W.H., 1965, Handbook of microwave ferrite materials (Academic Press, New York).

Wagner, C., 1935, Z. Tech. Phys. 16, 327.

Weber, G.H., 1970, IEEE Trans. Mag. 6, 533.

Weisz, R.S. and D.L. Brown, 1960, J. Appl. Phys. 31, 269S.

West, R.G., 1963, J. Appl. Phys. 34, 1113.

Wiechec, W., 1968, IEEE Trans. Mag. 4, 465.

Wijn, H.P.J., E.W. Gorter, C.J. Esveldt and P. Gelderman, 1954, Philips T. Tdsah 16, 124.

Wijn, H.P.J. and H. Van der Heide, 1957, Proc. IEE B104, 422.

Zener, C., 1948 [quoted in: C.S. Smith, 1940, Trans. AIME 175, 15].

chapter 4

MICROWAVE FERRITES

J. NICOLAS

Thomson-CSF
Laboratoire Central de Recherches
91401 Orsay
France

Ferromagnetic Materials, Vol. 2
Edited by E.P. Wohlfarth
© North-Holland Publishing Company, 1980

CONTENTS

1. Introduction*

1.1. What is a microwave ferrite?

A microwave ferrite can be defined simply as a high-resistivity, magnetic material used at frequencies comprised between 100 MHz and the highest microwave frequencies now in use, i.e. 500 GHz. In this frequency range, two categories of ferrite devices have been designed, namely:

The first category comprises non-reciprocal devices for which ferrites are almost irreplaceable and which have the greatest number of applications. These devices are essentially isolators or circulators which perform the following functions: (a) An isolator is merely a waveguide section in which the incident electromagnetic wave (from input to output) can propagate without any substantial attenuation, while the reverse waves are highly attenuated (from output to input). (b) An n-port circulator is a device having n inputs or outputs and operating as follows (fig. 1). An electromagnetic wave incident at port 1 emerges at port 2. A wave incident at port 2 emerges at port 3, and so on. A wave incident at port n emerges at port 1. Other propagation paths are inhibited. The devices of this type provide easy isolation between the different channels of a microwave circuit, which is essential, especially in the present techniques for radar and radio links.

The second category comprises reciprocal devices, e.g. electrically controlled phase-shifters. These circuits can use components other than ferrites, such as semiconductor diodes.

The above definition is perhaps too general. The ferrites of which the scalar permeability between 100 and 500 MHz is used, for instance in inductors, are considered as soft ferrites (for non-microwave applications) which are dealt with by Slick in chapter 3 of this volume. Thus, in practice, the boundary between the domains of soft ferrites and the microwave ferrites is rather vague when considering devices which are reciprocal to within a few hundred MHz. Nevertheless, as will be shown further on, the approach to the soft-ferrite and microwave-ferrite domains is extremely different.

* The MKSA International System of Units, as recommended by the International Electrotechnical Commission is used throughout this article. In this unit system, magnetic induction is written as $B = \mu_0 (H + M)$.

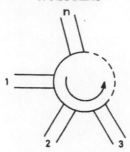

Fig. 1. n-port circulator.

As regards the history of microwave ferrite, the first microwave device using ferrites was described by Hogan from Bell Labs in 1952. This was a circulator based on the Faraday rotation effect, working at about 9 GHz.

All the applications of ferrites to microwaves since then have used the same basic principle. In all cases, the technique in microwave application uses a ferrite structure whose permeability is controlled by the gyromagnetic phenomenon. As is shown later, this permeability depends, firstly, on the values of magnetization and static applied magnetic field (basis for electrically-controlled phase-shifters) and, secondly, on the polarization of the electromagnetic waves relative to the direction of magnetization (basis for non-reciprocal devices).

Theory and general data on microwave ferrites can be found in the books of Lax and Button (1962), Von Aulock (1965), Sparks (1964), Pircher (1969), Landolt and Börnstein (1970), and in the papers of Hudson (1970), Deschamps (1970), Schlömann (1971), Nicolas et al. (1973), Nicolas and Désormière (1977), and Bolle and Whicker (1975) give an annotated survey of the literature of microwave ferrite materials and devices.

1.2. Basic physical principles

1.2.1. General
Let us consider a magnetic material assumed to be demagnetized and not subjected to a static magnetic field but only to an alternating field at a frequency f. When this frequency increases the permeability resulting from the displacement of the domain walls eventually disappears and that due to rotations of magnetization also vanishes. This is the normal condition for the ferrites that we will consider at microwave frequencies. Their permeability, in the conditions described above, only approximates that of the air. However, they can be rendered magnetically active through application of a static magnetic field, owing to a magnetic resonance effect whose principle and features will now be briefly reviewed.

1.2.2. Magnetic resonance
Let us consider an ellipsoid with axes Ox, Oy, Oz, consisting of a magnetic material (fig. 2). This ellipsoid is located in a static magnetic field H, uniform

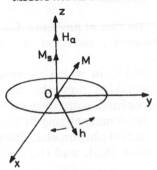

Fig. 2. Small ellipsoidal sample of magnetic material in a static magnetic field H and a microwave magnetic field h.

before introduction of the material, and with direction Oz. The field in the material H_{int} is the sum of H and of the internal demagnetizing field, which is uniform

$$H_{int} = H - N_z M_s \tag{1}$$

where M_s is saturation magnetization, and N_z is the demagnetizing factor in the direction Oz.

It is assumed that this static field is high enough to saturate the material whose magnetization M_s is oriented in the direction Oz. In addition, a microwave field, h, is applied perpendicular to Oz (hence located in the plane Ox, Oy) and with an angular velocity $\omega = 2\pi f$. It is assumed that the corresponding wavelength is very high compared to the length of the ellipsoid axes, so that the field is uniform throughout the sample volume. The corresponding field in the material is called h_{int} and is assumed to have a sufficiently low intensity with respect to H_{int}.

The magnetization vector M deviates slightly from the Oz axis and has x and y components, M_x and M_y. This gives

$$M = \chi_{int}(H_{int} + h_{int}) \tag{2}$$

in which χ_{int} is the magnetic susceptibility of the material.

This susceptibility can be explained by analyzing the movement of vector M. This may be achieved by applying to the magnetic moment carriers, e.g. the electrons, the kinetic moment theory, which yields

$$\frac{1}{\gamma}\frac{dM}{dt} = \mu_0 M \times (H_{int} + h_{int}) + \text{damping term} \tag{3}$$

in which γ is the gyromagnetic ratio, i.e. the ratio of the kinetic moment to the magnetic moment: $\gamma\mu_0 = 1.105 \times 10^5 g_{eff}$ (m/A s), where g_{eff} is the effective Landé factor or effective gyromagnetic factor, equal to 2 for the free electron, and μ_0 is the permeability of vacuum ($4\pi \times 10^{-7}$ H/m).

Different formulae have been proposed for a phenomenological description of damping, as explained for instance by Bloembergen (1956). It is now recognized that damping involves spin waves such that the elementary magnetic moments do not remain parallel to one another at any time. Consequently, eq. (3) is only a first approach which is, however, sufficient to enable the essence of the phenomena to be interpreted.

The solution of eqs. (2) and (3), in which an appropriate damping term has been chosen, is effected by taking only the time-related components into account. A susceptibility χ, defined with respect to the applied field h, and related to χ_{int}, allowing for the demagnetizing field, is considered because it is involved more directly in the actual phenomena. Thus, the components h_x and h_y of this field are related to those $(h_{int})_x$ and $(h_{int})_y$ of the internal field through formulae analogous to (1)

$$(h_{int})_x = h_x - N_x M_x \qquad (h_{int})_y = h_y - N_y M_y \qquad (4)$$

in which N_x and N_y are the x and y demagnetization factors.

1.2.3. Polder tensor

According to Polder and Wills (1949), when the material is magnetized to saturation in the Oz direction, the alternating components of the magnetization vector, M_x, M_y, M_z, are related to the components h_x, h_y, h_z of the periodic field through the formula

$$\begin{vmatrix} M_x \\ M_y \\ M_z \end{vmatrix} = \begin{vmatrix} \chi_{xx} & \chi_{xy} & 0 \\ \chi_{yx} & \chi_{yy} & 0 \\ 0 & 0 & 0 \end{vmatrix} \begin{vmatrix} h_x \\ h_y \\ h_z \end{vmatrix}. \qquad (5)$$

The susceptibility matrix (which has tensor properties) is called the Polder tensor. When the material has the form of an ellipsoid of revolution above the Oz axis, $\chi_{xy} = -\chi_{yx}$; the eigenvectors of the matrix are aligned with Oz and with the vectors rotating at the angular velocity of the microwave magnetic field, in the positive and negative senses of the plane xOy. Thus, calling h_+ and h_- the components of the microwave magnetic field rotating in the plane xOy, with

$$h_+ = h_x + jh_y \qquad \text{and} \qquad h_- = h_x - jh_y$$

the following can be written

$$M_+ = \chi_+ h_+ \qquad M_- = \chi_- h_-. \qquad (6)$$

In order to allow for the phase-shift between the magnetization and the field, χ_+ and χ_- are complex quantities

$$\chi_+ = \chi'_+ - j\chi''_+ \qquad \chi_- = \chi'_- - j\chi''_-. \qquad (7)$$

The corresponding permeabilities are also defined

$$\mu_+ = 1 + \chi_+ \qquad \mu_- = 1 + \chi_-. \qquad (8)$$

1.2.4. Characteristics at resonance

Figure 3 shows the curve of the observed variations of μ'_+, μ'_-, μ''_+, μ''_- versus the applied field, H, at the fixed frequency f. In this figure, the positive values of H

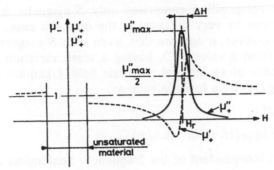

Fig. 3. Real and imaginary parts of permeability for circular polarization (μ_+ and μ_-) versus static magnetic field H, μ_+' and μ_+'' are ploted at $H > 0$, μ_-' is ploted at $H < 0$, μ_-'' equal practically zero at any field.

correspond to the positive polarization for h, and the negative values of H to the negative polarization for h. This results from the fact that a reversal of the field H amounts to a reversal of the direction of polarization of the circularly polarized wave in the plane perpendicular to H. The curves of μ_+' and μ_+'' have a shape which is characteristic of a resonance phenomenon whose main features are: (i) the resonance field H_r and (ii) the line width ΔH.

(i) The resonance field, H_r, which corresponds to a maximum value of μ_+'' is expressed by the Kittel (1948) formula (for any ellipsoid)

$$\omega = \gamma\mu_0\{[H_r - (N_z - N_x)M_s][H_r - (N_z - N_y)M_s]\}^{1/2} \qquad (9)$$

where N_x, N_y, and N_z are the demagnetization factors of the ellipsoid in the directions Ox, Oy, Oz ($N_x + N_y + N_z = 1$). If the ellipsoid is an ellipsoid of revolution about Oz ($N_x = N_y = N_t$), the resonance condition becomes

$$\omega = \gamma\mu_0[H_r - (N_z - N_t)M_s] \qquad (10)$$

and, in the case of a sphere, in the same way as in an infinite medium ($N_x = N_y = N_z$)

$$\omega = \gamma\mu_0 H_r. \qquad (11)$$

The formula (11) can also be written for specific units

$$f \text{ (MHz)} = 17.6\, g_{\text{eff}} H_r \text{ (kA/m)} = 1.4\, g_{\text{eff}} H_r \text{ (Oe)}$$

where g_{eff} is the effective gyromagnetic factor.

(ii) The linewidth ΔH is the mid-height width of the curve μ_+'' versus H.

Note: When the field H intensity is too low (less than a few tens of kA/m), the material cannot be magnetically saturated. The phenomena become more complex and no longer exactly follow the simple theory mentioned above. This area has been excluded from fig. 3.

In fact, g_{eff} closely approximates the value for the free electron condition

($g_{eff} = 2$) when the composition comprises only S-magnetic ions ($L = 0$); e.g. $Fe^{3+}Mn^{2+}$, Gd^{3+}. It can be very different in the opposite case, as will be seen. Strictly speaking, however, it appears that, even with S-magnetic ions only, g_{eff} can differ slightly from a value of 2, having a small variation with frequency, owing to the presence of an internal magnetic field. Okamura et al. (1952) has suggested replacing (11) by a formula such as

$$\omega = \gamma' \mu_0 (H_r + H_i) \tag{12}$$

or $f(MHz) = 17.6 g'_{eff} (H_r + H_i)$ (kA/m)

γ' and g'_{eff} are then independent of the frequency. Schlömann expressed H_i by the formula

$$H_i = -\tfrac{1}{2}(K_1/\mu_0 M_s) + \tfrac{1}{3} p M_s \tag{13}$$

in which K_1 is the 1st order anisotropy constant and p is the material porosity.

1.2.5. Classical expressions of χ_+ and χ_-

The phenomena illustrated in fig. 3 may be approximately represented by a Lorentzian solution of eq. (3), i.e. for a spherical sample and at a fixed frequency

$$\chi_+ = \frac{M_s}{H - H_r + j(\Delta H/2)} \qquad \chi_- = \frac{M_s}{H + H_r + j(\Delta H/2)} \tag{14}$$

and, separating the real and imaginary parts

$$\chi'_+ = M_s \frac{H - H_r}{(H - H_r)^2 + (\Delta H/2)^2} \qquad \chi'_- = M_s \frac{H + H_r}{(H + H_r)^2 + (\Delta H/2)^2}$$

$$\chi''_+ = M_s \frac{\Delta H/2}{(H - H_r)^2 + (\Delta H/2)^2} \qquad \chi''_- = M_s \frac{\Delta H/2}{(H + H_r)^2 + (\Delta H/2)^2}.$$

Similar formulae have also been given for an ellipsoid of revolution, whose axis is parallel with the static field. In these formulae, H is considered as always positive.

1.3. Characteristics related to the use of polycrystalline ferrites at microwave frequencies

1.3.1. General

In order to achieve the effect sought in a microwave ferrite device an electromagnetic wave and a piece of ferrite material interact. The propagation of electromagnetic waves is fully defined at a given frequency by:

Maxwell's equations.

Dielectric and magnetic properties of the materials involved.

Conditions at the limits resulting from the geometry of the system.

Only the properties of the material will be considered. The properties more particularly involved in the problem are:

The complex relative dielectric constant: $\epsilon = \epsilon' - j\epsilon''$. The dielectric losses, resulting from term ϵ'', (or the loss angle tangent, $\tan \delta_\epsilon = \epsilon''/\epsilon'$) have to be considered more particularly, for they constitute a spurious effect, resulting in direct insertion losses in the equipment, which have to be limited to the lowest possible level.

The magnetic characteristics defined by the elements in the Polder tensor are complex permeabilities expressed by the formula (8)

$$\mu_+ = \mu'_+ - j\mu''_+ \quad \text{and} \quad \mu_- = \mu'_- - j\mu''_-.$$

The terms μ''_+ and μ''_- are also the cause of unwanted magnetic losses. It is important to know how these losses vary with the static magnetic field H and also with the microwave field h (non-linear effects).

Magnetization is a factor in all the magnetic susceptibility terms [eq. (14)] and therefore determines the material efficiency. It also "sets" the value of the internal dipolar fields in the natural resonance effect (in the demagnetized state), which is of prime importance in the choice of a material for a given application. These various points are described in detail below.

1.3.2. Dielectric losses

Dielectric losses in microwave ferrites basically result from the existence of iron in two states of valence: trivalent and divalent. This results in an excess of electrons which may jump from one Fe ion to another and therefore cause some conduction and dielectric losses. This explains the necessity of removing any trace of divalent Fe ion from the composition. For this purpose, it is generally necessary to obtain a pure, homogeneous material with a single phase having exactly the proportion of ions given by the theoretical chemical formula. It is a technological problem inherent with each fabricated material. The present state of technology allows the fabrication of ferrites with (depending on the exact composition) dielectric loss angles ($\tan \delta_\epsilon$) at 10 GHz, of 10^{-4} and slightly above.

1.3.3. Magnetic losses at a low microwave power level ΔH_{eff}

As already mentioned, gyromagnetic phenomena are expressed, to a first approximation, by a Lorentzian equation [eq. (14)] and the curves of fig. 3. However, things are in fact more complicated.

For a given field H, an effective linewidth ΔH_{eff} may be defined by starting from the Lorentzian equation [similar to (14)]. This yields

$$\Delta H_{\text{eff}} = 2M_s \, \text{Im}(1/\chi_\pm) \tag{15}$$

where $\text{Im}(1/\chi_\pm)$ is the imaginary part of $1/\chi_\pm$. Experience has shown that ΔH_{eff} becomes independent of the field H (i.e., the effective relationship is a Lorentzian equation), far from resonance, especially for large fields which saturate the material above resonance. That has been well established independently by Patton (1969) and Vrehen (1969). Conversely, near the resonance field, the linewidth ΔH is broadened by effects studied in detail by Schlömann (1956, 1958, 1971). He gave an expression of ΔH which shows the principal causes of such broadening:

$$\Delta H = \Delta H_{int} + \Delta H_a + \Delta H_p \tag{16}$$

where ΔH_{int} is the intrinsic linewidth which one can consider as not essentially different from ΔH_{eff}.

ΔH_a is the broadening due to the magnetocrystalline anisotropy. According to Schlömann, when the saturation magnetization is high compared with the magnetocrystalline anisotropy (which usually applies in the case of microwave ferrites), the following line broadening can be admitted for a polycrystal

$$\Delta H_a = \frac{8\pi\sqrt{3}}{21}\frac{1}{M_s}\left(\frac{K_1}{\mu_0 M_s}\right)^2 G(\alpha) \qquad \Delta H_a = \frac{2.07}{M_s}\left(\frac{K_1}{\mu_0 M_s}\right)^2 G(\alpha) \tag{17}$$

where $\alpha = \omega/\omega_M$, $\omega_M = \gamma\mu_0 M_s$, and $G(\alpha)$ is given fig. 4.

ΔH_p is the broadening due to a porosity, p, (resulting from the existence, in the vicinity of the pores, of local demagnetizing fields). According to Schlömann this is expressed by

$$\Delta H_p = \beta p M_s \tag{18}$$

where $\beta = 8/\pi\sqrt{3} = 1.47$ for a perfectly magnetically saturated material.

The concepts of effective linewidth, ΔH_{eff}, and linewidth of main mode, ΔH, are of great interest from a practical standpoint. This may be shown, for instance, by considering a non-reciprocal device operating outside the resonance region, which is the most current situation. In this device, assume a static field H to be applied such that magnetic losses are low, i.e. μ'' very small, and μ'_+ and μ'_- sufficiently different (which is the basis of the non-reciprocity effect). A small linewidth ΔH_{eff} is the direct condition required for low magnetic losses, since μ'' is directly proportional to ΔH_{eff}, far from the resonance. A small linewidth ΔH also allows the choice of a field H best meeting the second condition (μ'_+ and μ'_- sufficiently different). As a matter of fact, a low ΔH facilitates the design of a device for a given performance, all other conditions remaining unchanged.

To obtain a low ΔH_{eff}, it is necessary to reduce the damping of the movement of magnetic moment carriers as much as possible. For this purpose, the relaxing magnetic ions, i.e. those which, owing to their electronic configuration, enable the magnetic energy of the electronic spins to be passed to the lattice in the form

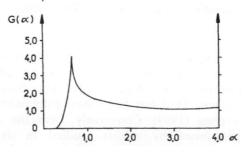

Fig. 4. $G(\alpha)$ versus $\alpha = \omega/\omega_M$ to calculate the broadening of resonance linewidth due to magnetocrystalline anisotropy for a polycrystalline sample (after Schlömann).

of heat (lattice vibrations). The spin–orbit coupling often plays an important part in this energy degradation process. In practice, a low ΔH_{eff} can be obtained simply by using only trivalent iron ions with very low relaxation as magnetic ions, because their electronic configuration, i.e. half-full outer shell 3d (hence 5 electron spins and no orbital moment) does away with spin–orbit coupling. It is true that ferrites that have only trivalent Fe ions as magnetic ions (e.g. the yttrium–iron garnet $Y_3Fe_5O_{12}$ or the lithium ferrite $Li_{1/2}Fe_{5/2}O_4$) have very low ΔH_{eff} (about 0.3 kA/m, 3 Oe). However, this property is not always compatible with other required characteristics, especially a good stability with temperature or good behaviour at high power. Consequently, in practice, compromises have to be found, as will be discussed further on.

According to eq. (16), a low ΔH can be obtained when ΔH_{eff} is low. Porosity also has to be reduced and this is a technological problem. Lastly, the magnetocrystalline anisotropy has to be low, which effectively determines the choice of the crystallographic structure and chemical formula.

Formula (16) does not take into account the surface roughness which plays a part similar to that of porosity in the broadening of the linewidth. In fact, this term may be appreciable only for very small samples used for measurement of ΔH. Therefore, the small linewidths $[\Delta H < 1.6 \, \text{kA/m} \, (20 \, \text{Oe})]$ have to be measured on small, perfectly polished spheres (diameter $\simeq 1$ mm).

1.3.4. Magnetic loss at a high microwave power level ΔH_k

All the foregoing statements apply when the magnetic field at microwave frequencies is sufficiently low. Above a given critical field, non-linear phenomena appear and result in additional magnetic losses which rapidly become prohibitive in practical devices. These non-linear effects are both first-order (which usually occur at fields H lower than the resonance field H_r), and second-order (which alter the resonance itself). The first-order effects have a lower threshold than the second-order effects and occur in a static magnetic field area which is often used in applications. These are the most disturbing effects in practice and will be dealt with in more detail in the following.

Non-linear effects are due, according to Suhl (1956), to the fact that electronic spins, the magnetic-moment carriers, do not remain parallel to one another in their movements, and that spin waves are produced. The critical microwave field h_c at which such effects appear is related to the applied static field H, as shown in fig. 5. After Suhl, theoretical and experimental work was done by several authors, for instance: Kittel (1959), De Gennes et al. (1959), Schlömann et al. (1960), Schlömann and Joseph (1961).

The critical field h_c passes through a minimum for a subsidiary field H_{sub} and tends toward a very high value for a limiting field H_{lim}. Beyond this limiting field, the first-order non-linear effects no longer exist. The Suhl theory is in agreement with experimental results. The existence of spin waves is related to damping of the spin movements. Spin waves are produced with more ease when the damping of this movement decreases. To express this damping, a spin wave linewidth ΔH_k is introduced, and this yields, for a spherical sample

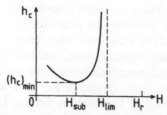

Fig. 5. Minimum critical signal field h_c to cause first-order non-linear effects (subsidiary resonance) as a function of static magnetic field H.

$$h_{c\,(min)} = \Delta H_k \frac{1 - (\omega_0/\omega)}{1 - (\omega_M/2\omega) + [1 + (\omega_M/2\omega)^2]^{1/2}} \frac{2\omega}{\omega_M} \qquad (19)$$

where $\omega_0 = \mu_0 \gamma H$, $\omega_M = \mu_0 \gamma M_s$, and h_c, in this formula, is the modulus of the field vector rotating in the plane xOy (circular polarization). In devices, such as circulators, using ferrite slabs, the critical field $h_{c(min)}$ is currently expressed by the formula

$$h_{c(min)} = 0.5 \Delta H_k (\omega/\omega_M). \qquad (20)$$

The limiting field is then

$$H_{lim} = (\omega/2\mu_0 \gamma) + N_z M_s. \qquad (21)$$

1.3.5. Role of magnetization

The role of magnetization is fundamental. In the first place, it is a factor in all the terms of the magnetic susceptibility, as shown by eq. (14). Magnetization therefore directly determines the efficiency of the material. Stability with temperature is also related directly to the variation of magnetization with temperature. One might think that to obtain the highest possible susceptibility term, i.e. the best efficiency of the material, it would be necessary in all cases to choose the highest possible magnetization. However, things are not as simple as that, owing to the natural resonance effect. This effect results from the following: In a demagnetized material, there are internal fields whose origin may be either the magnetocrystalline anisotropy, or the magnetic dipole interactions. The so-called "natural" resonance phenomenon exists in a material when the frequency of the alternating field meets the resonance condition (11), in which case H represents an internal field. In a material, the internal field varies from one point to another with a wide range of values. It follows that the natural resonance covers a wide range of frequencies. It is important to know the highest frequency at which the magnetic loss due to this resonance phenomenon exists. In the materials studied (excepting the hexagonal ferrites), the equivalent anisotropy field is most of the time low compared to the dipolar fields, related to the arbitrary orientation of crystallites with respect to one another, and to the effect of the magnetic domain walls. The highest value of these dipolar fields is M_s. Therefore, magnetic losses will occur in the material up to a frequency limit corresponding to a resonance field H_r, equal to or lower than M_s. This frequency limit is expressed by the formula

$$\omega_\ell = 2\pi f_\ell = \gamma \mu_0 M_s. \tag{22}$$

In view of the foregoing, the curve of μ_+'' can be considered with respect to the applied field H at a fixed frequency f for a given material, several cases will be considered:

(i) $f > f_\ell$: magnetic losses exist only near the resonance field H_r (between fields H_2 and H_3), as shown in fig. 6a.

(ii) $f \lesssim f_\ell$: the phenomena vary as shown in fig. 6b. There are losses at the low fields, which disappear for a field H_1. The losses due to the resonance proper reappear for a field H_2. The field interval $(H_1 H_2)$ for which losses are negligible decreases with frequency. This results from the fact that the zero-field loss increases when the frequency decreases and the resonance field H_r decreases with frequency.

(iii) $f \ll f_\ell$: if the frequency is sufficiently low, the interval $(H_1 H_2)$ during which losses are negligible, completely disappears and the losses at low field values overlap those due to resonance itself, as shown in fig. 6c. Magnetic losses become negligible above the field H_3.

The consequences of the foregoing in the choice of a material for the production of a given device, will now be assessed. As stated above, most devices use a static field such that magnetic losses are negligible. When feasible, a field lower than the resonance field is adopted in order to reduce the cost of the magnets. Thus, it is necessary to operate in condition (i) or, at least in condition (ii): if the operating frequency of the device is f, M_s must not exceed, according to eq. (22), the following value

$$M_s^{\max} = 2\pi f / \mu_0 \gamma. \tag{23}$$

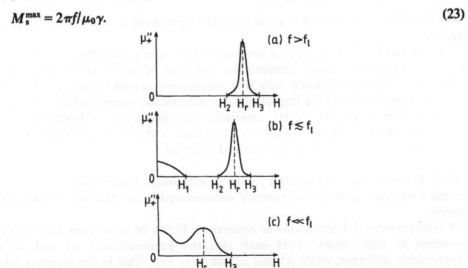

Fig. 6. Imaginary part of permeability μ_+'' as a function of static magnetic field for three values of frequency f compared with the value of $f_\ell = \gamma \mu_0 M_s / 2\pi$. (a) $f > f_\ell$, (b) $f \lesssim f_\ell$, (c) $f \ll f_\ell$.

It can be seen that magnetization has to be adapted to the operating frequency f and be substantially proportional to it. In addition, at a frequency of 1000 MHz, even with low-magnetization materials, the situation is as illustrated in fig. 6c and it is necessary to choose a field higher than the resonance field when $f \geqslant$ 1500 MHz. The previous condition is then no longer applicable.

1.3.6. Use of partially magnetized ferrites

It has been assumed initially (see section 1.2.2) that the ferrite was subjected to a static field H sufficient to saturate it. The foregoing therefore applies only if this condition is met. Let us now consider the case in which the fields H are low and the material unsaturated. In practice, the ferrite is almost always used under the conditions shown in fig. 6a, i.e. with no magnetic loss at low fields, and this will be the basic case considered here.

As a rule, if magnetization is always oriented along Oz, the susceptibility tensor resulting from eq. (6) becomes

$$\chi = \begin{vmatrix} \chi_+ & 0 & 0 \\ 0 & \chi_- & 0 \\ 0 & 0 & \chi_{zz} \end{vmatrix} \tag{24}$$

where $\chi_{zz} \neq 0$. In the fully demagnetized state, the susceptibility is scalar ($\chi_+ = \chi_- = \chi_{zz} = \chi_0$) and for ($\gamma\mu_0 M < \omega$) it is of the form

$$\chi_0 = -\tfrac{2}{3}(\gamma\mu_0 M_s/\omega)^2 \tag{25}$$

as shown independently by Nicolas (1967b) and Courtois and Deschamps (1967). Schlömann (1970) proposed a slightly different formula

$$\mu_0' = 1 + \chi_0' = \tfrac{2}{3}[1 - (\gamma\mu_0 M_s/\omega)^2]^{1/2} + \tfrac{1}{3} \tag{26}$$

which seems to constitute a good approximation over a fairly wide frequency range.

Green and Sandy (1974a) give empirical expressions for a first approximation of the permeability tensor elements for the case of any magnetization. These expressions actually contain only two parameters, $\mu_0\gamma M_s/\omega$ and $\mu_0\gamma M/\omega$. The latter parameter shows the importance, for electrically controlled devices, of the magnetization curve $M = f(H)$, especially for the shape of hysteresis loops. Green and Sandy (1974b) also give the threshold field of non-linear effects which limits the applications for an unsaturated material.

1.3.7. Different applications of a ferrite at microwave frequencies

Some cases among the most currently encountered in practice will be mentioned below.

(i) Non-reciprocal devices outside resonance: It can be seen from fig. 3 that it is possible to find values of H such that the permeabilities μ_+' and μ_-' are appreciably different, while μ_+'' and μ_-'' are very low. This is the property which is the most frequently used in the design of non-reciprocal devices. The applied field H may be either lower than the resonance field H_r, or higher than H_r; the first alternative is generally preferred, if only for economical considerations, owing to the smaller magnets required. In this case, the magnetization of the material does not exceed the value of the eq. (23), and in practice it is such that

$$M_s^{max}/2 < M_s < M_s^{max}.$$

However, the fact that, as shown by eq. (9), the resonance field decreases with frequency should be taken into account: in practice, below 1000 MHz, it is no longer possible, before reaching resonance, to find fields H such that μ'' is negligible, therefore fields H higher than H_r are then used and the preceding condition is no longer applicable.

(ii) Non-reciprocal devices at resonance: In some cases (e.g. in isolators) at resonance the material is polarized with a field H approximating H_r, and the difference between the magnetic losses (given by μ'') for the two directions of polarization of the electromagnetic wave is used directly.

(iii) Electrically controlled devices: In variable phase-shifters, either reciprocal or not, in latching circulators and other electrically controlled devices, use is made of the variations in permeability due to variations in magnetization, e.g. on the hysteresis loop. To ensure correct operation of the device, a low coercive field (e.g. approximately 100 A/m, or 1 Oe) and a very rectangular hysteresis loop are usually required.

2. Polycrystalline garnets for microwave applications

2.1. General

2.1.1. History
The magnetic garnets were discovered by Bertaut and Forrat from Grenoble University (1956). Following the work of Geller from Bell Laboratories (Geller 1960, Geller and Gilleo 1957) who designed and developed various types of substitution, the importance of garnets in microwave applications has been constantly growing. One can find data on different substituted garnets in the papers of Saunders and Green (1961), Harrison and Hodges (1961), Vassiliev et al. (1961), Geller et al. (1964), Smolenskii and Polyakov (1964), Vassiliev and Nicolas (1965), Nicolas and Lagrange (1970), Winkler et al. (1972), Sroussi and Nicolas (1974). Indeed, with magnetizations M_s of between 25 and 150 kA/m ($4\pi J_s$ between 300 to 1900 G), and small linewidths, these materials are suitable for the design of devices operating at frequencies anywhere between a few hundred MHz and 9000 MHz. They were derived from the yttrium–iron garnet $Y_3Fe_5O_{12}$, whose magnetization is $M_s = 142.5$ kA/m ($4\pi J_s = 1790$ G).

2.1.2. Preparation
The garnet microwave ferrites are manufactured using conventional, ceramic techniques. This manufacture is summarized for the yttrium–iron garnet. The other garnets and the spinel microwave ferrites are produced in a similar way.

J. NICOLAS

The raw materials, i.e. the ferric oxide Fe_2O_3 and the yttrium oxide Y_2O_3, taken in the form of fine powder, are weighed in the desired proportions and thoroughly mixed. The compound is heat treated (pre-sintering) at a temperature of approximately 1200°C. The powder obtained is milled to a grain size smaller than one micron. The milled powder, mixed with an organic binder, is granulated (e.g. by means of a spray dryer). These granules are pressure moulded at a pressure of 1 to 2×10^3 kg/cm². The piece thus formed is heat-treated at about 500–600°C, for elimination of the organic binders, and then sintered in an oxygen atmosphere, at about 1500°C, for 5 to 10 hours. A longer treatment enables a much lower porosity to be obtained (less than one percent).

An important feature is the necessity of a good stoichiometry, as shown in fig. 7. This figure shows the dielectric loss angle tangent (tan δ) and the linewidth ΔH measured at 10 GHz, obtained under given technological conditions (sintering at 1470°C), as a function of the iron content of the initial powders (before mixing and grinding). It can be seen in the figure that to achieve both low dielectric loss and a small linewidth, it is necessary to ensure the optimum stoichiometry, to within approximately 10^{-3}. For this purpose, account must be taken, with the greatest precision, of the iron introduced in the mixing and grinding operations and it is essential to perfectly control the regularity and reproducibility of all the technological operations.

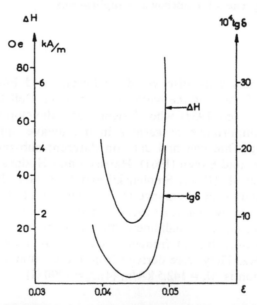

Fig. 7. Linewidth ΔH and dielectric losses tangent tan δ measured at 10 GHz as a function of initial iron defect ϵ (considered before grinding in this case) for a material having the formula $Y_3 Fe_{5(1-\epsilon)}O_{12}$. (after Miss M.C. Lalau from the Laboratoire Central de Recherches Thomson-CSF (Lalau 1978)).

2.1.3. General considerations on the selection of relevant data and its presentation

The physical and magnetic properties of yttrium–iron garnet, and of the garnets derived therefrom by various substitutions, are described in chapter 1 of this book, by M.A. Gilleo. Likewise, the chapter by S. Krupicka deals with spinel ferrites and that of H. Kojima and M. Sugimoto with hexagonal ferrites (see volume 3/4). This section has therefore been based on the following principles: (a) We have used only the diagrams or portions of diagrams which are effectively used, or seem to us the most suitable, for microwave applications. (b) We have given only the characteristics of interest for this technique. It should be noted that at the same time as giving values in SI units, we have also given some in cgs units. Thus $4\pi J_s$ represents the saturation magnetization in Gauss.

The characteristics appearing in the tables are usually those of materials available in the trade. They are not the optimum values obtainable in the laboratories, but those guaranteed by a manufacturer with the presently available techniques. ΔH, ΔH_{eff}, and ΔH_k, $\tan\delta$ are measured at $10\,\text{GHz}$. For these commercial materials, the curves of magnetization versus temperature within a technically significant interval are usually also given. Unless otherwise specified, the other results are taken from papers and enable the effects of the various parameters on the quantities involved to be evaluated.

2.2. Yttrium–iron garnet (YIG)

2.2.1. Characteristics at room temperature

Table 1 gives the rated characteristics at ambient temperature of two materials with a composition $Y_3Fe_5O_{12}$ having slightly different porosities. The designations are those given by the manufacturer (Thomson-CSF).

2.2.2. Variation of characteristics with temperature

(i) Saturation magnetization versus temperature is given in fig. 14.

(ii) Linewidth – Figure 8 shows the variations of ΔH with temperature, obtained by Simonet, in our laboratory, for two materials of composition $Y_3Fe_5O_{12}$, with two different porosities.

(iii) ΔH_{eff} – Koelberer and Patton (1977) give the variation in effective linewidth (ΔH_{eff}) at low temperature (as seen in fig. 9) for samples with different porosites. It can be seen that for dense materials, i.e. those of the type used in practical applications, ΔH_{eff} varies only slightly with temperature.

(iv) ΔH_k – The spin wave linewidth ΔH_k versus temperature is given, according to Saunders and Green (1961), in fig. 10, which also shows the variation of ΔH for the same material. In principle, ΔH_k, contrary to ΔH, or ΔH_{eff}, is not dependent on porosity.

TABLE 1

Characteristics of two materials having the chemical composition $Y_3Fe_5O_{12}$ (YIG) (from the Thomson-CSF list). The linewidths ΔH, ΔH_{eff}, ΔH_k and the dielectric characteristics ϵ_r and $\tan \delta_e$ are measured at 10 GHz.

Type	M_s kA/m	$4\pi I_s$ G	T_c °C	g_{eff}	ΔH kA/m	ΔH Oe	ΔH_{eff} kA/m	ΔH_{eff} Oe	ΔH_k kA/m	ΔH_k Oe	ϵ_r	$\tan \delta_e$ 10^{-4}
Y 10	142.5	1790	280	2.00	3.6	45	0.3	4	0.15	2	15.4	<2
Y 101	144.9	1820	280	2.00	1.6	20	0.3	4	0.15	2	15.4	<2

TABLE 2

Characteristics of yttrium–aluminium–iron garnets (Thomson-CSF)

Type	M_s kA/m	$4\pi I_s$ G	T_c °C	g_{eff}	ΔH kA/m	ΔH Oe	ΔH_{eff} kA/m	ΔH_{eff} Oe	ΔH_k kA/m	ΔH_k Oe	ϵ_r	$\tan \delta_e$ 10^{-4}
Y 35	95.5	1200	225	2.01	3.2	40	0.3	4	0.15	2	15.0	<2
Y 34	79.6	1000	210	2.01	3.2	40	0.3	4	0.15	2	14.9	<2
Y 39	63.7	800	195	2.01	3.2	40	0.3	4	0.15	2	14.7	<2
Y 38	60.5	760	190	2.01	3.2	40	0.3	4	0.15	2	14.6	<2
Y 37	54.1	680	180	2.01	3.2	40	0.3	4	0.15	2	14.6	<2
Y 33	49.0	615	175	2.01	3.2	40	0.3	4	0.15	2	14.6	<2
Y 30	45.0	565	160	2.01	2.8	35	0.3	4	0.15	2	14.5	<2
Y 32	33.4	420	135	2.01	2.8	35	0.3	4	0.15	2	14.5	<2
Y 31	29.5	370	125	2.01	2.8	35	0.3	4	0.15	2	14.2	<2
Y 36	23.1	290	115	2.01	2.4	30	0.3	4	0.15	2	14.1	<2

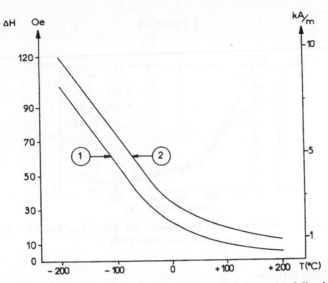

Fig. 8. Linewidth ΔH versus temperature for two YIG's which have the following densities: (1) $d_1 = 5.151$ g/cm³ (2) $d_2 = 5.128$ g/cm³. The X-ray density is 5.165 g/cm³. (after W. Simonet from the Laboratoire Central de Recherches Thomson-CSF (Simonet 1977)).

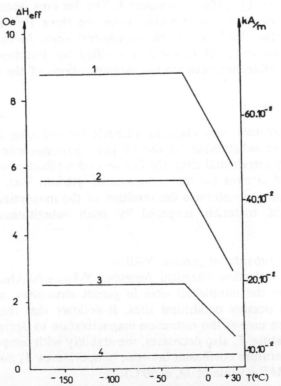

Fig. 9. High field effective linewidth ΔH_{eff} versus temperature for porous YIG at 9.94 GHz. (1) porosity = 16.3%, (2) porosity = 6.5%, (3) porosity = 1.9%, (4) porosity < 0.5%, (dense material)(after R.E. Koelberet and C.E. Patton, 1977).

261

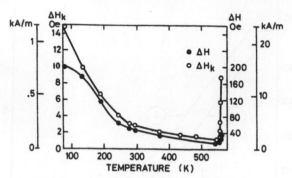

Fig. 10. Values of ΔH and ΔH_k versus temperature for polycrystalline YIG at 9.6 GHz (after Saunders and Green 1961).

2.3. Conventional garnets

2.3.1. Introduction
Different magnetization values can be obtained by substitutions in the yttrium garnet, as explained by Gilleo in chapter 1. The Fe ions occupy two types of crystallographic sites: for one molecule, there are three Fe^{3+} ions in the tetrahedral sites and two Fe^{3+} ions in the octahedral sites. In the theory of ferrimagnetism, given by Néel (1948) and verified by Pauthenet (1958), magnetization is the difference between the magnetizations of the magnetic sublattices

$$M_s = |M_{tetra} - M_{octa}|. \tag{27}$$

Actually, there are two very classical methods for reducing M_s: (a) The first method consists in substituting for the Fe ions, non-magnetic aluminium ions which will occupy tetrahedral sites. (b) The second method consists in substituting, in dodecahedral sites for the non-magnetic yttrium ions, gadolinium ions whose magnetization counteracts the resultant of the magnetization of Fe ions. The properties of materials prepared by such substitutions will now be examined.

2.3.2. Aluminium substituted garnets (YAlIG)
These have the following chemical formula: $Y_3Fe_{5-5y}Al_{5y}O_{12}$. The Al^{3+} ions preferably occupy the tetrahedral sites in garnet structures, although a small fraction of them occupy octahedral sites. It follows that the substitution of aluminium for iron causes the saturation magnetization to decrease (fig. 11). As the Curie temperature T_c also decreases, the stability with temperature is lower. An average temperature coefficient between temperatures T_1 and T_2 (°C) can be defined as (with M_s^{20} = value of M_s at 20°C):

$$\alpha_{T_2}^{T_1} = (M_s^{max} - M_s^{min})/(T_1 - T_2)M_s^{20}. \tag{28}$$

Figure 11 gives α_{-40}^{+85} versus y. This quantity provides an approximate idea of the

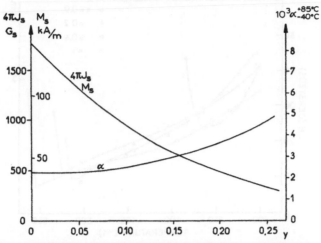

Fig. 11. Saturation magnetization at room temperature and mean temperature coefficient of magnetization (between −40 and +85 C) for garnets having the composition $Y_3Fe_{5-5y}Al_5yO_{12}$.

stability of materials with temperature. However, in practice, the temperature interval to be considered may be dependent upon the required application. The corresponding variations derived from the curves of figs. 14 and 15 then have to be taken into account. This also applies to the other materials which will be examined further on. The aluminium substituted garnets which are used in the devices are approximately within the interval $0 \leqslant y \leqslant 0.27$. Higher aluminium substitutions involve instability of characteristics versus temperature which is prohibitive in most cases. Figure 12 shows the variation of the linewidth ΔH with y, at room temperature, which has the advantage of remaining very low for all these materials. The variation of this linewidth with temperature, according to Harrison and Hodges (1961), is shown in fig. 13. As the magnetic ions are only Fe^{3+} ions, the effective linewidth ΔH_{eff} and the spin wave linewidth ΔH_k also remain very low in these materials. Typical characteristics of industrial materials shown in this diagram are given in table 2. The saturation magnetization of these materials versus temperature is illustrated in figs. 14 and 15.

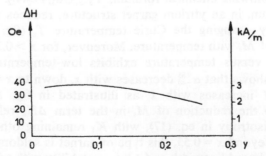

Fig. 12. Linewidth at room temperature of garnets having the composition $Y_3Fe_{5-5y}Al_5yO_{12}$.

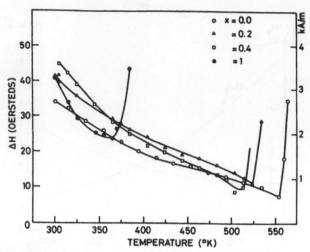

Fig. 13. Linewidth of $Y_3Fe_{5-5y}Al_5yO_{12}$ as functions of temperature (after Harrison and Hodges 1961).

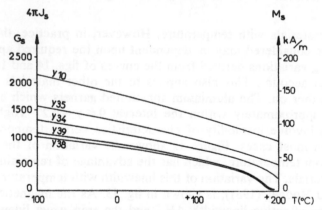

Fig. 14. Saturation magnetization versus temperature of some Thomson-CSF garnets having the composition $Y_3Fe_{5-5y}Al_5yO_{12}$ for $0 < y < 0.15$.

2.3.3. *Mixed yttrium–gadolinium garnets* (YGdIG)

These have the following chemical formula: $Y_{3-3x}Gd_{3x}Fe_5O_{12}$. The saturation of magnetic gadolinium, in an yttrium garnet structure, reduces the magnetization without practically changing the Curie temperature T_c (fig. 16). This results in a stabilization of M_s with temperature. Moreover, for $x > 0.3$ (approximately), the curve of M_s versus temperature exhibits low-temperature compensation points. Figure 16 shows that α_{-40}^{+85} decreases with x, down to $x = 0.4$. Conversely, the linewidth ΔH increases with x, as illustrated in fig. 17. This effect is essentially due to the reduction of M_s in the term ΔH_a related to linewidth broadening by anisotropy in eq. (17), with K_1 remaining otherwise practically constant with x. Beyond $x = 0.55$, this type of garnet is seldom used, because of the comparatively broad linewidth and a lower stability with temperature. Figure

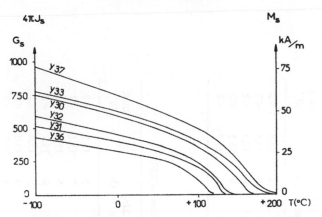

Fig. 15. Saturation magnetization versus temperature of some Thomson-CSF garnets having the composition $Y_3Fe_{5-5y}Al_{5y}O_{12}$ for $0.16 < y < 0.26$.

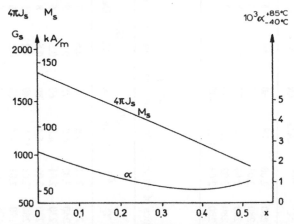

Fig. 16. Saturation magnetization at room temperature and mean temperature coefficient of magnetization (between -40 and $+85°C$) for garnets having the composition $Y_{3-3x}Gd_{3x}Fe_5O_{12}$.

17 also shows an increase of ΔH_k with x. Table 3 gives the characteristics of industrial materials and fig. 18 the variations of these with temperature.

The gadolinium ion is the only magnetic rare earth which can be used in the manner described above. The gadolinium ion is an S-ion ($L = 0$), it does not exhibit spin–orbit coupling causing rapid damping of the gyromagnetic movement (rapid relaxation) which increases linewidths ΔH_k, ΔH_{eff}, and ΔH to prohibitive proportions in terms of microwave application. The other rare earth magnetic ions can be used only at very low doping levels, as will be shown later.

2.3.4. Mixed yttrium–gadolinium garnet with aluminium substitution (YGdAlIG)

It can be seen that the aluminium and gadolinium substitutions in YIG decrease M_s, although having opposite effects on stability with temperature, and linewidth

TABLE 3

Characteristics of yttrium–gadolinium–iron garnets (Thomson–CSF)

Type	M_s kA/m	$4\pi J_s$ G	T_c °C	g_{eff}	ΔH kA/m	Oe	ΔH_{eff} kA/m	Oe	ΔH_k kA/m	Oe	ϵ_r	$\tan \delta_e$ 10^{-4}
Y 11	127.4	1600	280	2.00	4.8	60	0.4	5	0.2	3	15.4	<2
Y 12	112.9	1420	280	2.01	5.2	65	0.5	6	0.5	6	15.4	<2
Y 13	99.5	1250	280	2.01	6.0	75	0.6	8	0.6	8	15.4	<2
Y 14	87.5	1100	280	2.02	7.6	95	1.0	12	0.7	9	15.5	<2
Y 15	71.6	900	280	2.03	11.1	140	1.4	18	0.9	11	15.5	<2

TABLE 4

Characteristics of yttrium–gadolinium–aluminium–iron garnets (Thomson–CSF)

Type	M_s kA/m	$4\pi J_s$ G	T_c °C	g_{eff}	ΔH kA/m	Oe	ΔH_{eff} kA/m	Oe	ΔH_k kA/m	Oe	ϵ_r	$\tan \delta_e$ 10^{-4}
Y 71	81.2	1020	235	2.01	4.8	60	0.6	7	0.4	5	15.1	<2
Y 710	81.2	1020	240	2.02	6.0	75	0.7	9	0.6	7	15.1	<2
Y 77	75.6	950	230	2.01	4.8	60	0.5	6	0.4	5	15.0	<2
Y 78	63.7	800	220	2.00	6.4	80	0.6	8	0.6	8	15.1	<2
Y 708	63.7	800	260	2.04	11.1	140	1.2	15	1.2	15	15.3	<2
Y 74	53.3	670	190	2.01	4.8	60	0.5	6	0.5	6	15.0	<2
Y 72	43.0	540	175	2.01	4.8	60	0.5	6	0.5	6	14.7	<2
Y 75	31.8	400	160	2.03	5.2	65	0.5	6	0.5	6	14.4	<2
Y 76	31.0	390	160	2.02	4.0	50	0.5	6	0.5	6	14.3	<2

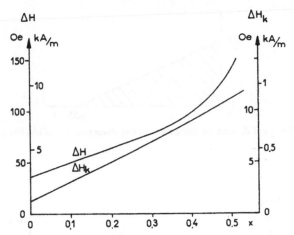

Fig. 17. Linewidth ΔH and spin wave linewidth ΔH_k at room temperature for garnets having the composition $Y_{3-3x}Gd_{3x}Fe_5O_{12}$.

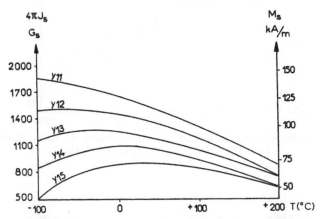

Fig. 18. Saturation magnetization as a function of temperature of some Thomson-CSF garnets having the composition $Y_{3-3x}Gd_{3x}Fe_5O_{12}$ $(0.1 < x < 0.5)$.

ΔH. An intermediate compromise may evidently be found in the diagram of the compositions: $Y_{3-3x}Gd_{3x}Fe_{5-5y}Al_{5y}O_{12}$, in which both types of substitutions are carried out at the same time. The useful part of this diagram is defined by the hatched area of fig. 19. Table 4 gives a general indication of the characteristics of the materials having this formula. The magnetizations of these materials versus temperature are given in figs. 20 and 21.

2.4. High magnetization garnets

By carrying out non-magnetic substitutions in the octahedral sites of the yttrium garnet, the magnetization at 0 K is increased, at least for low substitutions, in accordance with the theory of ferrimagnetism of Néel (1948). However the Curie

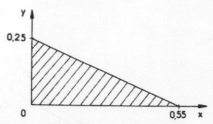

Fig. 19. Practical domain of utilization in the diagram: $Y_{3-3x}Gd_{3x}Fe_{5-5y}Al_5yO_{12}$.

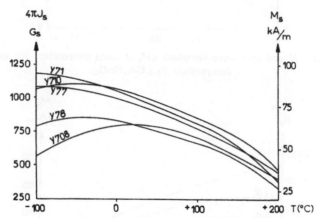

Fig. 20. Saturation magnetization as a function of temperature for some Thomson-CSF garnets having the composition $Y_{3-3x}Gd_{3x}Fe_{5-5y}Al_5yO_{12}$.

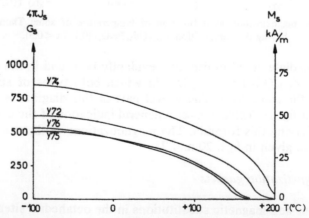

Fig. 21. Saturation magnetization as a function of temperature for some Thomson-CSF garnets having the composition $Y_{3-3x}Gd_{3x}Fe_{5-5y}Al_5yO_{12}$.

temperature is decreased by such substitutions. It follows from these two effects that at the room temperature, with this type of substitution, magnetization only reaches: $M_s = 155 \, kA/m$ ($4\pi J_s = 1950 \, G$). In fact, indium substitutions (In^{3+}) are made and lead to the compositions $Y_3Fe_{5-y}In_yO_{12}$, or zirconium substitutions (Zr^{4+}) are made (associated with Ca^{2+} substitutions to impose valence equilibrium with Fe^{3+}) which lead to the compositions $Y_{3-x}Ca_xFe_{5-x}Zr_xO_{12}$. Figure 22 shows the magnetization at room temperature versus x for the two chemical formulae, and the Curie temperatures T_c of these materials. Table 5 gives the characteristics of an industrial material of this type (indium substituted) and fig.

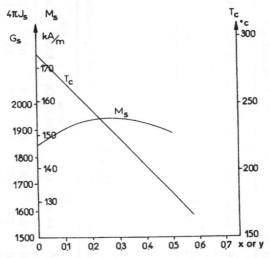

Fig. 22. Saturation magnetization at room temperature and Curie temperature T_c for garnets having the composition $Y_3Fe_{5-y}In_yO_{12}$ or $Y_{3-x}Ca_xFe_{5-x}Zr_xO_{12}$.

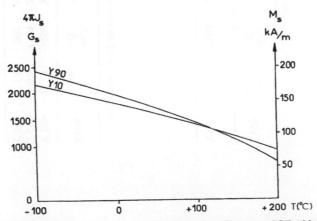

Fig. 23. Saturation magnetization versus temperature of a Thomson-CSF "high" magnetization garnet substituted by indium (Y 90) compared with YIG (Y 10).

TABLE 5
Characteristics of a high magnetization garnet (Thomson-CSF)

Type	M_s kA/m	$4\pi J_s$ G	T_c °C	g_{eff}	ΔH kA/m	ΔH Oe	ΔH_{eff} kA/m	ΔH_{eff} Oe	ΔH_k kA/m	ΔH_k Oe	ϵ_r	$\tan \delta_\epsilon$ 10^{-4}
Y 90	151.2	1900	240	2.02	3.6	45	0.3	4	0.15	2	15.0	<2

TABLE 6
Characteristics of narrow linewidth garnets (Thomson-CSF); ΔH is measured on optically polished spheres

Type	M_s kA/m	$4\pi J_s$ G	T_c °C	g_{eff}	ΔH kA/m	ΔH Oe	ΔH_{eff} kA/m	ΔH_{eff} Oe	ΔH_k kA/m	ΔH_k Oe	ϵ'	$\tan \delta_\epsilon$ 10^{-4}
Y 220	155.2	1950	205	2.01	0.8	10	0.2	2	0.1	1	15.5	<2
Y 215	120.0	1500	240	2.01	0.8	10	0.2	2	0.1	1	15.0	<2
Y 209	71.6	900	190	2.01	0.8	10	0.2	2	0.1	1	14.2	<2

23 shows the magnetization curve of this material versus temperature, compared with that of the YIG.

2.5. Low linewidth garnets

As demonstrated by several authors, in particular Winkler et al. (1972), it appeared that it was possible to produce polycrystalline garnets having very narrow linewidths ΔH, less than 0.8 kA/m (10 Oe) and even less than 0.4 kA/m (5 Oe), all with acceptable Curie temperatures for particular applications. To obtain such very narrow linewidths, the different linewidth broadening terms of eq. (16) have to be reduced to the greatest possible extent. These various contributions are described below:

A low intrinsic linewidth ΔH_{int} is obtained by using Fe^{3+} ions as magnetic ions.

A narrow linewidth ΔH_a involves a low anisotropy constant K_1. However, the anisotropy observed in garnets containing only Fe^{3+} ions, as magnetic ions, mostly results from the Fe^{3+} ions of the octahedral sites. Non-magnetic, octahedral substitutions therefore reduce K_1 to 0 K. The lowering of the Curie temperature by non-magnetic substitutions is also an important factor of the reduction of K_1 at room temperature, but this occurs to the detriment of the stability of the material with temperature, and a compromise has to be found in this field.

The linewidth broadening ΔH_p, due to porosity, falls to a very low value if porosity is reduced to one-thousandth, for instance. This can be achieved by a suitable choice of composition and careful optimization of technology. The materials that now appear the most significant concerning this point have the following compositions:

$$Y^{3+}_{3-x}Ca^{2+}_{x}Fe^{3+}_{5-x}Zr^{4+}_{x}O_{12} \qquad Y^{3+}_{3-2x}Ca^{2+}_{2x}Fe^{3+}_{5-x-y}In^{3+}_{y}V^{5+}_{x}O_{12}.$$

The In^{3+} or Zr^{4+} substitutions (associated with Ca^{2+} for valency compensation) have a similar effect, but the presence of Ca^{2+} seems to enable a very low porosity to be obtained by improving the sintering. A V^{5+} ion tetrahedral substitution gives a reduction in M_s, if required, while maintaining the Curie temperature substantially the same.

The characteristics of the first materials were given in the preceding section (fig. 22). Figure 24 completes this data by showing the variation with temperature of the linewidth ΔH of low-porosity materials of this diagram; the densities of the materials are marked on each curve. Practically, only the substitutions $0 < x < 0.4$ seem to be of interest, allowing for the lowering of Curie temperature T_c with x. The second diagram comprises octahedral substitutions (In^{3+}), which contribute to a reduction of magnetocrystalline anisotropy, and V^{5+} substitutions (associated with Ca^{2+} for valency compensation) which enable the magnetization to be reduced without substantial lowering of the Curie temperature. Figures 25 and 26 give the saturation magnetization at room temperature and the Curie temperature, T_c, of these materials. Linewidths of between 0.8 and 1.2 kA/m (10 and 15 Oe) have been measured by us for these materials, for the compositions

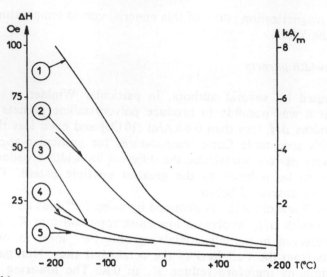

Fig. 24. Linewidth versus temperature of garnets having the composition $Y_{3-x}Ca_xFe_{5-x}Zr_xO_{12}$. (1) $x = 0$, density of material = 5.151, (2) $x = 0.2$, density of material = 5.094, (3) $x = 0.4$, density of material = 5.042, (4) $x = 0.5$, density of material = 5.011, (5) $x = 0.8$, density of material = 4.926 (after W. Simonet 1977).

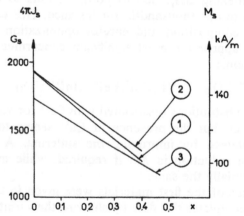

Fig. 25. Saturation magnetization at room temperature of garnets having the composition $Y_{3-2x}Ca_{2x}Fe_{5-x-y}In_yV_xO_{12}$. (1) y = 0.2, (2) y = 0.4, (3) y = 0.6 (after W. Simonet 1977).

corresponding to $y = 0.2$, and between 0.5 and 0.6 kA/m (4 to 7 Oe) for compositions corresponding to $y = 0.4$ and $y = 0.6$. Useful compositions appear to be situated, in practice, in the substitution intervals $0 < x < 0.8$, $0 < y < 0.4$. The characteristics of three industrial materials are given in table 6. The linewidths were measured on optically polished, spherical samples. The variation of the magnetizations of these materials with temperature is illustrated in fig. 27.

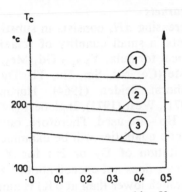

Fig. 26. Curie temperature of garnets having the composition $Y_{3-2x}Ca_xFe_{5-x-y}In_yV_xO_{12}$ (1) $y = 0.2$, (2) $y = 0.4$, (3) $y = 0.6$ (after W. Simonet 1977).

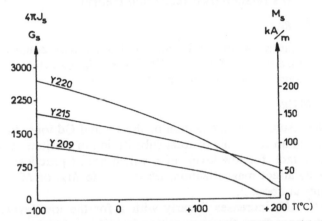

Fig. 27. Saturation magnetization versus temperature of Thomson-CSF materials having the composition $Y_{3-2x}Ca_{2x}Fe_{5-x-y}In_yV_xO_{12}$.

2.6. Garnets for high peak power

2.6.1. Introduction

In general, microwave polycrystalline ferrites cannot be used when the microwave field exceeds a critical field, beyond which non-linear effects appear. One of the solutions which may be used to raise the value of this critical field consists in increasing the spin wave linewidth, ΔH_k (see section 1.3.4). As seen in fig. 17, the value of ΔH_k can be raised by substituting Gd for Y. ΔH_k can be increased by other methods. The first one consists in obtaining small substitutions of relaxing (or doping) magnetic ions. These ions may be either magnetic rare earths (such as holmium or dysprosium), or cobalt ions. So the complex effects of cobalt will be studied. Firstly, the effect of doping with magnetic rare earths will be examined.

2.6.2. Rare earth doped garnets

A very simple way of increasing ΔH_k consists in substituting partly for yttrium, in the conventional garnets, a small quantity of relaxing rare earth ions. This gives the following chemical formula: $Y_{3-3x-3z}Gd_{3x}Me_{3z}Fe_{5-5y}Al_{5y}O_{12}$. The effect of various rare earths, Me (Ce, Nd, Sm, Eu, Tb, Dy, Ho, Er, Yb) has been studied by several authors: Seiden (1964), Hartmann–Boutron (1964a,b), Borghese and Roveda (1971), West (1971), Inglebert (1973). In practice, the most active rare earths – Dy or Ho – are used. Therefore, only the action of these two ions on the various magnetic quantities will be examined in the following:

Effect on M_s: The substitution of Dy or Ho for Y leads to a decrease in magnetization. However, as z does not exceed about 5×10^{-2}, the effect on M_s at room temperature (effect much lower than at 0 K) is almost negligible. The Curie temperature remains unchanged.

Effect on g_{eff}: g_{eff} is given for the mid-temperature range (i.e. between 200 K and $T_c - 10$ K) by Kittel's relation (fast relaxation theory)

$$g_{eff} = (M_s/M_a)g_s \tag{29}$$

g_s is the gyromagnetic factor for Fe ions and practically equal to 2. M_A is the resulting magnetization of Fe ions. This formula can be extended to the composition containing gadolinium ions. This yields

$$g_{eff} = (M_s/M'_A)g_s \tag{30}$$

M'_A then is the resultant magnetization of the Fe and Gd ions. Formula (29) has been well verified up to a scale of total substitution of yttrium by a relaxing rare earth ($z = 1$). In fact, for low levels of z which are of practical interest and do not substantially affect magnetization, M_s is near to M_A, or M'_A as appropriate, and g_{eff} remains at about 2.

Effect on ΔH_k: ΔH_k increases linearly with z for the low levels studied, in a manner which depends on the composition (x and y values).

Effect on ΔH: ΔH is a linear function of z. A factor of proportionality has been observed between the increase in ΔH and the increase in ΔH_k (with z), which is independent of the composition (x and y values).

Effect on ΔH_{eff}: ΔH_{eff} is a linear function of z, the value of this increase being of the same order of magnitude as ΔH.

Table 7 gives the increases in ΔH_k, ΔH, ΔH_{eff} corresponding to $z = 0.01$. It can be seen that for a given Curie temperature, the product of the increase, $\delta(\Delta H)$ [or $\delta(\Delta H_k)$], of ΔH (or ΔH_k) by the saturation magnetization is comparatively constant for a given doping ion

$$\delta(\Delta H) \cdot M_s = \text{constant.} \tag{31}$$

This constant decreases with the Curie temperature.

Consequently, increasing ΔH_k by relaxing ions, such as Dy or Ho, has the advantage of simplicity. This doping has practically no effect on magnetization or its variation with temperature. In addition, the required technology can remain similar to that applicable to the basic composition without doping, with dielectric

TABLE 7

Effect of Dy or Ho doping on properties of garnets having the composition $Y_{3-3x-3z}Gd_{3x}Me_{0.03}Fe_{5-3y}Al_{3y}O_{12}$ (Me = Dy or Ho)

Me	x	y	M_s kA/m	$4\pi J_s$ G	T_c K	ΔH kA/m	ΔH Oe	ΔH_{eff} kA/m	ΔH_{eff} Oe	ΔH_k kA/m	ΔH_k Oe	$\frac{\delta(\Delta H_{eff})}{\delta(\Delta H)}$	$\frac{\delta(\Delta H)}{\delta(\Delta H_k)}$	$\frac{\delta(\Delta H_{eff})}{\delta(\Delta H_k)}$	References
Dy	0	0	142	1780	560	2.23	28								Harrison and Hodges (1961)
Dy	0	0	142	1780	560	2.15	27			0.83	10.5		2.6		Borghese and Roveda (1971)
Dy	0	0	142	1780	560	2.07	26	2.16	27.2	0.68	8.5	1.04	3.1	3.2	Inglebert (1973)
Dy	0.1	0	127	1600	560	2.39	30			0.72	9		3.3		Inglebert (1973)
Dy	0.2	0	113	1420	560	2.79	35			0.80	10				Inglebert (1973)
Dy	0.3	0	97	1220	560	3.26	41	3.26	41	1	12.6	1	4.1	4.1	Inglebert (1973)
Dy	0.4	0	85.1	1070	560	3.30	41.5			1.03	13		3.2		Inglebert (1973)
Dy	0.5	0	71.6	900	560	3.74	47			1.03	13				Inglebert (1973)
Dy	0	0.1	79.6	1000	493	2.75	35	3.21	40.4	0.92	11.5	1.15	3.1	3.5	Inglebert (1973)
Dy	0	0.18	47.7	600	440	3.58	45			1.19	15		3		Inglebert (1973)
Ho	0	0	142	1780	560	3.66	46								Harrison and Hodge (1961)
Ho	0	0	142	1780	560	3.42	43			1.39	17.5		2.4		Borghese and Roveda (1971)
Ho	0	0	142	1780	560	3.36	46	4.22	53	1.45	18.2	1.10	2.4	2.8	Inglebert (1973)
Ho	0.3	0	98.7	1240	560	5.33	67	6.05	76	1.55	19.5	1.14	2.4	2.6	Inglebert (1973)
Ho	0	0	142	1780	560					0.34	4.3				Inglebert (1973)
Ho	0	0.14	63.7	800	465					0.75	9.4				Inglebert (1973)
Ho	0	0.24	25.1	315	400					1.99	25				Inglebert (1973)

losses remaining optimized. The drawback of this solution is an increase in ΔH and ΔH_{eff}. In practice the levels of doping with Ho or Dy are limited to that required to obtain ΔH_k approximately equal to 2.5 kA/m (some thirty oersteds) at maximum.

2.6.3. Effect of the cobalt ion in the polycrystalline garnet

Cobalt doped YIG

The ion Co^{2+} is substituted for the Fe^{3+} ion in an octahedral site. To have only divalent cobalt ions, it is necessary to simultaneously substitute an Fe^{3+} ion for an Si^{4+} or Ge^{4+} ion which also occupies an octahedral site. This leads, for instance, from the YIG to a formula of the type: $Y_3^{3+}Fe_{5-2u}^{3+}Si_u^{4+}Co_u^{2+}O_{12}$. The introduction of the relaxing ion Co^{2+} results in an increase of ΔH_k, as shown by fig. 28. The effect of these substitutions of Co and Si on M_s is in accordance with the theory of Néel (decrease of M_s) and it is low for the low values of u achieved in practice. In other respects, the effect of Co^{2+} is more complex than that of Dy^{3+} or Ho^{3+}, due to a considerable modification of the magnetocrystalline anisotropy.

Figure 29 shows the variations of ΔH and ΔH_{eff} which are no longer simply related. It also shows the variation of g_{eff} which decreases as a result of the variation in anisotropy. As the anisotropy due to the Co^{2+} ions is positive, it can compensate for the negative anisotropy due to the Fe^{3+} ions. By varying the amount of cobalt in the YIG it was possible to measure a maximum permeability point (at 10 kHz); this point also coincided with the point of minimum coercive force. Conversely, the effect on ΔH is less obvious.

Co^{2+} doped YGdlAlG

The same effects are found as in YIG, i.e. ΔH_k linearly increases for low levels of Co^{2+}, plus an increase of ΔH_{eff}. The effects of some anisotropy compensation,

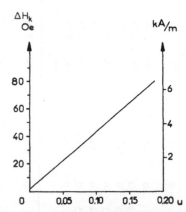

Fig. 28. Spin wave linewidth ΔH_k measured at room temperature at 9 GHz of garnets having the composition $Y_3Fe_{5-2u}Si_uCo_uO_{12}$ (after R.L. Inglebert from the Laboratoire Central de Recherches Thomson-CSF (Inglebert 1973)).

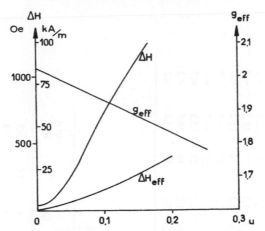

Fig. 29. Gyromagnetic factor g_{eff}, linewidth ΔH, and effective linewidth ΔH_{eff} of garnets having the composition $Y_3Fe_{5-2u}Si_uCo_uO_{12}$ (after Inglebert 1973).

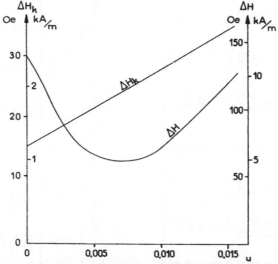

Fig. 30. Linewidth ΔH and spin wave linewidth ΔH_k measured at room temperature at 9 GHz on garnets having the composition $Y_{1.8}Gd_{1.2}Al_{0.225}Fe_{4.775}Co_uSi_uO_{12}$. Saturation magnetization of these materials is $M_s = 61$ kA/m ($4\pi J_s = 765$ G).

especially on ΔH, were also observed by Nicolas et al. (1970,1971). These effects increase with the equivalent anisotropy field of the initial composition. This applies to conventional garnets containing a large quantity of gadolinium. These phenomena are illustrated in fig. 30.

2.6.4. Examples of high peak power garnets

Table 8 gives examples of industrial materials doped with dysprosium or cobalt. The corresponding magnetization versus temperature curves are given in figs. 31 and 32.

TABLE 8
Characteristics of high-power garnets (Thomson-CSF)

Doping	Type	M_s kA/m	$4\pi J_s$ G	T_c °C	g_{eff}	ΔH kA/m	Oe	ΔH_{eff} kA/m	Oe	ΔH_k kA/m	Oe	ϵ_r	$\tan \delta_\epsilon$ 10^{-4}
Dy	D1	111.4	1400	270	2.00	8.8	110	2.7	34	1.3	16	15.5	<2
Dy	D5	85.2	1070	270	2.02	11.9	150	2.9	36	1.8	23	15.5	<2
Co	Y91	81.2	1020	240	2.02	4.8	60	1.3	17	1.1	14	15.2	<2
Dy	D2	71.6	900	270	2.01	14.7	185	2.0	25	1.9	24	15.5	<2
Co	Y86	66.1	830	270	2.03	7.6	95	2.7	34	2	25	15.5	<2
Co	Y94	62.1	780	250	2.02	6.0	75	1.1	14	1.8	23	15.3	<2
Dy	D3	47.0	590	175	2.00	6.8	85	1.3	16	1.5	19	14.5	<2
Dy	D4	46.2	580	170	2.00	11.1	140	2.7	34	2.6	33	14.4	<2

TABLE 9
Characteristics of Mn-Mg ferrites (Thomson-CSF)

Type	M_s kA/m	$4\pi J_s$ G	T_c °C	g_{eff}	ΔH kA/m	Oe	ΔH_{eff} kA/m	Oe	ΔH_k kA/m	Oe	ϵ_r	$\tan \delta_\epsilon$ 10^{-4}
U 21	191	2400	275	2.03	23.1	290	0.5	6	0.3	4	13.0	<3
U 20	167.2	2100	300	2.01	28.7	360	0.5	6	0.3	4	13.0	<3
U 33	127.4	1600	230	2.02	23.1	290	0.6	8	0.3	4	12.4	<3
U 32	111.4	1400	210	2.01	20.7	260	0.6	8	0.3	4	12.1	<3
U 30	89.9	1130	175	2.00	14.3	180	0.7	9	0.4	5	12.0	<3
U 31	71.6	900	120	2.02	10.3	130	0.7	9	0.4	5	11.7	<3

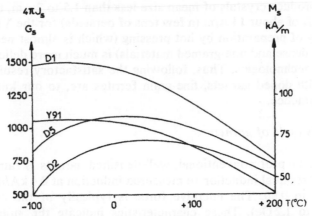

Fig. 31. Saturation magnetization versus temperature of some Thomson-CSF high-power garnets. Y 91 is a cobalt doped garnet. D 1, D 2 and D 5 are dysprosium doped garnets.

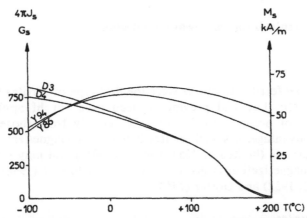

Fig. 32. Saturation magnetization versus temperature of some Thomson-CSF high power garnets. Y 86 and Y 94 are cobalt doped garnets. D 3 and D 4 are dysprosium doped garnets.

2.6.5. Fine-grain garnets

The materials discussed above were assumed to consist of crystals sufficiently large (at least 20 to 40 μm) to keep the effect of size on ΔH_k and ΔH_{eff} negligible. When grain size decreases sufficiently (less than a few μm) a substantial increase in the linewidths ΔH_k and ΔH_{eff} is observed. Koelberer and Patton (1977) showed in a very detailed study of the temperature dependence of the yttrium–iron garnet, that ΔH_k and ΔH_{eff} were dependent not only on grain size but also on porosity. This grain-size reduction can effectively be used for raising the threshold of non-linear effects and, in some cases, the compromise between increasing the threshold field (via ΔH_k) and increasing the magnetic losses (via ΔH_{eff}) may be slightly better than in doped materials. However, this method has various drawbacks. Firstly, the effect obtained is comparatively limited, since it is difficult to produce crystals of mean size less than 1.5 to 2 μm, this giving, for example, a ΔH_k of about 1 kA/m (a few tens of oersteds) for the YIG. Moreover, the technology of preparation by hot pressing (which is almost necessarily used to obtain both dense and fine-grained materials) is much more delicate and costly than standard technologies. Thus, following the satisfactory results obtained in most cases with doped garnets, fine-grain ferrites are, to our knowledge, very little used in practice.

2.7. Hysteresis cycle of garnets

The hysteresis loop of conventional, well-densified garnets is fairly rectangular with a ratio of residual induction to measured induction at 0.4 kA/m (5 Oe) of the order of 0.85 to 0.90. The coercive force is typically of the order of 40 to 100 A/m (0.5 to 1.2 Oe). These characteristics indicate the suitability of this ferrite for use in controlling hysteresis loops in switching devices involving residual magnetization (latching devices).

3. Polycrystalline spinels for microwave applications

3.1. General

Historically, spinel ferrite structures were the first type of ferrites to be used for microwaves (1952); they had previously been used for hf applications. The general methods of preparation do not differ greatly from those described in section 2. The advantage of spinel ferrites is that their magnetizations are higher than those of garnets (limited at 155 kA/m or 1950 G) and may reach 475 kA/m (5000 G). The magnetization is given by the Néel's theory (1948) as shown by Pauthenet (1951, 1952) and Gorter (1954).

3.2. Mn–Mg ferrites

The basic chemical composition is $Mn_xMg_yFe_zO_4$ with $x + y + z = 3$. Although they were the first ferrites used for microwave applications, these ferrites are fairly complex. The Mn^{2+} ion has the same electronic configuration as Fe^{3+}. However, part of the manganese can exist in the Mn^{3+} state. In addition, the metallic ions are distributed between tetrahedral and octahedral sites in a manner which depends upon the heat treatment applied; thus x, y and z are determined experimentally, in optimizing the characteristics, for each of the synthesis techniques used. Magnetizations of from $M_s = 160$ to 190 kA/m ($4\pi J_s = 2000$–2400 G), with Curie temperatures between 270 and 300°C, are obtained. The temperature coefficients of M_s are comparatively high ($\alpha_{-40}^{+85} \approx 2.6 \times 10^{-3}$). The linewidths ΔH due mainly to broadening by anisotropy, are of the order of 25 kA/m (300 Oe), while spin wave linewidths are low, of the order of 0.3 kA/m (3.5 Oe). These ferrites are therefore applicable to low-power microwave usage in the 7 to

15 GHz range, where temperature conditions are not severe. Useful compositions appear to be situated, in practice, in the intervals defined by

$$0.10 < x < 0.15 \qquad 0.4 < y < 0.5 \qquad 0.7 < z < 0.9.$$

Magnetization can therefore be reduced by substituting aluminium for part of the iron (usually between 0 and 10%) to the detriment of the Curie temperature and temperature coefficient. The hysteresis loop of this type of material is rectangular in shape: the coercive force field is fairly weak (0.1 kA/m approximately 1 Oe). Typical characteristics of industrial materials are given in table 9. The magnetization versus temperature curves of these materials are given in figs 33, 34.

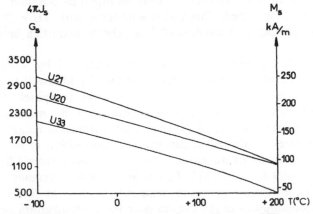

Fig. 33. Saturation magnetization versus temperature of some Thomson-CSF manganese–magnesium ferrites.

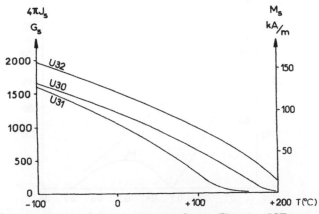

Fig. 34. Saturation magnetization versus temperature of some Thomson-CSF manganese–magnesium ferrites substituted by aluminium.

3.3. Nickel ferrites

The nickel ferrite $NiFe_2O_4$ has a Curie temperature of about 570°C and its
magnetization $M_s = 255$ kA/m ($4\pi J_s = 3200$ G) is therefore much more stable with
temperature ($\alpha_{-40}^{+85} \approx 0.43 \times 10^{-3}$) than those of the former materials. However, the
Ni ion, having an orbital magnetic moment, is a relaxing ion; it follows that the
Ni ferrite is not suitable for devices with very low insertion losses, but rather for
those required to withstand a certain peak power. The effect of the magnetocrystalline anisotropy on ΔH of the Ni ferrite is fairly high, but it can be
reduced by low substitutions of Co^{2+} whose anisotropy is very high and is of
sign opposite to that of the iron ions. This cobalt ion further increases ΔH_k. This
leads to a composition of the $N_{1-x}Co_xFe_2O_4$ type (with $x \approx 0.02$). However, for
physicochemical reasons, it is difficult to obtain a dense product with no
dielectric losses. In order to reduce these as much as possible, a small percentage of manganese is added. This yields a material with $\Delta H = 20$ kA/m (250 Oe),
$\Delta H_k = 2$ kA/m (25 Oe) and $\tan \delta_\epsilon < 10^{-3}$, all these quantities being measured at
9 GHz.

Materials with a lower magnetization are obtained by substitution of aluminium for a particular quantity of iron as shown by Maxwell and Pickart (1953).
The Al^{3+} ions preferably occupy octahedral sites in spinel structures, but a
certain quantity of Al^{3+} ions occupy tetrahedral sites in a manner which depends
upon the heat treatment. Figure 35 shows the saturation magnetization and the
Curie temperatures of the materials whose composition is $NiFe_{2-2x}Al_{2x}O_4$ and
which have been carefully annealed (very slow cooling from 1300°C). There is a
compensation point for $x = 0.315$. Figure 36 gives the gyromagnetic factors and
the linewidths measured at room temperature and at 9 GHz on the same
materials. Discontinuities in the curves near the compensation points, are due to

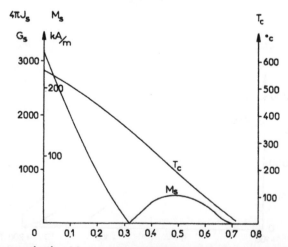

Fig. 35. Saturation magnetization M_s at room temperature and Curie temperature T_c of nickel spinels
having the composition $NiFe_{2-2x}Al_{2x}O_4$. These materials have been carefully annealed.

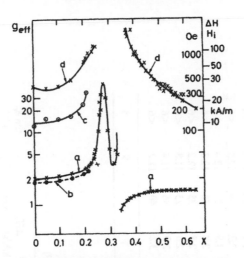

Fig. 36. Gyromagnetic resonance properties of nickel spinels having the composition $NiFe_{2-2x}Al_{2x}O_4$ which have been carefully annealed (a) gyromagnetic factor g_{eff} measured at 9.5 GHz, (b) gyromagnetic factor g'_{eff} resulting of the Okamura's formula $\omega = \gamma' \mu_0 \, (H_r + H_i)$, (c) internal field H_i, (d) resonance linewidth ΔH at 9.5 GHz.

sublattice effects described by Wangsness (1953). In fact, industrial materials, such as the nickel ferrites, contain: (i) small quantities of cobalt for adjusting the magnetocrystalline anisotropy and possibly increasing ΔH_k (power behaviour); (ii) small quantities of manganese to improve dielectric losses. Aluminium substitutions used in practice are limited roughly to the interval $0 < x < 0.2$ up to the compensation point and to a value approximately $x = 0.5$ after the compensation point.

Magnetization can be raised by partial substitution of zinc for the nickel, as expressed by the formula $Ni_{1-x}Zn_xFe_2O_4$ in a substitution interval $0 < x < 0.5$. The Curie point decreases with x (see fig. 37). These effects were described by Guillaud (1951). Typical industrial Ni ferrites for microwave usage are given in table 10. The material Z50 is a NiZn ferrite, while the others are aluminium-substituted ferrites. They are included in Thomson-CSF catalogues, except for the material TTZ-113 which is extracted from that of Trans Tech. The latter material is a typical Ni ferrite with a high aluminium substitution (x approximately equal to 0.5) for which the effective Landé factor is much lower than 2. The saturation magnetizations of these materials versus temperature are given in figs 38, 39, 40.

3.4. Lithium ferrites

The lithium ferrite $LiFe_5O_8$ or $Li_{1/2}Fe_{5/2}O_4$ has a magnetization of approximately $M_s = 285$ kA/m ($4\pi M_s = 3600$ G) with a high Curie temperature, approximately 645°C (see for instance Vassiliev 1962, Vassiliev and Lagrange 1966) so that its stability with temperature is similar to that of the Ni ferrite. It should be noted

TABLE 10

Characteristics of Ni ferrites. Z 50 ... N 29 are from the Thomson-CSF list. TT-2-113 is from the Trans Tech. list

Type	M_s kA/m	$4\pi J_s$ G	T_c °C	g_{eff}	ΔH kA/m	ΔH Oe	ΔH_{eff} kA/m	ΔH_{eff} Oe	ΔH_k kA/m	ΔH_k Oe	ϵ_r	$\tan \delta_e$ 10^{-4}
Z 50	382.1	4800	390	2.10	15.1	190	0.8	10	0.8	10	14.1	<6
N 25	254.7	3200	560	2.30	19.9	250	8.0	100	2.1	26	12.7	<6
N 40	246.8	3100	560	2.30	29.5	370	10.3	130	2.7	34	12.5	<6
N 28	218.9	2750	550	2.30	26.3	330	8.0	100	1.9	24	12.4	<6
N 41	199.0	2500	530	2.30	29.5	370	10.3	130	2.8	35	12.3	<6
N 26	187.1	2350	520	2.30	23.9	300	8.0	100	2.8	35	12.2	<6
N 27	175.1	2200	500	2.30	26.3	330	8.0	100	2.0	25	11.8	<6
N 42	151.2	1900	480	2.30	27.9	350	10.3	130	2.9	36	11.4	<6
N 29	111.4	1400	450	2.40	30.2	380	8.0	100	3.2	40	11.2	<6
TTZ-113		500	120	1.54		150					9	8

TABLE 11

Characteristics of lithium ferrites (Thomson-CSF)

Type	M_s kA/m	$4\pi J_s$ G	T_c °C	g_{eff}	ΔH kA/m	ΔH Oe	ΔH_{eff} kA/m	ΔH_{eff} Oe	ΔH_k kA/m	ΔH_k Oe	ϵ_r	$\tan \delta_e$ 10^{-4}
A 50	398	5000	450	2.06	13.5	170	0.3	4	0.25	3	15.3	<5
A 37	294.5	3700	565	2.08	31.8	400	0.3	4	0.25	3	16	<5
A 30	238.8	3000	555	2.08	35.8	450	0.3	4	0.25	3	16.4	<5
A 28	222.9	2800	540	2.08	35.8	450	0.3	4	0.25	3	16.6	<5
A 23	183.1	2300	505	2.08	35.8	450	0.3	4	0.25	3	16.8	<5
A 230	183⁻	2300	500	2.07	35.8	450			0.64	8	16.6	<5

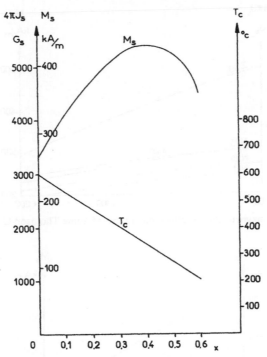

Fig. 37. Saturation magnetization M_s at room temperature (normalized to X-ray densities) and Curie temperature T_c of nickel spinels having the composition $Ni_{1-x}Zn_xFe_2O_4$ (after Pauthenet 1952).

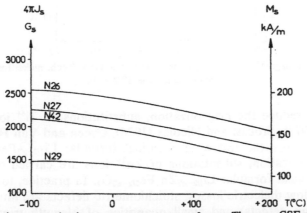

Fig. 38. Saturation magnetization versus temperature of some Thomson-CSF nickel ferrites.

that, like the yttrium–iron garnet, the Li garnet is a substance which only has trivalent iron ions as magnetic ions. This results in a very narrow spin wave linewidth ΔH_k, of the order of 0.25 kA/m (3 Oe), and hence, a narrow effective linewidth ΔH_{eff}. This makes the lithium ion ferrites more suitable for uses at low microwave power levels with low direct losses.

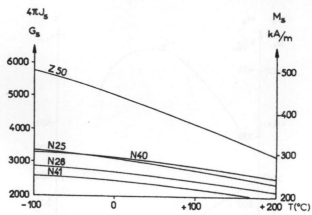

Fig. 39. Saturation magnetization versus temperature of some Thomson-CSF nickel ferrites.

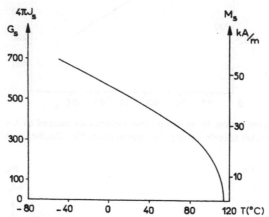

Fig. 40. Saturation magnetization versus temperature of a Trans Tech. nickel ferrite substituted by aluminium (type TT2-113).

In order to reduce the magnetization, substitutions of Ti^{4+} ions, studied by Baba et al. (1972), Sroussi and Nicolas (1976), Green and Van Hook (1977) are carried out, as per the following chemical formula: $Li_{1/2+y/2}Fe_{5/2-3y/2}Ti_yO_4$. To increase magnetization, substitutions of Zn^{2+} ions are carried out, as per the following chemical formula: $Li_{1/2-x/2}Zn_xFe_{5/2-x/2}O_4$. In practice, to obtain correct densities and low dielectric losses, stoichiometric defects are introduced (a few percent of iron defects) and small quantities of bismuth trioxide Bi_2O_3 are introduced (approximately 10^{-3} mol added to the above formulae) and of manganese (e.g. of the order of 0.04 mol of MnO_2 added to the same formulae). The characteristics, i.e. magnetization, Curie temperature and linewidth measured on these materials are given in figs. 41 and 42. The spin wave linewidth ΔH_k and effective linewidth of these materials are low, 0.2 to 0.3 kA/m (2 to 4 Oe). The hysteresis loops can be remarkably rectangular. The coercive force

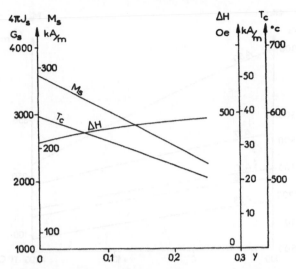

Fig. 41. Saturation magnetization M_s, resonance linewidth ΔH (at 10 GHz) at room temperature and Curie temperature T_c of lithium spinels having the composition $Li_{1/2+y/2}Fe_{5/2-3y/2}Ti_yO_4$ with additions of Mn and Bi ions.

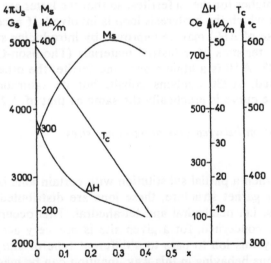

Fig. 42. Saturation magnetization M_s, resonance linewidth (at 10 GHz) at room temperature and Curie temperature of lithium spinels having the composition: $Li_{1/2-x/2}Zn_xFe_{5/2-x/2}O_4$ with additions of Mn and Bi ions.

fields H_c typically comprise between 0.1 and 0.2 kA/m (1 to 2 Oe). The ratio of the remanent induction B_r to that measured at a maximum field equal to $5H_c$ can reach 0.9 to 0.95. It should be noted that the Mn–Mg spinels, contrary to the Ni ferrites, also have rectangular hysteresis loops with characteristics approximating those of the Li ferrites. However, the stability of B_r with tem-

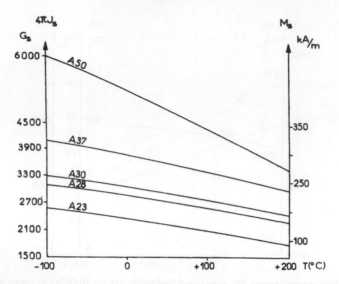

Fig. 43. Saturation magnetization versus temperature of some Thomson-CSF lithium ferrites.

perature is much higher for the Li ferrites, so that the latter are usually preferred for applications in which the hysteresis loop is involved. If necessary, the power behaviour (increase of ΔH_k) may be improved by introducing small quantities of cobalt. The characteristics of industrial materials (Thomson-CSF) are given in table 11 and fig. 43. A 50 is a lithium and zinc ferrite, the other materials being titanium substituted. A 230 contains cobalt, but its saturation magnetization versus temperature curve is practically the same as that of A 23.

3.5. Ionic distribution between crystallographic sites

3.5.1. General
It is known that when a partial substitution with certain ions is made in ferrites having a spinel or garnet structure, these ions are distributed in two different categories of sites, i.e. octahedral and tetrahedral. This occurs when the preference, of the ion concerned, for a given site is not very accentuated, i.e. the energy required by this ion to pass to a different type of site is comparatively low. Among the ions behaving in this way, mention can be made of Fe, Al, Mn, Cu, etc.

3.5.2. Distribution of Al and Fe ions in the nickel ferro-aluminates
By way of example, we will examine the phenomena which determine the distribution of Fe and Al ions in a case we have discussed in detail: that of the nickel ferro-aluminates, whose chemical formula is $NiFe_{2-2x}Al_{2x}O_4$. In these spinels, aluminium preferably occupies the octahedral sites, although part of it occupies tetrahedral sites. Maxwell (1953) showed that the equilibrium depends

on the heat treatment. The dynamic behaviour in reaching equilibrium was studied by Nicolas (1966, 1967a).

Controlling factors

The distribution of Fe and Al ions between the octahedral sites (or B sites), i.e. two per molecule, and the tetrahedral sites (or A sites), i.e. one per molecule, is determined by the value of B given in table 12. It may be considered that we have a reversible chemical reaction which can be written as follows

$$\underbrace{Fe^A + Al^B}_{1} \rightleftarrows \underbrace{Al^A + Fe^B}_{2}.$$

The speeds of reaction from 1 to 2 and 2 to 1 can be expressed by the following formulae

$$V_{12} = \left(\frac{dB}{dt}\right)_{12} = K_{12}(2x - B)(1 - B) \tag{32}$$

$$V_{21} = \left(\frac{dB}{dt}\right)_{21} = K_{21}B(1 - 2x + B). \tag{33}$$

The resultant speed will be

$$V = \frac{dB}{dt} = V_{12} - V_{21}. \tag{34}$$

Value of B at equilibrium

When the substance is brought, for an infinitely long period of time, to an absolute temperature T, the equilibrium ($V = 0$) reached is in accordance with a value $B = B_1$ expressed by

$$K = \frac{K_{12}}{K_{21}} = \frac{B_1}{1 - B_1} \frac{1 - 2x + B_1}{2x - B_1} = \exp(-W/kT) \tag{35}$$

where k is Boltzmann's constant. W is shown in electron-volts in fig. 44, at two different equilibrium temperatures 550°C and 605°C.

TABLE 12
Distribution of the cations among the crystallographic sites in the spinels having the composition $Ni\,Fe_{2-2x}\,Al_{2x}\,O_4$

Site A Tetrahedral	Site B Octahedral
	Ni
$(1 - B)$ Fe	$(1 - 2x + B)$ Fe
B Al	$(2x - B)$ Al

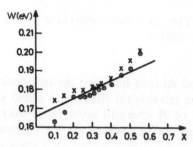

Fig. 44. Activation energy in electron-volts which controls the distribution of Fe and Al ions among the octahedral and tetrahedral sites in spinels having the composition $NiFe_{2-2x}Al_{2x}O_4$. ⊙ equilibrium at 550°C, x equilibrium at 605°C.

Dynamic behaviour in reaching equilibrium

Assuming that at an absolute temperature T and an initial time $t = 0$, the value of B is equal to B_0, different from the value of B_1 at equilibrium, the value of B will evolve in time from the value B_0 to the value B_1, when the temperature is kept constant at the value T, in accordance with the formula

$$z = \frac{B - B_1}{B - B_0} \frac{B_0 - B_2}{B_0 - B_1} = \exp(-t/\tau) \qquad (36)$$

with
$$B_2 = -\frac{2KX}{B_1(1 - K)}. \qquad (37)$$

The time constant τ can be written

$$\tau = \frac{1}{(B_1 - B_2)(1 - K)} \frac{1}{K_{21}} = \frac{(2x - B_1)(1 - B_1)}{2[x(1 - x) - (x - B_1)^2]} \frac{1}{K_{21}}. \qquad (38)$$

K_{21} can be considered at the frequency of permutation of the couples Al–Fe in the easiest direction of permutation. This quantity depends upon an energy of activation

$$K_{21} = f_0 \exp(-E/kT) \qquad (39)$$

where $E = 1.9 \mp 0.2$ eV. These relations satisfactorily explain the phenomena up to a temperature of approximately 600°C to 700°C.

It has been possible to show that the basic phenomenon was a cation diffusion by vacancies which have been created at higher temperatures: above 900°C. f_0 is thus a frequency proportional to the number of such cation vacancies. It has been verified that f_0 was strongly dependent upon the stoichiometry deviation and the conditions of preparation of the material. The value of f_0 found for substances prepared from stoichiometric proportions and sintered in air at 1350°C was $f_0 = 4 \times 10^7 \text{ s}^{-1}$.

4. Hexagonal ferrites for microwave applications

4.1. Introduction

Apart from those having garnet and spinel structures, the ferrites used essentially at microwave frequencies have a hexagonal structure. These materials have an axis of symmetry, c, and a high magnetocrystalline anisotropy. They are prepared by the ceramic technique similar to that already described, subject to a few variants. In particular, during pressing the grains are oriented by a magnetic field in order to align the c-axes of all the crystallites in the same direction. This yields a sample similar to a permanent magnet. When a microwave field is applied in a plane perpendicular to the magnetization direction, a gyro-resonance effect, which is expressed in a manner similar to that given in section 1, is observed. The resonance frequency f_r is related to the applied field (in the direction of magnetization) by a formula similar to (9), but allowing for an anisotropy field H_a, such that

$$f_r = (\gamma\mu_o/2\pi)\{[H_z + H_a + (N_x - N_z)M_z][H_z + H_a + (N_y - N_z)M_z]\}^{1/2}. \qquad (40)$$

Owing to the high values of H_a, which may be up to 2780 kA/m (35 kOe), resonance may occur at millimetric frequencies in the absence of a continuous field, H_z, or with a weak field, H_z, which is therefore easy to handle. This is the main advantage of this type of material.

What are the chemical formulae used for this type of material? These are essentially uniaxial, hexagonal materials of the M and W types described in the general chapter of Kojima and Sugimoto (see volume 3/4). As in the case of garnets and spinels, high resistivity and low dielectric losses are required for the microwave applications. This implies that all the Fe ions have to be in the trivalent state. When this condition is met, the compositions are chosen in order to obtain, with an adequate magnetization, an anisotropy field compatible with the operating frequency. Other conditions, such as a good stability with temperature, are also sought. Compositions of hexagonal ferrites which are suitable for microwave applications were given by some authors, especially Du Pré et al. (1958), Rodrigue et al. (1962), De Bittetto (1964), Taft (1964), Deschamps (1969), Dixon (1971), Okazaki et al. (1974), and Sweschenikow et al. (1976). We are going to summarize the main points of their results.

4.2. M-type hexagonal ferrites

These ferrites are derived from barium or strontium ferrites, whose formulae are $BaFe_{12}O_{19}$ and $SrFe_{12}O_{19}$, by partial substitutions of Al^{3+} ions (or Ga^{3+} or Cr^{3+} ions) for Fe^{3+} ions, which causes the magnetization to increase and the magnetocrystalline anisotropy field to increase; by substitutions of the groups $Ti^{4+}Me^{2+}$ or $Ge^{4+}Me^{2+}$ (with $Me^{2+} = Co, Ni, Zn \ldots$) for Fe^{3+} ions, which causes both the magnetization and the anisotropy field to decrease; by combined

TABLE 13

Characteristics of some hexagonal ferrites described in the literature as used at millimeter wave frequencies

References	Compositions	Types	M_s kA/m	$4\pi I_s$ G	H_a kOe	H_a kA/m	Frequencies of uses GHz	10^4 tan δ, at 9 GHz
Taft (1964)	$SrAl_xFe_{12-x}O_{19}$	M						
	x = 0	M	332	4180	19	1512	60	
	x = 0.4	M	262	3300	20	1590	64	10
	x = 0.8	M	193	2430	23.5	1870	70	10
	x = 1	M	165	2080	25	1990	73	10
	x = 1.6	M			31	2470	86	
	x = 1.9	M			34	2700	93	
Deschamps (1968)	$Ba_{0.02}TiNiAl_{1.2}$ $Fe_{8.6}Mn_{0.2}O_{19}$	M					36	
	$SrTi_{0.3}Ni_{0.3}Al_{0.3}$ $Fe_{10}Mn_{0.2}O_{19}$	M					72	
Okazaki et al. (1974)	$BaZn_{0.3}Ti_{0.3}Fe_{11.4}O_{19}$	M			13.7	1090	45	20
	$BaAl_{0.3}Fe_{11.7}O_{19}$	M			17.5	1390	55	20
	$SrNi_{0.3}Ge_{0.3}Al_{1.86}$ $Fe_{9.54}O_{19}$	M			27.3	2170	82	60
	$SrNi_{0.3}Ge_{0.3}Al_{2.3}$ $Fe_{9.1}O_{19}$	M			32.6	2595	96	80
Rodrigue et al. (1962)	$BaNi_{2-x}Co_xFe_{15.6}O_{27}$	W						
	x = 0.1	W			11	875	37	
	x = 0.2	W					36	
	x = 0.5	W			7	560	23	
Taft (1964)	$BaNi_2Al_xFe_{15.6-x}O_{27}$	W						
	x = 0.8	W	194	2440	15.2	1210	49	10
	x = 1.46	W	165	2080	17.9	1425	55	30
	x = 0.2	W			19.7	1570	62	

substitutions of both of the above types, which result in better stability of H_a with temperature.

4.3. W-type hexagonal ferrites

These are derived from barium ferrites, whose formula is $Ni_2BaFe_{16}O_{27}$, by partial substitution of Al^{3+} ions for Fe^{3+} ions, which causes the magnetization to increase and the anisotropy field to increase; by partial substitution of Co^{2+} ions for Ni^{2+} ions, which causes the anisotropy field to decrease, while the magnetization remains unchanged. Lastly, the W-type ferrites can be used to produce devices such as circulators with resonance frequencies lower than those of the M-type.

4.4. Typical applications

A few typical applications described by various authors are shown in table 13. It should be noted that some of these formulae exhibit Fe defects with respect to the stoichiometry or addition of Mn ions, which usually enable the dielectric characteristics to be improved.

5. Use of monocrystalline ferrites at microwave frequencies

Monocrystalline ferrites differ essentially from the polycrystalline ferrites in that they feature a linewidth ΔH of a lower order of magnitude: 30 A/m (0.4 Oe). This property is used in specific applications, such as, electronically tuned filters, and oscillators, as studied by De Grasse (1959), Patel (1961), Anderson (1962), Désormière (1963) and Comstock (1964). In such devices the materials are used in the form of a sphere approximately 1 mm in diameter. The samples behave like microwave cavities whose resonance frequency may be tuned by variation of the applied static magnetic field. The material most currently used is the yttrium garnet $Y_3Fe_5O_{12}$ which can be used at frequencies higher than 3.3 GHz. Below this frequency, non-linear, first-order effects appear and garnets with a lower magnetization, with galium substitution, whose formula is $Y_3Fe_{5-5z}Ga_{5z}O_{12}$ are generally used. Typical monocrystalline materials available in the trade are given in table 14.

TABLE 14

Characteristics of single crystal garnet spheres which are comercially available

Materials	M_s kA/m	$4\pi J_s$ G	ΔH A/m	Oe	Sphere diameter mm	Suppliers
Pure YIG	142	1780	24	0.3	0.8	Litton Airtron (USA) Cristal-Tec (France)
Gallium substi- tuted YIG	55.7 67.6	700 850	52 52	0.65 0.65	0.8 1.0	Litton Airtron (USA) Cristal-Tec (France)

6. Conclusion

We have endeavoured in this paper to describe a microwave ferrite. The physical phenomenon exploited is the gyromagnetic resonance that appears when the material is subjected to both a static magnetic field and to a microwave field. This effect is inherently non-reciprocal, which accounts for the tensorial (non-scalar) nature of the magnetic permeability. This non-reciprocity is widely used in various devices. The saturation magnetization, M_s, is a fundamental quantity, not only as a factor of efficiency, but also because of the existence of a natural resonance (in the absence of an applied static field) in the internal fields. Consequently, magnetization is the first quantity to be considered in the choice of material for a given application. As in any resonance effect, the linewidth is of prime importance. It is fundamentally related to the damping effects of spin movements, which seem possible to approach (without thoroughly understanding them) by considering either the linewidth off-resonance (ΔH_{eff}) or the non-linear effects related to the spin waves (ΔH_k). Conversely, the linewidth of the main mode (ΔH) observed at resonance, on the polycrystals, is broadened by the magnetocrystalline anisotropy effects, or by spurious effects related to faults and porosity.

Ferrites now used at microwave frequencies have spinel, garnet, or hexagonal structures. It was decided to describe only the chemical compositions which seem significant in the light of current techniques, while trying, at the same time, to go into the details of the composition intervals currently used. In line with this practical approach, the characteristics of a number of spinel and garnet type industrial materials have been given. Conversely, it should be noted that the examples given for hexagonal materials were taken from the literature, since to our knowledge, no material of this type has yet been marketed.

It is to be remarked that the field of microwave ferrites has greatly evolved during the last decade and this trend will probably continue during the coming decade. Will progress be made in any fields of theory: for instance for a better understanding of the magneto-crystalline anisotropy or of the relaxation mechanisms? It is difficult to predict that now. On the other hand it is more than likely that progress will be made in materials. Garnets will probably improve more; in some cases linewidths could be lessened by an ultimate reduction of porosity, which could be achieved by judicious choice of compositions and progress in technology. We may expect a development of lithium spinel ferrites and an improvement of their characteristics by a better knowledge and control of sintering. The development of uses in the millimetre frequency range, which is fairly slow but constant, would instigate more work in the field of hexagonal ferrites, and lead to their availability on the market.

References

Anderson, L.K., 1962, Stanford Microwave Laboratories. Techn. Rep. **880**.

Baba, P.D., G.M. Argentina, W.E. Courtey, G.F. Dionne and D.H. Temme, 1972, IEEE Trans. Magn. **MAG-8**, 83.

Bertaut, F. and F. Forrat, 1956, C.R. Hebd. Séan. Acad. Sci. **242**, 382.

Bloembergen, N., 1956, Proc. IRE. **44**, 10, 1259.

Bolle, D.M. and L. Whicker, 1975, IEEE Trans. Magn. **MAG-11**, 3, 907.

Borghese, C. and R.L. Roveda, 1971, J. Phys. **C1-32**, 150.

Comstock, R.L., 1964, IEEE Trans. on Microwave Theory and Techniques **MTT-12**, 6, 599.

Courtois, L. and A. Deschamps, 1967, C.R. Hebd. Séan. Acad. Sci. **264**, 1333.

De Bitetto, D.J., 1964, J. Appl. Phys. **35**, 12, 3482.

de Gennes, P.G., C. Kittel and A.M. Porthis, 1959, Phys. Rev. **116**, 323.

De Grasse, R.W., 1959, J. Appl. Phys. **30**, 1555.

Deschamps, A., 1969, Z. Angew. Phys. **26**, 2, 190.

Deschamps, A., 1970, Câbles et Transmission **4**, 357.

Désormière, B., 1963, Propriétés des monocristaux de grenat d'yttrium–fer et application à l'étude de filtres microondes accordables. Thèse 3è cycle. Université de Paris.

Dixon, S., Jr., 1971, J. Appl. Phys. **42**, 4, 1732.

Du Pré, F.K., D.J. De Bitetto, F.G. Brockman, 1958, J. Appl. Phys. **29**, 7, 1127.

Geller, S., 1960, J. Appl. Phys. **31**, 30S.

Geller, S. and M.A. Gilleo, 1957, Acta Cryst. **10**, 239.

Geller, S., H.J. Williams, G.P. Espinosa and R.C. Sherwood, 1964, Phys. Rev. **136**, A1650.

Gorter, E.W., 1954, Philips Res. Rev. **9**, 295.

Green, J.J. and F. Sandy, 1974, IEEE Trans. on Microwave Theory and Techniques **MTT-22**, 6, 641.

Green, J.J. and F. Sandy, 1974, IEEE Trans. on Microwave Theory and Techniques **MTT-22**, 6, 645.

Green, J.J. and H.J. Van Hook, 1977, IEEE Trans. on Microwave Theory and Techniques **MTT-25**, 2, 155.

Guillaud, Ch., 1951, J. de Phys. et le Radium **12**, 3.

Harrison, G.R. and L.R. Hodges, Jr., 1961, J. Am. Ceram. Soc. **44**, 214.

Hartmann-Boutron, F., 1964a, Phys. Kondens. Materie **2196**, 4, 80.

Hartmann-Boutron, F., 1964b, C.R. Hebd. Séan. Acad. Sci. **259**, 2085.

Hermosin, A., 1970, Contribution à l'étude de la mesure des caractéristiques diélectriques à 9 GHz. Thèse 3è cycle. Université d'Orsay.

Hogan, C.L., 1952, The Bell System Technical Journal, **31**, 1, 15.

Hudson, A.S., 1970, J. Phys. D (Appl. Phys.) **3**, 251.

Inglebert, R.L., 1973, Etude de la relaxation d'ondes de spins dans des grenats polycristallins par résonance ferrimagnétique et par pompages parallèle et perpendiculaire. Thèse 3è cycle. Université d'Orsay.

Inglebert, R.L. and J. Nicolas, 1974, IEEE Trans. Magn. **MAG-10**, 3, 610.

Kittel, C., 1948, Phys. Rev. **73**, 2, 155.

Kittel, C., 1959, Phys. Rev. **115**, 6, 1587.

Koelberer, R.E. and C.E. Patton, 1977, IEEE Trans. Magn. **MAG-13**, 5, 1230.

Lalau, M.C., 1978, Laboratoire Central de Recherches Thomson-CSF, private communication.

Landolt-Börnstein, M., 1970, Magnetic and other properties of oxides and related compounds, parts a and b, in numerical data and functional relationships in science and technology, vol. 4. (Springer-Verlag, Berlin, Heidelberg, New-York).

Lax, B. and K.J. Button, 1962, Microwave ferrites and ferrimagnetics. (McGraw-Hill, New York).

Maxwell, L.R. and S.J. Pickart, 1953, Phys. Rev. **92**, 1120.

Néel, L., 1948, Ann. Phys. **3**, 137.

Nicolas, J., 1966, Etude par un nouveau moyen d'un phénomène de diffusion dans des spinelles ferrimagnétiques. Thèse. Université de Paris, edited by Thomson-CSF B.P. 10 91401 Orsay (France).

Nicolas, J., 1967a, J. Phys. Chem. Solids **28**, 847.

Nicolas, J., 1967b, Ann. Radioélectricité **22**, 109.

Nicolas, J. and B. Désormière, 1977, Techniques de l'Ingénieur 6, E 248.

Nicolas, J. and A. Lagrange, 1970, Proc. Int. Conf. on Ferrites, Japan (University Park Press, London) p. 527.

Nicolas, J., A. Lagrange, R. Sroussi, 1970, IEEE Trans. Magn. MAG-6, 3, 608.

Nicolas, J., A. Lagrange, R. Sroussi, M. Hildebrandt and J. Llabrès, 1971, J. de Phys. Cl, 2–3, 32, 153.

Nicolas, J., A. Lagrange, R. Sroussi and R.L. Inglebert, 1973, IEEE Trans. Magn. MAG-9, 546.

Okamura, T., Y. Torizuka and Y. Kojima, 1952, Phys. Rev. 88, 1425.

Okazaki, T., H. Yutaka and Y. Akaiwa, 1974, Electronics and Communications in Japan, 57, 7, 188.

Patel, C.N., 1961, Stanford Electronics Laboratories Techn. Rep. 411, 1.

Patton, C.E., 1969, Phys. Rev. 179, 352.

Patton, C.E., 1972, IEEE Trans. Magn. MAG-8, 433.

Pauthenet, R., 1951, Aimantation spontanée des ferrites. Thèse Grenoble.

Pauthenet, R., 1952, Ann. Phys. 7, 710.

Pauthenet, R., 1958, Les propriétés magnétiques des ferrites d'yttrium et de terres rares de formule 5 Fe_2O_3 3 M_2O_3. Thèse Grenoble.

Pircher, G., 1969, Ferrites et Grenats. Phénomènes non-linéaires (Dunod, Paris).

Polder, D., H.H. Wills, 1949, Phil. Mag. 40, 99.

Rodrigue, G.P., J.E. Pippin, M.E. Wallace, 1962, J. Appl. Phys. Suppl. 33, 1366.

Saunders, J.H. and J.J. Green, 1961, J. Appl. Phys. 32, 161S.

Schlömann, E., 1956, IEEE Spec. Pub. Proc. Conf. Mag. and Mag. Mat. 91, 600.

Schlömann, E., 1958, J. Phys. Chem. Solids 6, 2–3, 242.

Schlömann, E., 1970, J. Appl. Phys. 41, 204.

Schlömann, E., 1971, J. Phys. 32, 2–3, 443.

Schlömann, E. and R.I. Joseph, 1961, J. Appl. Phys. 32, 3, 165S.

Schlömann, E., J.J. Green and U. Milano, 1960, J. Appl. Phys. 31, 386S.

Seiden, P.E., 1964, Phys. Rev. 133, 3A.

Simonet, W., 1977, Report no. 77, 70081 (Délégation Générale à la Recherche Scientifique et Technique, Paris).

Smolenskii, G.A. and V.P. Polyakov, 1964, Fiz. Tverd. Tela (USSR) 6, 2556 (in Russian); English transl. Sov. Phys.–Solid State (USA) 6, 2038.

Sparks, M., 1964, Ferromagnetic relaxation theory (McGraw-Hill, New York).

Sroussi, R. and J. Nicolas, 1974, IEEE Trans. Magn. MAG-10, 3, 606.

Sroussi, R. and J. Nicolas, 1976, Rev. Tech. Thomson-CSF 8, 1, 41.

Suhl, H., 1956, Proc. IRE 44, 10, 1270.

Sweschenikow, J.A., E.K. Merinow and B.P. Pollak, 1976, Electronik 26, 7, 262.

Taft, D.R., 1964, J. Appl. Phys. 35, 3, 2, 776.

Vassiliev, A., 1962, Ferrospinelles comprenant l'ion Li^+ et contribution à l'étude de leurs propriétés magnétiques. Thèse, edited by Thomson-CSF B.P. 10 91401 Orsay (France).

Vassiliev, A. and A. Lagrange, 1966, IEEE Trans. Magn. MAG-2, 707.

Vassiliev, A. and J. Nicolas, 1965, Z. Angew. Phys. 18, 557.

Vassiliev, A., J. Nicolas, M. Hildebrandt, 1961, C.R. Hebd. Séan. Acad. Sci. 252, 2529.

Von Aulock, W.K., 1965, Handbook of microwave ferrite materials (Academic Press, New York).

Vrehen, Q.H.F., 1969, J. Appl. Phys. 40, 1849.

Wangsness, R.K., 1953, Phys. Rev. 91, 5, 1085.

West, R.G., 1971, J.A.P. 42, 4, 1730.

Winkler, G., P. Hansen and P. Holst, 1972, Philips Res. Repts. 27, 151.

chapter 5

CRYSTALLINE FILMS FOR BUBBLES

A.H. ESCHENFELDER

IBM Corporation
San Jose, CA 95193
USA

Ferromagnetic Materials, Vol. 2
Edited by E.P. Wohlfarth

CONTENTS

1. Introduction

Magnetic bubbles are cylindrical domains of reverse magnetization in thin magnetic films. The films must have uniaxial anisotropy, normal to the film plane, large enough to overcome the demagnetizing forces due to the film geometry. The bubbles are stable over a range of magnetic bias fields and can be moved in the film by externally controlled magnetic field gradients. They are, therefore, useful for solid state devices (especially for information storage). It is not within the scope of this chapter to discuss the applications of bubbles. On the other hand, the purpose of this chapter is to identify the parameters of bubbles that are important for those applications, to tabulate the relationship of those practical parameters to the more fundamental magnetic and physical characteristics of the films, to discuss briefly the origin of phenomena that are unique to bubbles, and to indicate how materials can be selected to achieve particular bubble properties. More detail on these subjects and also on devices and applications can be found in the textbook by Eschenfelder (Eschenfelder 1980).

Gaussian units will be used throughout this chapter. Table A.1 in Appendix A is included to provide conversion from these Gaussian units to the alternative SI units.

2. Review of elementary bubble concepts

2.1. Static properties

Magnetic bubbles of diameter d exist in films of appropriate magnetic properties in a range of film thickness, h, and applied bias field, H_B, which is perpendicular to the plane of the film (see fig. 1). The film must have an anisotropy energy E_K, in its magnetization, M_S, such that the "easy" direction for the magnetization is perpendicular to the film plane. The bubble is actually defined by a domain wall of thickness $\delta = \pi\sqrt{A/K}$ where A is the "exchange constant" and K is the "anisotropy" constant of the bubble film. The domain wall has an energy per unit area, $\sigma_W = 4\sqrt{AK}$. Two important practical parameters defined for bubble materials are the "quality factor", Q, and the "characteristic length", l. These

299

parameters are defined in terms of the fundamental magnetic parameters as

$$Q = \frac{K}{2\pi M_S^2} \tag{1}$$

$$l = \frac{\sigma_W}{4\pi M_S^2} = \frac{4(AK)^{1/2}}{4\pi M_S^2}. \tag{2}$$

Q is significant because it denotes the relative strength of the anisotropy in preserving the perpendicular magnetization against the demagnetizing effect of the film configuration. For stable bubbles $Q > 1$. The characteristic length is useful in that other significant dimensions are conveniently described in units of l, e.g., d/l, h/l, δ/l.

The relationships between these static parameters have been extensively explored theoretically. A good point of departure for an understanding of how this is done is the paper by Thiele (1970). The results of such calculations will be given in useful form in section 3.

The bubble size and stability are determined by the combination of three components of energy: the wall energy necessary to sustain the bubble domain wall, the demagnetization energy saved by the presence of the bubble and the magnetic energy of the bubble in the externally applied field. As a result, bubbles are stable in a range of applied field, H, between two extremes: H_{CO}, which is high enough to cause the bubble to collapse and H_{SO} which is low enough that the bubble experiences an elliptical instability and explodes into a long serpentine stripe. $H_{CO}/4\pi M_S$ is called the collapse field in units of $4\pi M_S$ and $H_{SO}/4\pi M_S$ the stripeout field in the same units. The diameter of the bubble at collapse in units of the characteristic length is d_{CO}/l and at stripeout is d_{SO}/l. In general the bubble size varies by a factor of three over the stable range, corresponding to a range in $H_B/4\pi M_S$ of about 10%. The exact values depend on h/l as will be described later.

2.2. Bubble states

The twist of the magnetization through the domain wall can assume a variety of configurations. The simplest state is the "unichiral" state where the orientation of the magnetization in the very center of the bubble wall is either continuously clockwise or counterclockwise, as illustrated in fig. 2, visualizing the bubble of fig. 1 from above. This state is assigned a "state number" $S = 1$ where S is defined as

$$S = \oint d\phi/2\pi \tag{3}$$

the net number of revolutions of the magnetization in the center of the wall as we trace around the bubble counterclockwise.

A different state that will have lower energy in a sufficient in-plane field, H_{ip}, is illustrated in fig. 3. This bubble has $S = 0$ according to eq. (3).

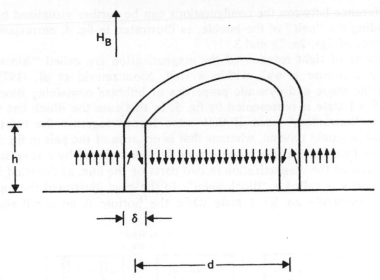

Figure 1. Basic configuration of a magnetic bubble.

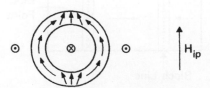

Fig. 2. Wall configuration for an $S = 1$ bubble.

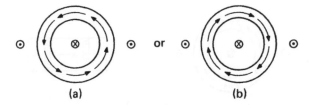

Fig. 3. Wall configuration for an $S = 0$ bubble.

The difference between the configurations can be further visualized by mentally unfolding the "pelt" of the bubble, as illustrated in fig. 4, corresponding to the examples of figs. 2a, 2b and 3.

The regions of tight rotation of the magnetization are called "Bloch lines" since they comprise a wall within a wall. Slonczewski et al. (1973) have described the static and dynamic properties of bubbles containing Bloch lines. Another $S = 1$ state is represented by fig. 5. In this case the Bloch line pair are said to be "unwinding" since if they were forced together the twist of the magnetization would unwind, whereas that is not true of the pair in fig. 3.

The stray field from a Bloch line can be further reduced by a reversal in the sense of twist of the magnetization in two parts of the line, as depicted in fig. 6. This singularity is called a "Bloch point". In the case illustrated the top of the bubble is essentially an $S = 1$ state while the bottom is an $S = 0$ state. The

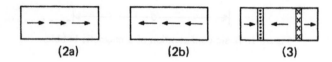

(2a)   (2b)   (3)

Fig. 4. Wall "pelts" corresponding to bubbles of figs. 2 and 3.

Fig. 5. Wall configuration for an "unwinding" $S = 1$ bubble.

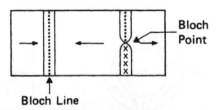

Bloch Line

Fig. 6. Wall "pelt" for an $S = \frac{1}{2}$ bubble showing a Bloch point.

combination is an $S = \frac{1}{2}$. In an in-plane field the Bloch point would be displaced from the center in order to favor the magnetization in the Bloch line that is oriented in the direction of the in-plane field. Then the bubble state will be some fraction between 0 and 1. The properties of Bloch points have been discussed by Slonczewski (1975).

Bubbles can exist in more complicated combinations characterized by integer or non-integer state numbers, by virtue of the presence of multiple Bloch lines with and without floating Bloch points. In addition the state may change continuously or discontinuously under the influence of applied fields. More understanding of bubble states and how they can be manipulated can be obtained from Slonczewski et al. (1973), Hsu (1974) and Beaulieu et al. (1976). Bubbles with many Bloch lines are called "hard" bubbles because of their unusual characteristics (Tabor et al. 1973). Hard bubbles persist to much higher bias fields before collapsing, compared to normal bubbles, and they move much more slowly in a given field gradient. Methods for the suppression of hard bubbles involve the creation on the bubble film surface of a "capping" layer which provides a low reluctance flux path. Three different methods for accomplishing this are: ion-implementation of the surface (Wolfe et al. 1973), deposition of a very thin film (~ 300 Å) of Permalloy onto the bubble film surface (Takahashi et al. 1973), and preparation of double layer bubble films where the second layer has the appropriate characteristics (Bobeck et al. 1973).

2.3. Dynamic properties

As depicted in fig. 7, a bubble will move in the presence of a gradient in the bias field with a velocity, V, and in a direction, θ, with respect to that gradient, provided that the gradient exceeds a minimum value associated with the coercive field of the film, H_C, and yet is not so large that saturation effects occur. The usual expression for this relationship is

$$V = \frac{\mu}{2}\left(\Delta H - \frac{8}{\pi} H_C\right) \qquad (4)$$

where ΔH is the difference in bias field across the bubble diameter [$(\partial H_z/\partial y)d$] and μ is the mobility.

In general, the bubble will not move in the direction of the field gradient because there is a gyrotropic force on a moving Bloch line, $F = G \times V$, where G

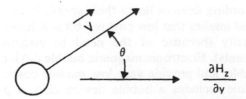

Figure 7. Bubble deflection in a magnetic field gradient.

is a vector parallel to the Bloch line with a sense depending on the sense of the rotation of the magnetization in the Bloch line, and V is the velocity with which the Bloch line is moving. In the case of an $S = 0$ bubble the forces on the Bloch lines cancel each other so that in this special case the bubble will move parallel to the field gradient. Bubbles of state $S \neq 0$ will be deflected at an angle θ.

The angle of deflection, θ, is defined as

$$\sin \theta = 4\mu S/\gamma d \tag{5}$$

where S is the "state number" defined in section 2.2 and γ is the gyromagnetic ratio of the film composition. Appropriate non-integral values of S allow this formula to apply to bubbles with Bloch points also.

The mobility, μ, is theoretically expected to be

$$\mu = \frac{\gamma}{\alpha} \frac{\delta}{\pi} = \frac{\gamma}{\alpha} \sqrt{\frac{A}{K}} = \frac{M_S}{\lambda'} \sqrt{\frac{A}{K}} \tag{6}$$

where α is the Gilbert damping parameter for the film composition. The "damping parameter" λ' is defined as $\alpha M_S/\gamma$ and is often used in the practical equations since it absorbs the variation of M_S and μ is then proportional to $(1/\lambda')\sqrt{A/Q}$. Since A does not vary appreciably over useful compositions and Q must be within a narrow range for device operation the practical variation of mobility is largely reflected by the λ' characteristic of the film composition.

The linear dependence of bubble velocity on field gradient is reproducibly observed for small field gradients. At higher magnitudes of field gradients complications occur, such as velocity saturation, overshoot, and bubble state changes. These effects occur due to distortions in the wall caused by the higher drive gradients and the influences of the film surface boundary conditions. They, therefore, depend on material parameters such as γ and Q as well as the geometric parameter l/h. An expression, derived by Slonczewski (1973), for the velocity at which these complications are expected is

$$V_0 = \frac{24\gamma A}{h\sqrt{K}} = \frac{12\gamma\sqrt{A}}{Q} \frac{l}{h}. \tag{7}$$

2.4. Properties required for practical bubble devices

Bubbles have been used in information storage devices where their electronic nature gives them an advantage over other forms of magnetic storage devices that involve some form of mechanical motion. The mechanical motion involved in the magnetic recording devices limits their speed, is the most frequent cause of device failure, and implies that low cost per bit is achieved only for devices of large storage capacity (because of the need to distribute the cost of the mechanical components). Electronic magnetic bubble devices have faster access, are even more reliable, and provide very low module costs.

The bubble module includes a bubble device chip, a permanent magnet to supply the bias field (H_B) needed to sustain the bubbles, and means for propagating the bubbles (e.g., coils to provide a rotating drive field). There are several device forms (e.g., Permalloy bar, contiguous disc, bubble lattice) and the properties required for devices will obviously depend somewhat on the particular form. Nevertheless we can discuss the general requirements, and then later in this chapter see how those requirements can be met with garnets, and in the next chapter how they can be met with amorphous materials.

Bubble films for devices must have appropriate: (a) bubble size, (b) Q, (c) dynamic properties, μ and V_0, (d) thermal variation of properties, (e) implantability. The following general statements can be made regarding these requirements:

a. Bubble size, d

The bubble size required is determined by the specific device configuration to be used and the lithography available. For example, some devices require a nominal bubble size equal to twice the minimum feature size of the device, so that 1 μm lithography leads to the need for 2 μm bubbles. The size of the bubble, varying as it does with bias field, actually varies somewhat as it proceeds through the device, experiencing variations in effective bias field due to such factors as Permalloy overlay patterns and applied drive fields. However, because of the latitude allowed between d_{CO} and d_{SO} ($\sim 3\times$) it is sufficient to achieve a material with the nominal bubble size and the burden is on the device design to ensure that the bubble experiences the correct average bias field with tolerable variations.

b. Q

If the Q is too low spurious bubbles are nucleated during device operation. If the Q is too high it becomes too difficult to nucleate wanted bubbles and also the V_0 will be unnecessarily lowered (since $V_0 Q \simeq$ constant). The Q required obviously depends on the particular device type. One estimate for the minimum Q required for Permalloy bar devices has been published by Kryder et al. (1976) as follows

$$Q_{min} \simeq 1.3 + 735/4\pi M_S.$$

Since $4\pi M_S$ is determined by the desired bubble size, Q_{min} is also related to the bubble size. The material should be designed to give a Q that exceeds Q_{min}, but by no more than is necessary for this to be so over the entire range of operating temperatures.

c. Dynamic properties, μ and V₀

From section 2.3 we see that μ and V_0 are related to some of the same factors as are other device related parameters, e.g., d and Q. Therefore μ and V_0 can be maximized, but only at the expense of these other parameters. Thus a compromise is necessary that balances the dynamic properties with the static properties. First the requirements on $4\pi M_S$, K, and A for d and Q should be satisfied, and then the bubble velocity optimized within that constraint in order to get the fastest possible device operation.

From eq. (4) we would be inclined to design for the highest possible mobility, μ, to get the highest velocity with a minimum requirement on drive field. However, we also see from (4) that the ratio $2V_0/\mu$ is the excess drive field, $[\Delta H - (8/\pi)H_C]$, that will produce dynamic instabilities. That excess field must not be so low that it is exceeded during the variations in effective field as the bubble traverses the device. Therefore the bubble material is usually designed to have a μ somewhat lower than it might otherwise be, in order to get a good operating speed but without suffering instability. In some devices the peak velocity of the bubble during its traverse of the device is $\geq 4\times$ its average velocity.

d. Thermal variation of properties

Practical devices must operate over a reasonable temperature range, e.g., 0–100°C. It is therefore necessary that the requirements for both the static and dynamic properties be satisfied over that entire temperature range. The variation of these practical properties is not obvious since they depend on quotients of the basic magnetic parameters ($4\pi M_S$, K, A, etc.). Except near a compensation temperature all of these basic parameters decrease as the temperature increases toward the Curie temperature. Thus the variation in the practical properties depends on the relative rates of decrease of the basic parameters. In general, A and K decrease somewhat faster than M_S^2, and the net result is that Q, d and H_{C0} usually all decrease with temperature, but more slowly than does $4\pi M_S$, while μ increases. In cases where a compensation temperature is involved, or very near to the Curie temperature, the thermal variation of the practical parameters can be very unusual.

The principal factor to be controlled is the required bias field. The bias field is typically supplied by a permanent magnet (e.g., barium ferrite) whose $4\pi M_S$ decreases approximately linearly at $-0.2\%/°C$ over the range 0–100°C. Thus the bubble films are usually tailored so that their $4\pi M_S$ decreases at the rate which will cause the required bias field to closely match that supplied by the permanent magnet.

e. Implantability

We have already seen in the last section that some treatment of the surface, usually by ion-implantation, is necessary to prevent hard bubbles. In addition some device structures are produced by patterned ion-implantation (e.g., contiguous disc devices, see Lin et al. 1977), and in others ion-implantation is used to produce a "cap" that is conducive to switching bubble states (e.g., bubble lattice devices, see Beaulieu et al. 1976). In essence, the ion-implantation is used to overcome the uniaxial anisotropy in the surface layer of the film. This is accomplished by creating a surface stress which introduces a counteracting in-plane anisotropy through a negative magnetostriction, λ_{111}. The thickness of the layer is primarily determined by the energy of the incident ions, the stress in that surface layer by the total dosage of those ions, and the in-plane anisotropy is produced by the stress through λ_{111}. Since the practical level of stress is limited, the primary requirement is for a sufficient λ_{111}.

The next section, 3, will describe the procedures for interrelating these practical bubble parameters with the fundamental magnetic and non-magnetic film parameters. In subsequent sections the range of available fundamental parameters for useful materials will be discussed and methods indicated for tailoring the materials to produce the desired combination of those parameters for practical purposes.

3. Relationship of practical bubble parameters to fundamental magnetic and non-magnetic parameters

3.1. Important practical bubble parameters

From the discussion in section 2 it can be seen that the bubble parameters that are important when considering the practical use of bubbles are: $d \equiv$ bubble size, $Q \equiv$ quality factor, $\mu \equiv$ mobility, $V_0 \equiv$ saturation velocity, $H_C \equiv$ coercive force, $H_{CO} \equiv$ collapse bias field at $d = d_{CO}$, $H_{SO} \equiv$ stripeout bias field at $d = d_{SO}$, $W_S \equiv$ natural demagnetized stripewidth.

3.2. Important fundamental magnetic parameters

The practical parameters of section 3.1 are functions of more fundamental magnetic parameters characteristic of the bubble film material and non-magnetic parameters characteristic of the film structure. The important primary magnetic parameters are: $M_S \equiv$ magnetization, $A \equiv$ exchange constant, $K \equiv$ anisotropy constant, $\lambda \equiv$ magnetostriction constant, $T_C \equiv$ Curie temperature, $\gamma \equiv$ gyromagnetic ratio, $\alpha \equiv$ Gilbert damping constant.

Convenient secondary parameters that are combinations of these and which are very useful, are defined as follows:

$$\delta \equiv \text{domain wall thickness} = \pi\sqrt{A/K} \tag{8}$$

$$\sigma_W \equiv \text{domain wall energy} = 4\sqrt{A/K} \tag{9}$$

$$l \equiv \text{characteristic length} = \sigma_W/4\pi M_S^2 \tag{2}$$

$$Q \equiv \text{quality factor} = (K/2\pi M_S^2) = (H_K/4\pi M_S) \tag{1}$$

$$H_K \equiv \text{anisotropy field} = 2K/M_S \tag{1a}$$

$$\lambda' \equiv \text{damping parameter} = \alpha M_S/\gamma. \tag{10}$$

3.3. Important non-magnetic film parameters

The parameters of the film structure of importance to the practical parameters are: $h \equiv$ film thickness (μm), $\Delta a_0 \equiv$ difference in lattice spacing between the substrate and the film $(a_0^s - a_0^f)$ (Å), $h, k, l \equiv$ film crystallographic direction perpendicular to the film plane.

The accurate measurement of film thickness is important because it is used in the determination of $4\pi M_S$ and l as well as in a number of the parameter conversion formulae. It is usually measured using an optical interference method (e.g. Pierce and Venard 1974) wherein the change in number of fringes, Δm, is measured at two or more wavelengths (λ_i) and the thickness, h, is derived from

$$h = \frac{\Delta m}{2(n_j/\lambda_j - n_i/\lambda_i)}$$

where n_i is the refractive index of the film at wavelength λ_i. It is necessary, for accuracy, to use the proper value of the refractive index. This index varies with composition and sometimes the use, for measurement of h on one composition, of the refractive index obtained from a different composition has led to errors in thickness of 10–20%. Values for the index as a function of wavelength and composition for garnet films can be extrapolated from the data of Wöhlecke and Suits (1977) and of McCollum et al. (1973).

3.4. Interrelating formulae

The relationship of Q and l to the more fundamental magnetic parameters are given by their defining expressions (1) and (2). It is convenient to rewrite these expressions in terms of $4\pi M_S$ in units of 1000 G, $4\pi M_S(10^3)$; K in units of 10^4 ergs cm^{-3}, $K(10^4)$; and A in units of 10^{-7} ergs cm^{-1}, $A(10^{-7})$ because these parameters have this order of magnitude in useful bubble materials. The defining equations then become, within 1%

$$Q = \frac{1}{4}\frac{K(10^4)}{[4\pi M_S(10^3)]^2} \qquad K(10^4) = 4Q[4\pi M_S(10^3)]^2 \tag{11}$$

$$l(\mu) = 0.016\frac{[A(10^{-7})K(10^4)]^{1/2}}{[4\pi M_S(10^3)]^2}. \tag{12}$$

3.4.1. Quality factor, Q
Figure 8 is a plot of eq. (11) and can be used to interrelate Q, $K(10^4)$ and $4\pi M_S(10^3)$. Given any two of these quantities the third can be readily determined.

3.4.2. Characteristic length, l
The characteristic length is more conveniently considered in relation to Q, in terms of Q/l and Q/l^2 which are, in turn, expressed in terms of the fundamental parameters as follows

$$\frac{Q}{l(\mu)} = 16\left(\frac{K(10^4)}{A(10^{-7})}\right)^{1/2} \tag{13}$$

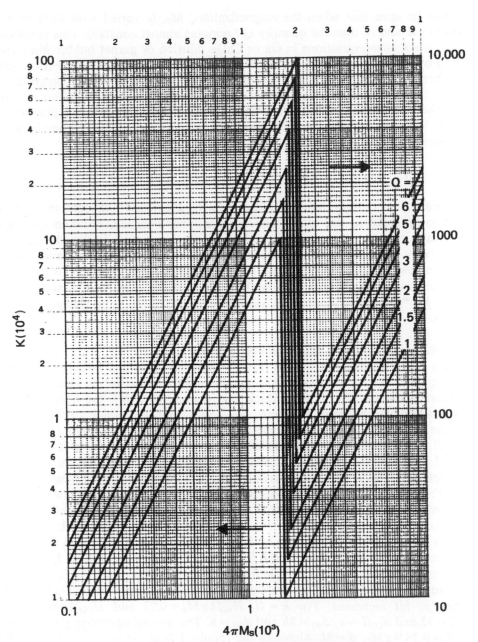

Fig. 8. $K(10^4\,\text{erg cm}^{-3})$ vs. $4\pi M_S(10^3\,\text{G})$ and Q.

$$\frac{Q}{[l(\mu)]^2} = 10^3 \frac{[4\pi M_S(10^3)]^2}{A(10^{-7})}. \tag{14}$$

It can be seen that when the magnetization, M_S, is varied with little or no change in K and A, then the quantity Q/l remains almost constant. This situation occurs when slight variations in Ga or CaGe dilution of garnet bubble films are caused by changes in growth temperature, as we shall see in a later section. On the other hand when the anisotropy, K, is varied with little or no change in M and A, then the quantity Q/l^2 remains almost constant. This situation occurs when the iron content of garnets is kept constant but the rare earth content is varied. We note in passing that constant Q/l implies constant domain wall thickness, δ, since from eq. (8)

$$\delta = \pi \sqrt{\frac{A}{K}} = \frac{\pi}{2}\frac{l}{Q} \qquad \delta(\mu) = \frac{\pi}{2}\frac{l(\mu)}{Q}. \tag{15}$$

Figure 9 depicts the relationships between Q, l, Q/l and Q/l^2 such that when any two are known, the others can be readily determined. Q/l and Q/l^2 are graphically related to the fundamental magnetic parameters in figs. 10 and 11, respectively. In using these figures we note that A varies very little and slowly with composition changes in the useful bubble materials so the lines of constant A are particularly helpful in showing how Q/l and Q/l^2 vary with changes in K and $4\pi M_S$, respectively, such as with rare earth or dilution changes in the garnets.

3.4.3. Range of applied field and bubble diameters

Thiele's (1970) calculations of the range of applied fields and bubble diameters for stable bubbles are presented both graphically for visualization and in a table for more accuracy. For a given thickness in units of l, h/l, the field must be between that which allows stripeout, $H_{SO}/4\pi M_S$, and that which causes bubble collapse, $H_{CO}/4\pi M_S$. Over that range the bubble diameter will range from d_{SO}/l to d_{CO}/l. Between the extremes Thiele showed that the bubble size will vary with applied field according to fig. 12. Thiele also showed that the preferred value for the thickness h is $4l$. The thicknesses usually employed in devices range from $4l$ to $9l$. From the figure it is seen that operation in the middle of the allowable field range will result in a bubble size that is roughly the average of the collapse and stripeout diameters. This also is close to the demagnetized stripewidth, W_S. The ratio of d_{SO}/d_{CO} is almost exactly 3.0 for $h = 4l$ and only slightly less than 3 for larger values of h. The values of the field limits vs. h/l are plotted in fig. 13 and of the corresponding bubble sizes in fig. 14. The allowable range in H_B is seen to be about 0.1 $4\pi M_S$ for all h/l and this range is centered around increasingly larger values as h/l increases. For $h = 4l$, $H_{SO}/4\pi M_S \simeq 0.23$ and $H_{CO}/4\pi M_S = 0.33$; $d_{SO}/l \simeq 12$ and $d_{CO}/l \simeq 4$; $d_{SO} \simeq 3h$ and $d_{CO} \simeq h$. The normal operating point would be $H = 0.28\ 4\pi M_S$, $d = 8l$. Although the required field increases by 65% if the thickness of the film is increased to $9l$, the bubble size only increases by 12%. The value of the demagnetized stripewidth in units of l, W_S/l, is also plotted in fig. 14.

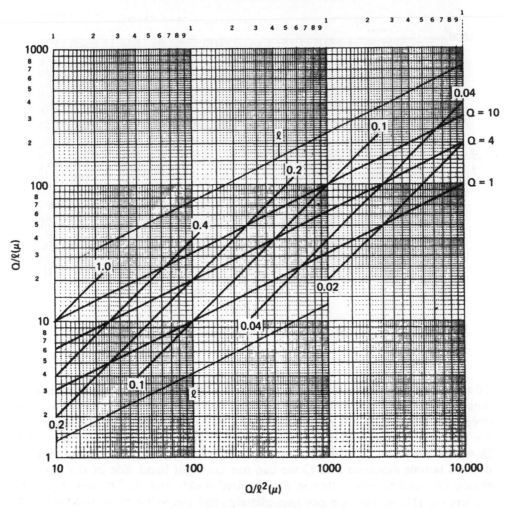

Fig. 9. $Q/l(\mu)$ vs. $Q/l^2(\mu)$, $l(\mu)$ and Q.

The values for the extremes of field and bubble size as well as the demagnetized stripewidth over the range of h/l from 3.0 to 9.0 are also given in tabular form for convenience in table 1.

These charts can be readily used for the conversion of parameters, e.g., deduction of the values of K and A from measured bubble properties or vice versa. This procedure is illustrated in Appendix B.

3.4.4. Summary chart

Figure 15 is a three-component chart that usefully summarizes the relationships that connect bubble size and Q to the fundamental magnetic parameters of the bubble film. The central section depicts eq. (14). Lines of constant $A = 2.0$ and

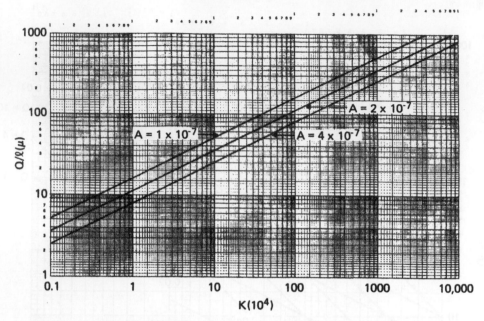

Fig. 10. $Q/l(\mu m)$ vs. $K(10^4 \, \text{erg cm}^{-3})$ and $A(10^{-7} \, \text{erg cm}^{-1})$.

4.0×10^{-7} are drawn because the A of most useful bubble materials lies between these two values. Thus $4\pi M$ is the strongest variable in determining Q/l^2. Q/l^2 is decomposed into Q and l on the right hand side of the chart. In addition scales for $8l$ and $9l$ are given, corresponding to the bubble diameter for a "thin" film ($h = 0.5d = 4l$) and a "thick" film ($h = 1.0d = 9l$) respectively. Thus given a desired bubble diameter and Q we can use the right hand side of the graph to obtain Q/l^2 and translate that in the required $4\pi M_S$ and A. The upper margin depicts eq. (11) so that we can immediately read the value of K required with the $4\pi M_S$ to obtain the desired Q. Points and interrelationships for a $2\,\mu m$ bubble size and $Q = 5$ in a $1\,\mu m$ thick film are shown for illustration yielding a $Q/l^2 = 80$ and a required $K = 4.8 \times 10^4 \, \text{erg cm}^{-3}$ for $4\pi M_S = 490 \, \text{G}$, $A = 3.0 \times 10^{-7} \, \text{erg cm}^{-1}$.

3.4.5. Mobility

The mobility is derived experimentally from the variation of bubble velocity with applied field gradient as in eq. (4). The expected value, based on the more fundamental magnetic parameters is given by eq. (6). Since λ' for the most commonly used materials has values of the order of 10^{-7}, the most convenient expressions are

$$\mu = 2500 \frac{4\pi M_S(10^3)}{\lambda'(10^{-7})} \left(\frac{A(10^{-7})}{K(10^4)} \right)^{1/2} \text{cm s}^{-1} \text{Oe}^{-1} \tag{16}$$

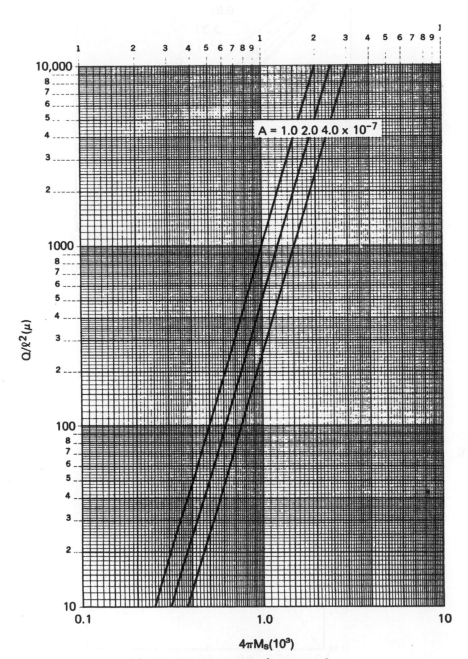

Fig. 11. $Q/l(\mu)$ vs. $4\pi M_S(10^3)$ and $A(10^{-7})$.

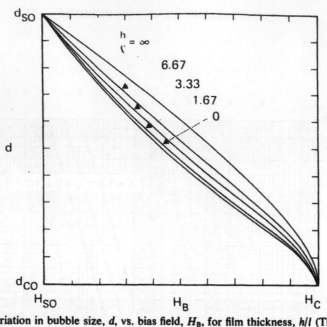

Fig. 12. Variation in bubble size, d, vs. bias field, H_B, for film thickness, h/l (Thiele 1970).

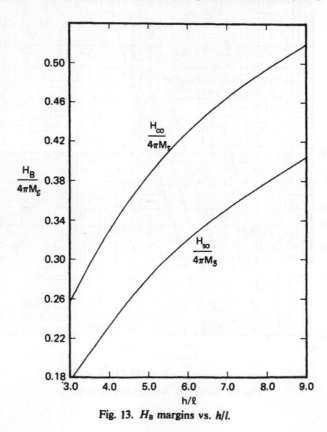

Fig. 13. H_B margins vs. h/l.

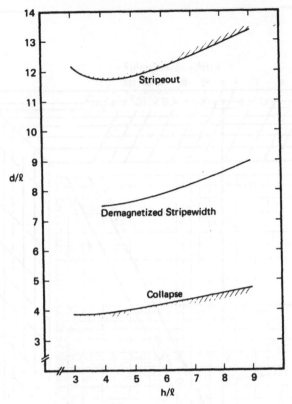

Fig. 14. Extremes of bubble size vs. h/l.

TABLE 1

Extreme of bias field and bubble size and demagnetized
stripewidth for various values of h/l

h/l	$H_{so}/4\pi M_s$	$H_{co}/4\pi M_s$	d_{so}/l	d_{co}/l	W_s/l
3.0	0.174	0.256	12.17	3.89	
3.5	0.206	0.295	11.86	3.89	
4.0	0.235	0.330	11.75	3.93	7.52
4.5	0.258	0.359	11.77	4.00	7.58
5.0	0.280	0.385	11.86	4.08	7.67
5.5	0.301	0.408	11.99	4.16	7.79
6.0	0.320	0.430	12.14	4.23	7.93
6.5	0.337	0.449	12.33	4.31	8.08
7.0	0.352	0.466	12.53	4.40	8.24
7.5	0.366	0.480	12.73	4.49	8.41
8.0	0.379	0.494	12.93	4.58	8.60
8.5	0.391	0.507	13.14	4.67	8.80
9.0	0.403	0.519	13.36	4.76	9.00

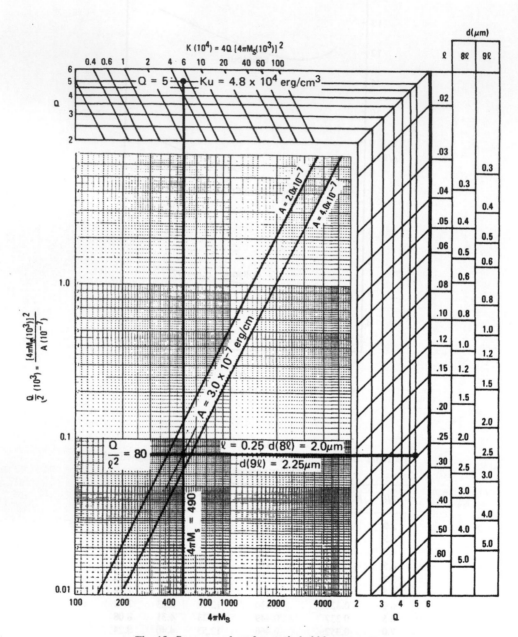

Fig. 15. Summary chart for static bubble parameters.

and using (11)

$$\mu = \frac{1250}{\lambda'(10^{-7})}\left(\frac{A(10^{-7})}{Q}\right)^{1/2} \text{cm s}^{-1}\text{Oe}^{-1}.$$ (17)

Thus for typical values of $A \simeq 3 \times 10^{-7} \text{erg cm}^{-1}$, $Q \simeq 4$, $\lambda' \simeq 10^{-7}$: $\mu \simeq 1000 \text{cm s}^{-1}\text{Oe}^{-1}$.

3.4.6. Saturation velocity

A similar reduction of (7) yields for the velocity at which we expect the simple mobility representation to break down

$$V_0 = 2400 \frac{\gamma(10^7)A(10^{-7})}{h(\mu)\sqrt{K(10^4)}} \text{cm s}^{-1}$$ (18)

or

$$V_0 = 38000 \frac{\gamma(10^7)\sqrt{A(10^{-7})}}{Q}\frac{l}{h} \text{cm s}^{-1}$$ (19)

Thus for typical values of $\gamma \simeq 1.8 \times 10^7 \text{Oe}^{-1}\text{s}^{-1}$, $A \simeq 3 \times 10^{-7} \text{erg cm}^{-1}$, $Q \simeq 4$, $h \simeq 6l$: $V_0 \simeq 5000 \text{cm s}^{-1}$.

4. Categories of crystalline bubble materials

As stated in section 2.1 a bubble material must have an anisotropy such that $Q > 1$. This anisotropy can arise from several different mechanisms. Some materials will have a sufficient natural crystalline anisotropy such as the hexagonal ferrites. Other materials, essentially cubic, such as the garnets, can have a sufficient anisotropy induced during the film growth process. There are three principal categories of crystalline materials that have received serious attention as bubble materials: garnets, orthoferrites, and hexaferrites. The earliest work involved orthoferrites; the most work has involved garnets; the future may include hexaferrites. Each of these categories admits wide substitution of both magnetic and non-magnetic ions, yielding a wide variation in the basic magnetic parameters and, hence, also in the practical bubble parameters. This variation is summarized in fig. 16.

Figure 16 contains two sets of coordinates. Fundamentally the materials are plotted according to their magnetization as the abscissa and their anisotropy field, $2K/M_S$, as the ordinate. In addition, lines of constant Q and approximate bubble diameter (assuming $A \simeq 4 \times 10^{-7} \text{erg cm}^{-1}$) are given in order to visualize the materials in terms of the practical bubble parameters as well as the fundamental magnetic parameters. Each category of materials will be discussed in later sections with references to the literature. Only the significant features will be given here to convey the overall picture.

An undiluted garnet bubble film has a $4\pi M$ of 1200–1800 G (see table 2) and if the rare earth mixture is correct, as will be seen in section 5.4 the anisotropy

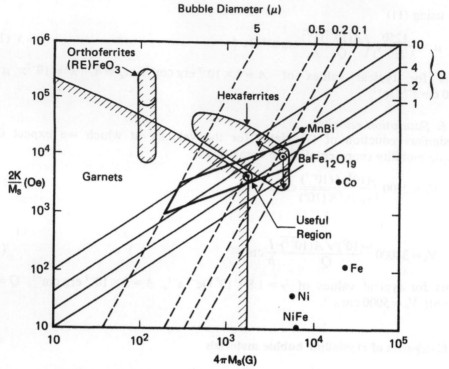

Fig. 16. Range of properties for several families of bubble materials.

TABLE 2
Properties of rare earth garnets, $(RE)_3Fe_5O_{12}$ (Davies and Giess 1975)

RE	$4\pi M$ (G)	T_C (K)	a_0 (Å)	$\lambda_{111}(10^{-6})$	$\lambda_{100}(10^{-6})$	$\lambda'(10^{-7} Oe^2 s)$
Sm	1675	578	12.529	−8.5	+21.0	12.0
Eu	1172	566	12.498	+1.8	+21.0	2.1
Gd	56	564	12.471	−3.1	0.0	0.52
Tb	198	568	12.436	+12.0	−3.3	48.0
Dy	376	563	12.405	−5.9	−12.6	26.0
Ho	882	567	12.376	−4.0	−3.4	42.0
Y	1767	553	12.376	−2.4	−1.4	0.52
Er	1241	556	12.347	−4.9	+2.0	7.0
Tm	1397	549	12.323	−5.2	+1.4	1.2
Yb	1555	548	12.302	−4.5	+1.4	4.2
Lu	1815	549	12.283	−2.4	−1.4	0.52

$K(10^4 \text{ erg cm}^{-3})$ will be 10–30 yielding a value for the anisotropy field of 2000–4000 Oe, a $Q = 2$ or slightly more, and a bubble size about 0.5 μm. $Sm_{1.2}Lu_{1.8}Fe_5O_{12}$ is an example which is plotted as a circle in fig. 16. The $4\pi M_S$ is reduced as the iron is replaced by a non-magnetic ion such as Ga or Ge and the K is reduced as the identity or ratios of the rare earth ions are changed. Thus the range of properties indicated by the cross-hatch for garnets can be obtained.

The hexaferrite, $BaFe_{12}O_{19}$, has $4\pi M_S = 4700$ G, $H_K = 17\,000$ Oe and this is also plotted as a circle. Actually, $BaFe_{12}O_{19}$ is a member of a broader family of hexaferrites which will be discussed later in section 7. Within this family the $4\pi M_S$ can be reduced by non-magnetic dilution of the iron and the K can be reduced by other magnetic substitutions for the iron. Thus a range of properties, as indicated, can be obtained in the hexaferrite category.

The orthoferrites have the formula $(RE)FeO_3$ and a range of properties is obtained depending on the specific rare earths, RE, used.

Bubble sizes smaller than 5 μm are desired for devices and the Q requirement varies somewhat with bubble size. The range of useful bubble size and Q is indicated by the heavy line. The figure indicates that bubbles are too large in the orthoferrites, that garnets provide ideal bubble materials down to ~ 0.4 μm and that hexaferrites may be useful for bubbles smaller than that. These three categories of bubble materials will now be discussed in greater detail with the greatest emphasis on garnets because these materails are the most widely studied and used.

5. Garnets

5.1. Introduction

Chapter 1 of this volume contains a description of the basic structure and magnetization of garnet bulk crystals. These materials can also be grown into single crystal films on suitable substrates for magnetic bubbles. In general, useful films will be compounds of the simple rare earth garnets (whose basic properties are listed in table 2, as listed by Davies and Giess 1975) but with the iron lattice diluted with a non-magnetic ion to reduce the magnetization to more useful values.

As in chapter 1, section 3, it is convenient to think of yttrium–iron garnet (YIG), $Y_3Fe_5O_{12}$, as the parent garnet for bubble films with non-magnetic dilution of the iron used to adjust the magnetization and substitution of magnetic rare earths used primarily to increase the anisotropy and secondarily to adjust the dynamic parameters. The magnetization variation is the same for bubble films as described in that section for bulk materials. Ga is the simplest dilutant for bubble films and has been used extensively; its influence is as described in chapter 1, section 3.2.2. However, because Ge has a stronger preference for the tetrahedral garnet sites with a smaller fraction going to the octahedral sites, less Ge is required than Ga to achieve an equivalent decrease in magnetization (Bonner et al. 1973). Therefore there is a smaller reduction in A and the Curie temperature, T_C, with Ge dilution and the variation of properties with temperature is consequently less for the same $4\pi M_S$ than with Ga dilution. For this reason Ge is often used in preference to Ga, especially for bubbles larger than one micron where the dilution is appreciable. Because Ge^{4+} is the natural ionization state it is necessary to also use an equivalent quantity of a non-magnetic divalent ion, such as Ca^{2+} to provide charge compensation. Most of the specific bubble materials referenced involve Ga^{3+} or $Ca^{2+}Ge^{4+}$ as dilutants. Other ions used include Al^{3+} and Si^{4+}.

While the magnetization vs. composition of garnet films for bubbles are the same as their equivalent bulk counterparts, the anisotropy of the films is quite different due to the specific nature and circumstances of the films. Therefore we will consider the anisotropy of garnet bubble films quite extensively in a subsequent section, 5.4. Because one component of the anisotropy is stress-induced if there is a mismatch between the garnet film and the substrate we will consider briefly the choice of substrate and the film growth next. After discussing the achievement of specific anisotropy we will relate that to the achievable range of garnet bubble film properties and specific compositions.

5.2. Film fabrication

Garnet bubble films can be fabricated by any technique for producing oriented single crystal thin films. The most popular method is liquid phase epitaxy, LPE, on oriented single-crystal substrates made from non-magnetic garnets. The gallium garnets are usually used for substrates with the rare earths of the bubble film chosen to provide lattice matching. $Gd_3Ga_5O_{12}$ (GGG) boules and prepared wafer substrates are readily available from a number of vendors. There is also an ample literature on the LPE growth technique (Giess and Ghez 1975, Blank et al. 1972). In general the composition of the grown film is different from that of the melt depending on the growth conditions (melt temperature, substrate rotation rate, etc.). Therefore an understanding of these effects is necessary in order to achieve a particular film composition. The following sections will relate to the actual composition of the film as determined by a technique such as electron microprobe analysis and produced by the use of the film growth techniques described in the literature.

5.3. Lattice mismatch

Any mismatch between the substrate lattice spacing and that of the film, $\Delta a = a_S - a_f$, produces stress in the film which can cause cracking if large enough and which will cause magnetic effects through the magnetostriction. To prevent cracking Δa should be kept less than $0.03 \times h^{-1/2}$ Å, where h is the film thickness in μm (Matthews et al. 1973). The influence of Δa on the anisotropy will be discussed in the next section.

Figure 17 (Tolksdorf 1975) depicts the lattice spacing, for iron garnets, gallium garnets and the influence of gallium dilution. It can be seen that GGG ($a_0 =$ 12.383 Å) is a very close match to YIG ($a_0 = 12.376$ Å). Other rare earths can be used in combinations which also match GGG, e.g., $Eu_1Tm_2Fe_5O_{12}$. The lattice spacings of the simple garnets are given in tabular form in table 2 and those for mixed garnets can be obtained by simple interpolation. Other substrates which are being increasingly used are $Sm_3Ga_5O_{12}$ (SGG: $a_0 = 12.437$ Å) and $Nd_3Ga_5O_{12}$ (NGG: $a_0 = 12.506$ Å) especially for garnets with a high proportion of the larger rare earths (Davies and Giess 1975).

Dilution with CaGe produces even less lattice spacing change than Ga. The lattice changes per atom substitution in the formula unit are $\Delta a \simeq -0.014$ Å/Ga and -0.003 Å/CaGe.

5.4. Magnetization

Chapter 1 describes how the magnetization of yttrium and rare earth iron garnets decrease with dilution of the iron lattice. The explanation of that will not be repeated here, but fig. 18 is included for convenience.

The figure shows the decrease in $4\pi M_S$ vs. dilution for $Y_3Fe_5O_{12}$ and for $Eu_1Tm_2Fe_5O_{12}$ when the dilutant is Ga or CaGe. The case of $Eu_1Tm_2Fe_5O_{12}$ is very similar to $Y_3Fe_5O_{12}$ except that the net magnetization is somewhat reduced because the magnetization of the rare earth sublattice opposes the net magnetization of the iron sublattices. These two represent extremes and most useful garnets will have intermediate rare earth content, and the basic undiluted

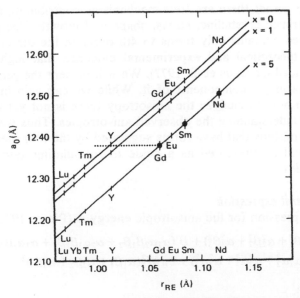

Fig. 17. Lattice spacing of $(RE)_3Ga_xFe_{5-x}O_{12}$ garnets (Tolksdorf 1975).

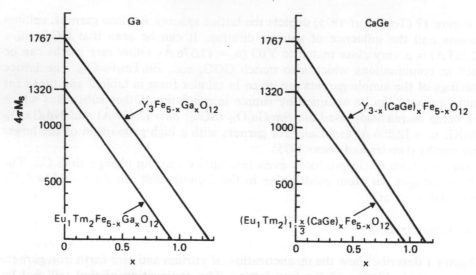

Fig. 18. $4\pi M_S$ vs. rare earth content and Ga or CaGe dilution x.

magnetization can be estimated by interpolating the values given in table 2. Then the decrease with dilutant can be constructed intermediate between the curves of fig. 18.

5.5. Anisotropy

5.5.1. Introduction

For a thin film garnet there are four mechanisms contributing to the magnetic anisotropy: intrinsic crystalline, stress, shape and growth. The general expression that is used contains only terms to 4th order in the direction cosines but there is both theoretical and experimental evidence that higher order terms should be included (Josephs et al. 1977). We will present the general expression and discuss the major components of it. While we can gain insight and clues from this as to how to analyze the anisotropy there is not yet a firm basis for quantitatively understanding the observed anisotropies. Thus we must resort to empirical relationships that have forms suggested by the four mechanisms. Such relationships will be presented as a guide to formulating garnets for specific bubble properties.

5.5.2. The general expression

The general expression for the anisotropic energy is (Pierce 1971)

$$E = A'(\alpha_1^2\beta_1^2 + \alpha_2^2\beta_2^2 + \alpha_3^2\beta_3^2) + B'(\alpha_1\alpha_2\beta_1\beta_2 + \alpha_2\alpha_3\beta_2\beta_3 + \alpha_3\alpha_1\beta_3\beta_1)$$
$$+ K_1(\alpha_1^2\alpha_2^2 + \alpha_2^2\alpha_3^2 + \alpha_3^2\alpha_1^2) + 2\pi M_S^2(\alpha_1\beta_1 + \alpha_2\beta_2 + \alpha_3\beta_3)^2 \tag{20}$$

where

$$A' = A + \tfrac{3}{2}\lambda_{100}\sigma_0 \qquad B' = B + 3\lambda_{111}\sigma_0$$

A, B are the growth anisotropy coefficients, λ_{100}, λ_{111} are the magnetostriction coefficients, σ_0 is the isotropic stress in the film, K_1 is the first order crystalline anisotropy constant, and the $2\pi M_s^2$ term is due to the film shape.

The direction cosines of the orientation of the magnetization with respect to the principal axes of the garnet are α_1, α_2, α_3 and the film is oriented so that the direction cosines of the normal to the film plane are β_1, β_2, β_3.

For most useful garnets the growth anisotropy is the dominant factor and the crystalline anisotropy the weakest, being however a perturbation that must be taken into account under some particular circumstances. The shape anisotropy is an important factor that puts a lower bound on the others in order to have $Q > 1$ and the magnetization oriented normal to the film for bubbles. The stress anisotropy is usually a second order correction except for large bubble sizes ($d \geq 5$ μm) where the growth anisotropy is purposely kept very low.

Because so much of the bubble work and, indeed, useful devices have utilized 111 oriented garnet films we will give most of our attention to that orientation. However, the 001 orientations seem capable of producing larger anisotropies and will probably become more attractive as the technology develops and submicron bubbles are required. In addition an increasing amount of work is being done with 110 orientations because an orthorhombic anisotropy can thereby be achieved where the easy axis is still normal to the film but there is a subsidiary anisotropy in the plane. This orientation has been useful for the study of anisotropy and its effects on other parameters (e.g., mobility). Some day this orientation may be used to achieve higher saturation velocities (Stacy et al. 1976).

When the orientation of the film normal is i, j, k the expression for the energy E_{ijk} reduces to

$$E_{111} = \frac{A'}{3} + \frac{B'}{3}(\alpha_1\alpha_2 + \alpha_2\alpha_3 + \alpha_3\alpha_1) + K_1(\alpha_1^2\alpha_2^2 + \alpha_2^2\alpha_3^2 + \alpha_3^2\alpha_1^2)$$

$$+ 2\pi M_s^2\left(\frac{\alpha_1 + \alpha_2 + \alpha_3}{\sqrt{3}}\right)^2 \tag{21}$$

$$E_{001} = A'\alpha_3^2 + K_1(\alpha_1^2\alpha_2^2 + \alpha_2^2\alpha_3^2 + \alpha_3^2\alpha_1^2) + 2\pi M_s^2\alpha_3^2 \tag{22}$$

$$E_{110} = \frac{A'}{2}(\alpha_1^2 + \alpha_2^2) + \frac{B'}{2}\alpha_1\alpha_2 + K_1(\alpha_1^2\alpha_2^2 + \alpha_2^2\alpha_3^2 + \alpha_3^2\alpha_1^2) + 2\pi M_s^2\left(\frac{\alpha_1 + \alpha_2}{\sqrt{2}}\right)^2. \tag{23}$$

If we arbitrarily let the energy in the direction of the film normal be zero, then the energy in other directions for the several film orientations is given in table 3.

Since $\alpha_1 + \alpha_2 + \alpha_3 = 0$ for the plane normal to 111, and therefore also $(\alpha_1 + \alpha_2 + \alpha_3)^2 = 0$, we can deduce that $E_{111} = \frac{1}{3}A' - \frac{1}{6}B' + \frac{1}{4}K_1$ for all directions in the 111 plane and there is consequently no in-plane anisotropy for 111 films. For the plane normal to 001, $\alpha_3 = 0$ and $E_{001} = K_1\alpha_1^2\alpha_2^2$ so the only in-plane anisotropy for 001 films is due to K_1. For films normal to 110, however, there is an in-plane anisotropy depending on the magnitudes of A' and B' as well as K_1.

TABLE 3
Anisotropy energy for various film orientations

Magnetization direction	Film orientation 111	Film orientation 001	Film orientation 110
111	0		
$1\bar{1}0$	$-\frac{1}{2}B' - \frac{1}{12}K_1 - 2\pi M_S^2$		$-\frac{1}{2}B' - 2\pi M_S^2$
$11\bar{2}$	$-\frac{1}{2}B' - \frac{1}{12}K_1 - 2\pi M_S^2$		
001		0	$-\frac{1}{2}A' - \frac{1}{4}B' - \frac{1}{4}K_1 - 2\pi M_S^2$
100		$-A' - 2\pi M_S^2$	
010			
010		$-A' - 2\pi M_S^2$	
110		$-A' + \frac{1}{4}K_1 - 2\pi M_S^2$	0

5.5.3. Stress anisotropy

The values of the magnetostrictive coefficients for the simple garnets are given in table 2. For practical purposes linear interpolation can be used to obtain values for mixed garnets. Dilution of the iron lattices by Ga, Ge, Al, etc., will decrease the values somewhat. Increase in temperature will also decrease the values. White (1973) justified the linear interpolation and also showed that the decrease with dilution is linear except for the contribution due to Eu, if present, which is quadratic because of the special nature of that ion. The λ_{111} in general decreases to about 50% when one of the Fe has been replaced by a dilutant.

Stress can be introduced locally in bubble device films due to patterns of ion implantation and/or overlay films, such as Permalloy, used to generate the device functions. This would be in addition to the uniform, isotropic stress, σ_0, which is due to lattice mismatch between the substrate and the film, as discussed in section 5.3. The mismatch stress can be related to the mismatch Δa by the following formula (Davies and Giess 1975)

$$\sigma_0 = \frac{E}{1 - \nu} \frac{\Delta a}{a_0}$$

where E is Young's modulus ($\sim 2 \times 10^{12}$ dyne cm^{-1}) and ν is Poisson's ratio (~ 0.3). Thus

$$\sigma_0 \approx 2.8 \times 10^{12} \frac{\Delta a}{a_0}.$$

Since only small $\Delta a/a$ can be tolerated ($\sim 0.2\%$) and the magnetostriction is of the order of -2×10^{-6} for many practical garnets, we see that the magnetic anisotropy due to lattice mismatch will generally be less than 10^4 ergs cm^{-3}. A good example of the linear variation of anisotropy with lattice mismatch, Δa, is given by the study of Giess and Cronemeyer (1973).

5.5.4. Growth anisotropy – general

The theoretical basis for growth anisotropy and the comparison with experimental data is extensively treated in the open literature (e.g., Rosencwaig and Tabor 1972). We will confine this section to summarizing the essential elements.

Experimental values for the non-cubic anisotropy of garnet films are in general substantially higher than can be attributed to stress. The largest values occur when the film contains a mixture of rare earths and is correlated with the size difference between those rare earths (Gyorgy et al. 1973). Examination of the garnet crystal structure reveals that the rare earth dodecahedral sites alternate with tetrahedral iron sites along strings that are oriented in the cubic directions but do not intersect. There are six magnetically inequivalent sites, each of which has a different orientation. The six sites and their local orthorhombic axes of symmetry are given by Rosencwaig and Tabor (1972) as in table 4.

If all of these sites were uniformly occupied by the rare earth ions the local anisotropies would be expected to average out. On the other hand, a preferential population of the sites when there is more than one kind of ion for these sites could lead to a net orthorhombic anisotropy of the form included in eq. (20). The values of A and B for this equation would depend on the identity of the ions and their preferential population ratios, and the latter can be expected to vary with the growth direction of the film as well as with the identity of the ions. Such a dependence has roughly been observed. Mixed garnets films involving Eu or Sm with smaller ions such as Y, Gd, Er, Tm, Lu usually have $A > 0$, $B < 0$ and $B/A \simeq 0.4$ (Hagedorn et al. 1973). This means that there is an easy axis normal to 111 films (suitable for bubbles) but that 001 films have the magnetization lying in the plane. Combination with Eu or Sm of ions that are larger does seem to reverse the sign of A leading to the stability of bubbles in some 001 films (Wolfe et al. 1976).

At the time of this writing no model has yet been devised that can be used to predict the values of A and B for particular rare earth combinations and film growth orientations. Indeed there is evidence that it may be necessary to include terms in the anisotropy of higher order in the direction cosines such as $B_1(\alpha_1\alpha_2\beta_1\beta_2\beta_3^2 + \alpha_2\alpha_3\beta_2\beta_3\beta_1^2 + \alpha_3\alpha_1\beta_1\beta_2^2)$ (Josephs et al. 1977).

TABLE 4

Symmetries of the garnet rare earth sites

Site	Principal axes		
	a	b	c
x_1	$01\bar{1}$	011	100
x_2	011	$01\bar{1}$	$\bar{1}00$
y_1	$\bar{1}01$	101	010
y_2	101	$\bar{1}01$	$0\bar{1}0$
z_1	$1\bar{1}0$	110	001
z_2	110	$1\bar{1}0$	$00\bar{1}$

Until further understanding is achieved it is necessary to rely on empirical models for the correlations of anisotropy with bubble film composition. One such model which reflects the features of the growth anisotropy mechanism is presented next for 111 oriented films.

5.5.5. Model for growth anisotropy of 111 oriented films

When viewed from the 111 direction the x_1, y_1 and z_1 sites present one configuration to an arriving rare earth ion and the x_2, y_2 and z_2 a different configuration. Thus when 111 films are grown with large ions like Eu or Sm as well as smaller ions like Y, Tm, Yb or Lu, the large ions preferentially populate one set of these sites, which induces a net anisotropy of the form shown in eq. (20) with $A = 0$ and $B < 0$. It is more convenient to discuss the growth anisotropy in terms of a positive constant, K_G, which represents the difference in energy between the "hard" in-plane orientation of the magnetization and the "easy" 111 orientation. It is seen from table 3 that $K_G = -\frac{1}{2}B'$.

A number of compositions which involve Eu or Sm and the smaller ions have been grown, measured, and their properties published. A few of the results are displayed in table 5.

Because the characteristic length is 0.05–0.06 these compositions are useful for bubble sizes of 0.4–0.5 μm. For larger bubble sizes these compositions are diluted with Ga or CaGe to lower the magnetization and with Y to lower K_G to keep the Q in a convenient range. Eschenfelder (1978) has published an empirical model that seems to correlate the anisotropy value with dilutions. The basic elements of that model are as follows:

(i) The K_G of these compositions are related to an intrinsic K'_j for each pair of rare earth ions modified by the product of the ionic abundances, $x_i x_j$. Thus, for SmLu $K'_j = 14$, yielding $K_G = 14 \times 1.2 \times 1.8 = 30$ for $Sm_{1.2}Lu_{1.8}Fe_5O_{12}$.

(ii) When multiple pairs of rare earths are involved with x_i of each large rare earth and x_j of each small ion, $K_G = \Sigma_{ij} K'_j x_i x_j$ where the sum is over all such pairs and the K'_j are given in table 6. Thus for $Sm_{0.8}Lu_{1.2}Y_{1.0}Fe_5O_{12}$

$$K_G(10^4) = (14 \times 0.8 \times 1.2) + (3.3 \times 0.8 \times 1.0) = 16$$

(iii) When Ga or CaGe are added to dilute the magnetization the K_G must be reduced by a factor A/A_0, reflecting the reduction in exchange energy, and plotted vs. dilution in fig. 19 (Krahn et al. 1978). CaGe should produce a

TABLE 5
Anisotropy of undiluted garnets

Composition	Source	$4\pi M$ (G)	$l(\mu m)$	$K_G(10^4\,erg\,cm^{-3})$
$Sm_{1.2}Lu_{1.8}Fe_5O_{12}$	Bullock et al. (1974)	1450	0.05	30
$Sm_{0.85}Tm_{2.15}Fe_5O_{12}$	Yamaguchi et al. (1976)	1380	0.05	19
$Eu_{1.0}Tm_{2.0}Fe_5O_{12}$	Giess et al. (1975)	1380	0.06	12

TABLE 6

K_j^i for various rare earth pairs

Large rare earth	Small ion			
	Lu	Yb	Tm	Y
Sm	14	12	10.5	3.3
Eu	9.5	8	6	1

result like x in fig. 19 whereas Ga will decrease A more rapidly because 10% is octahedral. Thus for $Eu_{1.0}Tm_{2.0}Ga_{0.6}Fe_{4.4}O_{12}$

$$K_G(10^4) = [6 \times 1.0 \times 2.0]0.6 \simeq 7.$$

(iv) The K_j^i correlate strongly with the difference in ionic radius of the two ions, Δr_{ij}, being close to $K_j^i(10^4) = 200 \ (\Delta r_{ij} - 0.05 \ \text{Å})$. Combinations of ions having size differences less than 0.05 Å seem to produce a negligible contribution to the growth anisotropy.

(v) These K_j^i, reflecting the relative ordering on the two sets of sites, vary somewhat with factors like growth temperature.

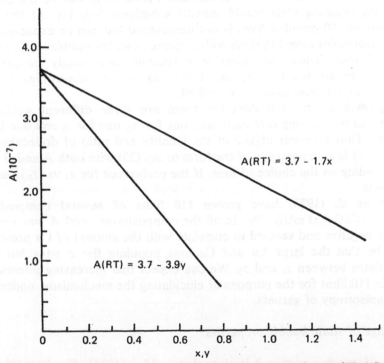

Fig. 19. Reduction in $A(10^{-7} \text{ erg cm}^{-1})$ due to tetrahedral dilution (x) in $\{Y_{3-x}Ca_x\}[Fe_2](Ge_xFe_{3-x})O_{12}$ and octahedral dilution (y) in $\{YLu\}[Sc_yFe_{z-y}](Fe_3)O_{12}$ (Krahn et al. 1979).

As understanding and additional data are acquired, better models will undoubtedly become available. Until that time this model provides some guidance in relating growth anisotropy to composition for useful garnet films.

5.5.6. Other orientations

For growth in the 100 direction the x_1 and x_2 sites present one configuration while a different configuration is presented by y_1, y_2, z_1 and z_2. Thus we can expect to see preferential occupation of these two sets and this leads to an anisotropy of the form shown in eq. (21). A will be positive or negative depending on the identity of the ions and might be a maximum when the ratio of ions is different from unity, in contrast to the case of 111 films. We might also expect a greater degree of ordering than for 111 films since the two configurations are more dissimilar for 100 than for 111.

A variety of films grown on 100 substrates do exhibit the expected form of K_G with the magnitude being substantially larger than typical for 111. An example is $Eu_2Y_1Fe_5O_{12}$ grown on 100 $Sm_3Ga_5O_{12}$ by Plaskett et al. (1972) where the anisotropy with no lattice mismatch would extrapolate to 30×10^4 with the 100 plane the preferred orientation of the magnetization. Thus A is positive and substantially larger than the negative B for 111. Plaskett also found that growth of this composition at low temperatures ($\sim 750^{\circ}C$ vs. $\sim 890^{\circ}C$) led to substantial incorporation of Pb from the melt flux and a reversal in sign of the anisotropy, so that the resulting films would support submicron bubbles. At this time the anisotropy of 100 oriented films is not understood but can be expected to allow larger anisotropies than 111 films with a correct sign for bubbles with the proper choice of ions. Thus such films may become increasingly popular as the technology progresses to very small bubbles where anisotropies larger than achievable in 111 orientations are desired.

For growth in the 110 direction there are three different configurations presented to the arriving rare earth ion; one for z_1, one for z_2 and one for x_1, x_2, y_1 and y_2. Thus a proper choice of the identity and ratio of dodecahedral ions should result in an anisotropy of the form of eq. (23) with both A and B of either sign depending on the choice of ions. If the preference for z_1 vs. z_2 is negligible, then $B \simeq 0$.

Wolfe et al. (1976) have grown 110 films of several compositions of $Eu_xLa_{3-x-y}(CaGe)_y(FeAl)_{5-y}O_{12}$. In all the compositions cited A was negative. B also was negative and seemed to correlate with the amount of Ca present. Thus it may be that the large La and Ca ions populate the z sites but only Ca differentiates between z_1 and z_2. We can expect that increasing interest will be shown in 110 films for the purpose of elucidating the mechanisms underlying the growth anisotropy of garnets.

5.6. Range of static bubble properties

From the foregoing section it is clear that $4\pi M \simeq 1400\,G$, $K \simeq 20 \times 10^4\,erg\,cm^{-3}$, and $A \simeq 4 \times 10^{-7}\,erg\,cm^{-1}$ can be achieved in several undiluted, 111 oriented garnet films. Such films yield $0.5\,\mu m$ bubbles with $Q > 2$. We also see that

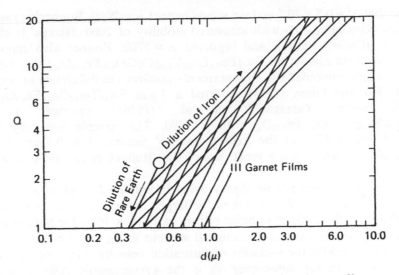

Fig. 20. Range of static bubble properties for 111 oriented garnet films.

dilution with either Ga or CaGe will decrease $4\pi M_S$ substantially with modest decreases in K and A at the same time. In addition the K can be further reduced by changing the rare earth couple or diluting them with Y. Using the variations of $4\pi M_S$, K and A with dilution that have been described we find that 111 garnets can provide bubble sizes and Q as illustrated in fig. 20. Dilution of the iron yields a decrease in $4\pi M_S$ with K remaining proportional to A so that by eq. (13) Q/l remains constant. Dilution of the rare earth decreases K with practically no change in $4\pi M_S$ or A so that by eq. (14) Q/l^2 remains constant.

5.7. Dynamic bubble properties

As we have seen, A for useful garnets is between 2 and 4×10^{-7} erg cm^{-1} and $Q = 4 \pm 2$. We find the values of λ' in table 2. We can use eq. (17) to predict the mobility μ. For 5 μm bubbles there is considerable dilution to get $4\pi M_S$ down around 200 G. Thus $A \simeq 2 \times 10^{-7}$, $Q \simeq 5$ and the dodechaderal ions are mostly Y with little Eu or Sm (as seen in the examples of table 7) so $\lambda' \simeq 1 \times 10^{-7}$ Oe^2 s. We therefore expect $\mu \simeq 1000$ cm s^{-1} Oe^{-1}. For smaller bubbles the mobility decreases because of the need to increase the Eu or Sm content in order to achieve the necessary anisotropy to preserve Q as $4\pi M_S$ increases. We see from table 2 that the use of Sm increases λ' and hence decreases μ rapidly and that Eu is preferable in this regard. The use of $Eu_{1.0}Tm_{2.0}Fe_5O_{12}$ for 0.5 μm bubbles would lead to an interpolated $\lambda'(10^{-7}) = 1.8$ and expected mobility of ~ 900. The use of $Sm_{1.2}Lu_{1.8}Fe_5O_{12}$ would lead to an interpolated $\lambda'(10^{-7}) = 5.1$ and expected mobility of $\simeq 300$.

Actual measurements agree pretty well with these expectations from basic properties. In the larger bubble region (6 μm) both Bonner et al. (1973) and Hiskes (1976) have reported measurements on samples with no magnetic rare earths so that $\lambda' \simeq 0.5 \times 10^{-7}$ and we would expect $\mu \simeq 2000$. Bonner's example is $Lu_{0.2}Y_{1.88}(CaGe)_{0.92}Fe_{4.08}O_{12}$ with measured mobility of 2000. Hiskes' is close to $Lu_{0.4}La_{0.1}Y_{1.5}(CaGe)_{1.0}Fe_{4.0}O_{12}$ and reported $\mu \simeq 2700$. Bonner also reported a $6 \mu m$ sample with almost no Eu ($Eu_{0.1}Lu_{0.3}Y_{1.64}(CaGe)_{0.96}Fe_{4.04}O_{12}$) that had $\mu \simeq 1400$. For smaller bubbles the measurements confirm the difference between Eu and Sm. Hu and Giess (1974) reported a $1 \mu m$ $Eu_{0.8}Tm_{2.2}Ga_{0.5}Fe_{4.5}O_{12}$ with $\mu \simeq 1000$ while Yamaguchi et al. (1976) reported a $1 \mu m$ $Sm_{0.4}Lu_{0.56}Y_{1.29}(CaGe)_{0.75}Fe_{4.25}O_{12}$ with $\dot{\mu} \simeq 600$. This sample has only 0.4 Sm compared with 0.8 Eu in the higher mobility sample. For $0.5 \mu m$ material, $Sm_{1.5}Lu_{1.5}Fe_5O_{12}$, which has a great deal of Sm, Bullock et al. (1974) reported a $\mu \simeq 200$.

Thus we see that the simple theory using interpolated λ' gives a reasonable approximation to the mobility. Films for larger bubbles can be made to have $\mu > 1000$ and by the judicious choice of rare earth ions, films for bubble sizes all the way down to $0.5 \mu m$ can be designed without going much below that level.

Now let us consider the variation of saturation velocity. From eq. (19) we see that a principal factor influencing V_0 is the gyromagnetic ratio γ, which is proportional to g. For most of the compositions used not containing Eu, $g \simeq 2$ and $\gamma \simeq 1.8 \times 10^7$. The introduction of Eu produces significant changes because the effective g of Eu is very large since it is proportional to the magnetic moment divided by the angular momentum and Eu has $J \simeq 0$ but a finite moment. With a combination of ions in the garnet each having magnetization $4\pi M_i$ and g_i the total effective g is

$$\frac{\Sigma \, 4\pi M_i}{g_{eff}} = \Sigma \, \frac{4\pi M_i}{g_i}. \tag{24}$$

LeCraw et al. (1975) showed that very large g_{eff} could be produced in $Eu_3Fe_{5-x}Ga_xO_{12}$ when the Ga dilution, x, was sufficient to reduce the net Fe magnetization to small values so that $4\pi M_{Fe}/g_{Fe}$ no longer overwhelms $4\pi M_{Eu}/g_{Eu}$. Because of the high value of g_{Eu}, $\Sigma \, [4\pi M_i/g_i]$ will be equal to zero at a slightly larger Ga content, x_1, than that for $\Sigma \, 4\pi M_i = 0$, x_0. The latter point is of course the magnetization compensation point where the net Fe magnetization equals the Eu magnetization. The former point, x_1, is called the momentum compensation point where the net Fe magnetization has to be less than the Eu magnetization by the ratio of the g-factors (hence $x_1 > x_0$). We see that the effective g can vary dramatically according to

Range of x	$\Sigma \, 4\pi M_i$	$\Sigma \, \dfrac{4\pi M_i}{g_i}$	g_{eff}
$x < x_0$	>0	$> \frac{1}{2} \Sigma \, 4\pi M_i$	$0 < g_{eff} < 2$
$x = x_0$	$=0$	>0	$=0$
$x_0 < x < x_1$	<0	>0	<0
$x = x_1$	<0	$=0$	∞
$x_1 < x$	<0	$< \frac{1}{2} \Sigma \, 4\pi M_i$	>2

Of course, the very large g_{eff} only occur for garnets which may be said to be over-compensated, i.e., $x_0 < x$, but, at the same time, g_{eff} and hence γ are reduced in all garnets which contain Eu even for $x < x_0$. Therefore the γ of Eu containing garnets is $< 1.8 \times 10^7\, Oe^{-1}\, s^{-1}$ and more so for larger Eu content and/or greater Fe lattice dilution.

LeCraw et al. (1975) found that with a special design towards momentum compensation they could achieve $V_0 = 30\,000\, cm\, s^{-1}$ with a $Q = 6.9$ ($V_0 Q = 200\,000\, cm\, s^{-1}$). To do this, however, they had to reduce $4\pi M_S$ to 218 G and the bubble size was $\simeq 5\, \mu m$. In garnets for smaller bubbles where larger values of $4\pi M_S$ are necessary and a close approach to momentum compensation is not possible γ will usually lie between 1.8×10^7 and $1.3 \times 10^7\, Oe^{-1}\, s^{-1}$. Therefore, from eq. (19) we expect

$$10\,000 < V_0 Q(cm\, s^{-1}) < 30\,000\, cm\, s^{-1}.$$

The measured values of practical films do fall in this range with Eu containing films being lower and thin films of SmLuYCaGe being higher.

Because the velocity saturation is basically due to instabilities in the bubble wall structure, saturation can also be inhibited and velocities substantially increased by other mechanisms which will stabilize that wall structure. Two such mechanisms are the application of an in-plane field (Vella-Coliero 1973, Bullock 1973, de Leeuw 1973) or the introduction of an in-plane anisotropy (Stacy et al. 1976, Wolfe et al. 1976).

5.8. Selection of garnet compositions for applications

We will now consider the procedures for selecting garnet compositions to match the requirements for devices as described in section 2.4.

a. Static properties

It is convenient to use fig. 21, which is the basic conversion chart (fig. 15) with two additions. The curves in the central portion of the chart reflect the decrease in A and $4\pi M_S$ of a basic garnet with $4\pi M_S = 1760\, G$ as either Ga or CaGe is added as a replacement of the iron. At the top the Kryder Q criterion is plotted to show the range of Q values acceptable for various bubble sizes.

To use the chart we first locate the desired bubble size and Q on the right side, making sure the Q is in the acceptable range. A line drawn horizontally from this point will intersect the lines for Ga and CaGe dilution. A vertical line through this intersection determines the required $4\pi M_S$ at the bottom of the chart and K at the top of the chart. We can then construct a composition having these values of $4\pi M_S$ and K plus matching the substrate lattice by using the guidelines of sections 5.4, 5.5, and 5.6, respectively.

As an example, suppose we want a SmLuCaGe composition for 3 μm bubbles with a Q of 5 in a "thick" film ($h \simeq d = 3\, \mu m$) on a GGG substrate. The chart construction is shown in fig. 22 and indicates $4\pi M_S \simeq 330\, G$ and $K(10^4) \simeq 2$. For $4\pi M_S = 330$ we need CaGe $\simeq 0.9$ from fig. 18. For this value of CaGe $A/A_0 \simeq \frac{2}{3}$.

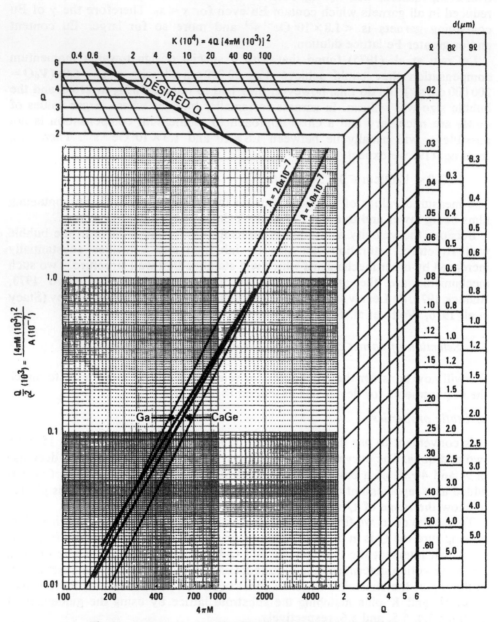

Fig. 21. Augmented summary chart for static bubble parameters.

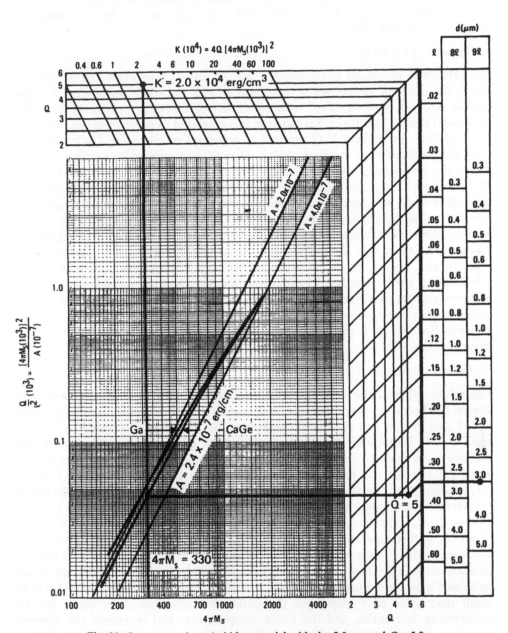

Fig. 22. Parameters for a bubble material with $d = 3.0\ \mu m$ and $Q = 5.0$.

Our formula should therefore be $Sm_x Lu_y Y_{2.1-x-y}(CaGe)_{0.9} Fe_{4.1} O_{12}$ and $\Sigma K_{ij} x_i x_j = \frac{3}{2} \times 2 = 3$. Then, $14xy + 3.3x(2.1 - x - y) = 3$ and to match GGG $0.05x - 0.03y = 0.003$. The solution is $x = 0.3$ and $y = 0.4$ and the formula is $Sm_{0.3} Lu_{0.4} Y_{1.4}(CaGe)_{0.9} Fe_{4.1} O_{12}$.

b. Dynamic properties

The principles discussed in section 5.7 and the eqs. (17) and (19) can be used to adjust the dynamic properties in conjunction with the static properties. For the bubble size we need to dilute with either Ga or CaGe to obtain the required $4\pi M_S$ and then either Eu or Sm must be incorporated to achieve the required K. We saw that the highest $V_0 Q$ product is achieved by choosing CaGe and Sm. This will ordinarily be the best choice. However, as we go to smaller bubbles where the Sm content becomes large the mobility can become too low because of the large λ' of Sm. At the same time the Fe magnetization is higher to provide the higher $4\pi M_S$ so that the incorporation of Eu does not affect the γ as much as when more dilution is used. In this case then some Sm can be replaced by Eu to achieve the right balance of V_0, and μ, as discussed in section 2.4. Additional factors to be taken into account as we go to smaller bubbles are that we may still want to use Sm vs. Eu because its higher λ_{111} is more compatible with ion-implantation (to be discussed later in this section) and the small bubble device patterns and larger drive fields imply that for equivalent operating speeds the mobility does not have to be as high as for the larger bubbles. Thus the tradeoff between Sm and Eu is made using these principles but adapted to the particular device situation.

c. Thermal variation

The relationship of the variation with temperature of bubble film properties to device requirements was discussed in section 2.4. The decrease in $4\pi M_S$, K and A and also in Q, d and H_{CO} with temperature for garnets is illustrated by Nielsen et al. (1974). Fortunately the $4\pi M_S$ of the useful garnet compositions, reflecting primarily the orderly decrease of the net Fe magnetization, comes close to matching that of the typical permanent magnets ($-0.2\%/°C$). There are two ways to fine-tune the magnetization variation with temperature to closely match the permanent magnet. Because more Ga has to be added than CaGe to produce a given $4\pi M_S$ (refer to fig. 18) the Curie temperature is lowered and the decrease of $4\pi M_S$ with temperature is more rapid. Thus for 3 μm bubbles Ga dilution tends to produce too much thermal decrease and CaGe not enough. For smaller bubbles with less Ga the thermal decrease will be just right. For the larger bubbles CaGe can be used and some Lu applied as the small rare earth. Lu is so small that some of it populates the octahedral iron sites, decreases the Fe subnetwork interaction and the Curie temperature, and increases the thermal decrease of magnetization. Thus the thermal decrease can be fine-tuned by a balance of Ga vs. CaGe and/or by the addition of some Lu as the small rare earth (Blank 1975).

A typical garnet designed to decrease 20% in H_{CO} from 0–100°C to match the barium ferrite may decrease $\sim 30\%$ in $4\pi M_S$, $\sim 50\%$ in $(4\pi M_S)^2$, $\sim 65\%$ in K and $\sim 65\%$ in A. In accord with our equations, the required bias field decreases $\sim 20\%$, matching the magnet, but bubble size also decreases $\sim 20\%$ and the $Q \sim 30\%$. Fortunately most device forms can tolerate this variation in bubble size, but the material clearly has to be designed to have a room temperature Q at least 30% larger than Q_{min}.

d. Implantability

As discussed in section 2.4 the primary requirement for implantability is for an adequate λ_{111}. The in-plane anisotropy required is of course proportional to the uniaxial anisotropy of the bubble film. Since K must increase as we go to smaller bubbles to maintain the Q with the higher $4\pi M_S$, the requirement for λ_{111} also increases for smaller bubbles. The variation of λ_{111} with composition was discussed in section 5.5.3. Suffice it to remark here that the introduction of Sm to achieve K_u is more compatible with the requirement for implantability than Eu since Sm increases λ_{111} whereas Eu decreases it. On the other hand, for smaller bubbles with high K, λ_{111} is not high enough for implantation to overcome K even when Sm is used and other techniques must be used for some practical devices (Lin et al. 1977). The most obvious technique is the growth of double layer films where the surface to be implanted is formulated to have a large λ_{111} and lower K that can be overcome by the implantation.

5.9. Typical compositions

Obviously, from the foregoing discussion, there are many different garnet compositions suitable for bubbles. Device oriented compositions involve Eu or Sm in combination with smaller ions on the dodecahedral sites. At the present time, data has been published on many more compositions for 3–6 μm bubbles than for smaller bubbles. As device structurres and available lithography make practical the use of smaller bubbles, surely many more small bubble compositions will be studied and the results published. A few examples of compositions covering the practical range of bubble sizes are listed in table 7. The most popular 5–6 μm compositions are listed. $Sm_{0.3}Lu_{0.3}Y_{1.48}(CaGe)_{0.92}Fe_{4.08}O_{12}$ with some slight variations in Sm and Lu content has been used almost universally for 3 μm bubble devices. Smaller bubbles, down to 0.4 μm will be achieved with dilution of the parent garnets shown for the 0.4–0.5 μm range.

6. Orthoferrites

Orthoferrites were the first materials to be investigated for magnetic bubbles. Interest shifted to the garnets, however, when studies of these materials led to the conclusion that they could only support bubbles much larger than useful for practical devices.

The orthoferrites have the formula $RFeO_3$ where R is a rare earth or yttrium.

TABLE 7

Typical garnet compositions

Bubble size range (μm)	Composition	Source	Properties			
			$l(\mu m)$	Q	$4\pi M_S$	$K(10^4)$
0.4–0.5	$Eu_{1.0}Tm_{2.0}Fe_5O_{12}$	Giess et al. 1975	0.06	1.7	1380	12
	$Sm_{0.85}Tm_{2.15}Fe_5O_{12}$	Yamaguchi et al. 1976	0.05	2.5	1380	19
	$Sm_{1.2}Lu_{1.8}Fe_5O_{12}$	Bullock et al. 1974	0.05	2.5	1450	30
0.8–1.2	$Eu_{1.0}Tm_{2.0}Ga_{0.6}Fe_{4.5}O_{12}$	Giess et al. 1975	0.15	3.0	900	9
	$Sm_{0.3}Tm_{2.75}Y_{1.25}(CaGe)_{0.75}Fe_{4.25}O_{12}$	Yamaguchi et al. 1976	0.10	3.1	504	3
	$Eu_{1.2}Lu_{1.8}Ga_{0.5}Fe_5O_{12}$	Hu and Giess 1974	0.12	4.0	770	9
2–3	$Sm_{0.3}Lu_{0.3}Y_{1.48}(CaGe)_{0.92}Fe_{4.08}O_{12}$	Blank 1975	0.35	4.8	320	2
	$Eu_{1.0}Tm_{2.0}Ga_{0.6}Fe_{4.4}O_{12}$	Giess et al. 1975	0.23	3.0	750	7
5–6	$Sm_{0.10}Y_{1.92}(CaGe)_{0.98}Fe_{4.02}O_{12}$	Blank et al. 1976	0.66	4.1	163	0.5
	$Sm_{0.38}Y_{2.62}Ga_{1.15}Fe_{3.85}O_{12}$	Nielsen 1976	0.63	8.8	200	1.5
	$Eu_{0.65}Y_{2.35}Ga_{1.2}Fe_{3.8}O_{12}$	Giess and Cronemeyer 1973	0.77	8.0	160	0.8

They have a distorted perovskite structure and so have an easy direction of magnetization along the "c" axis above a "reorientation" temperature and along the "a" below. For most of the orthoferrites the reorientation temperature is below room temperature so an easy axis normal to a film can be achieved that is suitable for bubbles.

The problem with the orthoferrites is that the magnetization is so low, e.g., $4\pi M_S \leq 150\,G$. Thus with an anisotropy of $K(10^4\,erg\,cm^{-3}) \simeq 80$ the Q is exceedingly high and the bubbles $\simeq 100\,\mu m$. To reach a more practical range the $4\pi M_S$ would have to be increased or the K reduced. There has been little success with $4\pi M_S$. K has been reduced by mixing orthoferrites so as to bring the reorientation temperature close to room temperature (e.g., $Sm_{0.55}Tb_{0.45}FeO_3$, Bobeck et al. 1969). This brought the size of the bubbles down to $20\,\mu m$, but this approach also produces unacceptable variability of the properties with temperature. It therefore appears that there is no way to make orthoferrites attractive for bubble applications unless the application is an unusual one that derives an advantage from larger bubbles.

7. Hexaferrites

Hexaferrites have the opposite problem to orthoferrites – the $4\pi M_S$ is so large that coupled with the natural anisotropy the bubbles are very small. In this case however both the $4\pi M_S$ and the K can be varied by substitution and these materials may provide bubbles smaller than $0.4\,\mu m$ which cannot be achieved with garnets.

The hexaferrites represent a family of oxides related to $BaFe_{12}O_{19}$. Within this family there are a variety of structures (M, W, Y, Z, etc.) that are built up of sections of different types (designated R, S and T) alternating along the c axis. R and T have a hexagonal close-packed arrangement of oxygen atoms while S is a cubic close-packed arrangement such as is found in spinels. The formula and layer structure for several of the more interesting types are as follows

M	$BaFe_{12}O_{19}$	RSR*S*
S	$Me_2Fe_4O_8$	SSSS
W	$BaMe_2Fe_{16}O_{27}$	RSSR*S*S*
Y	$Ba_2Me_2Fe_{12}O_{22}$	TSTS
Z	$Ba_3Me_2Fe_{24}O_{41}$	RSTSR*S*T*S*

Me represents a divalent transition metal and R* represents R rotated 180° about the c axis. Smit and Wijn (1959) provide an excellent discussion of these structures and their properties. These structures have $4\pi M_S \sim 2000–5000\,G$. M and W structures have the c axis as an easy axis of magnetization with $K(10^4\,erg\,cm^{-3}) \leq 330$. Y structures generally have $4\pi M_S$ lying in the basal plane because of the nature of the T section. These properties can be varied both by the selection of Me and also by dilution of the Fe by Ga, Al, etc., as in the case of garnets.

The "a" spacing is such that it should be possible to grow films epitaxially on a non-magnetic garnet or spinel of the proper lattice spacing and have the easy axis normal to the film for bubbles. However, all of the work reported to date has been on slices cut from bulk-grown crystals.

Kooy and Enz (1960) have shown that $BaFe_{12}O_{19}$ platelets have $4\pi M_S = 4700$ G and $K(10^4 \text{ erg cm}^{-3}) = 330$ which yields bubbles $\simeq 0.15$ μm with good Q ($\sim$3–4). It is not clear whether these bubbles will have adequate dynamic properties. Measurements on platelets gave $\mu < 10$ cm s^{-1} Oe^{-1} and $V_0 = 100$–600 cm s^{-1} (Bobeck 1970). Further insight into this must await the preparation of good films.

Further investigation will also demonstrate to what degree dilution and substitution in these materials will allow tailoring the bubble size and Q to fill the gap between the above values for $BaFe_{12}O_{19}$ and the lower limit of garnets. That will require both the reduction of $4\pi M_S$ (to increase d) and of K (to keep Q at a desirable level). Approaches with regard to achieving both of these have already been demonstrated.

Al has been used to successfully lower $4\pi M_S$ and produce larger bubbles (Van Uitert et al. 1970). It was possible to reduce $4\pi M_S$ to as low as 900. However, at the same time, as in the case of garnets, such dilution of the iron also causes a decrease in A, K, and Curie temperature, and an increase in the thermal variation of bubble properties. It remains to be seen, therefore, just how much reduction in $4\pi M_S$ is practical for useful bubbles. It would appear reasonable from the available data that bubbles as large as 0.4 μm could be achieved.

If K decreases with A from dilution we can expect Q/d to remain constant from eq. (13). Thus the increase from 0.15 to 0.4 μm bubble size would also increase the Q to about 10. Some additional reduction in K would therefore be desireable to reduce Q. Reductions in K of the type desired have been achieved by partially replacing Fe^{3+} with $Ti^{4+}Co^{2+}$ (Lotgering et al. 1961).

Much more work will undoubtedly be done on hexaferrites, especially as the prospects improve for being able to use submicron bubbles in actual devices. Much remains to be learned but there is reason to be optimistic that variations of these materials will become useful. It is not yet clear whether suitably oriented, single crystal films can be grown (Glass and Liaw 1978).

8. Conclusion

The elementary concepts of magnetic bubbles and charts, as well as formulae, for the conversion of important parameters have been presented. These apply to both the crystalline films for bubbles discussed in this chapter and the amorphous films to be discussed in chapter 6. The presentation of concepts has been necessarily brief; more details can be found in the references cited. The parameter conversion charts are intended to be useful for the practitioner of bubble films.

Garnet films are very useful for magnetic bubbles because their properties can be so widely varied. The principles described in section 5 can be used to formulate garnets suitable for particular bubble properties as long as the bubbles are larger than 0.4 μm. At that point 111 oriented garnets are limited in available anisotropy as well as magnetization. Additional anisotropy may be obtainable by converting to 100 orientation of the garnets.

The hexaferrites may yield useful bubbles considerably smaller than 0.4 μm. The variety of compositions and the ionic substitutions allowed evidently will provide the range of magnetization and anisotropy required. However it will be necessary to develop the techniques for growing suitable, very thin, oriented single crystal films.

Appendix A: Conversion of Gaussian to SI units

TABLE A.1

Parameter	Gaussian Formula	Units	SI Formula	Units	SI Gaussian Ratio
Magnetomotive force	$MMF = \dfrac{4\pi NI}{10}$	Gilbert	$MMF = NI$	A T	$10/4\pi = 0.796$
Magnetic field H	$\int H \cdot dl = MMF$	Oe	$\int H \cdot dl = MMF$	A m^{-1}	$10^3/4\pi$ A/m Oe
Flux ϕ	$\phi = \int B \cdot dA$	Maxwell	$\phi = \int B\, dA$	Weber	10^{-8} W/Mx
Flux density B	$B = H + 4\pi H$	Gauss	$B = \mu_0(H + M)^*$	Tesla	10^{-4} T/G
			$= (\mu_0 H + J)^{**}$	$\equiv$ Wb m^{-2}	10^{-4} Wb/m^2 G
Magnetization	$4\pi M$	Gauss	M^*	A m^{-1}	$10^3/4\pi$
Magnetic polarization	$4\pi M$	Gauss	$J = \mu_0 M^{**}$	Tesla	10^{-4} T/G
Volume susceptibility	$\chi = dM/DH$		$\kappa = dM/dH$		$1/4\pi$
Demagnetization factor	$H_d = -NM$		$H_d = -NM$		$1/4\pi$
Anisotropy	K	erg cm^{-3}	K	J m^{-3}	10^{-1}
Exchange	A	erg cm^{-1}	A	J m^{-1}	10^{-5}
Mobility	μ	cm s^{-1} Oe^{-1}	μ	m^2 A^{-1} s^{-1}	$4\pi \times 10^{-5}$
Energy product	$\dfrac{1}{8\pi} BH = \tfrac{1}{2} MH$	erg cm^{-3}	$\tfrac{1}{2} BH = \dfrac{\mu_0}{2} MH$	J m^{-3}	10^{-1}
Anisotropy field	$H_K = 2K/M$	Oe	$H_K = 2K/\mu_0 M$	A m^{-1}	$10^3/4\pi$
Q	$Q = H_K/4\pi M$		$Q = H_K/M$		
Wall energy density	$\sigma = 4\sqrt{AK}$	erg cm^{-2}	$\sigma = 4\sqrt{AK}$	J m^{-2}	10^{-3}
Characteristic length	$l = \sigma/4\pi M^2$	cm	$l = \sigma/\mu_0 M^2$	m	10^{-2}
Permeability	$\mu = dB/dH$		$\mu = dB/dH$		$4\pi \times 10^{-7} = \mu_0$
	$= 1 + 4\pi\chi$		$= \mu_0(1 + \kappa)$		$= 12.57 \times 10^{-7}$

*RMKS.
**MKSA

Appendix B. Use of the charts for conversion of bubble material parameters

1. Deduction of K and A from measured values of l, Q and $4\pi M_S$

Figure 8 gives the values for K immediately from the measured values of $4\pi M_S$ and Q. Then fig. 10 or fig. 11 can be used to obtain the value of A, but first it is necessary. to have the value of Q/l or Q/l^2, respectively. These values, in turn, can be obtained from fig. 9 once l is known. The value of l can be obtained from measured values of film thickness, h, and W_S or d_{CO} or $H_{CO}/4\pi M_S$.

2. Deduction of l from measured values of h and W_S or d_{CO}

Figure B.1a yields the value of l from either the measured demagnetized stripewidth, W_S, or the bubble size just before collapse, d_{CO}, plus h. Figure B.1b conveniently converts the measured h and W_S to the ratio h/W_S. Note that $d_{CO} \simeq 0.52 W_S$.

3. Deduction of l from measured values of H_{CO}, $4\pi M_S$ and h

Figure 13 can be used to obtain h/l from the measured value of $H_{CO}/4\pi M_S$ and then l can be calculated or read directly from fig. B.1b.

4. Deduction of Q from H_K and $4\pi M_S$

Q is usually determined from measurements of the anisotropy field, H_K and $4\pi M_S$. Figure B.2 gives this conversion directly.

5. Deduction of implied bubble d and Q from known basic magnetic parameters

Figure 11 yields Q/l^2 for known values of $4\pi M_S$ and A. Given the value for K the implied value of Q/l is given by fig. 10. Figure 9 can then be used to separate Q and l. l is converted into bubble size via fig. B.1a (remember that the mid-bias value of $d \simeq W_S$).

6. Deduction of required K for a given bubble Q and d

Figure B.1a is used to convert d to l and fig. 9 to convert Q and l to Q/l. Then the required value of K is read from fig. 10.

7. Deduction of required $4\pi M_S$ for a given bubble Q and d

Figure B.1a is used to convert d to l and fig. 9 to convert Q and l to Q/l^2. Then the required value of $4\pi M_S$ is read from fig. 11.

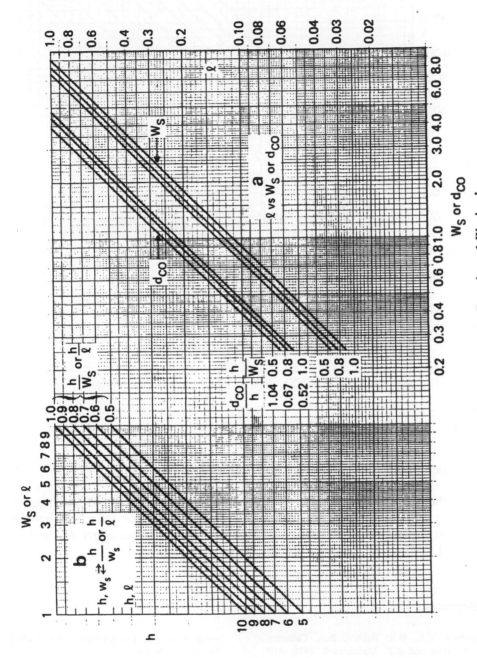

Fig. B.1. Conversion chart for the dimensions: ℓ, W_s, d_{CO}, h.

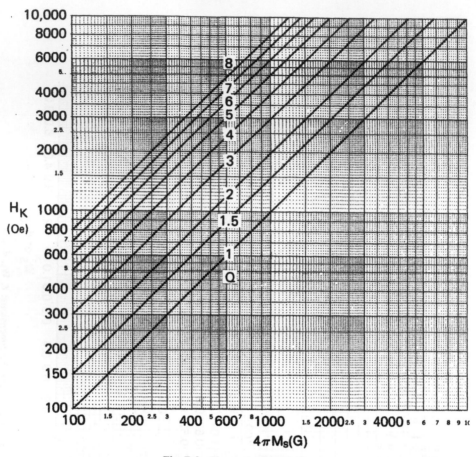

Fig. B.2. H_k vs. $4\pi M_S$ and Q.

8. Deduction of H_B required to stabilize a bubble of size d

Remembering that the mid-bias $d \simeq W_S$ we use fig. B.1 to obtain the required value for h/l. Then fig. 13 shows the required bias field, H_B, for a free bubble. Permalloy overlays will further stabilize the bubble so that larger H_B will be required in their presence as if the thickness were increased.

References

Beaulieu, T.J., B.R. Brown, B.A. Calhoun, T.L. Hsu and A.P. Malozemoff, 1976, AIP Conf. Proc. 34, 138.

Blank, S.L., 1975, Presented at the 1975 fall meeting of the Electrochem. Soc. Dallas, Texas, October 1975.

Blank, S.L., B.S. Hewitt, L.K. Shick and J.W. Nielsen, 1972, AIP Conf. Proc. 10, 256.

Blank, S.L., J.W. Nielsen, W.A. Biolsi, 1976, J. Electrochem. Soc. 123, 856.

Bobéck, A.H., 1970, IEEE Trans. MAG-6, 445.

Bobeck, A.H., R.F. Fischer, A.J. Perneski, J.P. Remeika and L.G. Van Uitert, 1969, IEEE Trans. MAG-5, 554.

Bobeck, A.H., S.L. Blank and H.J. Levinstein, 1973, AIP Conf. Proc. 10, 498.

Bonner, W.A., J.E. Geusic, D.H. Smith, L.G. Van Uitert and G.P. Vella-Coleiro, 1973, Mat. Res. Bull. 8, 1223.

Bullock, D.C., 1973, AIP Conf. Proc. 18, 232.

Bullock, D.C., J.T. Carlo, D.W. Mueller and T.L. Brewer, 1974, AIP Conf. Proc. 24, 647.

Davies, J.E. and E.A. Giess, 1975, J. Mat. Sci. 10, 2156.

De Leeuw, F.H., 1973, IEEE Trans MAG-9, 614.

Eschenfelder, A.H., 1978, J. Appl. Phys. 49, 1891.

Eschenfelder, A.H., 1980, Magnetic Bubble Technology (Springer Verlag, Heidelberg/New York).

Giess, E.A. and D.C. Cronemeyer, 1973, Appl. Phys. Lett. 22, 601.

Giess, E.A. and R. Ghez, 1975, Epitaxial Growth, part A (Academic Press, New York).

Giess, E.A., J.E. Davies, C.F. Guerci and H.L. Hu, 1975, Mat. Res. Bull. 10, 355.

Glass, H.L. and J.H.W. Liaw, 1978, J. Appl. Phys. 49, 1578.

Gyorgy, E.M., M.D. Sturge, L.G. Van Uitert, E.J. Heilner and W.J. Grodkiewicz, 1973, J. Appl. Phys. 44, 438.

Hagedorn, F.B., W.J. Tabor and L.G. Van Uitert, 1973, J. Appl. Phys. 44 432.

Hiskes, R., 1976, AIP Conf. Proc. 34, 166.

Hsu, T.L., 1974, AIP Conf. Proc. 24, 724.

Hu, H.L. and E.A. Giess, 1974, AIP Conf. Proc. 24, 605.

Josephs, R.M., B.F. Stein, H. Callen and W.R. Bekebrede, 1977. Paper 1–4 presented at 1977 Intermag. Conf., Los Angeles, California, USA.

Kooy, C. and U. Enz, 1960, Philips Res. Rep.

Krahn, D.R., P.E. Wigen and S.L. Blank, 1978, J. Appl. Phys. 50, 2189.

Kryder, M.H., C.H. Bajorek and R.J. Kobliska, 1976, IEEE Trans. MAG-12, 346.

LeCraw, R.C., S.L. Blank and G.P. Vella-Coleiro, 1975, Appl. Phys. Lett. 26, 402.

Lin, Y.S., G.S. Almasi and G.E. Keefe, 1977,

IEEE Trans. MAG-13, 1744.

Lotgering, F.K., U. Enz and J. Smit, 1961, Philips Res. Rep. 16, 411.

Matthews, J.W., E. Klokholm and T.S. Plaskett, 1973, AIP Conf. Proc. 10, 271.

McCollum, B.C., W.R. Bekebrede, M. Kestigian and A.B. Smith, 1973, Appl. Phys. Lett. 23, 702.

Nielsen, J.W., 1976, IEEE Trans. MAG-12, 327.

Nielsen, J.W., S.L. Blank, D.H. Smith, G.P. Vella-Coleiro, F.B. Hagedorn, R.L. Barns and W.A. Biolsi, 1974, J. Elec. Mat. 3, 693.

Pierce, R.D., 1971, AIP Conf. Proc. 5, 91.

Pierce, R.D. and W.B. Venard, 1974, Rev. Sci. Instrum. 45, 14.

Plaskett, T.S., E. Klokholm, H.L. Hu and D.F. O'Kane, 1972, AIP Conf. Proc. 10, 319.

Rosencwaig, A. and W.J. Tabor, 1972, AIP Conf. Proc. 5, 57.

Slonczewski, J.C., 1973, J. Appl. Phys. 44, 1759.

Slonczewski, J.C., 1975, AIP Conf. Proc. 24, 613.

Slonczewski, J.C., A.P. Malozemoff and O. Voegeli, 1973, AIP Conf. Proc. 10, 458.

Smit, J. and H.P.J. Wijn, 1959, Ferrites (John Wiley, New York) p. 177–211.

Stacy, W.T., A.B. Voermans and H. Logmans, 1976, Appl. Phys. Lett. 29, 817.

Tabor, W.J., A.H. Bobeck, G.P. Vella-Coleiro and A. Rosencwaig, 1973, AIP Conf. Proc. 10, 442.

Takahashi, M., H. Nishida, T. Kobayashi and Y. Sugita, 1973, J. Phys. Soc. Japan, 34, 1416.

Thiele, A.A., 1970, J. Appl. Phys. 41, 1139.

Tolksdorf, W., 1975, IEEE Trans. MAG-11, 1074.

Van Uitert, L.G., D.H. Smith, W.A. Bonner, W.H. Grodkiewicz and G.J. Zydzik, 1970, Mat. Res. Bull. 5, 445.

Vella-Coleiro, G.P., 1973, AIP Conf. Proc. 18, 217.

White, R.L., 1973, IEEE Trans. MAG-9, 606.

Wöhleke, M. and J.C. Suits, 1977, Appl. Phys. Lett. 30, 395.

Wolfe, R., J.C. North and Y.P. Lai, 1973, Appl. Phys. Lett. 22, 683.

Wolfe, R., R.C. Le Craw, S.L. Blank and R.D. Pierce, 1976, Appl. Phys. Lett. 29, 815.

Yamaguchi, K., H. Inoue and K. Asama, 1976, AIP Conf. Proc. 34, 160.

chapter 6

AMORPHOUS FILMS FOR BUBBLES

A.H. ESCHENFELDER

IBM Corporation
San Jose, CA 95193
USA

Ferromagnetic Materials, Vol. 2
Edited by E.P. Wohlfarth

CONTENTS

1. Introduction

Crystalline garnet films provide excellent properties for bubble devices, as described in chapter 5. On the other hand, there are limitations to the range of properties obtainable in garnets and, in addition, the crystalline substrates required are relatively expensive compared to those useful for other solid state devices. Amorphous intermetallic films, such as GdCoMo, offer an alternative that has a wider range of magnetic properties and that can be fabricated on less expensive substrates by other film deposition techniques (e.g., sputtering or evaporation). On the other hand, amorphous intermetallic films also have disadvantages compared to crystalline garnets (e.g., electrical conductivity). The purpose of this chapter is to describe the properties, important for bubbles, that are available in amorphous films, and to provide a comparison of those properties with those available in garnets. The basic requirements for the materials are the same as for garnets, as described in the previous chapter. A knowledge of those requirements and the formulae which interrelate practical bubble parameters and fundamental magnetic properties of the materials is assumed in the following discussion.

Gaussian units will be used throughout this chapter. Table A.1 in the Appendix is included to provide conversion from these Gaussian units to the alternative SI units.

2. General characteristics

The amorphous films that have been considered for magnetic bubbles have the general composition $[(RE)_{1-x}(TM)_x]_{1-y}Z_y$, where RE represents one of the rare earths such as Gd, Ho, Tb or a convenient non-magnetic replacement such as yttrium. TM represents one of the transition metals such as Mn, Fe, Co, Ni; Z is a third element such as Mo, Cu, Au or Cr, introduced to provide additional flexibility for tailoring the magnetic properties. For simplicity we will leave out the quantifying subscripts when referring to classes of these compositions. Hence GdCoMo will refer to the class of compositions $(Gd_{1-x}Co_x)_{1-y}Mo_y$.

Films of these compositions have been fabricated by both sputtering and evaporation. In both cases structure analysis by X-rays and electron beams indicates that atomic ordering is localized to within 15 Å so that the films are called amorphous (Herd and Chaudhari 1973).

Some films have an easy axis of magnetization perpendicular to the film plane ($K_u > 0$) whereas for others the magnetization naturally lies in the plane ($K_u < 0$). The variation is not simply understood in terms of composition or fabrication method. For instance, for bias sputtered GdCo $K_u > 0$ but evaporated GdCo has $K_u < 0$ and yet evaporated GdFe has $K_u > 0$ and behaves just the reverse of GdCo in several other respects. Those films which have $K_u > 2\pi M_S^2$ will support bubbles and, indeed, because of the high values of $4\pi M_S$ that can be produced, have yielded bubbles smaller than seen in garnet films.

The magnetization of these films can be easily varied over a wide range because it is due to the difference in the separate magnetizations of the transition metal and rare earth subnetworks. Because the subnetwork magnetizations vary with temperature in a different manner there is usually a compensation temperature, T_{comp}, at which the net magnetization is zero for a very uniform film. Chaudhari et al. (1973) showed how the net magnetization could be varied by changes in the subnetwork ratios for GdCo alloys and also showed how this produces substantial shifts in the compensation temperature. As a result room temperature values of magnetization from close to zero up to greater than 5000 G can easily be achieved. If the lower values, suitable for bubbles larger than 1 μm, are achieved by balancing the sublattices, then T_{comp} lies so close to room temperature that the magnetization varies rapidly with temperature, an undesirable trait for bubble devices. An alternative way of reducing room temperature $4\pi M_S$ with the compensation temperature displaced away from room temperature is to dilute the magnetization with a third ingredient, Z. An example from Hasegawa et al. (1975a) is given in fig. 1. This figure shows both the sublattice magnetizations and the net magnetization and we see that with Co/Gd = 0.85 and about 14 at.% Mo the magnetization, $4\pi M_S$ is approximately 600 G with little variation from 300–380 K. This means that $4\pi M_S$ would be appropriate for ~ 1 μm bubbles with little variation in $4\pi M_S$ over the temperature range from 20 to 100°C.

The third ingredient, Z, can effect the magnetization in two ways. It may be a simple non-magnetic dilutant such as Au, decreasing the density of magnetic moments and weakening the magnetic exchange. In addition, it may also donate electrons to the transition metal thus lowering the transition metal magnetic moment, such as Mo does in association with Co and Fe. We will discuss the effect of these dilutions in conjunction with GdCo.

A number of things must be taken into consideration that are related to the fabrication process. In bias sputtering, the ratio of Co/Gd in the film increases and so does the inclusion of Ar as the bias is increased. In addition, the presence of any oxygen in the deposition system will deplete Gd from the effective film composition because it combines with oxygen. For the same reason either film surface may act like a higher Co/Gd ratio, because some Gd has combined with some oxygen from the atmosphere or from the substrate. Thus it is necessary to do careful film composition analysis when trying to correlate film properties with composition.

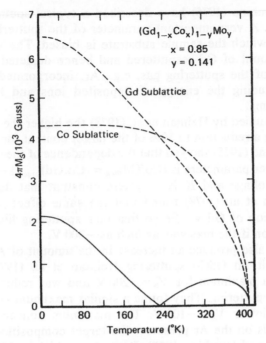

Fig. 1. Total and subnetwork $4\pi M_S$ vs. temperature for $x = 0.85$ and $y = 0.14$ (Hasegawa et al. 1975a).

It was with GdCo that Chaudhari et al. (1973) first discovered that such amorphous films could have K_u large enough to support bubbles. A considerable insight has been developed since then about this system, so it is appropriate to discuss that first, especially since in many ways this system is the prototype for others. We will in fact discuss the system GdFe in conjunction with GdCo because the contrasts between these two systems and the properties of mixtures of them are helpful for discussing the mechanisms responsible for the properties. We will then discuss GdCoMo because this is the system that has been most extensively explored for bubble applications. We will follow with other Z ingredients combined with GdCo such as Cu, Au and Cr and then briefly mention systems with other RE's and TM's.

3. GdCo–GdFe

3.1. Film fabrication

Films of GdCo, GdFe and intermediate alloys have been fabricated by both sputtering and evaporation. It is very clear that the parameters of the fabrication process produce variations in composition and microstructure in the deposited films which must be taken into account in the explanation of the magnetic properties.

Cuomo and Gambino (1975) have described a typical sputtering system and target preparation. A very important parameter of the sputtering process is the voltage, V_b, with which the sample substrate is biased. The value of this bias influences the amount of Gd resputtered and hence depleted in the deposited film, the amount of the sputtering gas, e.g., Ar, incorporated in the film, and other factors including the energy of deposited ions and hence the microstructure of the films.

In GdCo films studied by Heiman et al. (1978) the higher the bias the more Gd was depleted, with a reduction to 50% of the target content for a bias of −400 V. Cuomo and Gambino (1975) showed that the dependence of the Co/Gd ratio in the film on V_b and other parameters is $(Co/Gd)_{film} = (Co/Gd)_{target}[1 - KV_b/V_t]^{-1}$ where V_t is the target voltage and K is a system constant that depends largely on geometry. Heiman et al. (1978) found that the same effect produced an even more rapid depletion of Gd vs. Fe so that bias sputtering films of GdFe were essentially pure iron if the bias was as high as −300 V.

Increases in V_b also produce an increase in the amount of Ar included in the films up to a limit. In GdCo sputtering Heiman et al. (1978) observed that included Ar was a maximum for $V_b = -200$ V and was reduced at higher bias voltages due to resputtering. They found a similar maximum value for GdFe but it occurred at a lower $V_b = -100$ V. The maximum concentration of Ar incorporated depends on the Ar pressure, the target composition and the system geometry (Cuomo and Gambino 1977). It can be as high as 10–20% (Heiman et al. 1978).

A number of structural studies have detected slight differences in X-ray diffraction patterns for bias sputtered films vs. no bias (Wagner et al. 1976, Onton et al. 1976). These have been interpreted as indicating that the bias sputtered films have structural order of a different kind or over larger distances than the unbiased films. Onton et al. (1976) showed that their pattern for bias sputtered $GdCo_3$ was like that calculated for randomly oriented $GdCo_5$ clusters, broadened to correspond to cluster sizes of ~11 Å. Everyone seems to agree that structural differences may be significant magnetically but are nevertheless small and difficult to detect, let alone quantify, with available techniques. Esho and Fujiwara (1976) compared the measured decrease in density with Ar incorporation as a function of bias and concluded that whereas the density at low bias was what would be expected on the basis of the interpolated composition, the decrease in density at high bias would require some microvoids in addition to the measured Ar content.

It is generally conceded that films prepared by evaporation upon unheated substrates have microstructural anisotropies oriented in the direction of the incident beam (Heiman et al. 1975, Herd 1978). There is also evidence for such a structure in very thin bias-sputtered films, (Graczyk 1978) but not in thick bias-sputtered films (Cargill and Mizoguchi 1978). Thus for the GdCo and GdFe films to be discussed it is necessary to carefully distinguish not only whether those films have been deposited by evaporation or sputtering but also to understand the subtle compositional and structural variations produced by the particular process parameters.

3.2. Magnetization

As long as the compositional shifts during bias sputtering caused by Gd resputtering and the dilution by Ar incorporation ($\Delta 4\pi M_S = -13\%$ for each 5 at.% Ar, compounded) are taken into account, the values for $4\pi M_S$ of GdCo and GdFe films are the same for evaporated and sputtered films. See fig. 2

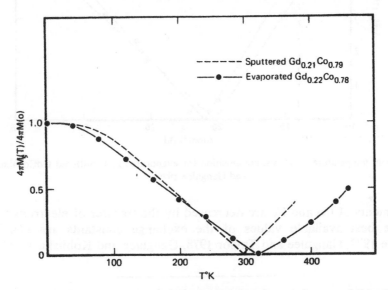

Fig. 2. Normalized $4\pi M_S$ vs. temperature for evaporated $Gd_{0.22}Co_{0.78}$ and sputtered $Gd_{0.21}Co_{0.79}$ films. (Taylor and Gangulee 1976).

(Taylor and Gangulee 1976). Figure 3 shows that

$$\text{for} \quad Gd_{1-x}Co_x: \quad 4\pi M_S(G) = |36800 - 46300x| \quad \text{(Cronemeyer 1973)}$$
$$\text{for} \quad Gd_{1-x}Fe_x: \quad 4\pi M_S(G) = |36800 - 49700x| \quad \text{(Taylor 1976)}$$

for $-0.08 < x - x_0 < +0.08$ where x_0 corresponds to compensation. Thus compensation occurs for $x_0 = 0.79$ in GdCo and $x_0 = 0.74$ in GdFe. An analysis of magnetization data with the molecular field approach yields the following values for the key parameters involved (Hasegawa 1974, Taylor and Gangulee 1977, Gangulee and Taylor 1978, Gangulee and Kobliska 1978).

	g	S
Gd	2.00	3.5
Fe	2.15	$1.071 - 0.201(X_{Gd}/X_{Fe})^{0.826}$
Co	2.22	$0.775 - 0.848(X_{Gd}/X_{Co})^{1.5}$

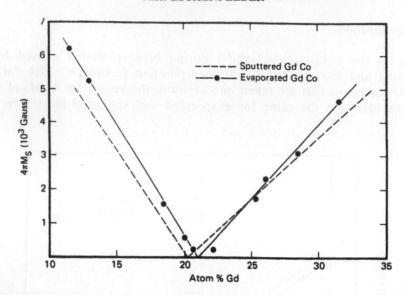

Fig. 3. Room temperature $4\pi M_S$ vs. composition for evaporated and sputtered GdCo films (Taylor and Gangulee 1976).

The moments of Co and Fe are decreased by the transfer of electrons from the Gd. The best available values of the exchange constants are (Taylor and Gangulee 1977, Gangulee and Taylor 1978, Gangulee and Kobliska 1978):

	$J(10^{-15} \text{ erg})$
CoCo	$+23.0 - 25.0(X_{Gd}/X_{Co})$
CoFe	$+16$
FeFe	$+4.85 + 2.01(X_{Gd}/X_{Fe})$
GdFe	$-3.42 + 3.36(X_{Gd}/X_{Fe})$
GdCo	$-2.52 - 6.17X_{Ar}$
GdGd	$+0.27$

We see that these too are altered by electron transfer and also somewhat by the inclusion of Ar from the sputtering process. The dominant interactions are evidently CoCo, FeFe and CoFe in mixed GdCo–GdFe. The GdGd coupling can for some purposes be ignored.

These J values give the strength of the coupling between individual pairs of spins according to: $W_{ex} = -2J_{ij}S_i \cdot S_j$ and the corresponding exchange parameter for the assemblage is

$$A = \sum_{i,j} 2|J_{ij}|S_iS_jP_{ij}a_{ij}$$

where $S_{i,j}$ are the average spin values for the members of each pair, a_{ij} is the average interatomic distance of the pair and P_{ij} is the relative population of the pair in the total assemblage.

The corresponding A values for GdCo are a function of the composition parameter, x, but range from 6.0×10^{-7} erg cm^{-1} for $x = 0.78$ to 6.5×10^{-7} erg cm^{-1} for $x = 0.84$. This is the composition range of practical interest over which $4\pi M_S$ ranges from near zero at compensation ($x \simeq 0.79$) to ~ 2000 G at $x = 0.84$. These magnitudes for A were worked out by Hasegawa (1974) using the interatomic distances determined by the structure study of Cargill (1973)

$$a_{CoCo} = 2.5 \text{ Å} \qquad a_{GdCo} = 3 \text{ Å} \qquad a_{GdGd} = 3.5 \text{ Å}.$$

These "a" values agree roughly with those obtained by Wagner et al. (1976) from refined interference functions by X-ray transmission and reflection

$$a_{CoCo} = 2.50 \text{ Å} \qquad a_{GdCo} = 2.95 \text{ Å} \qquad a_{GdCo} = 3.45 \text{ Å}.$$

Values quoted by Gangulee and Kobliska 1978, and Gangulee and Taylor 1978, are

$$a_{CoCo} = 2.50 \text{ Å} \qquad a_{GdCo} = 3.04 \text{ Å} \qquad a_{GdGd} = 3.58 \text{ Å}$$
$$a_{FeFe} = 2.50 \text{ Å} \qquad a_{GdFe} = 3.03 \text{ Å}$$
$$A_{CoFe} = 2.50 \text{ Å}.$$

3.3. Anisotropy

3.3.1. General

The magnetic anisotropy of films for GdCo and for GdFe is due to some form of ordering in these nominally amorphous materials and this ordering is related to the way the film has been built up and, hence, to the details of the film deposition process. In addition, even for the same preparation the anisotropies of the two systems are quite opposite. Evidences for these two statements include:

(i) Bias sputtered films of GdCo have $K_u > 0$ whereas zero-biased sputtered GdCo films have $K_u < 0$ (Chaudhari et al. 1973).

(ii) GdCo films evaporated over a broad range of Gd/Co ratio and with widely varying substrate temperature have $K_u < 0$ in contrast to bias sputtered GdCo films (Heiman et al. 1975).

(iii) Evaporated GdFe films have $K_u > 0$ in contrast to evaporated GdCo films (Heiman et al. 1975).

(iv) Sputtered GdFe films have $K_u > 0$ for zero bias which becomes $K_u < 0$ with increased bias, the very opposite of sputtered GdCo (Katayama et al. 1977, see fig. 4).

(v) The anisotropy axis in evaporated films conforms to the direction of the incident atomic beam (Heiman et al. 1975) whereas the anisotropy axis in sputtered films is always normal to the film plane even if the biased substrate is canted with respect to the source target.

The ordering that is responsible for the anisotropy can be on an atomic scale (such as a preferential orientation of pairs of like atoms normal to the film plane)

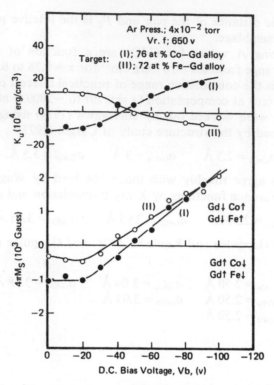

Fig. 4. K_u and $4\pi M_S$ vs. bias voltage for sputtered GdCo and GdFe films (Katayama et al. 1977).

or on a microscopic structural scale (such as columnar inhomogenieties oriented normal to the film plane that extend through the thickness of the film but have a very small lateral dimension). Cargill and Mizoguchi (1978) considered both types and concluded that for both of them the induced magnetic anisotropy would have the form

$$K_u \propto k \left[\frac{M_A}{X_A} + \frac{M_B}{X_B} \right]^2$$

where $k = pX_A X_B$ for pair ordering, and $k = V(1-V)(\Delta x)^2$ for columnar composition fluctuations. These parameters have the following meaning: $M_A, M_B =$ the average magnitude of the subnetwork magnetizations, e.g., Gd, Co. $X_A, X_B =$ the fractional compositions, e.g., Gd, Co. $p =$ a parameter describing the anisotropy in orientation of unlike pairs. $V =$ the volume fraction of the compositional inhomogenieties. $\Delta x =$ the variation in the composition parameter.

A comparison of their calculations with actual data for bias sputtered $Gd_{0.11}Co_{0.67}Mo_{0.16}Ar_{0.06}$ indicated that the observed anisotropy could be accounted for either by the pair ordering mechanism with $p \simeq -0.015$ or by the compositional inhomogeneity mechanism with $V \simeq 0.5$ and $\Delta x \simeq 0.1$. Both are plausible. Such a small p is hardly detectable, but a compositional inhomogeneity of this magnitude should be readily detected. Indeed evidence of such microstructure has been reported for evaporated GdCo films but has not yet been found in bias-sputtered GdCo films.

Taylor (1976) has suggested that microstructure is quite probable in evaporated films because of the unidirectionality of the incident beam, but that the highly energetic deposition of atoms, coupled with resputtering, rare gas bombardment and inclusion, etc., would inhibit columnar growth in sputtered films. If pair ordering is the mechanism for magnetic anisotropy in sputtered films (as originally proposed by Gambino et al. 1974) then it would appear that biasing somehow produces changes in that ordering.

Now let us consider the quantitative measurements of anisotropy in, first, evaporated films and, then, sputtered films of GdCo/GdFe.

3.3.2. Evaporated films

The anisotropies of evaporated films have been reported for GdCo (Taylor and Gangulee 1976) GdFe (Taylor 1976, Gangulee and Taylor 1978) and alloys $Gd_x(Fe_{1-y}Co_y)_{1-x}$ (Taylor and Gangulee 1977). The anisotropy has been described by the formula: $K = \Sigma\, C_{ij}M_iM_j$. The M_i, M_j are the subnetwork magnetizations and the corresponding C_{ij} are

C_{ij}	CoCo	CoFe	FeFe	GdFe	GdCo	GdGd
	0.66	2.81	0.95	1.58	2.22	0.93

Since the magnetization of the rare earth subnetwork is oppositely oriented to that for the transition metal subnetworks, Gd–Co and Gd–Fe coupling tends to induce in-plane magnetization whereas RE–RE or TM–TM coupling (like pairs) contributes to a perpendicular orientation.

The net result is that:

(i) GdCo has $K_u < 0$ except for very high Co or Gd contents where the like pairs finally overwhelm the strong GdCo contribution

$$K = -0.63 M_{Co}^2\left[1 - 0.57\frac{M}{M_{Co}} - 1.48\left(\frac{M}{M_{Co}}\right)^2\right].$$

(ii) GdFe has $K_u > 0$ because the like pairs dominate.

$$K = +0.30 M_{Fe}^2\left[1 - 0.93\frac{M}{M_{Fe}} + 3.10\left(\frac{M}{M_{Fe}}\right)^2\right].$$

(iii) For a moderate content of Gd ($x \simeq 0.2$–0.35) K_u shifts from positive to negative as the Co content becomes comparable to or exceeds the Fe content. For low Gd content (~0.2) K_u of GdFe will be increased by the introduction of small amounts of Co before reversing and going negative because of the large influence of CoFe pairs. This is illustrated in fig. 5.

(iv) This system allows a convenient independent variation of K and $4\pi M_S$ since $4\pi M_S$ depends primarily on x with only slight variations with changes in y (Co/Fe) whereas K has a substantial change with y for a given x. This is

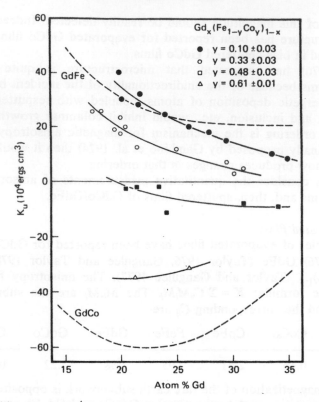

Fig. 5. Anisotropy as a function of $Gd_x(Fe_{1-y}Co_y)_{1-x}$ films (Taylor and Gangulee 1977).

illustrated in table 1. For $Q > 1$ it is necessary to keep the Gd content low and the Fe content high.

3.3.3. Sputtered films

Gangulee and Kobliska (1978) also evaluated the variation of magnetic aniso-tropy as a function of the subnetwork magnetizations for bias-sputtered GdCoMoAr films. When films with the same metals content but prepared under different biases so that the Ar content was different were examined the data of table 2 were obtained.

A number of conclusions can be drawn:

(i) The C_{ij}'s vary considerably, as does K_u, with bias and Ar content, even if the conditions are chosen so as to preserve the same metals ratios.

(ii) The C_{ij} values and K_u depend too much on preparation parameters to be simply related just to composition as in the case of evaporated films. (Note the different values vs. z and even between $z = 0.08$ and $z = 0.09$.)

(iii) It appears that each C_{ij} tends to increase with increasing Ar content and, presumably, bias.

(iv) Just as with evaporated GdFe, in all of the examples of table 2 the like pairs dominate over the influence of the unlike pairs so that the net $K > 0$.

TABLE 1

Room temperature magnetization $(4\pi M_S)$ and anisotropy (K_u) of Gd_x $(Fe_{1-y}Co_y)_{1-x}$ amorphous alloys (data from Taylor and Gangulee 1977)

	$4\pi M_S$(G)		K_u (10^4 erg cm^{-3})	
x / y	0.21	0.31	0.21	0.31
0.06	1600	3150	+28	+15
0.13	1900	3000	+32	+11
0.23	2250	3000	+22	+8
0.34	2150	3250	+12	+6
0.49	2300	3200	−2	+3
0.62	2200	3600	−3.5	−8

TABLE 2

Coupling constants, C_{ij}, of bias-sputtered amorphous films $(Gd_{0.10}Co_{0.73}Mo_{0.17})_{1-z}Ar_z$ (Gangulee and Kobliska 1978)

Ar(z)	C_{CoCo}	C_{GdCo}	C_{GdGd}	K_u (10^4)
0.02	0.17	0.37	0.42	+0.75
0.08	0.48	1.10	1.08	+3.11
0.09	0.58	1.29	0.77	+1.37
0.16	1.16	2.74	1.69	+2.54

(v) Comparing the C_{ij} for the sputtered GdCo vs. the evaporated, the principle difference appears to be in the relative magnitude of C_{GdCo}. In all the sputtered examples of table 2 C_{GdCo}/C_{CoCo} is within 5% of 2.26 even though the individual values vary substantially with z, whereas for the evaporated GdCo, $C_{GdCo}/C_{CoCo} = 3.36$.

The anisotropy of GdFe films is opposite to that of GdCo in the case of sputtered films, just as in the case of evaporated films, at least for the example shown in fig. 4.

Heiman et al. (1978) studied the anisotropy of GdCo and GdFe sputtered films as a function of bias. They found that for $Gd_{0.22}Co_{0.78}$ K_u became increasingly positive as the bias was increased, in agreement with Gangulee and Taylor (1978) and Katayama et al. (1977) but that it peaked for a bias of −200 V at a value $K_u \approx 60 \times 10^4$ erg cm^{-3} and was progressively smaller for higher bias fields. This peak at −200 V corresponded to the maximum Ar incorporation. The variation in K_u with bias is shown in fig. 6 along with the variation of $4\pi M_S$. It is clear that some of the increases in this example are due to the increase in Co content of the film due to Gd resputtering as reflected in the increase in $4\pi M_S$. However,

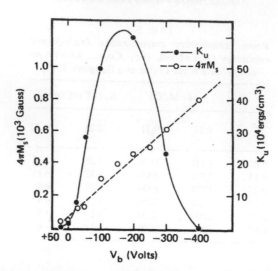

Fig. 6. $4\pi M_S$ and K_u as a function of substrate bias, V_b, for film deposited from a target of $Gd_{0.22}Co_{0.78}$.

Burilla et al. (1978) took care to adjust their deposition parameters so as to maintain the same film composition and showed that even with this compensation the magnetic anisotropy increases with bias and peaks near −200 V.

Heiman et al. (1975) found that the anisotropy of $Gd_{0.24}Fe_{0.76}$ sputtered films changed rapidly from positive to negative upon the application of bias, in agreement with the findings of Katayama et al. (1977), but in this case the anisotropy does not peak with bias but reaches a saturation level at a bias of ~ −200 V. This correlated well with the depletion of Gd to essentially pure iron films and a corresponding saturation in Fe content. The saturation value of $4\pi M_S$ was ~5000 G and of $K_u \simeq -40 \times 10^5$ erg cm^{-3}. Here too, then, we find that the anisotropy of the bias sputtered films behaves oppositely to evaporated films (increasing $|K|$ for increasing Fe content but of opposite sign).

3.3.4. Summary
In general, the anisotropies of films of both GdCo and GdFe in the interesting composition ranges have opposite sign and the same thing is true of bias sputtered and evaporated films. In other words, the sign of K_u is as follows:

	GdCo	GdFe
bias sputtered	+	−
evaporated	−	+

In all cases the K_u can be decomposed into components proportional to the products of the subnetwork magnetizations. The resulting C_{ij} serve well to describe the variation of K_u with composition and temperature in the case of the evaporated films. For bias sputtered films the C_{ij} are also useful but they are very sensitive to the particulars of the sputtering process, e.g., bias. The change in sign of the K_u reflects a fairly subtle shift in balance between the influence of the C_{ij} corresponding to unlike pairs (TM–RE) (contributing to $K_u < 0$) and those corresponding to like pairs (TM–TM or RE–RE) (which contribute to $K_u > 0$). It has been shown that such a dependence of anisotropy on subnetwork magnetizations can arise from several mechanisms (anisotropic microstructures as well as pair ordering). It is reasonable to expect that the microstructure in evaporated films is quite different from that in bias sputtered films and that the structure of bias sputtered films would be a function of the magnitude of the bias and the resulting atomic resputtering as well as gas inclusion. Indeed X-ray analysis of both kinds of films indicates that there are such differences. Therefore the C_{ij} can be expected to change depending on the compounding of the several mechanisms as influenced by the fabrication process. Thus the C_{ij} are useful but more study will be necessary to understand the precise variation of their magnitude with atomic species and fabrication process.

3.4. Practical properties

Since $A \simeq 6 \times 10^{-7}$ erg cm^{-1} we need $4\pi M_S \simeq 1200/d$ and $K(10^4$ erg cm$^{-3}) \simeq 25/d^2$ for magnetic bubbles with a Q of 4 and a diameter of $d\mu$m. Thus for bubbles 0.5 μm and larger we need $4\pi M_S < 2400$ G. We see from the data already displayed in this section that it is necessary to restrict the GdCo and GdFe compositions to those which have compensation close to room temperature to obtain $4\pi M_S$ values this low (fig. 3). If we do that we find that values of the anisotropy adequate to get quite useful combinations of Q and d can be obtained for sputtered GdCo and also for evaporated GdFe films. However, as indicated

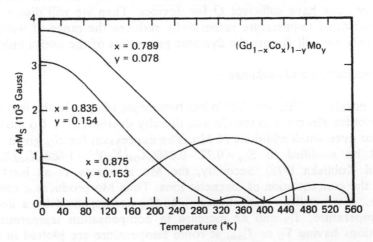

Fig. 7. $4\pi M_S$ vs. temperature for several examples of $(Gd_{1-x}Co_x)_{1-y}Mo_y$ (Hasegawa et al. 1975a).

in fig. 2, the value of $4\pi M_S$ varies very rapidly with temperature near compensation so such films would not really be useful in devices where properties must remain fairly constant over a moderate variation in operating temperature. In order to get a reasonably stable $4\pi M_S$ as a function of temperature it is necessary to select compositions for which room temperature is approximately midway between the compensation temperature and the Curie temperature, and then to find some way to dilute the magnetization down to the values required for magnetic bubbles. This has been done successfully by using Mo as such a dilutant as shown in fig. 7. Therefore, the preparation and properties of $(Gd_{1-x}Co_x)_{1-y}Mo_y$ films will be discussed next.

4. GdCoMo

4.1. Introduction

As shown in fig. 7 it is possible to obtain a fairly flat temperature variation of magnetization near room temperature in films of $(Gd_{1-x}Co_x)_{1-y}Mo_y$, where $x \simeq 0.87$ and $y \simeq 0.15$. The values of magnetization and anisotropy of such bias-sputtered films result in magnetic bubbles of $\sim0.5\ \mu m$ diameter with a Q sufficient for bubble devices. Thus GdCoMo films would appear to be especially suitable for magnetic bubbles. The question is: over what range of bubble sizes can useful properties be achieved?

The approach we will take to answering this question will be as follows: First we will construct the range of $4\pi M_S$ and A values that can be obtained by variation of x and y on a chart like fig. 8, which has been used in the section on garnet bubble films (chapter 5, section 3.4.4). This will show that a broad range of bubble sizes is possible if the anisotropy is sufficient. Therefore we will next consider the magnitude of K that can be achieved. This will reveal what range of bubble sizes can have sufficient Q for devices. Then we will discuss how the desired degree of temperature insensitivity reduces the range of useful bubble sizes. Finally we will discuss the dynamic properties of the useful bubbles.

4.2. Magnetization and exchange

The incorporation of Mo into GdCo has two major effects. In the first place, the Mo contributes electrons to the Co and thereby decreases the Co moment very rapidly for even small additions of Mo. The expression for S_{Co} given in section 3.2 must be modified to $S_{Co} = 0.775 - 0.848(X_{Gd}/X_{Co})^{1.5} - 1.688(X_{Mo}/X_{Co})$ (Gangulee and Kobliska 1978). Secondly, the Mo also acts as an inert dilutant, reducing the concentration of magnetic ions. Thus Mo produces a reduction in exchange, A, as well as magnetization, $4\pi M_S$, accompanied by a decrease in Curie temperature, T_C, and an increase in compensation temperature, T_{comp}. Compositions having T_C or T_{comp} at room temperature are plotted in a ternary diagram in fig. 9. Compositions lying above the T_C line are paramagnetic at room temperature because of an excess of Mo. Compositions to the right of the T_{comp}

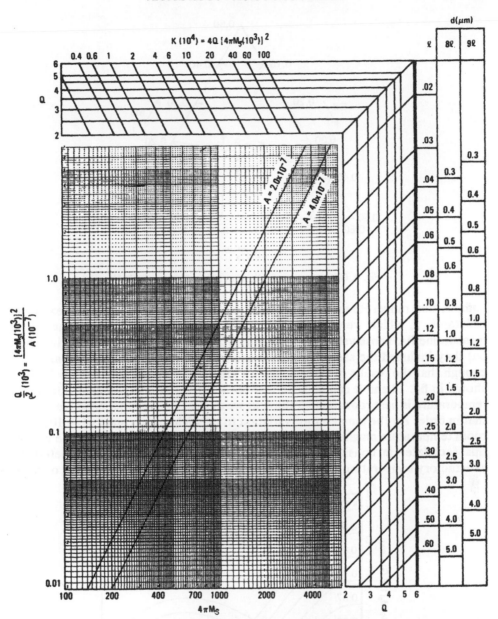

Fig. 8. Blank summary chart for static properties.

line have enough Co moment so that T_{comp} lies below room temperature. Since we are interested in compositions where room temperature is approximately midway between T_{comp} and T_C, in order to get a flat temperature characteristic, we are most interested in compositions in the vicinity of $x = 0.80-0.90$ and $y \simeq 0.10-0.16$.

Hasegawa (1975) calculated the variation of magnetization vs. temperature in this

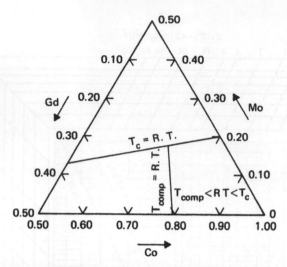

Fig. 9. GdCoMo ternary diagram (Chaudhari et al. 1975).

composition range. He used a mean field model with values for S_{Co} and exchange constants, J, which were obtained from an analysis of his own magnetization measurements. Those values differ slightly from the more recent results given here but they included the same rapid decrease in S_{Co} with Mo content and are close enough that the results of the calculation are useful. The results for $y = 0.16$ are shown in fig. 10 and it can be seen that the least variation in magnetization with temperature is near room temperature, as we desire. On the other hand we also see that in order for the temperature variation to be slight over a span of $\sim 100°C$ the room temperature magnetization, $4\pi M_S$, must be greater than 1200 G. We will use this fact later in considering the acceptable range of bubble parameters for satisfactory temperature stability. The variation in exchange constant, A, over this

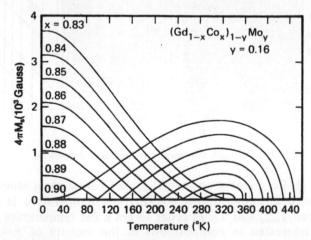

Fig. 10. $4\pi M_S$ vs. T for $y = 0.16$ and various values of x (Hasegawa 1975).

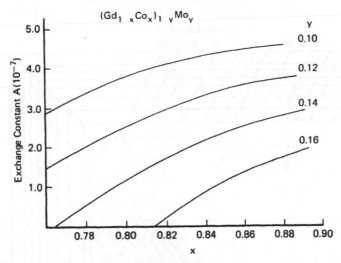

Fig. 11. Exchange constant A (10^{-7} erg cm^{-1}) vs. x and y (Hasegawa 1975).

same range of composition is displayed in fig. 11.

Now we can combine the values of room temperature $4\pi M_S$ and A on our bubble chart as in fig. 12. The effect of Mo in decreasing the Co moment as well as decreasing A is apparent.

It is important to remember that in the sputtering process some rare gas, e.g., Ar, and some oxygen may be incorporated in the film depending on the sputtering conditions. Ar will be especially included if large bias voltages are used to increase K_u for the benefit of Q. Ar has the effect of diluting the magnetic ions thereby reducing $4\pi M_S$ and A. Oxygen will combine with some of the Gd resulting in an increase in the effective value of x and in addition the resulting Gd_2O_3 will act as a dilutant. Thus some adjustment of the GdCoMo composition grid on fig. 12 must be made depending on the Ar and O_2 inclusion. It has been found that Ar inclusion decreases $4\pi M_S$ ~13%/5% Ar.

Comparing fig. 12 with fig. 13 we see that the values of $4\pi M_S$ and A available in GdCoMo overlap those available from garnets, and in addition allow lower values of A (and therefore lower $4\pi M_s$ for a given bubble size) and also yield smaller bubble sizes if sufficient anisotropy is available. We will next consider the available anisotropy.

4.3. Anisotropy and Q limitations

We have discussed the anisotropy of GdCo films in section 3.3. There is some empirical data but no systematic insight, at the time of writing, as to how that picture is modified by the incorporation of Mo. To a first order the same principles apply, modified as we would expect them to be by the effects of Mo as a reducer of the Co moment and as a dilutant. Anisotropy is generally reduced by the addition of Mo for equivalent Co/Gd ratio. It continues, as in GdCo, to be

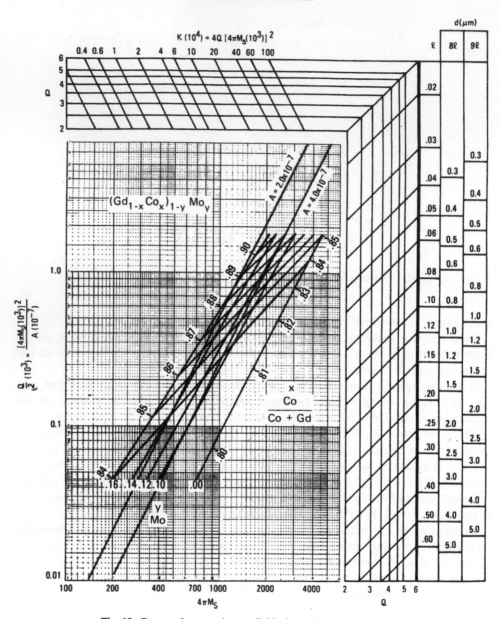

Fig. 12. Range of properties available from $(Gd_{1-x}Co_x)_{1-y}Mo_y$.

increased, in the composition range of interest to us, by increased Co/Gd ratio and by increased sputtering bias.

There is not much published data on the variation of anisotropy with Mo content. Chaudhari and Cronemeyer (1976) did try to analyze this for a few GdCoMo films that were bias sputtered onto glass substrates. They measured K_u by FMR as a function of temperature and made a fit of the data to our familiar expression: $K_u = \Sigma \, C_{ij}M_iM_j$. The C_{ij} they obtained were negative and did vary

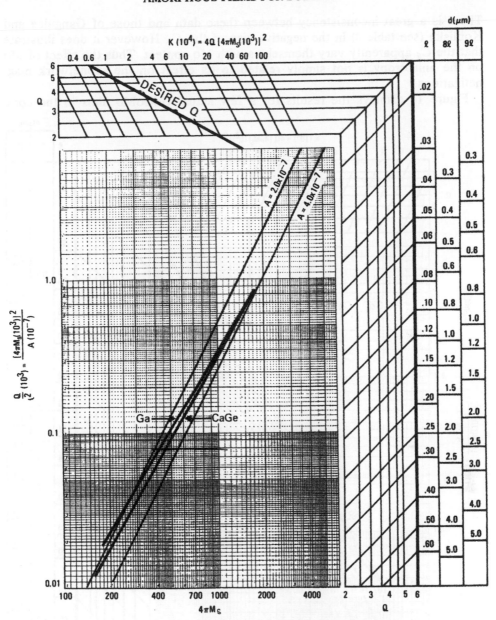

Fig. 13. Augmented summary chart for static bubble parameters.

appreciably with Mo content as follows for $(Gd_{1-x}Co_x)_{1-y}Mo_y$:

x	y	C_{GdCo}	C_{CoCo}	C_{GdGd}
0.74	0.11	−1.43	−0.617	−0.558
0.71	0.16	−0.483	−0.288	−0.041

There is a great inconsistency between these data and those of Gangulee and Kobliska (see table 2) in the negative sign of the C_{ij}. However it does illustrate that the C_{ij} apparently vary themselves with Mo content. Thus the effect of Mo on the anisotropy is not simply related to its effect on the subnetwork magnetizations.

Figure 14 displays the results for a few actual GdCoMo films in the com-

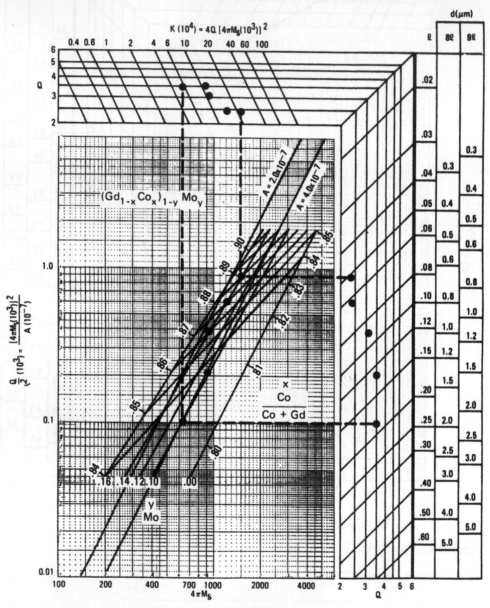

Fig. 14. Examples of GdCoMo films.

position range of interest. It is evident that anisotropies are sufficient to produce Q's greater than the Kryder device criterion (see chapter 1, section 5.8a) for a range of bubble sizes including some smaller than obtainable in garnets. However no one has yet reported anisotropies significantly greater than those displayed in fig. 14. If we assume that the magnitude of anisotropy is limited as indicated in fig. 14, (which would seem reasonable based on available data) then we see that the range of useful bubbles is limited by anisotropy as indicated on the right side of fig. 14. Thus bubbles as small as 0.2–0.25 μm may be achieved with device quality Q but smaller useful bubbles will require the discovery of ways to increase the anisotropy in these compositions.

4.4. Thermal variation

We now want to consider the thermal variation of the magnetic properties. The same requirements apply as in the case of garnets. We want as stable a bubble size and Q as possible. In addition, the collapse field should decrease ~0.2%/°C for most applications in order to track the decrease in field from the permanent magnet in the device. Most applications require an operating range from ~0°C to ~100°C although a smaller range will sometimes be satisfactory. As in the case of garnets, K and A decrease approximately linearly toward the Curie temperature so the Curie temperature should not be too low. In addition, we want room temperature to be approximately midway between compensation and Curie temperature in order to have a slightly decreasing magnetization with temperature for 0° to ~100°C as in fig. 7 with $x = 0.875$ and $y = 0.153$. If the magnetization rises with temperature, not only would the collapse field increase, getting seriously out of balance with the decreasing bias field, but also the increasing temperature coupled with decreasing K and A would produce drastic decreases in Q and d.

The compensation and Curie temperatures are plotted as a function of composition in fig. 15 (Hasegawa 1975a). This plot suggests the following composition limits for: (a) $T_C > 450$ K (the Curie temperature at least 150° above room temperature); (b) $\Delta T > 300$ K (the difference between compensation and Curie temperatures being at least 300° so the magnetization curve can be reasonably flat); (c) $T_m < 300$ K (the midpoint being below room temperature so that the magnetization decreases with temperature, rather than increasing).

| y | x | | | |
	$T_C > 450$ K	$\Delta T > 300$ K	$T_m < 300$ K	Net
0.10	>0.78	>0.83	>0.87	>0.87
0.12	>0.82	>0.85$^-$	>0.88$^-$	>0.88$^-$
0.14	>0.86	>0.86$^+$	>0.87$^+$	>0.87$^+$
0.16	>0.90	>0.89$^-$	>0.85	>0.90

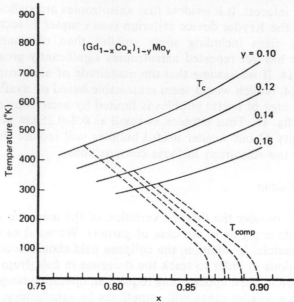

Fig. 15. Curie temperature T_C and compensation temperature T_{comp} vs. x and y (Hasegawa et al. 1975a).

These limits are plotted in fig. 16 and the compositions which satisfy all the criteria are shaded. These satisfactory compositions are, in turn, replotted on our bubble diagram in fig. 17. We see that the requirement that T_C be greater than 450 K is tantamount to requiring $A > ~2.5 \times 10^{-7}$.

We can see an example of how sensitive the temperature behavior is to composition by considering the case of $y = 0.154$, $x = 0.835$ in fig. 7. As we expect from fig. 16, even though the Mo content is about the same as for the quite satisfactory $y = 0.153$, $x = 0.88$, just 0.04 decrease in x yields a curve whose midpoint is only slightly above room temperature but whose $T_C \ll 450$ K and $\Delta T \ll 300$ K.

Kobliska et al. (1977) calculated the compositions, $4\pi M$ and T_C values for $T_m = 290$ K using a mean field model. They also took into account the Ar included in the film during sputtering. They found that for $T_m = 290$ K and $T_C = 450$ K the following results

Ar	Mo	Co/(Gd + Co)	$4\pi M_S$
10%	0.155	0.92	1700
15%	0.14	0.92⁻	1500
20%	0.125	0.91	1300

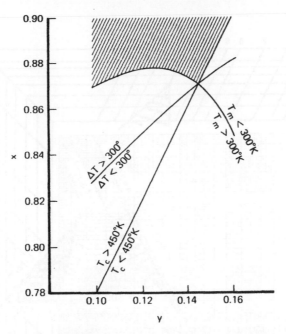

Fig. 16. Critical temperature limits as a function of x and y for $(Gd_{1-x}Co_x)_{1-y}Mo_y$.

The corresponding point in fig. 16 is $y = 0.145$, $x = 0.87$, $4\pi M_S \simeq 1200$ G. The higher value of x in Kobliska's results is due to the fact that the magnetization curve, being the difference between the Co and Gd sublattices, is skewed so that the peak in magnetization lies about 2/3 of the way from the compensation temperature to the Curie temperature, as indicated in figs. 1 and 7. A peak at room temperature therefore requires $T_{comp} \simeq 0$ K. From fig. 15 we see that the corresponding composition would be $y = 0.16$–0.17, $x = 0.90$–0.92. For such a composition fig. 12 would predict $4\pi M_S \simeq 2300$ G. Kobliska's $4\pi M_S$'s are lower, reflecting the 13% decrease compounded per 5% Ar, mentioned before (section 4.2). Similarly his compositions require less Mo as Ar is added since Ar accomplishes some of the decrease in A and T_C.

4.5. Static bubble properties

From the foregoing we see that bubbles with device quality Q can be achieved in GdCoMo sputtered films from very large sizes down to $\sim 0.2\ \mu$m. However we are restricted to bubble sizes less than $0.6\ \mu$m if we require reasonably stable properties over a temperature range of 0–100°C. The useful bubbles under these conditions are included in the shaded region on the right of fig. 17 and we see there are three limitations: we cannot use bubbles outside the T limit if we want a 0–100°C operating range; we cannot produce bubbles beyond the K limit unless we find a way to achieve larger anisotropy; we cannot use bubbles outside the device limit because their Q is not sufficient.

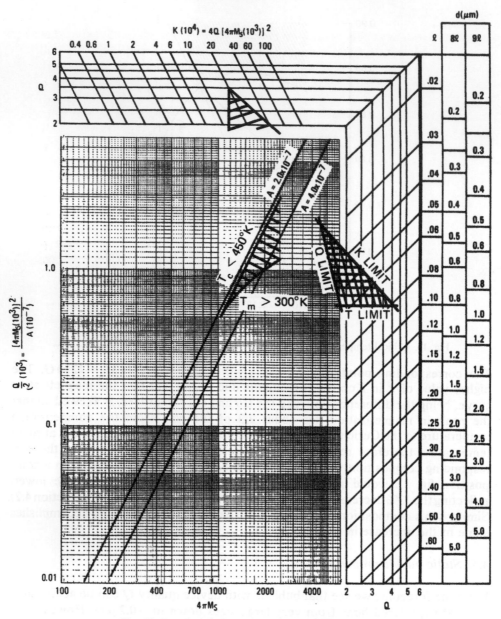

Fig. 17. Practical limits for amorphous GdCoMo films.

4.6. Dynamic properties

The dynamic parameters of typical $1\,\mu$m GdCoMo are: $\alpha \simeq 0.2$, $g \simeq 3$, $\lambda' \simeq 4 \times 10^{-7}$s Oe^2. This leads to a calculated $\mu \simeq 1.9\,\text{ms}^{-1}\,\text{Oe}^{-1}$. Measured $\mu \simeq 0.5 - 1\,\text{ms}^{-1}\,\text{Oe}^{-1}$. Thus mobility is an order of magnitude lower than for $1\,\mu$m garnets.

At the same time the coercive field is an order of magnitude higher, being several oersteds for GdCoMo vs. tenths of oersteds for typical garnets. Nevertheless GdCoMo films could be fast enough in typical bubble devices since the bubble experiences relatively high drive fields. Most of the actual measurements of bubble velocity have been made on 2 μm bubble GdCoMo films. In this case compensation is closer to room temperature and g factors are increased by 2–3×. Indeed measured mobilities are ~4× higher for 2 μm GdCoMo bubbles compared to those of 1 μm.

Kryder et al. (1977) reported a variety of dynamic bubble measurements for 2 μm GdCoMo. g was measured by FMR to be 7.1. The larger g factor compounded with the larger l value would explain a 4× increase in mobility. For bubble propagation in a field gradient Kryder et al. found $\mu \simeq 2\,\text{ms}^{-1}\,\text{Oe}^{-1}$ with a threshold field gradient of $\Delta H_c = 6.4\,\text{Oe}$. Bubbles were propagated around 63 bit HI bar loops with 1 μm lines and 0.5 μm gaps and 8 μm periodicity with no errors in 10^9 steps at 1 MHz. This is an average velocity of 8 m s^{-1} and the peak velocity is usually at least four times the average. Bubbles were also propagated around the 180 μm periphery of solid permalloy squares by a rotating in-plane magnetic field at 1 MHz. This corresponds to a velocity of 180 m s^{-1}. The bubble stretches out during this propagation. The expansion velocity of bubbles under a chevron expander was also determined to be ~100 m s^{-1} with low rotating in-plane fields of 45–60 Oe. The Slonczewski V_0 was calculated to be 160 m s^{-1}. Thus in the case of 2 μm GdCoMo velocities are high enough for 1 MHz operation of devices and such frequencies may well also be possible for submicron GdCoMo bubbles.

4.7. Other properties

a. Metallic characteristics

Since these amorphous materials have a metallic characteristic, in contrast to garnets, consideration must be given to potential problems due to corrosion or electrical conductivity. Corrosion can undoubtedly be avoided by using protective coatings. On the other hand, the electrical conductivity of the films leads to more stringent requirements on the superimposed insulating spacer, between the bubble film and the device metallurgy. Sputtered SiO$_2$ films, properly deposited, can reliably withstand electrical fields up to $6 \times 10^5\,\text{V cm}^{-1}$. While this may be adequate for 1 μm bubble devices with 1000 Å spacers (6 V), it may not be adequate for use with the very small bubbles which we have been discussing (<0.5 μm) (Bajorek and Kobliska 1976).

b. Optical characteristics

Plane polarized light reflected from the surface of GdCoMo films will become elliptically polarized with a polar rotation proportional to the net magnetization parallel to the light. The effect is produced in the surface "skin-depth" which is a few hundred angstroms. It is therefore sensitive to any surface contamination of the film and can be used to study such surfaces. Measurements can be made of the free surface or through a transparent substrate. The free surface can be ion-milled and overcoated with glass to eliminate the effects of surface oxidation both at the free surface and to a different degree at the substrate interface.

The polar rotation for pure Co is $\theta_{Co} = -0.33°$ and for pure $Gd \simeq +0.02°$. GdCoMo films in the composition range of our interest exhibit $\theta \simeq -0.12°$ (Malozemoff et al. 1977). If the total rotation is considered to be primarily due to Co with corrections due to the presence of other constituents then $\theta = -0.33° + \Sigma_i a_i X_i$ is a reasonable approximation for these Co rich compositions, where X_i is the atom fraction of the constituent and the a_i are: $a_{Mo} = +1.42°$; $a_{Gd} = +0.43°$; $a_{Ar} = +0.27°$; $a_O = +0.07°$. Thus we see that Mo has an unusually large effect on θ. The large reduction in θ caused by Mo essentially reflects the reduction in Co moment it produces. To the extent that the polar rotation is due almost exclusively to the Co subnetwork in our range of compositon, the variation in the Co subnetwork with temperature can be separated from that of the total magnetization by measuring the saturation rotation as a function of temperature (Argyle et al. 1975). The ellipticity, ϵ, of pure $Co \simeq -0.05°$ and of pure $Gd \simeq +0.02°$. The ellipticity for typical GdCoMo films is $-0.04°$. Both θ and ϵ increase with wavelength in the range of 4000 Å–7000 Å.

c. Implantability

Because of the amorphousness of these films the influence of ion implantation is not as obvious as in the case of single crystal garnet films. However this has been investigated and 2 MeV Ar ions in dosages $\sim 10^{14}$ ions/cm^2 produce substantial reductions in the anisotropy of sputtered GdCoMo films with only slight changes in the magnetization and compensation temperature (Mizoguchi et al. 1977).

d. Thermal annealing

Insofar as the anisotropy is due to structural ordering, either on an atomic scale or in microstructure, it might be reduced or even destroyed by thermal annealing. To be useful the film would have to withstand temperatures of 200–300°C since such temperatures are typically used in several device fabrication and packaging process steps.

In fact, GdCo and GdFe films do suffer an appreciable decrease in anisotropy when heated at 200°C for several hours, whereas GdCoMo is very much better in this respect. Even in the case of GdCoMo this would have to be taken into account in designing the device process and an anisotropy built into the initial bubble film large enough to accomodate the decrease that would take place during the processing.

If the film is heated in air there will be some oxidation of Gd so that the effective Co content is increased, especially in the surface. This can lead to a slight decrease in T_{comp} and an increase in $4\pi M_S$. However with reasonable care there is no appreciable change in $4\pi M$ or T_{comp} in GdCo, GdFe or GdCoMo. Heiman et al. (1978) found that 200°C for 4 hours produced no change in $4\pi M_S$, T_{comp}, composition as measured by microprobe, or in X-ray diffraction pattern. On the other hand K dropped 60% for GdCo and GdFe but practically not at all for GdCoMo. The addition of Mo clearly stabilizes GdCo against thermal anneal but even with GdCoMo there is some change in K when heated near 250°C. The anisotropy stabilizes after a few hours at a reduced value; the larger the Mo content, the smaller the reduction (Katayama et al. 1978).

The reason for the stabilization by Mo is not understood. Heiman et al. (1978) suggest that it is related to the good bonding of Mo to both Gd and Co as evidenced by the many stable compounds Mo forms with these atoms. The addition of Cu, which is insoluble in Co, results in GdCoCu films that suffer as much, if not more, anisotropy decrease upon thermal annealing as GdCo. Any mechanism proposed to explain the annealing results would have to explain the fact that the anisotropy only decreases so far and then stabilizes against further anneal at that temperature, but will decrease further if reannealed at higher temperature.

4.8. Preparation

GdCoMo films can be sputtered using a single target of appropriate composition (e.g., Cuomo and Gambino 1975) or with multiple targets of the elemental constituents (e.g., Burilla et al. 1976). The single target system has been the most extensively used and reported. We have already mentioned that the composition varies with bias voltage and sputtering gas pressure (e.g., Cuomo and Gambino 1975) and that sputtering gas is included in the film (e.g., Cuomo and Gambino 1977). A summary of the interaction of the magnetic properties with fabrication parameters is as follows:

(i) Voltage bias on the substrate is necessary to achieve substantial anisotropy. The bias voltage, V_b, ranges typically from -80 V to -160 V and the target voltage, V_t, is typically -1200 V.

(ii) V_b introduces Gd resputtering so that the Co/Gd ratio in the film is greater than in the target according to

$$(Gd/Co)_{film} = (Gd/Co)_{target}[1 - k(V_b/V_t)]$$

where k depends on the sputtering chamber geometry and is close to unity for geometries that give good uniformity, but may range up to about 4.

(iii) The Mo content does not seem to be appreciably altered by V_b but the amount of sputtering gas included increases with V_b up to a maximum for $V_b \simeq -200$ V.

(iv) Lower sputtering gas pressure allows greater Gd resputtering.

These facts are substantiated by the data of Bajorek and Kobliska (1976) with the following as example for an Ar pressure of 45 μ.

	Atomic %				
	Gd	Co	Mo	Ar	Gd/Co
Target	17.5	67.5	15	0	0.26
$V_b = 0$	17.5	67.5	15	0	0.26
$V_b = -80$ V	13	68	16	2	0.19
$V_b = -160$ V	8	65	16	10	0.12

The uniformity of properties as a function of the geometry has been measured by Kobliska et al. (1975). The ratio of the separation to diameter of the electrodes must be less than 0.1 for good uniformity.

These amorphous films do not require a single crystal substrate as the garnets do. On the other hand just ordinary glass won't do either. Both specially prepared glass, such as used for photolithographic masks, or Si wafers, such as used for semiconductor devices, provide the necessary smoothness and non-contamination. Si substrates are preferable because of their greater thermal conductivity (Bajorek and Kobliska 1976).

4.9. Practical application

Experimental major–minor loop bubble storage devices using 2 μm bubbles have been actually fabricated and operated on GdCoMo amorphous films (Kryder et al. 1974). The device overlay patterns were fabricated with 1 μm electron beam lithography and this device had a bit density greater than any bubble device on garnets fabricated at that time. All device functions operated with reasonable operating margins. This device provided insight into the scaling rules for device parameters, such as required drive field, as the bit density is increased by the use of smaller bubbles and finer lithography.

Even though the 2 μm devices on GdCoMo worked adequately at room temperature the tolerance to temperature variation was unacceptable. As already discussed, the temperature characteristic can be expected to be adequate for bubble devices only for bubbles $\leqslant 0.6$ μm.

5. GdCoZ

Since the addition of Mo to GdCo is effective for achieving more flexibility in tailoring the magnetic properties, it is worthwhile exploring the incorporation of other third ingredients in $(Gd_{1-x}Co_x)_{1-y}Z_y$, e.g., $Z = Cu$, Au, Cr, etc. All of these have ternary diagrams similar to fig. 9 with the room temperature compensation line, T_{comp}, and Curie temperature line, T_C, in a somewhat different place (Chaudhari et al. 1975). In each case the T_C line terminates at the same place on the GdCo boundary as for Mo and on the CoZ boundary close to the composition for which bulk CoZ alloys have T_C at room temperature. The T_{comp} line proceeds from $x_{comp} = 0.79$ on the GdCo boundary approximately as a straight line as in GdCoMo but to higher or lower x_{comp} with increasing y depending on the identity of Z. Values of x_{comp} which result in compensation at room temperature with $y = 0.11$ are

GdCoMo = 0.825   GdCoCu = 0.785   GdCoAu = 0.760   GdCoCr = 0.820.

Thus x_{comp} increases vs. y with Mo and Cr, remains almost constant with Cu, and decreases with Au. This is the result of the balance of two competing effects: (a) donation of electrons from Z to Co lowers the Co moment so that more Co is needed to compensate the Gd; (b) dilution by Z weakens the exchange coupling leading to reduced alignment of the already weakly aligned Gd subnetwork so that the Gd moment decreases and less Co is needed to compensate the Gd. With Cu the two effects are evidently just about balanced whereas there is excessive electron donation with Mo and Cr but very little with Au.

The difference between Cu and Mo is illustrated in fig. 18. Minkiewicz et al. (1976) studied compositions represented by the shaded region. We notice im-

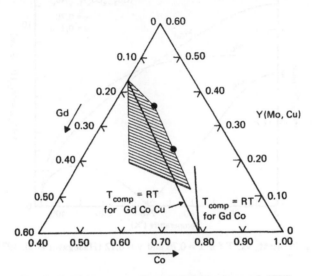

Fig. 18. GdCoY ternary diagram (Minkiewicz et al. 1976).

mediately that substantially more Cu is added than Mo without reducing the Curie temperature to room temperature. GdCoMo films with >20% Mo have T_C depressed below room temperature as shown in fig 9. GdCoCu films on the Co dominated side of the T_{comp} line have been made which support one micron bubbles with good Q. For instance, films of $Gd_{0.14}Co_{0.63}Cu_{0.23}$ have $4\pi M_S \simeq$ 1000 G, $K(10^4\ \mathrm{erg\ cm^{-3}}) \simeq 30$, $T_{comp} \simeq -115°C$ and $Q \simeq 8$ while $Gd_{0.13}Co_{0.51}Cu_{0.36}$ have $4\pi M_S \simeq 750$ G, $K(10^4\ \mathrm{erg\ cm^{-3}}) \simeq 16$, $T_{comp} \simeq -75°C$ and $Q = 7$. The locations of these two compositions are indicated by the dots on the diagram. The 23% Cu films have somewhat higher Co/Gd ratio producing the higher K, and lower T_{comp}. These K values are larger than obtained in GdCoMo films.

These increased K values are particularly important. We remember that with GdCoMo we can get a reasonably flat $4\pi M_S$ vs. T at $\sim 100°$ above room temperature for $4\pi M_S > 1200$ G but that the available anisotropy limits the Q. Higher K values, as achieved in GdCoCu, could alleviate this constraint. The earlier discussion on anisotropy showed that the most influential factor on the anisotropy is the Co moment. Since the lesser dilution of the Co moment by Cu results in higher K, Au might well be even more effective in doing so. The question arises whether the GdCoAu system also offers good $4\pi M_S$ vs. T characteristics like GdCoMo but higher K. In fact, this is true. The $4\pi M_S$ vs. T curves of GdCoAu films can be quite similar to those of GdCoMo because the ratios of exchange constants are similar (Hasegawa et al. 1975b). However the anisotropy for equivalent magnetization is higher. We can compare fig. 19 for GdCoAu with $x = 0.76$, $y = 0.12$ to fig. 1 for GdCoMo with $x = 0.85$, $y = 0.14$.

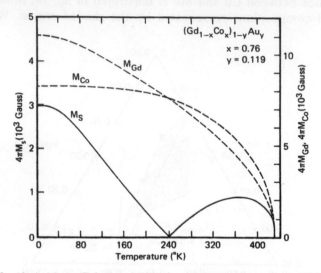

Fig. 19. $4\pi M_S$ vs. T for $x = 0.76$ and $y = 0.12$ (Hasegawa et al. 1975b).

The variation of $4\pi M_S$ with T is very similar but because of the lowered Co moment by Mo, the GdCoMo film requires a higher Co/Gd ratio to produce the same results. At the same time, even though the GdCoMo film has more Co it has a lower anisotropy because of the decreased Co moment. For the GdCoAu film $K(10^4 \, \text{erg cm}^{-3}) \approx 27$ while for the GdCoMo film $K(10^4 \, \text{erg cm}^{-3}) < 10$. The corresponding Co moments are 0.8 for the GdCoAu and 0.4 for the GdCoMo (Hasegawa et al. 1975b cf. Hasegawa et al. 1975a).

The difficulty with both GdCoCu and GdCoAu is that in both cases the anisotropy deteriorates when the films are subjected to temperatures $\geq 200°C$. A dilutant is needed that can keep the Co moment higher and, at the same time, provides films that are as stable, or more so, than GdCoMo.

GdCoCr films are very stable against thermal anneal. Films subjected to 250°C for 6 days did not change more than 5% (Schneider 1975). At the same time, since Cr, as observed above, dilutes the Co moment to about the same degree that Mo does, the magnetic features are about the same as for GdCoMo. With $x \approx 0.83$ and $y = 0.20$ $4\pi M_S$ has a reasonable flatness with T from 0° to 100°C. Unfortunately as we might expect from the Co moment dilution the anisotropy is not larger than for comparable films of GdCoMo. A drawback to these films is that they are very sensitive to residual oxygen in the sputtering chamber. Only films made in chambers evacuated to 10^{-8} Torr prior to admission of the Ar gas have low coercive force (~ 1 Oe). Films made in chambers evacuated to only 10^{-6} Torr were contaminated by 3–13 at.% oxygen and the coercive force increased to 10 Oe.

In summary, no other third constituent, Z, has yet been identified which provides a combination of the desirable properties better than Mo. Those properties include: $4\pi M_S$ values suitable for useful bubbles that do not vary appreciably from 0°C to 100°C and sufficient anisotropy to obtain adequate Q with stability in the face of device processing conditions.

6. Other RE–TM combinations

Certainly the most extensively investigated amorphous films for bubbles have been GdCo and GdFe with additions of Mo, Cu, Au and Cr. However, there has also been some fabrication and measurements on films of other RE–TM combinations. These have included:
(i) Other TM with Gd: e.g., GdAu and GdNi (Durand and Poon 1977) and GdCu (Heiman and Kazama 1978).
(ii) Other RE with Co: e.g., RE = Dy and Ho (Roberts et al. 1977) and RE = Nd (Taylor et al. 1978).
(iii) Other RE with Fe: e.g., RE = Tb, Dy, Ho, (e.g., Heiman et al. 1976) and RE = Nd (Taylor et al. 1978).
(iv) Other RE with Cu where RE = Tb, Dy, Ho (e.g., Heiman and Kazama 1978).

Some of these, especially in ternary combination with GdCo or GdFe, have magnetic properties in the range we might expect by extrapolation and support magnetic bubbles. None of them, however, has so far provided superior bubble properties to those compositions already discussed. All of these compositions are interesting for illuminating the mechanisms of magnetization. The Gd_xZ_{1-x} alloys, where Z is non-magnetic and Gd is an S state atom without local anisotropy, exhibit variations in magnetization expected from "percolation" theory where ferromagnetism disappears when the concentration of Gd is less than $x \simeq 0.33$ because of the formation of clusters which no longer have continuous nearest neighbor coupling to each other. Thus spin-glasses are characteristic of compositions having the Gd concentration near $x \leqslant 0.33$ and compositions having considerably more Gd are ferromagnets with T_C increasing with greater Gd content. When the rare earth is Tb, Dy or Ho, which have varying degrees of coupling of their moments to the local "crystal field", there is an added complication due to local anisotropy. In these cases the magnetization is less than would be provided by fully aligned moments and even this value is reached only after exceeding a large coercive force (Heiman and Kazama 1978). This lack of full alignment of the rare earth subnetwork may also occur when the rare earths with strong crystal field interaction are combined with the magnetic transition metals (e.g., Nd_xCo_{1-x} and Nd_xFe_{1-x}, Taylor et al. 1978).

7. The usefulness of amorphous films for bubbles

The principal advantage of amorphous films vs. crystalline films for bubbles would lie in the lower cost of the raw materials, film substrate and film deposition process. We have seen that the substrate requirement for cleanliness and smoothness is such that ordinary glass cannot be used. Nevertheless glass such as is available for photomasks or Si substrates which are common to the semiconductor device industry are adequate and would be less expensive than garnet single-crystal substrates.

Subsidiary advantages might be found in the flexibility of the materials and the fabrication process. For example, higher values of $4\pi M_S$ are available than with garnets, making possible smaller useful bubbles. These smaller bubbles may become useful when the rest of the bubble device technology advances to the point where bubbles smaller than $0.6 \mu m$ can be accommodated. To extend to bubbles smaller than $0.2 \mu m$ it will be necessary to discover how to introduce higher uniaxial anisotropy. In addition the composition of the films can be readily changed by changing the parameters of the deposition process, whether sputtering or evaporation. This suggests that multilevel films with graded composition can be made when that is useful.

The primary disadvantage of amorphous films is the difficulty in achieving a satisfactory variation of properties with temperature. In the garnets the net magnetization is primarily the difference between the two iron subnetworks, each of which has a good temperature characteristic. Thus through dilution of one subnetwork vs. the other, the magnetization can be reduced all the way to zero without destroying the good temperature characteristic. In the amorphous films, however, the net magnetization is the difference between two subnetworks with very different temperature characteristics. Therefore when the magnetization is reduced by either changing the RE/TM ratio or diluting the TM by something like Mo, Au or Cu, the temperature characteristic varies appreciably. The net result is that both materials are useful for small bubbles where the dilution is small, but only the garnets have a satisfactory temperature characteristic for larger bubbles. Amorphous materials appear to be restricted to the range below $0.6 \mu m$ unless a broad temperature range of operation is not required. On the other hand, garnets do not extend below $\sim 0.4 \mu m$ so that amorphous materials may provide the extendability to very small bubbles that the garnets cannot provide.

Additional complications in the use of amorphous films arise from their metallic nature. Thus special insulating or passivating layers may be necessary to accommodate the electrical conductivity and chemical reactivity of the amorphous films.

Appendix: Conversion of Gaussian to SI units

TABLE A.1

Parameter	Gaussian Formula	Units	SI Formula	Units	SI Gaussian Ratio
Magnetomotive force	$MMF = \dfrac{4\pi NI}{10}$	Gilbert	$MMF = NI$	A T	$10/4\pi = 0.796$
Magnetic field H	$\int H \cdot dl = MMF$	Oe	$\int H \cdot dl = MMF$	A m^{-1}	$10^3/4\pi$ A/m Oe
Flux ϕ	$\phi = \int B \cdot dA$	Maxwell	$\phi = \int B\,dA$	Weber	10^{-8} W/Mx
Flux density B	$B = H + 4\pi H$	Gauss	$B = \mu_0(H+M)^*$ $= (\mu_0 H + J)^{**}$	Tesla $=$ Wb m^{-2}	10^{-4} T/G 10^{-4} Wb/m^2 G
Magnetization	$4\pi M$	Gauss	$M\ddagger\S$	A m^{-1}	$10^3/4\pi$
Magnetic polarization	$4\pi M$		$J = \mu_0 M^{**}$	Tesla	10^{-4} T/G
Volume susceptibility	$\chi = dM/DH$		$\kappa = dM/dH$		$1/4\pi$
Demagnetization factor	$H_d = -NM$	Gauss	$H_d = -NM$		$1/4\pi$
Anisotropy	K	erg cm^{-3}	K	J m^{-3}	10^{-1}
Exchange	A	erg cm^{-1}	A	J m^{-1}	10^{-5}
Mobility	μ	cm s^{-1} Oe^{-1}	μ	m^2 A^{-1} s^{-1}	$4\pi \times 10^{-5}$
Energy product	$\dfrac{1}{8\pi}BH = \tfrac{1}{2}MH$	erg cm^{-3}	$\tfrac{1}{2}BH = \dfrac{\mu_0}{2}MH$	J m^{-3}	10^{-1}
Anisotropy field Q	$H_K = 2K/M$ $Q = H_K/4\pi M$	Oe	$H_K = 2K/\mu_0 M$ $Q = H_K/M$	A m^{-1}	$10^3/4\pi$
Wall energy density	$\sigma = 4\sqrt{AK}$	erg cm^{-2}	$\sigma = 4\sqrt{AK}$	J m^{-2}	10^{-3}
Characteristic length	$l = \sigma/4\pi M^2$	cm	$l = \sigma/\mu_0 M^2$	m	10^{-2}
Permeability	$\mu = dB/dH$ $= 1 + 4\pi\chi$		$\mu = dB/dH$ $= \mu_0(1+\kappa)$		$4\pi \times 10^{-7} = \mu_0$ $= 12.5 \times 10^{-7}$

*RMKS.
**MKSA

References

Argyle, B.E., R.J. Gambino and K.Y. Ahn, 1975, AIP Conf. Proc. 24, 564.

Bajorek, C.H. and R.J. Kobliska, 1976, IBM J. Res. Dev. 20, 271.

Burilla, C.T., W.R. Bekebrede and A.B. Smith, 1976, AIP Conf. Proc. 34, 340.

Burilla, C.T., W.R. Bekebrede, M. Kestegian and A.B. Smith, 1978, J. Appl. Phys. 49, 1750.

Cargill, G.S., 1973, AIP Conf. Proc. 18, 631.

Cargill, G.S. and T. Mizoguchi, 1978, J. Appl. Phys. 49, 1753.

Chaudhari, P. and D.C. Cronemeyer, 1976, AIP Conf. Proc. 29, 113.

Chaudhari, P., J.J. Cuomo and R.J. Gambino, 1973, IBM J. Res. Dev. 17, 66; Appl. Phys. Lett. 22, 337.

Chaudhari, P., J.J. Cuomo, R.J. Gambino, S. Kirkpatrick and L-J Tao, 1975, AIP Conf. Proc. 24, 562.

Cronemeyer, D.C., 1973, AIP Conf. Proc. 18, 85.

Cuomo, J.J. and R.J. Gambino, 1975, J. Vac. Sci. Technol. 12, 79.

Cuomo, J.J. and R.J. Gambino, 1977, J. Vac. Sci. Technol. 14, 152.

Durand, J. and S.J. Poon, 1977, IEEE Trans. MAG-13, 1556.

Esho, S. and S. Fujiwara, 1976, AIP Conf. Proc. 34, 331.

Gambino, R.J., P. Chaudhari and J.J. Cuomo, 1974, AIP Conf. Proc. 18, 578.

Gangulee, A. and R.J. Kobliska, 1978, J. Appl. Phys. 49, 4169.

Gangulee, A. and R.C. Taylor, 1978, J. Appl. Phys. 49, 1762.

Graczyk, J.F., 1978, J. Appl. Phys. 49, 1738.

Hasegawa, R., 1974, J. Appl. Phys. 45, 3109.

Hasegawa, R., 1975, J. Appl. Phys. 46, 5263.

Hasegawa, R., B.E. Argyle and L-J Tao, 1975a, AIP Conf. Proc. 24, 110.

Hasegawa, R., R.J. Gambino and R. Ruf, 1975b, Appl. Phys. Lett. 27, 512.

Heiman, N. and N. Kazama, 1978, J. Appl. Phys. 49, 1686.

Heiman, N., A. Onton, D.F. Kyser, K. Lee and C.R. Guarnieri, 1975, AIP Conf. Proc. 24, 573.

Heiman, N., K. Lee and R.I. Potter, 1976, AIP Conf. Proc. 29, 130.

Heiman, N., N. Kazama, D.F. Kyser and V.J. Minkiewicz, 1978, J. Appl. Phys. 49, 366.

Herd, S.R., 1978, J. Appl. Phys. 49, 1744.

Herd, S.R. and P. Chaudhari, 1973, Phys. Stat. Sol. (a) 18, 603.

Katayama, T., M. Hirano, Y. Koizumi, K. Kawanishi and T. Tsushima, 1977, IEEE Trans. MAG-13, 1603.

Kobliska, R.J., R. Ruf, and J.J. Cuomo, 1975, AIP Conf. Proc. 24, 570.

Kobliska, R.J., A. Gangulee, D.E. Cox and C.H. Bajorek, 1977, IEEE Trans. MAG-13, 1767.

Kryder, M.H., K.Y. Ahn, G.S. Almasi, G.E. Keefe and J.V. Powers, 1974, IEEE Trans. MAG-10, 825.

Kryder, M.H., L-J Tao and C.H. Wilts, 1977, IEEE Trans. MAG-13, 1626.

Malozemoff, A.P., J.P. Jamet and R.J. Gambino, 1977, IEEE Trans. MAG-13, 1609.

Minkiewicz, V.J., P.A. Albert, R.I. Potter and C.R. Guarnieri, 1976, AIP Conf. Proc. 29, 107.

Mizoguchi, T., R.J. Gambino, W.N. Hammer and J.J. Cuomo, 1977, IEEE Trans. MAG-13, 1618.

Onton, A., N. Heiman, J.C. Suits and W. Parrish, 1976, IBM J. Res. Dev. 20, 409.

Roberts, G.E., W.L. Wilson, Jr. and H.C. Bourne, Jr., 1977, IEEE Trans. MAG-13, 1535.

Schneider, J., 1975, IBM J. Res. Dev. 19, 587.

Taylor, R.C., 1976, J. Appl. Phys. 47, 1164.

Taylor, R.C. and A. Gangulee, 1976, J. Appl. Phys. 47, 4666.

Taylor, R.C. and A. Gangulee, 1977, J. Appl. Phys. 48, 358.

Taylor, R.C., T.R. McGuire, J.M.D. Coey and A. Gangulee, 1978, J. Appl. Phys. 49, 2885.

Wagner, C.N.J., N. Heiman, T.C. Huang, A. Onton and W. Parrish, 1976, AIP Conf. Proc. 29, 188.

chapter 7

RECORDING MATERIALS

G. BATE

*Verbatim Corporation**
Sunnyvale, CA 94086
USA

*The chapter was written while the author was employed by the IBM Corporation.

Ferromagnetic Materials, Vol. 2
Edited by E.P. Wohlfarth
© North-Holland Publishing Company, 1980

CONTENTS

1. Introduction

The objective in recording is to store information so that it may be recovered with little or no modification; the output signal must resemble the input signal as closely as possible. Magnetic recording is the preferred method of storage when the information must be read immediately after it is written (e.g., in instant replay, or for verification) or when the information is to be processed by a machine. The storage medium is cheap, it requires no development and, since the writing process is essentially one of reversing electron spins, it is very fast and apparently infinitely reversible. The information to be stored must first be encoded so that it can be represented as a pattern of magnetization. In analog recording, continuous variations in the amplitude (audio recording) or the frequency (video, instrumentation recording) must be stored. And to minimize distortion, linearity is required between the recorded signal and the state of magnetization. In digital recording, the information is encoded as a train of pulses. The time relationship between the pulses at the output should be the same as the relationship at the input. Digital recording is used at present in the storage of machine-readable information in computers and in the Pulse-Code-Modulation (PCM) method of instrumentation recording. Because of its inherently high signal-to-noise ratio (SNR) and because by the use of error correction techniques it can be made extremely reliable, it appears that digital recording will also eventually be widely used in both video and audio recording.

To read the information, flux from the magnetized regions on the storage medium must be coupled (usually inductively) to the reading head. Large signals call for a large amount of flux and thus we require that the storage medium have a high remanent magnetization. The coercivity should be high enough to prevent accidental erasure of the information by stray fields and high enough to reduce the effects of self-demagnetizing fields within the magnetic pattern. These requirements are those of a permanent magnet material having a remanent magnetization in the range 100–1500 emu/cc (0.126–1.88 Wb/m^2) and a coercivity between 250 and 1200 Oe (19.95–95.5 kA/m). The question of the detailed dependence of recording performance on magnetic properties will be deferred to section 3.6.

In practice, recording surfaces are made of fine single-domain particles immersed in a plastic binder, or of continuous, deposited films; in both cases a substrate provides the mechanical strength. Both kinds of recording surface will be discussed in the ensuing sections with thin film surfaces being treated first.

With very few exceptions the magnetic units used in discussion of recording materials are cgs. These units will be mentioned first in this section and the corresponding mks quantities and units will then be given in parentheses.

2. Hard films

2.1. Metal films

In digital recording the functional parameter by which progress is most commonly judged is the bit density – the number of bits-per-unit-length of a track. To record and reproduce reliably at high density we require that each recorded bit be stored as a discrete transition of magnetization and that interference between neighboring transitions be minimized during writing, storage, and reading. This in turn requires that the magnetic coating be thin, but to get a large signal from a thin coating implies the need for a reasonably high intensity of magnetization, M_s. The magnetization must not be too high, however, or demagnetization will be troublesome. To resist demagnetization and accidental erasure the coercivity, H_c, must be high but not so high that the writing and erasure heads have difficulty in switching the magnetization.

Metal films can be made to have magnetic properties covering an extremely wide range, for example, $M_s \leqslant 2000$ emu/cc (2.51 Wb/m^2) and $H_c \leqslant 2000$ Oe (159 kA/m). Furthermore, they can, if necessary, be made only a few hundred Ångstroms thick. In comparison, particulate recording surfaces can seldom be made thinner than 1μm. In addition to these desirable attributes the hysteresis loops of metal films can be made to have extremely steep sides; the distribution of switching fields is very narrow. Since metal films show such flexibility in magnetic and non-magnetic properties it would appear that the ideal recording surface for any conceivable application could be made in thin film form. Yet, with the exception of some drums and a few video disks, thin film recording surfaces are not found among today's recording surfaces.

Matching their impressive list of advantages is an equally impressive list of disadvantages. Their thinness usually means poor wear resistance and a susceptibility to serious damage in the event of contact with the head. Metals are inherently more reactive than oxides and so the thin films must be protected against corrosion. When metal films are applied to flexible substrates, it appears that either the adhesion is poor and so is the wear resistance, or the adhesion is good but stresses in the film causes cupping of the tape. Finally, when making particulate recording surfaces the magnetic properties of the particles are checked before coating. The changes that occur during coating are slight and predictable and the properties can be consistent. With thin films, on the other hand, the magnetic properties obviously cannot be known until the film is deposited and precise knowledge and control of a large number of processing variables is necessary if a consistent product is to be made.

At present the practical difficulties of making thin film recording surfaces are severe and yet so great is their potential that we must expect that the problems will eventually be overcome. There are four principal methods for making thin films. They are: electrochemical deposition, chemical deposition, evaporation, and sputtering.

2.1.1. Electrochemical deposition

Films of the ferromagnetic elements may be made by electroplating. The metallic ions are reduced from an aqueous solution at the cathode by means of electrons from an external supply. The process can be continuous, is of low cost, and can be used to plate two metal ions simultaneously if the deposition potentials of the two metals are similar. This is the case for iron, cobalt, and nickel and so binary and ternary alloy films may be formed. Recording surfaces require coercivities higher than 250 Oe (19.9 kA/m) and it is found empirically that to obtain high coercivities the films must contain a large amount of cobalt. Co–Ni is used rather than Co–Fe presumably because the deposition potentials of cobalt and nickel are very close together and possibly with the expectation that the presence of nickel would enhance the corrosion resistance of the films. Highest coercivities are obtained with films containing 60–95% cobalt.

Bonn and Wendell (1953) introduced the basic plating bath for cobalt–nickel films. The electrolyte contained nickel and cobalt chlorides, amonium chloride [to promote film uniformity (Koretzky 1963)] and sodium hypophosphite. The pH was critical and was kept at 4.0; the bath temperature was 50°C. Films having a thickness of about 0.1μm and containing approximately 85% Co were obtained. Sodium hypophosphite was used to add phosphorus to the film in quantities far beyond the solubility limit. It appears that the phosphorus, which is normally present in amounts of 1–3%, is segregated at the crystallite boundaries and that its principle role is in determining the size of the crystallites. The dependence of the coercivity of the film on the hypophosphite concentration of the bath is linear for small amounts (Freitag and Mathias 1964) but for more than about 5 g/l, it increases less rapidly (Koretzky 1963, fig. 1). Other parameters which influence grain size are the bath composition, pH, temperature and the current density. Similarly the magnetic properties of earlier plating baths were found by Koretzky (1963) to be improved by addition of KCNS to the bath. Radiographic techniques revealed that sulfur was incorporated in the films.

The deposition parameters which must be controlled in order to obtain films having predictable magnetic properties are: bath composition, pH, amount of additives, current density, temperature of the bath, and the roughness of the substrate. The composition of the films depends mainly on the bath composition but also on the hypophosphite concentration and on the temperature of the bath. At high temperatures electrodeposition is supplemented by autocatalytic deposition which, in acid solutions, preferentially deposits nickel. With increasing temperature, a bath having a constant composition will produce films of

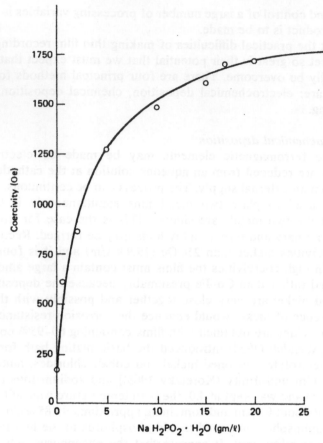

Fig. 1. Coercivity vs the hypophosphite concentration of the bath for a film of Co–Ni–P deposited electrochemically (Koretzky 1963).

decreasing Co/Ni ratio and thus decreasing saturation magnetization. Sallo and Olsen (1961) prepared films of Co–Ni–P and Sallo and Carr (1962, 1963) made Co–P films using the Bonn and Wendell (1953) method. In both experiments the film had coercivities up to 1000 Oe (79.6 kA/m) depending on deposition conditions. The coercivity increased with increasing concentrations of nickel, cobalt, ammonium chloride and sodium hypophosphite and with increasing current density. The magnetic hardness of the films was attributed to the structure of small crystallites in the shape of rods or platelets, separated by non-magnetic phosphides of cobalt and nickel. Kefalas (1966) measured the coercivity of Ni–Co films as a function of the deposition parameters; bath temperature and Ni^{2+}/Co^{2+} ratio, current density and thickness, and found optimum values for each. Judge (1972) observed that electrodeposited films of cobalt may have hcp and fcc structures with hcp predominating in alkaline solutions. In Co–Ni films the structure depended on composition and mixed hcp and fcc phases were commonly found in the range 70–80% cobalt. The higher coercivity films were generally those having a mixture of phases. Phosphorus is soluble in bulk cobalt only to 0.35 at.% but large amounts can be included in thin films. Kanbe and Kanematsu (1968) made films containing up to 24 at.% by electrodeposition. They found that the fcc structure became dominant with increasing amounts of phosphorus.

Fisher (1962) investigated the interrelation of stress, hcp/fcc content, and coercivity in electrodeposited Co–P films and concluded that the coercivity was controlled primarily by the crystalline anisotropy, with stress playing a secondary role. However, his films were of much lower coercivity (30–100 Oe, 2.39–7.96 kA/m) than those discussed above. In films of higher coercivity the effect of stress is expected to be of lesser importance.

The saturation magnetization-per-unit-mass of the films is usually 5–20% below the bulk value for an alloy of the same composition. The films not only have high squareness (often >0.9) but they are also isotropic in the plane of the film. The graph of torque vs applied field is a straight line and the maximum value of the anisotropy field is about 6500 Oe (33.6 kA/m). Structure determination of films thinner than 1μm can be difficult when the diffraction lines are few, broad, and ill-defined, implying a crystallite size less than 100 Å. Thus it appears that the most probable explanation of their hysteresis properties is that the films behave like an assembly of SD particles which are separated by the phosphorus-rich layer. The particles appear to be very strongly interacting and the interaction confers stability and isotropy on the magnetic properties.

Luborsky (1970) investigated the effect on the magnetic properties of Co and Co–Ni films of adding Cr, Mo, W from Group VIB, and P, As, Sb, and Bi from Group V. An analysis of grain size, crystal texture σ_s and H_c showed that the added elements, which were present as metastable solid solutions in cobalt, all acted in the same way. The high coercivity and squareness arose from a strongly-interacting array of SD crystals in which the role of the additives was to control crystal size and separation. Gerkema et al. (1971) investigated the effect of metallic underlayers between the substrate and the electrodeposited Co–P layer. They found that the underlayer determines to a large extent the crystallite size and orientation in the Co–P. Using the same plating conditions but with underlayers of Pt, Au, Ag, Sn, the coercivities were 160, 300, 630, 830 Oe (12.7, 23.9, 50.1, 66 kA/m). Similar but smaller effects were obtained by Tadokoro et al. (1975) who immersed the substrate in solutions of Au, Ag, Pd, Cr, etc. salts before electroplating Co–Ni. Films of Co–Ni–P are usually made with 60–95% cobalt in order to obtain high coercivities. Yamanaka et al. (1975), however, found that when the Ni/Co weight ratio was much lower (between 2 and 9) the noise level was very much reduced without much reduction in signal level.

The electrodeposited films discussed above were applied to metallic substrates. Wenner (1964) showed that it is possible to use electrodeposition on non-metallic substrates. His method was to apply first by autocatalytic deposition, a conducting film of Ni–P which was then used as the cathode for the electro-deposition of the magnetic film. Improved magnetic properties were claimed (Akad 1974) by the addition of sodium ethylenediamine tetraacetate to the electrolyte solution accompanied by the superimposition of strong, low-frequency pulses of polarity opposite to that of the plating current. The difficulties in using the Wenner approach to make tape, stem basically from the large number of processing variables which must be controlled. In addition to those variables related to the deposition of the magnetic layer, there are additional variables associated with the conducting layer. The function of the layer is not simply to carry the plating current. It is now the substrate on which the magnetic film must grow and its crystal structure will, as we have seen, determine the structure of the magnetic layer.

2.1.2. Autocatalytic deposition

Films of the ferromagnetic metals, particularly nickel and cobalt, may be prepared by a deposition process in which the potential required for the reduction of the metallic ions is provided by a reducing agent in the bath rather than by an external power supply. The newly-deposited metal then catalyzes the further deposition and so the reaction proceeds autocatalytically. Before the reaction can start, the substrate must be catalyzed and stannous and palladium ions are commonly used for this purpose. Cobalt or nickel chloride or sulfate provide the metal ions and the reducing agent is a hypophosphite or a borohydride. The metal ions and the reductant must co-exist at the catalytic surface but if the concentration of the metal ions becomes too high the absorption of the reductant is inhibited and the plating rate is decreased. Some method of controlling the metal ion concentration must be used, for example, by arranging that the metal ion exists in solution as a complex ion which dissociates at a convenient rate to provide an equilibrium amount of the free metal ion. The nickel ion Ni^{2+} does not exist alone in water but is bonded by shared electrons to eight molecules of water arranged around the Ni^{2+} ion with octahedral symmetry. The complex is formed by substituting a ligand for the water molecules and the concentration of the metal ions in solution may be controlled by choosing the proper ligand e.g., ammonia, citrate or tartrate groups. Ammonia forms a relatively weak bond and thus builds up a considerable free metal ion concentration. Cobalt, nickel, and iron possess incomplete 3d shells which readily accept electrons when the ions are chemisorbed on the surface of the catalyst to form more metal atoms which in turn catalyze the reaction. Limiting deposition rates are reached when the reaction on the catalytic surface, which has the lower activation energy at low rates, gives way at high rates to a homogeneous reaction throughout the bath. When this occurs the metal is deposited as fine particles which sink to the bottom of the bath. The homogeneous reaction is sometimes used to make metal particles for recording surfaces (see section 3.5.2). Magnetic films having coercivities up to 1800 Oe (143 kA/m) and saturation magnetizations up to 80% of that of bulk alloy have been made autocatalytically. The films most commonly made by this method are of Co–P and contain about 2–5% phosphorus.

TABLE 1

Baths for deposition of Co–P films on polyethylene terephthalate substrates (after Judge et al. 1965b)

Sulfate		Chloride	
$CoSO_4$	0.11 M	$CoCl_2$	0.13 M
NaH_2PO_2	0.19 M	$NH_4H_2PO_2$	0.18 M
$(NH_4)_2SO_4$	0.50 M	$(NH_4)_2H(citrate)$	0.155 M
$Na_3(citrate)$	0.12 M		

Judge et al. (1965b) gave examples of baths (see table 1) which were used to deposit Co–P films on polyethylene terephthalate substrates which had been made hydrophilic by treatment in first, hot chromic acid solution and then hot sodium hydroxide solution. The substrate was then pre-sensitized in $SnCl_2$–$PdCl_2$ solutions and plated in one of two baths. The temperature was 80°C and the pH was kept at 8.7 with NH_4OH; no agitation was used. In these baths the complexing group was the citrate radical which is more strongly bonded to the metal than is ammonia. The complex ion dissociates at the same rate as plating occurred so that the free metal ion concentration remains constant.

Since polynuclear complexes are not normally formed, at least one ligand is needed for each reducible metal ion. Nickel chloride or sulfate could be substituted for cobalt but Ni–P films are paramagnetic when thin and have coercivities up to only 60 Oe (4.8 kA/m) when thicker. Ni–P films made in acid baths are paramagnetic regardless of thickness.

Cobalt–phosphorus films are immediately ferromagnetic. Judge et al. (1965b) found that there were differences in properties between films prepared in the sulfate and chloride baths but the differences were slight. The values of σ_s were 111 emu/g (139 μWb m/kg) for the sulfate bath and 116 emu/g (146 μWb m/kg) for the chloride bath and were independent of the thickness of the deposit. The calculated values, assuming that the phosphorus content of the films (5.8 ± 0.35% sulfate and 5.0 ± 0.2% chloride) existed as nonferromagnetic Co_2P, were 120 emu/g (151 μWb m/kg) and 125 emu/g (157 μWb m/kg) respectively. Blackie et al. (1967) found that the coercivity of their films depended strongly on the bath pH but not on its temperature, while Ito et al. (1970) found H_c to depend on both pH and temperature. The effect of variations in bath conditions is found in the thickness, the crystallite size and orientation, and the phosphorus content. These in turn determine the magnetic properties of the films. The thickness dependence of coercivity and squareness of films reported by Judge et al. (1965b) are shown in fig. 2. Region (a), where the thickness was less than 200 Å, was composed of growing superparamagnetic particles. In region (b), 300–1000 Å, most of the particles were single-domain and had high coercivities; and in region (c) the crystallites have grown large enough to behave as multi-domains. A similar dependence of H_c in autocatalytically deposited films was found by Fisher (1966), who showed that adding Zn to Co–P increased the

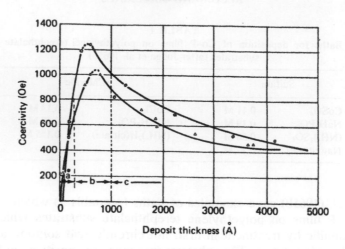

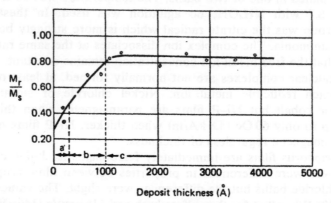

Fig. 2. Coercivity and reduced remanence vs film thickness for films of Co–P deposited from the chloride (△) and sulfate (○) baths (Judge et al. 1965b).

coercivity, and by Aspland et al. (1969), who plotted coercivity against particle diameter measured by dark field electron microscopy. Judge et al. (1965b) had noted earlier the correspondence between film thickness and the size of the crystallites, and Fisher (1972) confirmed this and showed that the particle diameter was proportional to the square root of the film thickness regardless of deposition rate.

Morton and Schlesinger (1968) and Sard (1970) found that the sensitization of the substrate created islands of palladium or tin. Nucleation of the Co–P film took place on the islands and a continuous film was obtained only after the islands had grown together. Clearly the nature of the substrate must be an important factor in such a process not only in determining the size and orientation of the crystallites but also in correcting any mismatch between the thermal expansion coefficients of the substrate and the film. Stone and Patel

(1973) described the use of several such layers. Their aluminum substrate was overcoated first with zinc, then nickel, then gold, and finally with a magnetic layer of Co–P. We see from fig. 2 that there is a marked thickness dependence of coercivity and it may not be possible to satisfy simultaneously the dual requirements of magnetic moment (proportional to thickness) and coercivity. Fisher and Jones (1972) suggested that by using the coercivity to determine the necessary film thickness, a composite film consisting of identical laminae could be built up. Each magnetic layer would be separated by a thin non-magnetic layer of Ni–P and the lamination would continue until the desired moment was reached.

The Co–P films discussed by Bate et al. (1965) contained only the hcp phase with a stacking fault frequency of about 0.13 and crystallite sizes from 200–700 Å. The lower coercivity films ($H_c < 330$ Oe, 23.9 kA/m) had the hexagonal c-axes predominantly normal to the film plane. In the higher coercivity films ($H_c \sim 1000$ Oe, 79.6 kA/m) the c-axes were found to be randomly distributed in the film plane and these films looked much smoother under the electron microscope. Films of intermediate coercivity (300–700 Oe, 23.9–55.7 kA/m) showed a random distribution of c-axes in and out of the film plane. When electroless nickel films are used as a substrate, Morton and Fisher (1969) found that Co–P films had an [0001] orientation perpendicular to the plane when the film thickness was small but that as the thickness increased the [10$\bar{1}$0] orientation became predominant. Frieze et al. (1968) found that all films began by having a ⟨0001⟩ orientation. For films having less than 2% phosphorus this orientation became more pronounced with increasing film thickness. However, if the phosphorus content was 2–5%, beyond a certain thickness the ⟨0001⟩ orientation gave way to ⟨10$\bar{1}$0⟩. Gălugăru (1970) found that it was possible to induce a uniaxial anisotropy into films of Co–P deposited electrolessly by heating them in a magnetic field. Films, 1200 Å thick and containing 9% P when heated to 670 K in a field of 6030 Oe (480 kA/m), developed a uniaxial anisotropy whose constant was $K_u = 2 \times 10^4$ erg/cc (2×10^3 J/m³). Attempts to induce magnetic orientation by applying fields during the deposition have usually been unsuccessful, but Aonuma et al. (1974) reported the results of a successful experiment. They used the chloride bath with hypophosphite, boric acid, and sodium citrate as complexing agent. A field of 600 Oe (47.8 kA/m) was applied and found to produce an orientation of hexagonal c-axes lying parallel and in the plane. The squareness along and across the orientation directions were 0.75 and 0.65 which does not reveal a high degree of orientation.

Ōuchi and Iwasaki (1972) measured the temperature dependence of coercivity in the range 20–400°C for electrolessly deposited films of Co–P containing different amounts of phosphorus. The rate of change of H_c increased with increasing P-content. The film containing 6% phosphorus showed a decrease in H_c from 480–200 Oe (38.2–15.9 kA/m) as the temperature rose from 20°C to 400°C.

All the films prepared by autocatalytic deposition which have been discussed so far were of Co–P. Judge et al. (1965a) showed that by depositing Co–Ni–P films it was possible to obtain higher coercivities than were obtained with Co–P. The highest coercivities of about 1400 Oe (111.4 kA/m) were found at 70–90% Co. The saturation magnetization, not surprisingly, decreased as nickel was added. Depew and Speliotis (1967) investigated the effect of adding iron to Co–P films and found that the coercivity fell sharply with increasing iron content up to 10%. Thus; by adding 10% Fe, the coercivity fell from 800–1400 Oe (63.7–111 kA/m) depending on thickness, to about 100 Oe (7.96 kA/m).

Predictably, many variations to the basic electroless bath have been proposed. For example, Shirahata et al. (1975b) found that the addition of a small amount of barium chloride brought about an increase of coercivity from 660–935 Oe (52.5–74.4 kA/m). The solution of palladium chloride used to sensitize the substrate is normally very weak (and long-lasting) compared with the stannous chloride solution. Ohiwa et al. (1975) found that by increasing the Sn/Pd ratio still more from 3 to between 4 and 8, the signal-to-noise ratio (SNR) of the tapes was improved by 10 dB. Traditionally in the plating field many strange ingredients are added to the bath as levelling agents or "brighteners." One such additive recommended by Nakao et al. (1975) as a means of enhancing the coercivity of the films from 480–800 Oe (38.2–63.7 kA/m), was Seneca extract. It is otherwide known as "rattlesnake root" and is found in cough medicines. By causing the plating solution to flow over the substrate at speeds between 5 and 25 m/min Hayashida et al. (1975, 1976) obtained films of Co–P and Co–Ni–P having improved SNR by virtue of a 5 dB reduction in their noise levels. The films contained about 2–5% phosphorus and had coercivities of about 1000 Oe (79.6 kA/m). They were about 1000–1500 Å thick and showed a marked orientation of the crystallites perpendicular to the substrate.

Polymeric substrates must be made hydrophilic before sensitization and the drastic treatment needed to do this (hot chromic acid then hot sodium hydroxide solutions) breaks the polymer bond so that the material on the surface of the substrate, while able to bond the Sn^{2+} and Pd^{2+}, has a reduced mechanical strength. The failure mode of a tape is then commonly, the loss of adhesion between the plated film and the substrate. Fisher et al. (1963) proposed to overcome this problem by means of an adhesive layer applied to the substrate. The loss of mechanical strength at the interface between the substrate and the film is one of many practical problems associated with making smooth, non-corroding, hard-wearing magnetic films. Whether it is better to use electrode-position or autocatalytic deposition is not at all clear. The apparent simplification gained by eliminating electrical contacts, current densities, etc., is balanced by the more complex chemistry of the electroless bath.

2.1.3. Vacuum deposition

Films of the ferromagnetic elements prepared by vacuum deposition are normally continuous, highly conducting, and of coercivity much less than 100 Oe (7.96 kA/m) – too low for use in recording surfaces. In order to increase the coercivity some way must be found to give the film a single-domain structure. In the electrodeposited and electroless films of Co–Ni–P and Co–P, phosphorus-rich material served this purpose by isolating the crystallites. This approach could not be used in films made by evaporation because of the different vapor pressures of the metal and phosphorus. Three methods have been described by which high coercivity, high magnetization films can be prepared by vacuum deposition. The methods are: evaporation at oblique incidence, evaporation at normal incidence, and low deposition rates, and evaporation at normal incidence over a layer of chromium which was also applied by vacuum deposition. Only the first and last methods have yielded coercivities high enough to be attractive in modern recording surfaces.

Speliotis et al. (1965a) described the preparation of thin films of Fe, Co or Ni by evaporation onto a variety of substrates at different angles of incidence, α. Substrate temperatures were usually in the range 20–50°C and evaporation rates of up to 50 Å/s were achieved by electron bombardment heating. The films had thicknesses ranging from 100–4000 Å and within each sample the thickness gradient was held to 5% by using a distance from source to substrate of at least 30 cm. The crystallite size was found to be strongly dependent upon the angle of incidence of the evaporant atoms and increased monotonically from 20–30 Å at normal incidence to 80–100 Å at 65° and about 1000 Å at 75°. Not surprisingly, the films showed anisotropic properties and the behavior of the coercivity (measured parallel and perpendicular to the plane of incidence) as a function of α is particularly characteristic of these films. It is shown in fig. 3 for iron films. Similar behavior was found for cobalt films. In the case of nickel, the form of the curves is the same but the highest value of coercivity reached was about 300 Oe (23.9 kA/m) compared with $H_c \sim 1000$ Oe (79.6 kA/m) for iron and cobalt. Morrison (1968) observed that although changing the angle of deposition had the largest effect on magnetic properties, other factors, e.g. deposition rate, substrate temperature, contamination of the evaporant source by the tungsten filaments, and film thickness, were thought to play some part in determining the film properties. Welch and Speliotis (1970) studied iron films and found a weak dependence of coercivity on thickness with a peak at about 400 Å. They also found that for deposition angles below 65°, the coercivity measured perpendicular to the plane of incidence was slightly higher than that measured parallel to the plane. For deposition angles greater than 65°, the curves crossed and the coercivity (parallel) rose rapidly to 1000 Oe (79.6 kA/m) at 85°. This corresponded to a uniaxial anisotropy coefficient $K_u = 2.95 \times 10^6$ erg/cc (2.95×10^5 J/m³). Kamberska and Kambersky (1966) explained the angular dependence of coercivity as follows. At low deposition angles, long chains of crystals are formed which run perpendicular to the plane of incidence. The structure is caused by a mechanism in which the first nuclei cast a shadow on the substrate and thus growth is encouraged along the chains rather than across them. At deposition angles greater than about 65°, particles tend to grow in columns along the direction of incidence. Kamberska and Kambersky were able to show that the magnetoelastic anisotropy constant was 10^4 erg/cc (10^3 J/m³). That is about an order of magnitude smaller than the measured K_u. Denisov (1969) measured the remanence coercivity as a function of α (in addition to the coercivity) and found

G. BATE

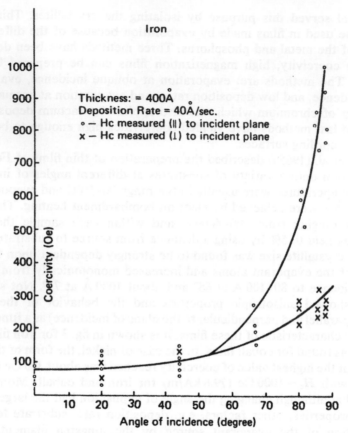

Fig. 3. Coercivity of iron films vs the angle of incidence (α) during deposition (Morrison 1968).

that H_r decreased with increasing angle up to $\alpha = 50°$ and thereafter increased in step with H_c. He also measured the variations of H_c and H_r with the angle, θ, between the direction of incidence and the direction of measurement. The angular variation of H_c suggests a coherent mechanism of magnetization reversal. Kondorsky and Denisov (1970) showed that the dependence of H_c on α shown in fig. 3 is only found when the substrate temperature is low. At higher temperatures a more complex dependence on α was found and is shown in fig. 4. Their interpretation of results on films of small α agrees with previous (self-shadowing) theories. However, Kondorsky and Denisov concluded that the high-results were best explained by the Néel anisotropy in the surface layers of adjoining crystallites. Münster (1971) developed a model based on a regular arrangement of oriented acicular particles and showed that the dependence of H_c on α shown in fig. 4 followed from the theory.

Keitoku (1973) measured the temperature dependence of torque curves on

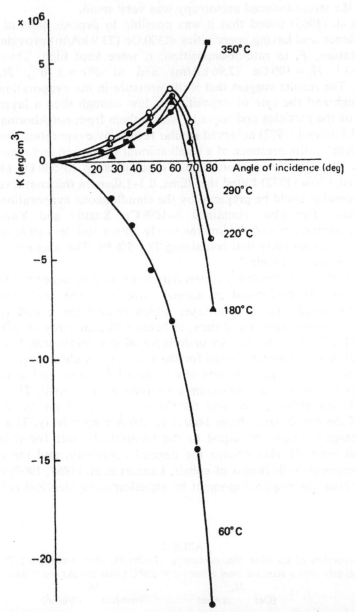

Fig. 4. Uniaxial anisotropy constant vs the angle of incidence (α) during deposition with the temperature of the substrate as parameter. The film thickness was 200 Å (Kondorsky and Denisov 1970).

cobalt films evaporated at $\alpha = 0$ to $75°$. He concluded that for $\alpha < 60°$, the anisotropy was magnetocrystalline, but for $\alpha > 60°$ shape anisotropy dominated. In all cases the stress-induced anisotropy was very small.

Davies et al. (1965) found that it was possible to deposit films of cobalt at normal incidence and having coercivities <300 Oe (23.9 kA/m) provided that the ratio of pressure, P, to rate-of-deposition, r, were kept high. Thus at $P/r = 10^{-7}$ torr$(Å/s)^{-1}$ $H_c = 100$ Oe (7.96 kA/m) and at $P/r = 3.10^{-5}$, $H_c = 300$ Oe (23.9 kA/m). The results suggest that if the pressure in the evaporation chamber is high enough and the rate of deposition is low enough then a layer of oxide could form on the particles and serve to isolate them from neighboring particles. Rodbell and Lommel (1972) achieved similar results by evaporating films of iron, cobalt, or nickel in the presence of a small amount of oxygen and then annealing the films at 50–600°C for a short time. Coercivities up to 530 Oe (42.2 kA/m) were reported. Clow (1972) found that films, 0.1–1.0 μm in thickness, suitable for magnetic recording could be prepared by the simultaneous evaporation of cobalt and chromium. The films contained 3–10% Cr. Suzuki and Yazaki (1975) evaporated platinum and cobalt simultaneously onto a stainless steel substrate at 250°C. They made an alloy film containing $77 \pm 5\%$ Pt. The magnetic properties of this film are given in table 2.

The third method by which high-coercivity films may be deposited by vacuum evaporation also is performed at normal incidence. The technique was to evaporate the cobalt onto a underlayer which caused the cobalt to grow in magnetically discrete regions and thus, to behave like an array of SD particles. Lazzari et al. (1967) found that an underlayer of chromium was best and was quite insensitive to the material used for the substrate. A chromium thickness of just over 300 Å and a deposition rate of less than 1 Å per second were found to be necessary to develop the maximum coercivity in the cobalt. The coercivity could be chosen between 200 and 600 Oe (15.9–47.8 kA/m) by varying the thickness of the cobalt layer from 1400 Å to 200 Å respectively. The measured saturation magnetization was equal to the theoretical value for cobalt within experimental error. Having chosen the desired coercivity, and thereby determined the appropriate thickness of cobalt, Lazzari et al. (1967, 1969) claimed to be able to obtain the required moment by superimposing identical cobalt layers

TABLE 2

Magnetic properties of an alloy film containing $77 \pm 5\%$ Pt, after evaporating Pt and Co simultaneously onto a stainless steel substrate at 250°C (after Suzuki and Yazaki 1975)

	H_c		M_r		S
	(Oe)	(kA/m)	(emu/cc)	(Wb/m^2)	
As deposited	380	30.2	191	0.24	0.58
After annealing 2 h at 550°C	2300	183	454	0.57	0.81

separated by layers of chromium as thin as 100 Å. From the temperature dependence of H_c they concluded that the dominant anisotropy was magnetocrystalline. Daval and Randet (1970) studied the growth of cobalt films on chromium underlayers and found that to produce films of high coercivity it was necessary to have large hcp grains in which the c-axes were mainly in the plane of the film. Since there was favorable matching between $\langle 11\bar{2}0 \rangle$ of cobalt and (100) of chromium, it was possible, as Daval and Randet found, to grow crystallites of cobalt (having the c-axes lying in the plane of the film) epitaxially on chromium. The $(11\bar{2}0)$ plane of cobalt has a rectangular lattice with constants of 4.07 Å and 4.34 Å which also match reasonably well the cubic lattice constant of 4.078 Å of gold. Ahn et al. (1973) used single crystal substrates overcoated with 2000 Å of gold and 500 Å Co–Ni by vacuum deposition and a background pressure of $2–4 \times 10^{-7}$ torr. The deposition rates were 30 Å/s for gold and 5 Å/s for Co–Ni. Without the gold underlayer films having a low coercivity, usually less than 100 Oe (7.96 kA/m), were obtained. With the gold underlayer, coercivities up to 1300 Oe (103.5 kA/m) were found; the coercivity being higher as the ratio hcp:fcc in the Co–Ni film increased. Ahn et al. (1973) believed that in addition to encouraging epitaxis, the gold underlayer provided a surface of high mobility to the cobalt and nickel atoms and thus enhanced the nucleation and growth of the hexagonal phase.

2.1.4. Sputtering and ion plating

Sputtering is done in a vacuum chamber inside which are an anode, a cathode, and, between them, the substrate on which films are to be deposited. The chamber is first evacuated and then filled with a flowing inert gas (argon or krypton) to a pressure of 10^{-1} to 10^{-4} torr. A potential difference of about 1000–5000 V is applied across the anode and cathode, either dc or rf. The electrodes, which are usually water-cooled are often in the form of flat plates. The pressures used are very much higher than those generally encountered in vacuum deposition and a glow discharge occurs in the inert atmosphere. Under bombardment by the positive ionized gas molecules, the cathode slowly disintegrates and the material is ejected from it either as free atoms or as atoms combined with residual gas molecules. The momentum transfer model of the process suggests that atoms are ejected from the cathode with velocities at least two orders of magnitude higher than those found during evaporation. At pressures of about 10^{-1} mm, the transport mechanism is diffusion controlled while at pressures around 10^{-4} mm, the mechanism is one of molecular flow. If the substrate–cathode distance is less than the mean free path then particles arrive at the substrate with high energies (10–30 eV). The substrate is also subjected to electron bombardment since the electrons in the glow discharge are accelerated to the anode. This serves to clean the freshly-deposited surface by removing molecules of adsorbed gases and the effect can be enhanced by applying a negative potential (~ 100 V relative to the anode) to the film during deposition.

The properties of the sputtered films depend on the gas used in the discharge. While argon is often used in the deposition of pure metals and alloys Kay (1961) reported that, by mixing argon and oxygen and controlling the partial pressure of the latter, different oxidation states and allotropic forms could be obtained selectively, for example γ-Fe_2O_3 rather than α-Fe_2O_3. The deposition rates and the film characteristics were found to depend markedly on the currents, voltages and pressures used and also on the geometry of the substrate, the electrodes and the glow discharge zones. Kay obtained films of pure α-iron with no apparent crystallite or magnetic orientation. The films had coercivities of about 250 Oe (19.9 kA/m) and $S > 0.95$.

Rogalla (1969) made two-phase Co–Cu thin films by sputtering 25Co–75Cu or 50Co–50Cu. In the as-sputtered state the films had the structure of a metastable fcc solid solution. After annealing at 500–700°C the films were found to consist of two phases, a copper matrix and fine particles of fcc 89Co–11Cu which showed single domain behaviour. Coercivities up to 280 Oe (22.3 kA/m) and squareness between 0.7 and 0.9 were found. Berndt and Speliotis (1970) sputtered alternate layers of cobalt and tungsten from separate targets. Cobalt films thicker than 30 Å were ferromagnetic and showed a peak in coercivity between 50 and 150 Å. The coercivity increase sharply with increasing substrate temperature and reached 750 Oe (59.7 kA/m) at 500°C. The high-coercivity films showed a crystallite size of about 100 Å and were fcc. Heller (1974) sputtered iron in an inert gas–oxygen atmosphere and, by changing the partial pressure of oxygen, he was able to deposit alternate layers (100–250 Å) iron and iron oxide. Recording surfaces ranging in coercivity from 400 to 700 Oe (31.8–55.7 kA/m) were made by this method.

In ion-plating the potentials used in dc sputtering are reversed and the substrate is placed near the cathode. The target, which is the anode, is heated as in vacuum deposition and 10–15% of the evaporated metal atoms are ionized in the glow discharge field. The metal ions are then accelerated toward the cathode and strike the substrate. It is found that the adhesion of the deposited film to the substrate is much greater than in conventional vacuum deposition since surface atoms on the substrate are continually being sputtered away and re-deposited. It is also possible to apply a uniform coating to substrates of a complex shape by ion plating. The ions tend to follow the field lines but because of the higher background pressure, they are scattered and can reach sites that would be hidden to atoms travelling a straight path from source to substrate.

Shirahata et al. (1975a) described the use of ion plating to make magnetic tape. They deposited a film of 75Co–20Ni–5Cr in a field of 1000 Oe at 0.01 torr of helium. The film was 0.22μm thick and had a coercivity of 320 Oe (25.5 kA/m) and a squareness of 0.55. When a cylindrical permanent magnet was used as cathode the squareness increased to 0.84. Takao and Tasaki (1976) also used ion plating to deposit Co–Ni films and found that the highest coercivities (about 500 Oe, 39.8 kA/m) were obtained at the composition 75Co–25 Ni, at a thickness of 0.1 μm and an argon pressure of 0.01 to 0.04 torr. The films consisted of crystallites whose average size was less than 300 Å surrounded by voids so that SD behavior was obtained.

In summary the advantages of vacuum deposition are that it is the oldest and best understood of the vacuum chamber deposition techniques and that the substrate can be kept relatively cool by having a large source–substrate distance. Sputtering permits the simultaneous deposition of several elements (including non-metals) from separate targets and one can clean the substrate by sputter etching. Ion plating gives good adhesion by virtue of a more gradual transition from the substrate material to the film material. This comes from the sputtering, followed by the redeposition of atoms of the substrate and also of the atoms and ions of the film material. Complicated shapes may be coated by this technique.

2.1.5. Undercoats and overcoats

The thinness of deposited magnetic films which is such a great advantage when recording at high densities becomes a liability when wear takes place. In some cases the failure mechanism is a breakdown in adhesion between film and substrate as was shown by Morrison (1968) using a two-layer film of cobalt over nickel. He observed that the ratio of cobalt to nickel does not change even when the films were almost worn away and surface wear would have produced a relative reduction of cobalt to nickel. Adhesive failure on the surface of polymeric substrates is clearly possible when a drastic hydrolyzing treatment has been given as in the case of autocatalytic deposition of a magnetic or a conducting film. It is also true that vacuum deposited or sputtered films of inactive metals show poor wear (Piskacek and Vavra 1968, MacLeod 1971, Judge 1972) but that the wear performance can be improved by putting an active layer e.g. of Ag, Al, Cu, Ni, Cr or W between the substrate and the magnetic film.

The abrasion resistance of thin films may be improved by vacuum-depositing or plating an overlayer of rhodium (Piskacek and Vavra 1968, MacLeod 1971). Tadokoro and Aonuma (1974) electroplated $0.2\mu m$ of copper and then $0.02\mu m$ of rhodium on their electrodeposited Co–Ni films, to improve the abrasion resistance. Overcoats of polymeric materials have also been used to give abrasion resistance. Siakkou et al. (1973) protected films of Fe–Ni–Co prepared by vacuum deposition at oblique incidence by coating with polyurethanes. Ito (1974) used a polyamide with an aliphatic acid as lubricant on Co–P films and Clow and Gilson (1976) applied polyisobutylene to Co–Cr films. Other materials which have been applied to improve abrasion resistance include tungsten carbide, applied in thicknesses up to $0.25\mu m$ by vacuum deposition (BASF 1974) and molybdenum disulfide, incorporated in electrodeposited films of Ni–Co by Sata et al. (1975). Hoyama and Suzuki (1973) electroplated a $0.6\mu m$ thick film of 85Fe–15Co and then soaked the film in an alkali solution of cobaltous and ferrous sulfates which was then warmed in a stream of oxygen to form an oxide film. The film was stabilized by being heated at 300°C and was claimed to give increased abrasion resistance.

Many attempts have been made to solve the key problem of reducing the sensitivity of metal films to corrosion and abrasion but it is not clear that a thoroughly satisfactory solution has been found. Until that happens metal films will offer only the tantalizing promise of superior recording surfaces.

2.1.6. Films having perpendicular anisotropy

The thin films discussed above by virtue of their geometry encourage the remanent magnetization to lie in the film plane. A completely different structure was proposed by Brownlow (1968). This consisted of an aluminum substrate anodized in warm sulfuric acid to give a porous film of Al_2O_3 in which the micropores extend from the outer surface to the metal. Using the substrate as cathode Brownlow then electroplated cobalt from solution into the pores and thus created a film which consisted of an array of elongated particles having their long axes perpendicular to the substrate. A similar method was used by Ishii and Uedaira (1974) and by Kawai and Ueda (1975). The structure was observed in transmission electron micrographs of thin sections and was also evident from hysteresis loops measured perpendicular to the film plane. Coercivities ranged from 500 to 1100 Oe (39.8 to 87.6 kA/m). Higher remanent magnetization was obtained by Kawai (1975) by gradually lapping away the outer layers of the oxide films to leave thin smooth surfaces suitable for use as magnetic disks. The corrosion resistance could be enhanced by immersing the films in boiling water which had the effect of sealing the pores.

The magnetic recording performance was measured out to 14 000 bpi (551 bits/mm) by Kawai and Ishiguro (1976). Further details of preparation and properties were given by Kawai et al. (1975).

A different method of preparing films having their easy axis of magnetization perpendicular to the film plane was described by Iwasaki and Yamazaki (1975). They used rf sputtering of Co–Cr onto a polyimide film. X-ray diffraction patterns of their films showed only the hcp (00.2) line. The role of chromium was to reduce M_s so that $H_k > 4\pi M_s$; M_s depended linearly on the chromium content. The coercivity measured perpendicular to the film plane was 1500 Oe (119.4 kA/m) and in the plane, 500 Oe (39.8 kA/m). The recording properties of these films were reported by Iwasaki and Nakamura (1977) out to 45 000 flux changes per inch (1772 fc/mm).

The interest in films having the easy axis perpendicular to the film plane is that it offers the possibility of decoupling bit density and demagnetization. In conventional films having the direction of magnetization. The resulting increase in demagnetization factor calls for a reduction in film thickness or an increase in coercivity or a decrease in remanent moment. In films having perpendicular anisotropy the direction of magnetization is perpendicular to the film plane and consequently an increase in bit density results in a reduction of the demagnetizing factor. It is also possible that the contribution to loss of signal that arises from the separation between the head and the medium may also be reduced with this geometry.

In order to take advantage of these interesting films, it is necessary to redesign the write and read heads. This question was recently addressed by Iwasaki and Nakamura (1977).

2.2. Oxide films

Continuous thin films made from the oxides of the ferromagnetic metals are unreactive and consequently their composition and properties are much more stable than those of the metal films. Because of the high temperatures involved in their preparation, oxide films are unsuitable for use with most flexible substrates. However, they adhere well to metals, glasses, and ceramics and are usually more wear resistant than are the metal films. Oxide films have two advantages over recording surfaces made of oxide particles immersed in a polymeric binder. The thinnest particulate coating which can be made by conventional techniques has a thickness of about 1μm but continuous oxide films may be made very much thinner than this. Furthermore, since the oxide loading is 100% in the films and is seldom more than 40% (by volume) in particulate coatings, the reduced film thickness (and improved signal resolution) can be achieved without sacrifice of magnetic moment.

Four methods have been used to make oxide films. They are (1) the vacuum deposition of an iron film followed (or accompanied) by its oxidation (2) the chemical deposition of compounds of iron which are then oxidized (3) chemical vapor deposition of the oxide and (4) reactive sputtering.

Schneider and Stoffel (1973) evaporated iron at the rate of about 30 Å/s onto a substrate which was held at 220°C. An oxygen partial pressure of 5×10^{-5} torr was maintained. The resulting films were found, by X-ray diffraction, to be of magnetite and to have magnetic properties which were independent of film thickness in the range 500–4000 Å. The saturation magnetization of the films was close to the bulk value for magnetite of 485 emu/cc (0.609 Wb/m^2) at room temperature. Coercivities were generally in the range 400–500 Oe (31.8–39.8 kA/m). In a modification of this method, Schneider et al. (1973) altered the deposition conditions intermittently by interrupting the flow of oxygen or by increasing the deposition rate when the gettering action of the iron caused the pressure to drop and a pure iron layer to be formed. In this way they were able to make films composed of alternating layers of magnetite and iron having $H_c = 420$ Oe (33.4 kA/m) and $M_s = 800$ emu/cc (1 Wb/m^2). By a similar process layered films of Co/CoO were made and gave $H_c = 430$ Oe (34.2 kA/m) and $M_s = 950$ emu/cc (1.19 Wb/m^2).

In the iron–oxygen system hematite, α-Fe$_2$O$_3$ is the stable compound, below 560°C, in the presence of excess oxygen and magnetite, Fe$_3$O$_4$, is stable when there is excess iron. Feng et al. (1972) used this fact in the preparation of thin films of magnetite. They first evaporated ($< 10^{-6}$ torr) a film of iron then oxidized it to hematite in oxygen at 450–550°C. The hematite was then covered with a second layer of pure iron. The layered film was annealed at 350–400°C in a vacuum better than 10^{-7} torr and the hematite was converted to a magnetite film which was thicker by a factor of 2.34 than the initial iron film. X-ray diffraction and α-particle back scattering revealed that the final film was magnetite. The saturation magnetization was about 87% of the bulk value. On silicon substrates coercivities were between 300 and 400 Oe (23.8–31.8 kA/m) but the films were rough (0.1μm) as a result of the oxidation step; Ahn et al. (1976), and Nose et al. (1970) were able to make films of γ-Fe_2O_3 by evaporating from particles of γ-Fe_2O_3 which were dropped onto a heated tungsten wire (1470–1700°C). The vacuum was 5×10^{-5} torr before and 2×10^{-4} torr during evaporation and the rate was 150 Å/min. Films of pure γ-Fe_2O_3 of thicknesses 600–3500 Å were obtained.

Films of iron prepared by evaporation can also be oxidized in solution. Inagaki et al. (1975a, b, c) evaporated iron at a pressure of 10^{-5} to 10^{-7} torr and a rate of 10 Å/s on an alumina substrate at 300°C. The film, 0.75μm thick, was then heated at 90°C in a solution of sodium hydroxide containing sodium nitrite as oxidizing agent. The magnetite film was sintered for three hours at 1200°C in air and converted to hematite, then heated for a similar time in hydrogen at 300°C and finally re-oxidized at 200°C to give a magnetite film. The magnetic properties were sensitive to the initial annealing temperature (see table 3).

The process of the formation of oxides by oxidation of metallic films was studied by Engin and Fitzgerald (1973). They used electron diffraction and found that pure iron films oxidized at temperatures between 300 and 700°C, formed hematite and a spinel, while above 900°C only hematite was formed. Between 700 and 900°C the diffraction patterns were indistinct and could not be identified. For cobalt films at 200°C, CoO and Co_3O_4 were formed and at temperatures greater than 300°C only Co_3O_4 was found. Films of $CoFe_2$ oxidized between 300 and 650°C formed a mixture of Co_3O_4, α-Fe_2O_3 and a spinel. Above 650°C only cobalt ferrite was found.

It is possible to make continuous films of magnetite or of γ-Fe_2O_3 by the same sequence of steps which are commonly used to prepare particles or iron oxide. Melezoglu (1972) sprayed an aqueous solution of ferric chloride and hydrochloric acid onto a glass or ceramic substrate heated to 350–700°C. A coating of hematite was formed and then reduced in hydrogen at 400°C to form magnetite. The final step was the re-oxidation to γ-Fe_2O_3 by heating in oxygen at 250°C.

Comstock and Moore (1974) made hematite and cobalt-modified hematite films by spin-coating a diluted solution of the metal nitrates on a glass or titanium substrate and then heating in air to 300°C. The films were converted to ferrite by

TABLE 3

Magnetic properties of a thin film of oxidized iron, as a function of the initial annealing temperature (after Inagaki et al. 1975a,b,c)

Annealing temperature (°C)	H_c		M_s		S
	(Oe)	(kA/m)	(emu/cc)	(Wb/m²)	
1000	400	31.8	382	0.480	0.63
1100	480	38.2	342	0.430	0.58
1200	550	43.8	302	0.380	0.61

being heated in a wet hydrogen atmosphere. Magnetite films whose coercivities ranged from 500–588 Oe (39.8–46.8 kA/m) were obtained. The saturation magnetization reached 353 emu/cc (0.443 Wb/m^2) and the squareness was 0.82.

Oxide films have also been made by chemical vapor deposition followed by oxidation. Metal compounds were chosen that do not decompose when vaporized; examples are ferrocene $Fe(C_5H_5)_2$ iron pentacarbonyl (Halaby et al. 1975). On heating ferrocene powder to at least 150°C in nitrogen the molecules are transported to a glass ceramic or metal substrate which was heated to 450°C. This temperature is high enough to cause the ferrocene to decompose and a film of iron is built up. The film is oxidized to hematite and in the next step, the conversion of the hematite film to magnetite, the partial pressure of oxygen in the atmosphere is critical. An iron film 1900 Å thick was heated in air at 450°C for an hour to oxidize the iron to hematite. Then the hematite was converted in an oxidation–reduction atmosphere of H_2, H_2O and N_2 at 525°C.

Steck et al. (1976) also decomposed iron pentacarbonyl onto a heated substrate (150–300°C) to form an adherent crystalline layer of hematite which they then immersed in an ether solution of palladium acetylacetonate. After evaporating the solvent the hematite film was reduced in a stream of hydrogen and water vapor at 280°C, to form a black magnetite film 0.7μm thick. The coercivity of the film was 313 Oe (24.9 kA/m) and the squareness was 0.47.

Similar films of $Co_xFe_{3-x}O_4$ were made by Chen and Murphy (1975) using iron pentacarbonyl and cobalt nitrosyltricarbonyl. A film 3000 Å thick of a nonmagnetic cobalt–iron oxide was grown on a heated substrate and then reduced to $Co_xFe_{3-x}O_4$. It is also possible to disperse a powder, for example, magnetite, in screening oil and to coat the dispersion on substrate which is then heated to the vaporizing temperature of the oil leaving a metal oxide layer. Herczog et al. (1975) then heated the glass substrate to its softening point (~700°C) for a time long enough to allow the film to bond to the glass.

Most recently, reactive sputtering has been used to make ferrite thin films. Inagaki et al. (1976) used a target of Fe–Al–Co and an atmosphere of argon and oxygen. The total pressure was 2×10^{-2} torr and stable films of hematite were obtained when the partial pressure of oxygen exceeded 2.5×10^{-3} torr. A sputtering rate of less than 34 Å/min was used to prepare films whose thickness was 0.1 to 0.25μm. The hematite film was reduced in hydrogen at 300°C to magnetite. Films containing 10 at.% Al and 1.5 at.% Co were found to give better magnetic properties than were obtained with pure magnetite. They reported the magnetic properties as follows: $H_c = 500$ Oe (39.8 kA/m), $M_r = 207$ emu/cc (0.26 Wb/m^2) and squareness, $S = 0.75$. The hardness of the aluminum substrate was improved by anodizing it in sulfuric acid before sputtering.

Cycling the oxygen partial pressure between 10^{-4} and 10^{-3} torr Heller (1976) was able to sputter alternate layers of Fe and α-Fe_2O_3 respectively. The resulting films had either a layered structure or were homogeneous depending on the cycle length. A film composed of eight iron layers and eight layers of Fe_2O_3, each layer being 150 Å thick, had a coercivity of 525 Oe (41.8 kA/m).

A method of epitaxially sputtering iron oxide at temperatures as low as 200°C

TABLE 4

Properties of some magnetically hard films

Metal	σ_s (emu/g)	σ_r/σ_s	H_c (Oe)	Film thickness, structure and orientation
Electrodeposition				
Co, 2%P	100	0.5	400	1–30 μm, platelets 3 × 0.5 μm, (110) ∥ film plane
Co, 3.5%P	99	0.5	1300	1–30 μm, platelets 3 × 0.5 μm, (002) ∥ film plane
Autocatalytic				
Co, 2–3%P	112	0.8	<330	<300 Å, hcp crystallites 200–700 Å, c-axis ⊥ film plane
Vacuum deposition				
Fe	200	0.85–0.90	~1000	~500 Å, hcp crystallites 200–700 Å, c-axis random in plane
		0.90–0.95	270	~500 Å, bcc crystallites 80–100 Å, easy-axis ⊥ incident plane
			>800	~500 Å, bcc crystallites 1000 Å, easy-axis ∥ incident plane
Sputtering				
Co, 13% Cr	60	0.90	1500	8000 Å, hcp crystallites 300 Å, c-axis ⊥ film plane
Oxide				
Fe₃O₄	41–68	0.80–0.82	500–588	1000–1500 Å, inverse spinel
(Fe₀.₈₈Co₀.₀₁Al₀.₁₁)₃O₄	58–73	0.75	550–400	0.6–1.5 μm, inverse spinel, 400–300 Å

was described by Ahn et al. (1976). They first sputtered bcc vanadium or chromium onto the substrate at 200°C. The metal film had a thickness of $0.2\mu m$ and a (110) texture which encouraged the formation of magnetite (or of γ-Fe_2O_3) during the subsequent sputtering from an iron oxide source. The films had a thickness of about $0.1\mu m$ and were magnetically isotropic with $H_c = 500$ Oe (39.8 kA/m). Ahn et al. claimed that their sputtering technique required less critical control of the atmosphere than is necessary when reactive sputtering is used.

The advantages of oxide thin films are that it is possible to make much thinner coatings than can be obtained by means of oxide particles in a polymeric binder. Furthermore, the film is pure oxide whereas in particulate coatings the maximum volume packing fraction is about 40%. The high temperature involved in the preparation of oxide films is a serious disadvantage and restricts their possible use to rigid disks having metallic, glass, or ceramic substrates.

The properties of representative metal films and oxide films are summarized in table 4.

3. Particles

The objective is to make particles with a coercivity high enough to successfully resist demagnetization, yet not so high that magnetization cannot readily be reversed by the field from the writing or erasing heads. At the same time the magnetization should be in the same range as the ferromagnetic metals to provide adequate flux at the reading head. In practice, the coercivities are confined to the range 250 Oe to 1200 Oe (19.9–95.5 kA/m) with most tapes and disks having coercivities between 250 and 600 Oe (19.9–47.8 kA/m). This means that the particles generally are single-domains. But it must be remembered that all manufacturing processes yield a range of particle shapes and sizes and that while most of the particles may be single-domain there will inevitably be some particles small enough to be superparamagnetic and some multidomain particles in any sample. The aim of the particle maker is to maximize the number of single-domain particles and usually most particles will be larger than a few hundred Ångstroms but smaller than a micrometer. Since, even for single-domain particles, the extrinsic magnetic properties such as coercivity depend upon the size of the particle, this must be controlled to lie within the closest possible limits. It is a further requirement that particles should be dispersed (or dispersible) and not bonded or sintered together. The composition, structure, and magnetic properties of the particles must be stable with respect to time and the environment (temperature, pressure, and contaminants) and the surface chemistry of the particles must be compatible with that of the binder.

Particles suitable for use in tapes and disks have been made of iron oxides (γ-Fe_2O_3, Fe_3O_4), cobalt-modified iron oxides, barium ferrite (for special applications), chromium dioxide, and a variety of metal particles. The methods of preparation and the structure and properties of these particles will now be described.

3.1. Iron oxide particles

The earliest particulate magnetic tapes were made of iron oxide and today it is still the most widely used magnetic material for disks and tapes. It is very cheap and can be made into recording surfaces which perform well in audio, video, instrumentation, or digital applications. When iron oxide is prepared in the form of acicular particles of γ-Fe$_2$O$_3$, of length approximately 0.1 to 0.7 μm, and axial ratio usually 3:1 to 10:1, the shape anisotropy confers a coercivity which is in the range 250–400 Oe (15.9–31.8 kA/m). The coercivity and a saturation magnetization of about 400 emu/cc (0.503 Wb/m^2) lead to a tolerable demagnetization and thus to a useful output at all but the highest recording densities.

3.1.1. Preparation of iron oxide particles

a. The basic reactions

The most common method by which iron oxide particles are prepared involves five steps. These are:
1. The preparation of small acicular seeds of α-FeO)OH,
2. The growth of α-(FeO)OH on the seeds,
3. The dehydration of α-(FeO)OH to α-Fe$_2$O$_3$,
4. The reduction of α-Fe$_2$O$_3$ to Fe$_3$O$_4$,
5. The oxidation of Fe$_3$O$_4$ to γ-Fe$_2$O$_3$.
(Steps 3 and 4 are usually combined).

Naturally occurring Fe$_3$O$_4$ (magnetite) and γ-Fe$_2$O$_3$ (maghemite) particles are usually equiaxed and this is not surprising when we remember that their crystal structures are basically cubic. The first two steps in the process were originally used to make yellow pigment of the orthorhombic α-(FeO)OH. As this led to the needle-shaped particles needed for recording materials, the only remaining task was the conversion of α-(FeO)OH (goethite) to γ-Fe$_2$O$_3$ without substantially changing the shape and size of the particles.

The yellow pigment can be made by a process described by Penniman and Zoph (1921) and by Camras (1954). Ferrous sulfate is put in agitating tank with water and excess dilute sodium hydroxide solution is added. The solutions are continuously agitated so that a fresh surface is continually exposed to the atmosphere. Colloidal seed crystals of α-(FeO)OH are precipitated:

$$4\text{FeSO}_4 \cdot 7\text{H}_2\text{O} + \text{O}_2 + 8\text{NaOH} \rightarrow 4\alpha\text{-(FeO)OH} + 4\text{H}_2\text{SO}_4 + 30\text{H}_2\text{O}$$

The seeds are then used as nuclei in the growth of larger (length 0.1–1 μm) crystals of synthetic goethite. The progress of the seed growth is marked by colour changes in the mixture; dark blue to green to brownish yellow to yellowish tan, at which point the growth is complete. The seed material, more ferrous sulfate, and pieces of iron are mixed with water in an agitating tank and heated to 60°C. Air is bubbled through the mixture for several hours and more goethite grows on the colloidal nuclei to produce pale yellow pigment particles of goethite. The size and shape of these particles are very similar to those of the end product, γ-Fe_2O_3. The sulfuric acid produced in the process reacts with the iron to produce more ferrous sulfate. The product has a composition and structure which are predominantly those of the mineral goethite (the most important iron ore after hematite, α-Fe_2O_3). However, Van Oosterhout (1965) showed that structurally there is actually a whole range of synthetic goethites which have more or less a defective structure when compared with the natural mineral. Goethite was considered to be paramagnetic, (Selwood 1956) and to obey Curie's law ($1/\chi \propto T$). But Hrynkiewicz and Kulgawchuk (1963) showed by Mössbauer experiments, and Van Oosterhout (1965) showed by this method and by the temperature dependence of susceptibility, that the material is actually antiferromagnetic with $T_N \sim 170°C$.

Yamamoto (1968) prepared four samples of α-(FeO)OH; one by air oxidation of a suspension of metallic iron, and the other three by dissolving ferric nitrate, $Fe(NO_3)_3 \cdot 9H_2O$, in water and precipitating with sodium hydroxide. Samples of the brown precipitate were hydrolyzed at 80, 95 and 130°C and the amorphous precipitate was converted into α-(FeO)OH. The Néel temperature was determined by measurements on the internal field from Mössbauer spectra and on the temperature dependence of susceptibility. It was found that the T_N was dependent upon particle size ($T_N = 122°C$ for $0.6\mu m$, 92°C for $0.2\mu m$). Yamamoto (1968) suggested that the change in interactions brought about by the proportionately larger surface area of the smaller particles might be the cause of this dependence.

The third and fourth steps in the production of particles of γ-Fe_2O_3 are usually combined. They are the dehydration of the goethite particles to hematite and the reduction to magnetite.

$$2\alpha\text{-(FeO)OH} \rightarrow 2\alpha\text{-}Fe_2O_3 + H_2O$$
$$3\alpha\text{-}Fe_2O_3 + H_2 \rightarrow 2Fe_3O_4 + H_2O$$

The reduction was completed when no more water vapor was evolved. Once again the equations represent a very simplified form of the chemistry. For example, Van Oosterhout (1965) believed that the iron oxide at this stage contains more Fe_2O_3 than is represented by the formula $(FeO) \cdot Fe_2O_3$.

The fifth and final step in preparing iron oxide particles is the careful oxidation of magnetite to γ-Fe_2O_3 (maghemite). A balance must be found between too low an oxidation temperature (long reaction time) and too high a temperature which would lead to the formation of α-Fe_2O_3. Camras (1954) reported that if particles of γ-Fe_2O_3 were again reduced to Fe_3O_4 and then reoxidized at about 200°C, a powder having improved magnetic properties would be obtained: $H_c = 400$ Oe (32 kA/m) cf. 260 Oe (21 kA/m) before re-reduction and reoxidation. Figure 7a (see section 3.1.4) shows an electron micrograph of particles of γ-Fe_2O_3 prepared by the standard method.

b. Modifications to the basic reactions

The properties of the final product depend strongly on the size and shape of the particles which as we have seen, are determined at the first two steps; the preparation of the seed particles and the growth on those particles. It is, therefore, not surprising that these steps are the ones most commonly modified in attempts to make particles with superior properties (generally higher coercivity, and better uniformity in size and greater dispersibility). Thus, Umeki (1975) added nickel salts to ferrous sulfate to help in controlling the particle size, Matsui et al. (1975) added $Cr_2(SO_4)_3$ to the $FeSO_4$ used during the growth step and produced particles containing 0.2% Cr ($H_c = 450$ Oe, 36 kA/m), $\sigma_s = 43.5$ emu/g (54.6 μ Wb m/kg). Woditsch et al. (1975b) described the addition of zinc sulfate and phosphate irons in the first step ($H_c = 337$ Oe, 26.8 kA/m).

The third and fourth steps in the process for manufacturing particles of iron oxide involve dehydration and then reduction at high temperatures. It is here that sintering of the particles is most likely to occur unless steps are taken to prevent it. The modern approach is to coat the particles of α-(FeO)OH with one of a wide range of chemicals. These may be inorganic: Baronius et al. (1966) used sodium silicate; Horiishi et al. (1974) added $Cr_2(SO_4)_3$ solution to an aqueous dispersion of goethite and produced particles coated with 0.3 at.% Cr ($H_c = 395$ Oe, (31.4 kA/m), $\sigma_s = 78.5$ emu/g (98.6 μ Wb m/kg). More commonly, organic additives are used. For example, Woditsch et al. (1975a) and Amemiya and Asada (1976) used silicone oil; Mizushima et al. (1975a, b, c, d, e) used silane, phosphoric acid ester surfactants, an aqueous solution of sorbitol-fatty acid esters, sodium stearate or an alkyl titanate. Taniguchi (1975) recommended coating the goethite particles with aqueous polyvinyl alcohol before reduction to magnetite. Köster et al. (1976b) treated their particles with barium hydroxide solution and an alkyl phenol. In each case the authors claim improved dispersibility or coercivity or both as a consequence of their treatments.

Leutner and Schoenafinger (1975) annealed the goethite at 500–800°C before its reduction to magnetite and then annealed the magnetite for 20 min at 450°C under nitrogen and obtained roughly a 10% enhancement of coercivity and remanence. Yoshimura et al. (1976) found that temperature control during reduction was less critical if the operation was performed in a glass tube which was open at one end.

In addition to the problem of sintering which occurs at the high temperatures used during the dehydration and reduction stages, the rapid evolution of large amounts of water vapor causes the particles to develop a porous structure. This is very noticeable in the electron micrographs of many iron oxide samples. Several workers have attempted to moderate the severity of these steps. Umeno (1975) performed the reduction by bubbling nitrogen through ethyl alcohol and then passing it over the particles while keeping the furnace at 350°C. The γ-Fe_2O_3 particles made from the resulting magnetite had a coercivity of 430 Oe, 34 kA/m and $\sigma_s = 76.0$ emu/g (95.5 μ Wb m/kg). Sato et al. (1976a) also used an inert gas but in addition they mixed the acicular goethite particles with a solution of starch. Sato et al. (1976b) also used carbon monoxide to achieve more than 60% of the reduction and completed the process with a stream of hydrogen. Horiishi and Takedoi (1975d) produced acicular particles of magnetite with improved magnetic properties, $H_c = 435$ Oe (34.6 kA/m), by increasing the ratio of ferrous to ferric ions by dispersing the magnetite particles in an aqueous solution of ferrous salts or ferrous hydroxide and heating to 95°C for two hours.

The fifth and final step consists of gently oxidizing the magnetic particles to γ-Fe_2O_3. Too high a temperature causes the formation of α-Fe_2O_3 and temperatures no higher than 250°C are usually recommended. Matsui et al. (1975) advised heating the magnetite in a mixture of nitrogen and oxygen so that the partial pressure of oxygen could be controlled. Woditsch et al. (1972) reoxidized under nitrogen at 250–300°C and produced particles with remarkably high coercivity, 488 Oe (38.8 kA/m), $\sigma_r = 36.0$ emu/g (45.2 μ Wb m/kg). These highly-acicular particles were made from goethite which had been grown from conventionally prepared goethite seeds in the presence of phosphate ions. Camras (1954) reported that if particles of γ-Fe_2O_3 were reduced to magnetite and then reoxidized at a temperature of about 200°C, a powder having a coercivity of about 400 Oe (31.8 kA/m) was formed. Kajitani and Saito (1975) used 2-ethyl-hexyl aldehyde as the reducing agent in the second reduction step. Cyclical oxidation and reduction procedures of this kind are believed to be used in preparing oxides for recording applications, however the details of these processes are often closely-kept secrets. For example, Horiishi and Takedoi (1975e) dispersed particles of γ-Fe_2O_3 in an aqueous solution of ferrous salts (or ferrous hydroxide) at 50–100°C, pH = 8.0–14.0, in a non-oxidizing atmosphere and then re-oxidized to γ-Fe_2O_3. The first cycle produced an increase in the coercivity of the particles from 370 Oe to 420 Oe (29.4–33.4 kA/m) and the cycle could be repeated. The end product had an atomic ratio Fe^{3+}:Fe^{2+} of 1:0.23, that is, the composition was almost midway between Fe_2O_3 and Fe_3O_4 (= $FeO.Fe_2O_3$).

c. Iron oxide particles from synthetic lepidocrocite

The reactions described above use synthetic goethite, α-(FeO)OH as a first product. A different approach uses synthetic lepidocrocite γ-(FeO)OH as a starting material. Baudisch (1933a) described the preparation of gamma ferric oxide monohydrate by dissolving chemically pure iron (from the decomposition of the carbonyl) in hydrochloric acid and then adding an excess of the bases aniline or pyridine. Air was bubbled through the solution and caused the precipitation of γ-(FeO)OH. The dehydration step was carried out by gently heating the solution at 150–280°C and the end product was γ-Fe_2O_3, Baudisch (1933b). Similarly, Bratescu and Vitan (1969) used an aqueous solution of ferrous sulfate, a mixture of hydrochloric acid and nitric acid as oxidants, and ammonia as the precipitant. The black precipitate was heated to 60°C yielding a yellow-brown powder of γ-Fe_2O_3.

The method of making synthetic lepidocrocite first rather than synthetic goethite is now used to make particles of γ-Fe_2O_3 which look quite unlike the standard product. The particles are more acicular, they have far fewer dendrites and consequently show higher coercivity, and are considerably easier to disperse and to orient. They also are noticeably free from "blow holes". Ayers and Stephens (1962a, b) showed that if a mixture of ferrous chloride, water and sodium hydroxide were strongly agitated at 27°C, a colloidal seed was formed which could be grown into platelets of γ-(FeO)OH when zinc chloride was added as a means of inhibiting the growth of α-(FeO)OH. The crystalline particles were uniform in size (about 1μm) and were transparent. Using a similar process, Bennetch et al. (1975) bubbled an oxygen-containing gas through a more concentrated ferrous chloride solution mixed with an aqueous alkali to provide vigorous agitation at a temperature of 27°C. The agitation was continued until the pH reached 3 to 3.5. This is apparently critical to the desired product as is the ratio of the mass of the product to that of the seed (preferably about 2:1). As the seed particles grow, their acicularity becomes less pronounced, thus, by keeping the growth ratio much lower than the usual 5:1, a more acicular product is obtained. Particles of lepidocrocite, whose length to width ratio was 20:1 to 50:1, were obtained and surface treated with a mixture of morpholine and a hydrophobic aliphatic monocarboxylic acid (e.g. coconut oil fatty acid) at 80°C. This important treatment was done to prevent agglomeration and sintering of the particles during the dehydration and reduction steps. The monomolecular layer acts inherently as a reducing agent and so reduction in hydrogen gas is unnecessary and more perfect particles are formed. Many different monocarboxylic acids suitable as surface treatment agents were disclosed by Greiner (1970). Uehori et al. (1974) also used a solution of a fatty acid in acetone to treat particles of α-Fe_2O_3 before its reduction. The finished oxide had a coercivity of 400 Oe (31.8 kA/m) and $M_r = 268$ emu/cc (0.337 Wb/m^2). The particles were 2μm in length, had an average axial ratio of 9.3:1, and were densified in a ball or roller mill.

The critical nature of the high temperature steps (dehydration and reduction) in the successful preparation of recording materials is shown by the difficulties experienced by earlier workers in converting synthetic lepidocrocite into iron oxides with useful properties. Sykora (1967) described unsuccessful attempts to dehydrate lepidocrocite; he found that the product was significantly inhomogeneous, not very acicular and had a wide distribution of particle lengths. He concluded that the only practical way to make oxides for recording applications was the standard method. It was of course just these problems which were successfully overcome by Bennetch et al. (1975) by their use of reduction by means of a monomolecular layer. Imaoka (1968) also studied by X-ray diffraction and by magnetic measurements, the preparation of γ-Fe_2O_3 by dehydration of γ-(FeO)OH at temperatures from 200 to 400°C. He found that there was a pronounced change in the crystallite size (determined by broadening of the X-ray lines) as the γ-(FeO)OH (180 Å) was transformed. Initially, the crystallite size of the γ-Fe_2O_3 was 40 Å and grew to 95 Å when the dehydration temperature was increased to 400°C; no two-phase structure was observed at any time. Imaoka's (1968) results supported Sykora's (1967) contention that the magnetic properties of the particles were not good enough for recording purposes. Imaoka (1968) found $M_r/M_s \leqslant 0.35$ and $H_c \leqslant 134$ Oe (10.7 kA/m). In view of these properties, his explanation that the first particles produced were superparamagnetic seems reasonable but an alternative explanation is that the particles became agglomerated into multi-domains.

d. Berthollide oxides and other high-coercivity iron oxides

The standard method of making fine particles of γ-Fe_2O_3 is to precipitate seeds of synthetic goethite, α-(FeO)OH and then to use those seeds as nuclei in the growth of larger particles of goethite. The method results in acicular particles whose coercivity is usually in the range 250–300 Oe (19.9–23.9 kA/m). With the objective of increasing the coercivity many modifications to the standard method have been described in recent years and some examples of these modifications will now be discussed.

In the final step of the standard method, the oxidation of Fe_3O_4 to γ-Fe_2O_3, there is always the danger that α-Fe_2O_3 may be formed. There must also be the possibility of the incomplete conversion of magnetite and so particles ostensibly of pure γ-Fe_2O_3 may well contain small amounts of α-Fe_2O_3 and of Fe_3O_4. The transformations were studied by Imaoka (1968) who used magnetic and X-ray diffraction measurements. In the transformation γ-$Fe_2O_3 \rightarrow \alpha$-$Fe_2O_3$ he found that the coercivity of the sample increased steadily from 260 to 310 Oe (20.7 to 24.7 kA/m), while the saturation magnetization decreased from about 100 emu/cc (0.126 Wb/m^2) to near zero. The ratio M_r/M_s remained throughout at 0.4 to 0.5. In the reduction of γ-Fe_2O_3 to Fe_3O_4, the coercivity increased from 370 to 390 Oe (29.4 to 31 kA/m) and then decreased to about 240 Oe (19.1 kA/m) as the magnetization increased. Imaoka interpreted his results in terms of heterogeneous reactions and stresses which caused each particle to contain many crystallites (Bando et al. 1965). Each particle proceeds during growth through the stages; supermagnetic $\rightarrow$ single-domain $\rightarrow$ multi-domain. Camras (1954) had found that particles of γ-Fe_2O_3 reduced to Fe_3O_4 and then reoxidized at about 200°C had considerably improved magnetic properties; for example the coercivity before the reduction–oxidation cycle was 260 Oe (21 kA/m) and 400 Oe (32 kA/m) afterwards. Similar results were reported by Horiishi and Takedoi (1975e), who found that the end product had a ratio of Fe^{2+} : Fe^{3+} of 0.23 : 1, a composition that is almost halfway between Fe_2O_3 and Fe_3O_4. Imaoka (1968) had previously found that the peak in coercivity was achieved at Fe^{2+} : Fe^{3+} of 0.14 : 1.

Compounds in which the ratio of positive ions to negative ions is not fixed are called "Berthollide compounds" and thus the oxides of iron discussed above in which the composition is variable between Fe_2O_3 and Fe_3O_4 are described as "Berthollide oxides". The highest coercivities found today in commercial iron oxides are usually Berthollide oxides. Coercivities up to almost 400 Oe (31.8 kA/m) are offered and even higher values have been reported in the literature. For example, Horiishi and Takedoi (1975d) dispersed particles of γ-Fe_2O_3 in a solution of ferrous sulfate and reduced to a composition between Fe_2O_3 and Fe_3O_4; the reported coercivity was 420 Oe (33.4 kA/m). Kajitani and

Saito (1975) found that reducing γ-Fe$_2$O$_3$ to Fe$_3$O$_4$ with an aldehyde and then reoxidizing gave a coercivity of 405 Oe (32.2 kA/m). Berthollide oxides are also used as starting material in the preparation of cobalt-impregnated iron oxides. This topic will be discussed in section 3.2.4.

It has also been possible to make particles of small diameter and high acicularity without the dendrites and voids which are so common in the older particles of iron oxide. The new particles are more easily dispersed and oriented. Indeed the highest orientation ratios now being reported are obtained with iron oxide particles rather than chromium dioxide particles as was the case prior to about 1975. In addition to the use of Berthollide oxides the techniques employed usually involve the treatment of the product precursors either at the stage of the oxyhydroxide or at the magnetite stage.

The oxyhydrides, goethite or lepidocrocite are often coated with organic compounds (Bennetch et al. 1975, Uehori et al. 1974) to reduce the tendency to agglomeration during the later stages. Amemiya and Asada (1976) treated their goethite particles with silicone oil containing methyl ethyl ketone and ultimately obtained particles of γ-Fe$_2$O$_3$, having a coercivity of 480 Oe (38.2 kA/m). Woditsch et al. (1972) grew their goethite particles in the presence of phosphoric acid. The end product was γ-Fe$_2$O$_3$ particles with a coercivity of 488 Oe (38.8 kA/m). Itoh and Satou (1975b, 1976) also introduced the phosphate ion and obtained the remarkable value of 600 Oe (52.5 kA/m) for the coercivity of the resultant oxide. Unfortunately on heating the product to 100°C and then keeping the powder in a dessicator for 20–30 days at room temperature, the coercivity gradually decreased to the same value as shown by particles not treated with sodium metaphosphate. However, the experiment is useful in that it suggests that the condition of the surface of the particle is important in determining the coercivity.

Pingaud (1972, 1974) stabilized iron oxide particles against transformation to α-Fe$_2$O$_3$ by adding ions of sodium or calcium to the goethite. The transition temperature was increased to 700°C and Pingaud also found that the particles had a coercivity of 370 Oe (29.5 kA/m).

The treatment of the particles at the magnetite stage has been directed mainly at eliminating the pores which are so noticeable in electron micrographs. The argument appears to be that pores encourage an incoherent mode of magnetization reversal and thereby reduce the coercivity. Leutner and Schoenafinger (1975) annealed their magnetite particles and found that the coercivity of the final Fe$_2$O$_3$ particles was increased from 352 to 381 Oe (28 to 30.3 kA/m). Horiishi and Takedoi (1975a) treated magnetite particles with ferrous sulfate solution and increased its coercivity from 410 to 435 Oe (32.6 to 34.6 kA/m).

Three other variants on the standard process were described by Sato et al. (1976a), Sato et al. (1976b) and Sato et al. (1976c). They involved (1) dehydrating the goethite mixed with a carbohydrate, (2) reducing the hematite to more than 60% Fe$_3$O$_4$ in carbon monoxide and then completing the reduction in hydrogen and cooling in a stream of nitrogen, and (3) heating the goethite in sodium hydroxide solution. In each case the magnetite was oxidized to γ-Fe$_2$O$_3$ and the coercivity of the powders was

Process	Fe_3O_4		$\gamma\text{-}Fe_2O_3$	
	(Oe)	(kA/m)	(Oe)	(kA/m)
1	415	33.0	370	29.5
2	440	35.0	390	31.0
3			410	32.6

Horiishi and Fukami (1975) precipitated $Fe(OH)_2$ with ferrous sulfate and sodium hydroxide and then treated the precipitate with ammonium bicarbonate to produce $FeCO_3$ which was then converted by warming in air to $\alpha\text{-}(FeO)OH$. The rest of the process was standard and gave particles with coercivities

Fe_3O_4	375 Oe	29.9 kA/m
$\gamma\text{-}Fe_2O_3$	380 Oe	30.2 kA/m

Yada et al. (1973) reported measurements on new high-coercivity particles of $\gamma\text{-}Fe_2O_3$ (450 Oe, 35.8 kA/m), and described the effect of milling time on coercivity but unfortunately they gave no information on the method of preparation other than that the objectives were to make particles of uniform, regular shape and with fewer pores and dendrites.

These examples show very clearly that every aspect of the standard method has been the subject of intense investigation in the pursuit of ever higher coercivities. In contrast the lepidocrocite process for producing highly-acicular, dendrite-free particles of $\gamma\text{-}Fe_2O_3$ (Bennetch et al. 1975) has so far yielded coercivities which while they are higher than those of the earlier standard iron oxide particles, are considerably below the values obtained by modifications to the standard process. Thus Bennetch et al. (1975) claim $H_c = 310$ Oe (24.7 kA/m) and Gustard and Wright (1972) only found $H_c = 290$ Oe (23.1 kA/m) for this material.

We can expect continued efforts to improve the shape and the uniformity of iron oxide particles and to enhance their coercivity. The motivation is that iron oxide particles seem to be free from the problems of abrasivity, corrosion and stress-, temperature- and time-dependence of magnetic properties, which are troublesome in the cobalt-modified iron oxides, chromium dioxide and the metal particles.

e. Modified acicular particles

A variation in the starting material was proposed by Dovey et al. (1954) and Dovey and Pitkin (1954) who began with finely divided iron oxalate or iron formate. Since the ensuing reactions, like those described above, are pseudomorphic, it was important to select particles of the oxalate or formate that were uniform and less than $1\mu m$ in diameter.

The first step was to heat the oxalate or formate at about 450°C in an atmosphere of gases that would neither oxidize nor reduce appreciably the ferrosoferric oxide (Fe_3O_4) formed at this stage. Suitable atmospheres were mixtures of steam, carbon dioxide and nitrogen; this part of the process took several hours, the exact time depended on the amount and the distribution of material in the reaction vessel. The completion of the decomposition was marked by a color change from light green or yellow to black.

The final step, involving the controlled oxidation of $Fe_3O_4 \rightarrow \gamma\text{-}Fe_2O_3$ has already been described. Dovey and Pitkin (1954) found it convenient to mix the cooled Fe_3O_4 with enough water to dampen it thoroughly and then to dry the dampened oxide in air in an oven maintained at a temperature between 100 and 200°C. They found that the water had the effect of preventing an unwanted rise in temperature during the oxidation and also appeared to have a catalytic effect so that oxidation took place at a lower temperature than was the case for dry Fe_3O_4.

A process whereby magnetite particles might be grown directly on goethite seeds was described by Fukuda et al. (1962a, b). The seeds could be grown by the method described by Camras (1954). To these seeds were added a mixture of ferric chloride and ferrous sulfate in such proportions that the ionic ratio $Fe^{2+} : Fe^{3+} = 1 : 2$ was the same as that of magnetite. The mixture was then heated in a sealed vessel containing nitrogen at 50°C. Sodium hydroxide solution was dripped into the vessel over a period of five hours while the mixture was vigorously agitated. The temperature was then raised to 80°C and held there for one hour. During this time magnetite shells grew around the goethite nuclei to form spindle-shaped particles of an average length of $0.7\mu m$ and width of $0.15\mu m$. The particles were either made directly into recording media or were first converted to $\gamma\text{-}Fe_2O_3$. It is interesting that seeds other than goethite were also used. For example, acicular particles of zinc oxide having adsorbed colloidal magnetite, or indeed any needle-shaped particle having a thin coating of iron oxide, can be used. Improved magnetic properties (H_c, M_r and the slope of the initial magnetization curve) were claimed but it is not known whether the particles have ever been used in commercial tapes or disks. Spindle-shaped particles were also made by Koshizuka et al. (1975) by treating the ferrous sulfate with ammonium or sodium carbonate.

f. Non-acicular particles of iron oxide

Square platelets. The methods of Ayers and Stephens (1962a, b) for making plate-like particles of length:width about 10:1 and width:thickness about 5:1 have already been described. Goto and Akashi (1962) described a method for making square, plate-like particles of $\gamma\text{-}Fe_2O_3$. Sodium hydroxide (in the presence of nitrogen) was added to an aqueous solution of ferrous sulfate. Ferrous hydroxide was precipitated and heated in the mother liquor for about five hours. This produced flattened and substantially square particles. The size of these particles was, on average, $1 \times 1 \times 0.1\mu m$; the shape being ascribed to layered structures of Fe^{2+} and OH^- ions in which there was a weak bond between the OH^- layers.

Two methods were used to convert the platelets to $\gamma\text{-}Fe_2O_3$; the first was the conventional process:

$$Fe(OH)_2 \rightarrow \alpha\text{-}Fe_2O_3 \rightarrow Fe_3O_4 \rightarrow \gamma\text{-}Fe_2O_3$$

The second method involved heating the particles in a closed vessel at 300°C.

In the coating process to make disks or tapes, drying occurs and causes a reduction of the coating thickness. During this process the platelets are pulled into planes parallel to the substrate but are randomly oriented about the normal to the substrate. Thus, we should expect isotropic magnetic properties in the plane of the coating and, therefore, isotropic recording properties. This would clearly by an advantage in flexible disks where residual orientation produced by the mechanics of the coating process produces an undesirable modulation of the signal with conventional acicular particles. The origin of the magnetic hardness of these particles is not clear since they approximate oblate spheroids which should have no hysteresis; Stoner and Wohlfarth (1948). The use of flat platelets (of diameter 1–40μm and diameter: thickness = 10–20:1 was recently described by Hirota (1976) who claimed that they produced coatings of high packing density and uniformity and of good magnetic properties.

Hexagonal platelets. Okamoto and Baba (1969a, b) described how hexagonal platelets could be prepared. They also introduced the possibility that an iron oxyhydroxide might have magnetic properties which would make it suitable for some high-density recording applications. We have already seen how synthetic goethite [α-(FeO)OH] and synthetic lepidocrocite [γ-(FeO)OH] can be made as intermediate compounds in the preparation of Fe_3O_4 and γ-Fe_2O_3; Okamoto and Baba (1969a, b) made δ-(FeO)OH. A mixture of an aqueous solution of ferrous sulfate and strong sodium hydroxide precipitated ferrous hydroxide in the alkali solution. The dispersed precipitate was heated under non-oxidizing conditions in an autoclave filled with nitrogen for one hour at 100°C to promote grain growth of the precipitated particles. After cooling, the ferrous hydroxide particles were rapidly oxidized with hydrogen peroxide at 45 to 50°C and the reddish-brown precipitate was washed, dried and identified by X-ray analysis as δ-(FeO)OH. The mean particle size was 0.6μm and the magnetic properties quoted were $H_c \sim 500$ Oe (39.8 kA/m) and a saturation moment density of 25 emu/g (31.4μWb m/kg). In comparison, if the heat-treatment step was omitted the properties were $H_c = 55$ Oe (4.4 kA/m), $\sigma_s = 22$ emu/g (27.6μWb m/kg) for particles whose average size was 0.007μm.

The relationships between the different phases of the iron hydroxide system were summarized by Bernal et al. (1959), who asserted that δ-(FeO)OH is produced by the very rapid oxidation of ferrous hydroxide. Greenblatt and King (1969) measured the Mössbauer spectra of magnetic hydroxides of iron that had been precipitated by porous silica and concluded that γ-(FeO)OH could be converted to δ-(FeO)OH by heating in air to 350°C.

To make oxide particles, Okamoto and Baba (1969a, b) precipitated δ-(FeO)OH as described above except that the heat-treatment step was omitted. The particles were reduced at 300°C in a stream of nitrogen containing a slight amount of hydrogen and water vapor to give magnetite which was then reoxidized to produce γ-Fe_2O_3. This was obtained in the form of particles having a hexagonal, plate-like appearance in the electron microscope and a mean size of $0.07\,\mu m$.

Delta ferric oxyhydroxide was first prepared by Glemser and Gwinner (1939), who examined the structure by X-ray diffraction powder methods and believed it to be δ-Fe_2O_3. They prepared it by reacting a strong solution of sodium hydroxide with a solution of a ferrous salt such as ferrous sulfate. The resulting flocculent precipitate was oxidized with hydrogen peroxide. The brown powder is somewhat unstable and must be carefully dried at a temperature below 80°C to avoid decomposition to α-Fe_2O_3. Bernal et al. (1959) found that δ-(FeO)OH was completely transformed to α-Fe_2O_3 by heating at 200°C.

Cubic particles. These were the first iron oxide particles to be used for magnetic recording but their coercivity was far too low [75 to 150 Oe (6.0 to 12.0 kA/m)] when compared with that of modern particles. The method of preparation will be described since with the same modifications, it can be used to prepare cubic particles in which 2–10% of the iron ions are replaced by cobalt. The cobalt-substituted iron oxide particles have high coercivities because of the high magnetocrystalline anisotropy introduced with the cobalt.

Krones (1955) precipitated ferrous hydroxide from ferrous sulfate with sodium hydroxide and oxidized the precipitate to magnetite to 70 to 90°C with sodium nitrate which was chosen because its oxidizing potential was insufficient to form α-Fe_2O_3. The magnetite powder was oxidized in air at 250° to 300°C to give cubic particles of γ-Fe_2O_3. It was possible to exercise some control over the size of the particles (within the range 0.05 to $0.3\,\mu m$) by varying the conditions of precipitation such as concentration, pH and temperature. Although the particles appear as deformed spheres in the electron microscope, the length:width is probably greater than the value of 1.1:1 at which Osmond (1954) deduced that the magnetocrystalline anisotropy and shape anisotropy were equal in γ-Fe_2O_3.

g. Additional methods

Methods of preparing particles of γ-Fe_2O_3 which did not transform to α-Fe_2O_3 at temperatures below 600°C were described by Pingaud (1972) and by Adams et al. (1975). The technique was to include between 0.1 and 5 wt% of sodium or calcium ions in the γ-Fe_2O_3. Adams et al. (1975) believed that the cations enter the lattice and stabilize it. Itoh and Satou (1975a) prepared particles whose composition was between that of Fe_3O_4 and γ-Fe_2O_3 by incorporating phosphorus and sodium into the particles. Adding 0.25 mole P per mole Fe_3O_4 produced particles whose coercivity was 600 Oe (47.8 kA/m).

A preparation method which allowed very close control over the size distribution of the particles was described by Horiishi and Fukami (1975). An aqueous solution of ferrous sulfate was reacted with sodium hydroxide to form the colloidal hydroxide. The colloid was then treated with ammonium bicarbonate to precipitate $FeCO_3$ which was converted first to α-Fe_2O_3 (by heating in air at 790°C), then to Fe_3O_4 (330°C in steam) and finally to γ-Fe_2O_3 (280°C in air). Mean particles diameters in the range 0.1 to $1\,\mu m$ with a standard deviation of less than 3 nm were claimed.

Finally, what must be the cheapest method of making particles of iron oxides suitable for magnetic recording uses was described by Nakano (1974). He began with the scale obtained in rolling steel. This was washed, separated magnetically, powdered in a ball mill and mixed with an oxidizing agent, ammonium nitrate. The powder, now converted to Fe_3O_4, had a mean particle diameter of $0.06\mu m$ and a coercivity of only 20 Oe (1.59 kA/m). However, after heating the powder to 170°C for 5 h a coercivity of 359 Oe (28.6 kA/m) was obtained with a remanence of 100 emu/cc (0.126 Wb/m²).

h. Reactions

The principal methods of preparation and reactions of iron oxides are shown in fig. 5.

Precipitation of the seed. The importance of small variations in the conditions under which the initial seed material is precipitated is shown by a comparison of the processes described by Camras (1954), Ayers and Stephens (1962a, b) and Bennetch et al. (1975). All use ferrous sulfate (Bennetch et al. prefer ferrous chloride but claim that their process works with the sulfate) and all precipitate with sodium hydroxide solution while bubbling air vigorously through the mixture. Camras produced goethite [α-(FeO)OH], and Bennetch et al. made highly acicular particles of lepidocrocite. Camras held his mixture at 60°C while the others used 27°C. Furthermore the concentrations of ferrous sulfate and sodium hydroxide which Camras used were at least twice those recommended by Ayers and Stephens and by Bennetch et al. We must suppose that these differences are enough to result in the formation of goethite in one case and lepidocrocite in the others. It is more difficult to discover differences between Ayers' and Stephens' method and that of Bennetch et al. The differences appear to be that Bennetch et al. used a concentration of ferrous salt which was twice that used by Ayers and Stephens; both used the same concentration of sodium hydroxide. Bennetch et al. also stressed the importance of close control of pH.

Krones (1955) also used ferrous sulfate and sodium hydroxide and precipitated ferrous hydroxide which was then used as seed material for the growth of cubic

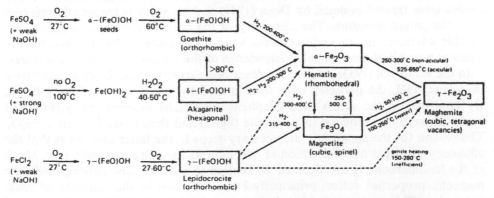

Fig. 5. The principal methods of preparation and the reactions of iron oxides.

particles. However, Krones did not bubble air through his mixture. The effect of strong agitation is to bring the reagents in constant contact with air and thus to form the seed colloid in a higher state of oxidation. This happens because the ferrous ions Fe^{2+} are very unstable in the presence of air and are usually transformed into a mixture of Fe^{2+} and Fe^{3+}. Had Krones oxidized his ferrous hydroxide very rapidly, δ-$(FeO)OH$ would probably have resulted (Bernal et al. 1959).

Another method of preparing acicular crystals of goethite was described by Derie et al. (1975). They precipitated goethite from ferrous sulfate by the addition of a mixture of sodium hypochlorite and sodium carbonate. An excess of $NaOCl$ was desirable and the reaction yielded the best product when the reaction temperature was between 60 and 75°C. The preparation of iron oxy-hydride containing elements such as Co, Mn, Cu, Cr, etc. was possible by precipitation from solutions of the appropriate combinations of metal sulfates.
Particle growth.   Ayers and Stephens (1962a, b) added zinc chloride to the seed particles and pieces of iron during the growth of their lepidocrocite platelets. The purpose of the zinc chloride was to inhibit the formation of goethite but the mechanism was not specified. Bennetch et al. (1975) omitted the zinc chloride but still produced lepidocrocite. Their critical parameter was to limit the ratio of product : seed to 2 : 1 so that the acicularity of the particles might be preserved by the slow growth rate. Leitner et al. (1973) believed that the use of large quantities of air at this stage helped to prevent dendrites.

Marcot et al. (1951) showed that small ($<1\mu$m) yellow acicular crystallites of ferric oxide monohydrate could be grown from ferrous sulfate in quite strong alkaline solutions. They claimed that the use of alkaline conditions substantially eliminated the anions (Cl^- or SO_4^{2-}) which would otherwise combine chemically with the particles and lead to their agglomeration during drying. Marcot et al. (1951) also used compounds like silicon dioxide which are capable of forming complexes with $(FeO)OH$. When added to the reagents in quantities of 0.1% to 5% the additives act as crystal growth directors and significantly increase the crystallinity of the product.
Reduction reactions.   The oxidation, reduction, and annealing behavior of iron oxides were treated in detail by Doan (1941) and his work is frequently referred to in the patent literature. The chemistry of the reactions is quite complex, but notable advances in our understanding have taken place within the last twenty years as a result of more detailed knowledge of the various iron oxide structures.

In the Baudisch (1933a, b) process described earlier for the direct precipitation of γ-$(FeO)OH$, the particles were transformed by gentle heating into γ-Fe_2O_3. By contrast, in the Ayers' and Stephens' (1962a, b) process the γ-$(FeO)OH$ particles were first reduced to magnetite (fig. 5) and then reoxidized to γ-Fe_2O_3. The reason for the apparently unnecessary steps in the latter process is that the efficiency of direct transformation of γ-$(FeO)OH$ to γ-Fe_2O_3 is quite low. Much of the lepidocrocite is transformed on heating to α-Fe_2O_3, and consequently the magnetic properties suffer, principally by a reduction in the intensity of mag-netization. This is supported by the results of Watanabe (1970) who prepared lepidocrocite by the oxidation of a colloidal solution obtained from ferrous iodide and sodium hydroxide. The needle-shaped crystallites, heated at 300°C for three hours, were converted to γ-Fe_2O_3 particles which resembled very closely in appearance the particles produced from synthetic lepidocrocite by Bennetcch et al. (1975); i.e. length 0.4–1.0μm, width 0.05–0.2μm. There were, however, significant differences in the magnetic properties. The remanent and saturation magnetization were only about 80% of conventional γ-Fe_2O_3 and the coercivity was 120 Oe (cf. 330–365 Oe, Bennetch et al. 1975). The results are consistent with the formation of a mixture of γ-Fe_2O_3 and α-Fe_2O_3. Imaoka (1968) studied the transformation by differential thermal analysis and X-ray diffraction and found that it occurred between 50 and 300°C. Berkowitz et al. (1968) found that when they prepared γ-Fe_2O_3 by heating lepidocrocite, the resulting particles were composed of crystallites whose average size was less than 100 Å. Furthermore, the magnetization and the coercivity were lower than was found in γ-Fe_2O_3 produced by the standard method and having a crystallite size greater than 100 Å.

The progression from particles of α-(FeO)OH to γ-Fe_2O_3 was studied by Olsen and Cox (1971) who used electron microscopy and diffraction. They found that some sintering of the particles occurs during the reduction from α-Fe_2O_3 to Fe_3O_4; this was revealed by the formation of dendrites and of polycrystalline agglomerates. A significant reduction in the sintering could be achieved by reducing the reduction temperature, however, at a price; it caused an increase in the porosity of the particles. These results were supported by Fagherazzi et al. (1974) who varied the reduction temperature between 350–600°C and studied the changes in microstructure by transmission electron microscopy and by small-angle X-ray scattering. At the higher reduction temperatures a reduction in the specific surface was observed (that is, agglomeration occurred) and the 80–300 Å micropores disappeared. The formation of agglomerates is very undesirable considering the use of the particles. Van der Giessen (1974) remarked that agglomerates were responsible for an increase in the noise level of the recording surface (since they cause an increase in the surface roughness and in the magnetic inhomogeneity below the surface) and a decrease in the signal level due to a wide switching field distribution, a reduced orientability, and again, greater surface roughness.

Finally a surprising result was obtained by Skorski (1972) who observed that the rate of reduction of α-Fe_2O_3 to Fe_3O_4 could be changed by the application of a magnetic field. He claimed that in a field of 500 Oe (39.8 kA/m) and at 300°C, reduction was 65% complete after 30 min in the field but only 19% complete without the application of a field.

Oxidation reactions. In the oxidation of Fe_3O_4 to γ-Fe_2O_3 a compromise must be reached between too low a temperature, which leads to uneconomically long processing times, and too high a temperature, which results in the undesirable formation of the barely-ferromagnetic α-Fe_2O_3. Camras (1954) recommended the use of temperatures between 200 and 250°C and this advice has apparently been followed in all the subsequent preparation methods. Camras showed that a sample oxidized at 380°C had only half the remanent moment of one oxidized at 285° and oxidizing at 665°C resulted in no remanent moment, i.e. pure α-Fe_2O_3 was formed. Klimaszewski and Pietrzak (1969) used ferromagnetic resonance to study the oxidation of Fe_3O_4. They found that the transformation to γ-Fe_2O_3 began at 100°C and is complete at 250°C and that the formation of α-Fe_2O_3 started at 250°C and finished at 500°C. The highest temperature used by Klimaszewski and Pietrzak was 1300°C. At about this temperature Parker (1975) reported that the evaporation of oxygen becomes appreciable and a transformation back to magnetite takes place.

Imaoka (1968) studied the transformation γ-$Fe_2O_3 \rightarrow Fe_3O_4$ by differential thermal analysis and X-ray diffraction and found that the range of temperatures over which it took place depended on the shape of the particles. For acicular particles the range was 560–650°C, while for non-acicular particles the transformation occurred between 250 and 300°C. Gustard and Scheuele (1966) found that acicular particles of γ-Fe_2O_3 were partially converted to α-Fe_2O_3 at temperatures of 525–650°C. Bando et al. (1965) showed that transformation temperature increased as the crystallite size of the γ-Fe_2O_3 particles increased. Further details of the transformation were elucidated by Kachi et al. (1970) (electron microscope) and by Morrish and Sawatzky (1970) (Mössbauer). Colombo et al. (1967) and Gazzarini and Lanzavecchia (1969) believed that the oxidation and reduction of iron oxides proceed by the diffusion of ferrous ions with the kinetics of the process depending strongly on the presence of vacancies in the lattice. Protons in the precipitated magnetite act as a stabilizer of the lattice-with-vacancies which is thus easily oxidized to γ-Fe_2O_3.

Water appears to play a very important role in the conversion of Fe_3O_4 to γ-Fe_2O_3. David and Welch (1956), for example, showed that "specimens of magnetite which gave gamma ferric oxide on oxidation invariably contained appreciable percentages of water, while specimens prepared under dry condition oxidized with great difficulty, never yielding the gamma oxide." Furthermore, they concluded that the water (0.5–1.0%) combined in the γ-Fe_2O_3 lattice may exercise a stabilizing effect since it cannot be removed without changing the structure of the material to α-Fe_2O_3. Healey et al. (1956) investigated the surface of a ferric oxide powder by calorimetry and found that two-thirds of the surface could absorb water under conditions of relative pressure at which a monolayer would be expected. The remainder of the surface chemisorbed water which could only be released by heating at a temperature of about 450°C.

The affinity of iron oxide particles for water led to the use of cationic dispersing agents such as amines, amine salts or imides, in the expectation that they would be attracted to the particles even more strongly than was water. The standard method for making iron oxide particles was originally developed for the manufacture of paint pigments. The problem of agglomeration and dispersion is important in that application also and Marcot et al. (1951) recommended the use of water-insoluble carboxylic acids, e.g., oleic, ricinoleic, naphthenic, etc. Wolff (1964) advocated heating the particles with Werner compounds of chromium and carboxylic acid. The chromium cations were chemically bonded to the oxygen ions on the particles and the carboxylic acid radicals had an affinity for the binder material thus allowing better dispersion and higher particle packing densities. More recently, Hartmann et al. (1975) described the addition of styrene and maleic acid as aids in the dispersion and orientation of iron oxide particles.

Further evidence of the role of water in the oxidation of magnetite to γ-Fe_2O_3 was provided by Elder (1965). He found that small particles ($<1\mu$m) of magnetite could be partially converted to γ-Fe_2O_3 by heating at 250°C but much larger particles ($\geqslant 25\mu$m) could only be converted to α-Fe_2O_3. The effect was attributed to the stabilizing effect of water absorbed or hydrated on the surface of the fine particles. Elder found that oxidation was incomplete when performed at temperatures below 250°C, but if the temperature was above 450°C then a α-Fe_2O_3 resulted. The action of the adsorbed or combined water on small particles of magnetite and the relatively small surface area of larger particles helps to explain why large single crystals of γ-Fe_2O_3 have not been grown. Gazzarini and Lanzavecchia (1969) stated that γ-Fe_2O_3 is readily obtained only if the particles are strained, for example, by being ground or if water is contained in the magnetite (as distinct from water being present in the oxidizing gas).

Colombo et al. (1964) concluded from thermogravimetric and differential thermal analysis and from X-ray diffraction, that the mechanism of oxidation of magnetite to maghemite is a two-stage process. In the first stage a solution of γ-Fe_2O_3 in Fe_3O_4 always occurs. Oxygen is absorbed and the Fe^{2+} ions are oxidized to Fe^{3+}. Then ferrous ions diffuse from the inside toward the surface, the rate of diffusion depending on the degree of crystalline imperfection. Gazzarini and Lanzavecchia (1969) further showed that if even a minute amount of hematite is present as an impurity in the magnetite there also occurs an autocatalytic growth of hematite. This growth initially co-exists with the formation of maghemite but ultimately predominates so that the whole particle becomes α-Fe_2O_3. On the other hand, if some lattice sites are vacant or if water is present in the lattice then the formation of maghemite is encouraged. A diagram showing the most important of the reaction discussed above is given in fig. 5.

3.1.2. Structure of iron oxides

Magnetite has the typical inverse spinel structure $Fe^{3+}[Fe^{2+}Fe^{3+}]O_4$; that is, it is ferrous ferrite; the square brackets include the ions situated on the octahedral (B) sites. The lattice constant of the face-centered cubic unit cell is 8.3963 Å and the space group Fd3m (1957, ASTM Index). There are eight molecules of Fe_3O_4 in each unit cell – 32 atoms of oxygen and 24 cations, of which 16 occupy the octahedral sites and 8 the tetrahedral sites. Magnetite in the form of fine particles is usually deficient in Fe^{2+}. Some of the ferrous ions have been oxidized to ferric and the composition lies between Fe_3O_4 and γ-Fe_2O_3 (Dunlop 1973).

On oxidation to Fe_2O_3 the oxygen sub-lattice remains substantially the same as in magnetite but the lattice constant is slightly reduced to 8.33 Å (ASTM index). The cubic unit cell of γ-Fe_2O_3 then contains $\frac{8}{9}$ of the number of iron ions in magnetite ($10\frac{2}{3}$ molecules of Fe_2O_3) and a defect spinel lattice is obtained with the formula

$$Fe_8^{3+}[(Fe_{13(1/3)}^{3+}\Delta_{2(2/3)})]O_{32}$$

where Δ represents a cation vacancy (Verwey 1935, Hägg 1935, Kordes 1935). Braun (1952) noticed that some specimens of γ-Fe_2O_3 (containing water molecules) gave very nearly the same X-ray diffraction lines as those obtained from ordered lithium ferrite $Fe_8[Li_4Fe_{12}]O_{32}$. He concluded that in those γ-Fe_2O_3 specimens the same ordered structure occurred with $1\frac{1}{3}$ Fe ions and $2\frac{2}{3}$ vacancies randomly distributed on the sites of the four lithium atoms. Thus the formula became

$$Fe_8^{3+}[(Fe_{1(1/3)}^{3+}\Delta_{2(1/3)})Fe_{12}^{3+}]O_{32}.$$

The ideal ordering ratio, lithium-type ions (octahedral Fe ions of 1:3) could be achieved by incorporating water molecules so that the limiting formula became

$$Fe_8[H_4Fe_{12}]O_{32}.$$

Van Oosterhout and Rooijmans (1958) prepared very pure gamma ferric oxide by decomposing pure ferrous oxalate dihydrate in an atmosphere of steam and nitrogen followed by oxidation in a mixture of air and nitrogen at 250°C. They were unable to detect by X-ray diffraction even the strongest line of alpha ferric oxide. They did find, however, a number of additional reflections, all of which could be indexed by assuming a tetragonal unit cell having $c/a = 3$ and $a = 8.33$ Å. Assuming that the structure was similar to that of spinel, they concluded that the new unit cell must contain 32 molecules of Fe_2O_3 with the lattice vacancies arranged on a superlattice, and that the space group is $P4_1$ (or $P4_3$). Thus one third of a complete (tetragonal) unit cell has the formula

$$Fe_8^{3+}[(Fe_{1(1/3)}^{3+}\Delta_{2(2/3)})Fe_{12}^{3+}]O_{32}.$$

In 1962, Aharoni et al. verified a suggestion by Kojima (1954) that γ-Fe_2O_3 was produced as an intermediate phase during the reduction in hydrogen of α-Fe_2O_3 to Fe_3O_4. They measured the saturation magnetization, the Mössbauer spectra, and the Curie temperature at stages in the reduction process. The result is rather difficult to accept since it involves the conversion at high temperature of the more stable α-Fe_2O_3 to the less stable γ-Fe_2O_3. To explain this, Aharoni et al. (1962) supported Braun's (1952) suggestion that γ-Fe_2O_3 consists of a solid solution of two materials

$$Fe_8[(Fe_{1(1/3)}\Delta_{2(2/3)})Fe_{12}]O_{21}$$

and

$$Fe_8[H_4Fe_{12}]O_{32}.$$

They showed that when γ-Fe_2O_3 was heated, hydrogen was evolved in amounts of about 70% of that required to give the formula $Fe_8[H_4Fe_{12}]O_{32}$.

However, Schrader and Büttner (1963) measured by counter diffractometer the X-ray spectra from samples of γ-Fe$_2$O$_3$ prepared by a number of different methods. They found that the whole system of peaks, including the very weak ones, corresponded to the tetragonal system of Van Oosterhout and Rooijmans (1958) ($a_0 = 8.33$ Å, $c/a = 3$). They concluded that γ-Fe$_2$O$_3$ exists in only one uniform phase and that it was unnecessary to invoke a stabilizing phase of, for example, hydrogen ion spinel. Strickler and Roy (1961) deduced that the hydrogen is not present in the structure of maghemite and is probably held as water.

Experimental evidence that the vacancies were largely restricted to the octahedral sites came from the neutron diffraction results of Ferguson and Hass (1958) and of Uyeda and Hasegawa (1962). The ratio of intensities of (400) and (440) reflections was found experimentally to be 0.443. The calculated value was 0.482 assuming a random distribution of vacancies, 0.580 assuming a preferential distribution over tetrahedral sites, and 0.433 assuming a preferential distribution over octahedral sites.

Because of the instability of γ-Fe$_2$O$_3$ at high temperatures it is difficult to grow single crystals of the material. Takei and Chiba (1966) were able to grow single-crystal films epitaxially on a (100) face of a magnesium oxide crystal by the decomposition of ferrous bromide with water vapour and oxygen at 680°C. Deposition took place at about 0.3μm/h and the maximum thickness obtained in the spinel structure was 1μm. The films were identified, by X-ray analysis, as single crystals having a cubic spinel-type structure with a lattice constant $a_0 = 8.35 \pm 0.01$ Å. Even after exposures (about 100 h) in the diffraction camera, no diffraction spots corresponding to (110), (210) or (320) reflections were found and Takei and Chiba, therefore, ruled out the presence of a superlattice in their films. Normally γ-Fe$_2$O$_3$ would be entirely transformed to α-Fe$_2$O$_3$ at the temperature used for the deposition of their films. Takei and Chiba suggested that in the case of epitaxy the $\gamma \rightarrow \alpha$ transition is hampered by the substrate.

Bernal et al. (1959) commented that "the significance of the conditions which lead to the various modifications of γ-Fe$_2$O$_3$ is still obscure." But despite the disagreement over the presence or absence of stabilizing ions or phases, the basic structure of γ-Fe$_2$O$_3$ is generally accepted – that is, a defect spinel with the cation vacancies arranged on a tetragonal superlattice.

The other compounds whose structures are of interest are the strongly ferrimagnetic δ-(FeO)OH and hematite which is formed as an intermediate product in every commercially-important method for preparing γ-Fe$_2$O$_3$ particles. The X-ray results of Francombe and Rooksby (1959) and of Bernal et al. (1959) indicate that the structure of δ-(FeO)OH is hexagonal. That is, the structure of the oxyhydroxide is similar to that of ferrous hydroxide, Fe(OH)$_2$, from which it is made by rapid oxidation. The arrangement of the ions is such that hexagonally close-packed layers of oxygen/hydroxyl are interleaved with Fe^{3+} ion layers. Francombe and Rooksby concluded that about 20% of the Fe^{3+} is on the tetrahedral sites.

The structure of hematite is rhombohedral but almost indistinguishable from hexagonal. Two-thirds of the octahedral interstices within the hexagonally close-packed oxygen lattice are occupied by ferric ions (Creer et al. 1975).

A summary of the structures of the relevant iron–oxygen compounds is included in fig. 5.

3.1.3. Transformations of iron oxides and oxyhydroxides

Early summaries of these transformation were given by Welo and Baudisch (1925, 1933) and the more recent work has been discussed by Francombe and Rooksby (1959), Bernal et al. (1959) and Dasgupta (1961). As Bernal expressed it: "The common feature of the group of iron oxides and hydroxides is that they are composed of different stackings of close-packed oxygen/hydroxyl sheets, with various arrangements of the iron ions in the octahedral and tetrahedral interstices".

Consequently many of these materials exhibit topotaxy which is the transformation of one solid crystalline phase into another while preserving some of the original crystalline planes and directions. For example, the [100], [010], and [001] axes of orthorhombic γ-(FeO)OH became, after transformation, the [001], [110] and [1$\bar{1}$0] of the cubic (with tetragonal vacancies) γ-Fe$_2$O$_3$. Similarly, orthorhombic α-(FeO)OH can be transformed to the related hexagonal α-Fe$_2$O$_3$ by removing the hydroxyl sheets and some of the oxygen in strips running parallel to the c-axis of the oxyhydroxide lattice, to form water. In this transformation the [100], [010] and [001] axes of goethite become the [111], [110] and [112] axes of hematite.

During the oxidation of synthetic magnetite to γ-Fe$_2$O$_3$ and then to α-Fe$_2$O$_3$ (which are further examples of topotactic transformations) crystals of α-Fe$_2$O$_3$ grow with their [001] axes parallel to the [111] of γ-Fe$_2$O$_3$ and the [1$\bar{1}$0] axes of hematite are parallel to the [110] of γ-Fe$_2$O$_3$. Natural magnetite apparently oxidizes only to hematite, unless small particles are prepared by grinding in water (Elder, 1965) before the oxidation is carried out.

The dehydration-reduction reaction of α-(FeO)OH to Fe$_3$O$_4$ is the basic method for making acicular particles of magnetite and γ-Fe$_2$O$_3$ and is thought not to be topotactic but, as we have seen, the acicularity of the particles is roughly preserved. This reaction does involve a notable weight loss and density change (4.35–5.197 g/cc); therefore, it is not surprising that the voids which are so apparent in γ-Fe$_2$O$_3$ particles prepared by this method, are formed at this step. A reduction of particle size occurs at this stage.

The final oxidation of Fe$_3$O$_4$ to γ-Fe$_2$O$_3$ involves, in comparison to the transformations which we have just considered, a slight change in structure and is clearly topotactic. A slight reduction in density (5.197 to 5.02 g/cc) indicates an increase in particle volume.

3.1.4. Morphology of γ-Fe$_2$O$_3$ particles

Early work on the morphology of acicular particles of γ-Fe$_2$O$_3$ showed considerable disagreement. Osmond (1953) believed that the particles were composed of small crystallites which all had their [111] axes parallel to the long axis.

Campbell (1957) used selected-area electron diffraction and after examining a large number of particles, concluded that each particle had as its long axis one of the low order directions ⟨111⟩, ⟨211⟩ etc., but that no one set of directions predominated. Van Oosterhout (1960) believed that Campbell's observation did not rule out a ⟨110⟩ orientation. Hurt et al. (1966) found by electron diffraction that about 80% of the particles were polycrystalline and showed a preference for the long axis to be [110]. Their single crystal particles also had a [110] orientation. Then using dark-field electron microscopy, Gustard and Vriend (1969) concluded that 40% by volume of their γ-Fe_2O_3 particles had [110] as their long axis. Köster (1970) measured the temperature dependence of susceptibility of assemblies of γ-Fe_2O_3 particles and he also concluded that the easy axes were ⟨110⟩. The transformation from α-(FeO)OH to γ-Fe_2O_3 was studied by electron microscopy and diffraction by Olsen and Cox (1971) who found that the single crystal nature of α-(FeO)OH was maintained through each step. However, if high temperatures were used during the reduction step from α-Fe_2O_3 to Fe_3O_4 then polycrystals with dendrites appeared. They found a preferred [110] orientation in the γ-Fe_2O_3 which had been reduced at low temperature but no preferred orientation in the branched and agglomerated particles reduced at high temperatures. Panter, Eliasberg and Yakobson (1971) used electron micrsocopy and electron and X-ray diffraction to study two kinds of γ-Fe_2O_3 of different origin. The Russian Type-6 particles were found to be composed of many crystallites sharing a common [110] orientation while the 3M, MTA21052 particles (probably resembling the "low-temperature reduction" particles of Olsen and Cox) were single crystals.

Thus it appears that the early confusion concerning the orientation of the long axis arose because of different preparation conditions. A low-temperature reduction preserves the single crystal nature of the α-Fe_2O_3 and α-(FeO)OH particles, while reducing to Fe_3O_4 at very high temperature causes a polycrystalline structure to appear which may or may not have a [110] orientation. A possible advantage in having particles which are single crystals with [110] orientation (or polycrystalline with a common [110] orientation) is that the magnetocrystalline easy axis is also [110] and thus shape and crystalline anisotropy would combine to yield higher coercivities. On the other hand, the high temperatures during the reduction step, although they may result in disoriented and agglomerated polycrystals, do produce particles without micropores according to Fagherazzi et al. (1974).

Typical particles of γ-Fe_2O_3 prepared by the traditional method [α-(FeO)OH $\rightarrow \alpha$-$Fe_2O_3 \rightarrow Fe_3O_4 \rightarrow \gamma$-$Fe_2O_3$] appear in the electron microscope as shown in fig. 7a. Such particles were measured by Sykora (1967) and his histograms are shown in fig. 6. Surprisingly similar results were later published by Robinson and Hockings (1972). They measured the lengths of about 100 particles from each sample and found that the lengths were normally distributed. However, if the particles were milled (or "densified") the distribution became log normal. Reference to densification of the particles frequently occurs in the literature. The process consists of the agitation of the particles in a ball mill with the

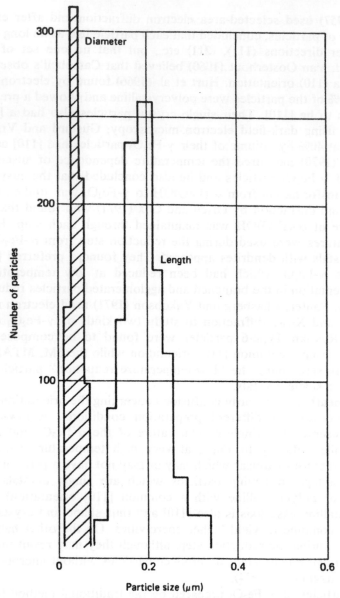

Fig. 6. Histograms of the lengths and diameters of acicular particles of γ-Fe$_2$O$_3$ (Sykora 1967).

objectives of breaking up the agglomerates and removing dendrites both of which hinder the preparation of smooth, uniform coatings of high packing density (and therefore, high signal level). Excessive milling is undesirable since small, less acicular particles would result and the process is usually monitored by measuring the oil absorption and the coercivity of the particles.

More recently high-acicular particles of γ-Fe$_2$O$_3$ having a clean outline (free of dendrites and micropores) have been made by the method of Ayers and Stephens (1962a, b) and Bennetch et al. (1975). The former obtained well-defined platelets of length $\sim 1\mu$m, length:width 10:1 and width:thickness ~ 4:1.

Using the same precipitation method Bando et al. (1965) obtained platelets of length 2.5μm width 0.2 to 0.5μm and thickness 0.01 and 0.02μm. The particles prepared by Bennetch et al. (1975) were needle-shaped and more than 70% were reported to have a length:width > 10:1 and a length of up to 2μm. Figure 7b is an electron micrograph of these particles. Berkowitz et al. (1968) found that when particles were prepared by dehydration of γ-(FeO)OH at temperatures below 400°C the crystallites within the particles were smaller than 100 Å in contrast to a crystallite size of more than 100 Å when the particles were prepared by the standard method.

3.1.5. Magnetic properties of the iron oxides

It is both convenient and appropriate to divide the magnetic properties into *intrinsic* properties, those fundamental properties which belong to the material per se and *extrinsic* properties, which depend largely on the state of preparation of the material. The saturation intensity of magnetization M_s and the Curie temperature θ_c are examples of intrinsic magnetic properties. The coercivity H_c and the remanent intensity of magnetization M_r are extrinsic properties since they depend upon, among other things, the shape and size of the oxide particles. The magnetic properties are summarized in table 5.

a. Intrinsic properties

Saturation intensity of magnetization. Values of σ_s are used more frequently than those of M_s since weighing is easier and less subject to errors caused by temperature changes or the presence of cavities than is the determination of the volume of a powder sample.

Magnetite has, as we have seen, the simple inverted spinel structure, which can be represented magnetically as

A sites	B sites
Fe^{3+}	$[Fe^{2+}Fe^{3+}]$
$\longrightarrow$	$\longleftarrow$ $\longleftarrow$
$5\mu_B$	$4\mu_B$ $5\mu_B$

where μ_B represents the Bohr magneton. Since the moments on the tetrahedral (A) sites and octahedral (B) sites are opposed, the resultant magnetic moment of the molecule is just that of the Fe^{2+} ion, $4\mu_B$. Smit and Wijn (1959) give 4.1μ_B/molecule as the experimental value.

The saturation moment-per-unit-mass, σ_{s0} at 0 K, can be calculated from $\sigma_{s0} = (5585/M)n_B$, where n_B is the number of Bohr magnetons per ion, and M is the molecular weight. For magnetite, $M = 231.6$ and $n_B = 4$, so that $\sigma_{s0} = 97$ emu/g (122μWb m/kg). Smit and Wijn give $\sigma_s = 92$ emu/g (116μWb m/kg) as the value at room temperature. The average value for σ_s of a number of powder samples of synthetic magnetite measured in the author's laboratory on a suscep-

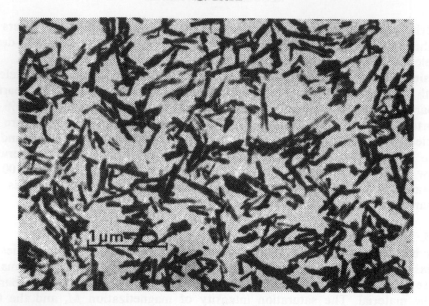

(a)

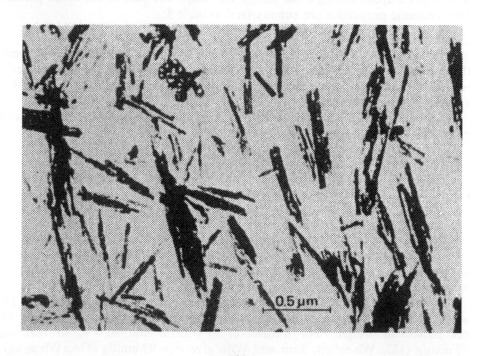

(b)

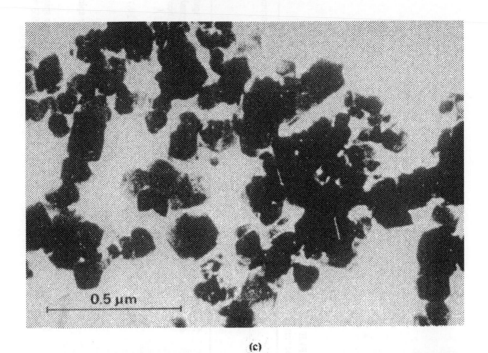

(c)

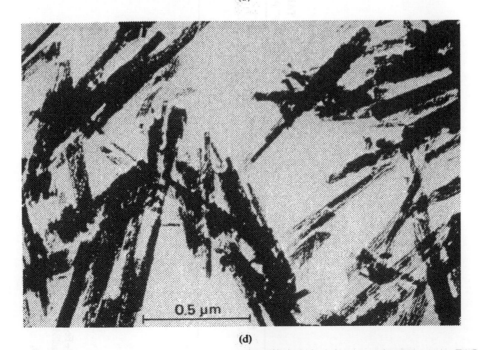

(d)

Fig. 7. Electron micrograph of particles; (a) γ-Fe$_2$O$_3$ prepared from α-(FeO)OH, (b) γ-Fe$_2$O$_3$ prepared from γ-(FeO)OH, (c) Cobalt-substituted γ-Fe$_2$O$_3$, (d) chromium dioxide, CrO$_2$.

TABLE 5
Magnetic properties of particles

Oxides	Intrinsic				Extrinsic				Structure
	σ_s (emu/g)	θ (°C)	K_1 (ergs/cc)	λ_s	ρ (g/cc)	H_c (Oe)	σ_r (emu/g)	Shape, size (μm)	
γ-Fe_2O_3	74	590	-4.64×10^4	-5×10^{-6}	5.07$_4$ (bulk) 4.60 (particles)	75–150	34	Equiaxed 0.05–0.3	Inverse spinel on a tetragonal superlattice $a = 8.33$ Å $c/a = 3$ $P4_1$ (or $P4_3$)
						250–320	37	Acicular; $l/w = 7$ $l = 0.2$–0.7 (110)	
Fe_2O_3-Fe_3O_4 Berthollide	~80	590	-3×10^4	$+7 \times 10^{-6}$		300–450	~40	Acicular; $l/w = 7$ $l = 0.2$–0.7	Inverse spinel
Fe_3O_4	84	585	-1.1×10^5	$+18 \times 10^{-6}$	5.197 (bulk)	305–335	42	Acicular; $l/w = 7$ $l = 0.2$–0.7	Inverse spinel $a = 8.3963$ Å Fd3m

Material									
$Co_xFe_{3-x}O_4$									
$x = 0.1$	87	580	$+1\times10^6$	$+18\times10^{-6}$	100		44	Equiaxed 0.2	Inverse spinel $a = 8.395 \pm 0.005$ Å
$x = 1.0$	65	520	$+1.8\times10^6$	-110×10^{-6}	980		33	1.0	Inverse spinel
$Co_xFe_{2-x}O_3$									$a = 8.354$ Å
$x = 0.04$	70	525	$+1\times10^6$	-15×10^{-6}	400	4.67	35	Cubic 0.05–0.08	Cubic
$x = 0.06$	62	525			515–600		31	0.05–0.08 18 m²/g	
$\gamma\text{-}(Fe_{1.96}Co_{0.04})_2O_3$	80.2		$+1\times10^6$	-15×10^{-6}	480	4.67	46.5	Acicular; $l/w = 10$ $l = 0.2$–0.5 $\langle100\rangle$	$\langle100\rangle$
Impregnated $(Co, Fe)_2O_3\text{-}(Co, Fe)_3O_4$	73		$k_u \sim 1.5\times10^6$		470–550		36	Acicular; $l/w = 7$ $l = 0.2$–0.5	
	60				580–700		30		
CrO_2	73	117	$+2.5\times10^5$	$+1\times10^{-6}$	470	4.88–4.95	36	Acicular; $l/w = 10$ $l = 0.28$ 23 m²/g	Tetragonal $c = 2.9182$ Å $a = 4.4218$ Å $D_{4h}^{14}\text{-}P4_2/mnm$
	72	117	$+2.5\times10^5$	$+1\times10^{-6}$	650	4.88–4.05	35	Acicular; $l/w = 17$ $l = 0.48$ 26 m²/g	
$BaO\cdot6Fe_2O_3$	68	450	$+3.3\times10^6$		715–5150	5.28	34	Platelets diam. = 0.08–0.15 diam./thickn. = 15	Hexagonal $c = 23.2$ Å $a = 5.88$ Å

tibility balance was 84 emu/g (106μWb m/kg). Aharoni et al. (1962) found experimentally $\sigma_s = 82$ emu/gm (103μWb m/kg) at 20°C. Using the value of 5.197 g/cc (5197 kg/cm) (ASTM Index) for the density of magnetite and Smit and Wijn's values of σ_s, the saturation intensity of magnetization per-unit-volume is found to be 478 emu/cc (0.601 Wb/m²).

One third of the tetragonal unit cell of the defect-spinel structure of γ-Fe_2O_3 can be represented magnetically as:

A sites   B sites
Fe_8^{3+} $[(Fe_{1(1/3)}^{3+}\Delta_{2(2/3)})Fe_{12}^{3+}]$

$\xrightarrow{\hspace{2cm}}$ $\xleftarrow{\hspace{1cm}}$ $\xleftarrow{\hspace{1cm}}$

$(8 \times 5)\mu_B$ $(1\frac{1}{3} \times 5)\mu_B$ $(12 \times 5)\mu_B$

Thus the resultant moment for this structural unit is $(\frac{16}{3} \times 5)\mu_B$ and so the moment-per-molecule of γ-Fe_2O_3 is $\frac{3}{32} \times \frac{16}{3} \times 5\mu_B = 2.5\mu_B$. It is interesting that if the limiting hydrogen ion spinel of Braun (1952) is used we get the same result and so the question of whether this structure is relevant to γ-Fe_2O_3 does not affect the value of σ_s

A sites   B sites
Fe_8^{3+} $(H_4Fe_{12}^{3+})$

$\xrightarrow{\hspace{2cm}}$ $\xleftarrow{\hspace{1.5cm}}$

$(8 \times 5)\mu_B$ $(12 \times 5)\mu_B$

The resultant moment-per-molecule is $4 \times \frac{5}{8} = 2.5\mu_B$ or $1.25\mu_B$ per iron atom. Henry and Boehm (1956) found experimentally a value of $1.18\mu_B$/atom at 4.2 K; from this, σ_{s0} can be calculated as 82.3 emu/g (103μWb m/kg). If the theoretical value of $1.25\mu_B$/atom is used, then $\sigma_s = 87.4$ emu/g (110μWb/m/kg). Brown and Johnson (1962) showed both experimentally and theoretically that

$$(\sigma_s, 290\text{ K})/(\sigma_s, 0\text{ K}) = 0.900$$

thus the values of σ_s at room temperature became 74.1 emu/g for the experimental value of $1.18\mu_B$/iron atom and 87.7 emu/g (110μWb m/kg) for the theoretical value of $1.25\mu_B$/iron atom. The average value of σ_s for a large number of powder samples obtained from different sources and measured in the author's laboratory was 74 emu/g (93μWb m/kg). This agrees well with the value of 73 emu/g (92μWb m/kg) found experimentally by Aharoni et al. (1962).

If we take the density of γ-Fe_2O_3 as 5.074 g/cc (5074 kg/cm, ASTM index) and the theoretical value of σ_s, the saturation intensity of magnetization-per-unit-volume is 400 emu/cc (0.503 Wb/m²) at 20°C, when the theoretical value of $1.25\mu_B$/iron atom is used. If the experimental value of $1.18\mu_B$/iron atom is taken then $M_s = 378$ emu/cc (0.475 Wb/m²). The discrepancy between theory and experiment is perhaps best explained by the Mössbauer studies of Morrish and Clark (1974) in terms of the non-collinearity of the magnetizations on the A and B sublattices. They deduced this from the presence of lines 2 and 5 of the 6 line pattern (field parallel to beam); the intensity of these lines should be zero for a collinear structure. The average angle between the A and B magnetizations and the applied field direction was 12° this angle increased with decreasing particle size and became 31° to 32° for supermagnetic particles. If we assume an angle of 12° then the theoretical value of $1.25\mu_B$ becomes $1.22\mu_B$.

The approach to saturation magnetization for fine particles of magnetite was measured by Parfenov et al. (1966). They found that the measured magnetization, M, between fields of 10 kOe and 20 kOe (796–1.592 kA/m) followed the formula: $M = M_s - bH^{-1}$, where b was a constant for each example and which decreased with increasing particle size. Measurements by Takei and Chiba (1966) on their epitaxially-grown single-crystal films of γ-Fe_2O_3, followed by an extrapolation according to $(1 - T^{3/2})$ to 0 K, gave $n_B = 1.45$. At 20°C the saturation magnetization-per-unit-volume, $M_s = 369$ emu/cc (0.463 Wb/m²) was determined by resonance measurements.

Saturation magnetization is usually thought of as being the intrinsic magnetic property par excellence. Yet Berkowitz et al. (1968) found a marked dependence on crystallite size for γ-Fe_2O_3. Their results are shown in fig. 8. Although the particle size was approximately the same in all samples (average length <1μm, axial ratio 5 to 7), the crystallite size, as determined by X-ray line broadening, ranged from 50 to 700 Å. No strain broadening of the lines was detected. All the γ-Fe_2O_3 samples with crystallite sizes greater than 100 Å were obtained from

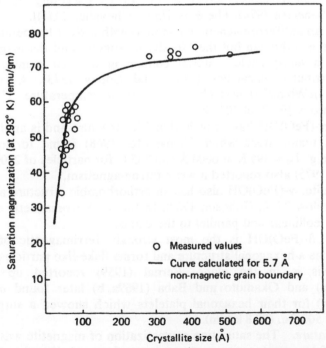

Fig. 8. Saturation magnetization of particles of γ-Fe_2O_3 vs their crystallite size (Berkowitz et al. 1968).

firms supplying oxides for magnetic recording. The samples having crystallize sizes less than 100 Å were prepared by the dehydration of γ-(FeO)OH at temperatures below 400°C.

Berkowitz et al. (1968) found $\sigma(293 \text{ K}) = 74 \text{ emu/g}$ ($93\mu \text{Wb m/kg}$) for the largest crystallite sizes but this decreased to 34 emu/g ($43\mu \text{Wb m/kg}$) for particles in which the crystallite size was about 50 Å. After ruling out, by X-ray diffraction, the possibility that their results were due to the presence of less magnetic oxides or oxyhydroxides of iron, they theorized that the dependence of σ_s on crystallite size was caused by magnetostatic interaction between the crystallites across non-magnetic boundaries, whose thickness they inferred to be about 6 Å. Additional evidence that the saturation magnetization of ferrites could be an extrinsic property was provided by Berkowitz and Lahut (1973) who found that the presence of a surfactant layer on particles could reduce σ_s by more than 50%. The magnetization could be increased to almost the bulk value by removing the surfactant. They explained the effect in terms of a chemical bonding of the organic molecules to the ferrite which decreased the magnetization of the surface layers of the particles.

Hematite, α-Fe_2O_3 has a rhombohedral, almost hexagonal structure made up of distorted hexagonally close-packed layers of oxygen atoms with layers of ferric ions between them. Alternate layers of ferric ions have antiparallel spins. The canted antiferromagnetic structure gives a weak parasitic ferromagnetism [0.1 emu/g ($0.13\mu \text{Wb m/kg}$), Creer et al. 1975), 0.4 emu/g, ($0.50\mu \text{Wb m/kg}$), (Stacey and Banerjee 1974). The θ_c is 675°C (Chevalier 1951)].

ϵ-Fe_2O_3, is an antiferromagnetic compound with a Néel temperature of 480 K. The structure is unknown but the Mössbauer spectra and the magnetic properties suggest a fine-particle material containing a large number ($\sim 10\%$) of randomly-distributed vacancies (Dezsi and Cooey 1973). A remanence of 0.3 emu/g ($0.4\mu \text{Wb m/kg}$) was observed at room temperature. The compound transforms into α-Fe_2O_3 at 1040 K.

Goethite, α-(FeO)OH has an orthorhombic structure and is antiferromagnetic with a Néel temperature which Yamamoto (1968) found to depend on the particle size e.g. $T_N = 395$ K at 6000 Å and 365 K for particles of 2000 Å diameter. Creer et al. (1975) also reported a weak ferromagnetism.

Lepidocrocite, γ-(FeO)OH also has an orthorhombic structure and is antiferromagnetic below 73 K, (Johnson 1969). In the antiferromagnetic state the Fe^{3+} moments are collinear and parallel to the c-axis.

Akaganite, δ-(FeO)OH is the most strongly ferrimagnetic of the oxyhydroxides. It has a hexagonal structure and forms flake-like particles in which the preferred axis is in the plane. Bernal (1959) reported up to 19 emu/g ($24\mu \text{Wb m/kg}$) and Okamoto and Baba (1969a, b) later found $\sigma_{s0} = 25$ emu/g ($31\mu \text{Wb m/kg}$) for their hexagonal platelets which showed a surprisingly high coercivity of 500 Oe. (39.8 kA/m).

Curie temperature. The saturation magnetization of magnetite was measured as a function of temperature in an atmosphere of pure helium by Aharoni et al. (1962). From these results they concluded that the Curie temperature $\theta_c = 575$°C.

They repeated the experiment in an atmosphere of oxygen and found a value of $\theta_c = 585°C$ which is commonly quoted in the literature e.g. by Gorter (1954). The X-ray diffraction pattern for the samples that had been heated to this temperature showed, however, that the material had been partially oxidized to α-Fe_2O_3. In contrast, the samples heated in helium during the measurement of the σ_s-T curves remained as Fe_3O_4. Smith (1956) also measured θ_c of single crystals of magnetite in a protective atmosphere of CO and CO_2 and found $\theta_c = 575.25°C$.

The obvious difficulty in determining the Curie temperature of γ-Fe_2O_3 is that the material is rapidly converted to α-Fe_2O_3 on being heated to temperatures above 400°C. One approach to this problem is to prepare a series of compounds in which γ-Fe_2O_3 is combined chemically with known small amounts of a compound that is structurally similar to iron oxide, in an attempt to stabilize the structure against the transformation. It has been known for some time (Sellwood 1956) that γ-Fe_2O_3 could be stabilized in this way. The method is then to determine the Curie temperature from the σ_s-T curves as a function of the amount of the added compound, and then to extrapolate the curve of θ_c vs additive amount to find θ_c for unadulterated γ-Fe_2O_3. Michel et al. (1950) used up to 7% Al_2O_3 and deduced the Curie temperature of γ-Fe_2O_3 to be 591°C. Taking a more direct approach Aharoni et al. (1962) measured σ_s vs T in an atmosphere of pure helium and concluded that $\theta = 590°C$.

A much higher value was obtained by Brown and Johnson (1962) who fitted a Néel molecular field model, neglecting interaction of the types A–A and B–B between the A and the B sublattices of the spinel, to experimental results of σ_s-T which were made at temperatures below the γ to α transition.

They found that the best fit between theory and experiment was obtained when $\theta_c = 747°C$. Banerjee and Bartholin (1970) also obtained $\theta_c = 747°C$ by extrapolation; in their case H/σ was plotted against σ^2 and fitted to $-\alpha + \beta\sigma^2$. The intercept and the slope β were replotted against temperature and they found α below 738.8 K extrapolated as a straight line to a Curie point of 1020 K (747°C).

At the other extreme of the published results, Takei and Chiba (1966) measured $\theta = 470°C$ on their epitaxially grown single crystal films of γ-Fe_2O_3.

Anisotropy constants. Bickford (1950) used a ferromagnetic resonance absorption method to determine the magnetocrystalline anisotropy constant K_1 of both actual and synthetic crystals of magnetite. He obtained the result that at room temperature, $K_1 = -1.10 \times 10^5$ erg/cc (-1.10×10^4 J/m^3) and $g = 1.12$. Later Bickford et al. (1957) used the torque method on synthetic single crystals and again found $K_1 = -1.10 \times 10^5$ erg/cc (-1.10×10^4 J/m^3). It was also possible from these measurements to make an upper-limit estimate of the second-order anisotropy constant, which they gave as $K_2 = -2.8 \times 10^5$ erg/cc (-2.8×10^4 J/m^3). Smith (1956) used a vibrating-coil magnetometer to measure the magnetization curves of a magnetite single crystal along the principal cubic direction and deduced that the room temperature value of K_1 was -1.2×10^5 erg/cc (-1.2×10^4 J/m^3). McNab et al. (1968) reported that the easy axes of magnetite are along the $\langle 111 \rangle$ directions and the hard ones are along the $\langle 100 \rangle$ directions, in agreement with the other results.

The determination of the anisotropy constants for γ-Fe_2O_3 presents problems. The usual method, cutting samples from a single crystal along the principal planes and then measuring the magnetization curves at different angles to the crystal axes, cannot be used since it has not yet been found possible to grow single crystals that are large enough. However, the first-order anisotropy constant for powdered γ-Fe_2O_3 can be determined by indirect methods. The anisotropy constant for γ-Fe_2O_3 powders was first determined by Birks (1950) from measurements on the complex permeability as a function of wavelength in the microwave region. He deduced that $|K_1| = 4.7 \times 10^4$ erg/cc (4.7×10^3 J/m³). He also measured samples of magnetite and found that $|K_1| = 1.3 \times 10^5$ erg/cc (1.3×10^4 J/m³) which is within 20% of the values measured later by Bickford (1950) and Bickford et al. (1957) on single crystals of magnetite. This suggests that Birk's method may also give a reasonably accurate value of $|K_1|$ for γ-Fe_2O_3.

More recently Valstyn et al. (1962) studied the ferromagnetic resonance spectra of three powder samples of different particle shapes and sizes (spherical, $d \sim 0.12 \mu m$; acicular, $l \sim 0.6 \mu m$ and $0.1 \mu m$). To obtain agreement between theory and their measurements they found it necessary to use a value of $|K_1| = 2.5 \times 10^5$ erg/cc (2.5×10^4 J/m³), which is larger by a factor of five than Birk's result. Later Morrish and Valstyn (1962) reported a value of $K_1 = -3.0 \times 10^4$ erg/cc (-3.0×10^3 J/m³). Köster (1970) measured the temperature dependence of susceptibility of assemblies of γ-Fe_2O_3 particles and deduced that the easy axes were $\langle 110 \rangle$. Takei and Chiba (1966) also used resonance measurements to determine the magnetocrystalline anisotropy constant of their single-crystal films of γ-Fe_2O_3. The plane of their sample was (001) and they found that the easy direction of magnetization was $\langle 110 \rangle$ and the hard direction $\langle 100 \rangle$. The constant K_1 was found to have the value -4.64×10^4 erg/cc (-4.64×10^3 J/m³), an excellent agreement with the earlier measurement on powders by Birks (1950) and probably the most reliable value for K_1.

b. Extrinsic properties

Critical sizes for single-domains. Particles large enough to support domain walls (separating regions having distinctly different directions of magnetization) can reverse their magnetization by wall motion. This process, in materials used in magnetic recording surfaces, occurs in much smaller fields than are required to rotate the magnetic moments in single domains. Thus the coercivity of multi-domain (MD) particles is usually smaller than that of single-domain (SD) particles as was found experimentally by Morrish and Watt (1957) and by Dunlop (1972). However, the differences are often not large and the magnitude of coercivity cannot usually by itself be used as proof of SD or MD behavior. For example the SD particles studied by Dunlop (1972) had a range of coercivity of 325 to 875 Oe (25.9–69.7 kA/m) while his MD particles gave coercivities of 92–205 Oe (7.3–16.3 kA/m). Amar (1957) discussed the difficulties involved in a theoretical treatment of multi-domain particles and more recently Craik (1970) has reviewed the problem. Multi-domains, single-domains, pseudo-single-domains and superparamagnetic particles can all be found in rocks and it is under the subject of rock magnetism that work on MD particles is most frequently published e.g., Parry (1975), Stacey and Banerjee (1974). This topic will not be pursued further here since the particles used in recording surfaces are at present, intended to be single-domains.

The theory of the magnetic properties of fine particles has been comprehensively reviewed by Wohlfarth (1959a) and Kneller (1969). The present discussion will be restricted to those parts of the theory which are necessary to understand the behavior of recording materials. Since the extrinsic magnetic properties of particles depend strongly upon their shape and size we shall consider this question first.

Morrish and Yu (1955) calculated the size at which particles change between SD and MD behavior by comparing the energy of the SD particle with the energy of MD configurations in which the external flux is reduced but to which the domain wall energy must be added. To simplify the calculation they neglected magnetocrystalline energy and considered only magnetostatic, exchange and domain wall energies. They found that particle shape was important in that acicular particles could be larger than spherical ones and still remain SD. They also predicted that the critical size for γ-Fe$_2$O$_3$ should be slightly larger than for Fe$_3$O$_4$ particles. However, Kneller (1969) disagreed with this approach and observed that the key question is not at what size does the SD particle have the smaller energy, but rather at what size is it impossible to nucleate a non-uniform magnetization.

Interactions between particles also affect the critical size. Kondorskii (1952) showed that using the energy minimization method if a particle is SD when isolated it will also be SD when close to other particles. However a particle which is MD when isolated may behave as an SD particle at high packing densities. Some evidence of this was found by Morrish and Watt (1957). One of their samples showed a coercivity which was at first independent of packing density as expected for MD particles, but at greater concentrations the coercivity decreased as the packing density increased.

Dunlop (1972) studied the anhysteretic remanence of synthetic magnetite particles of different shapes and sizes. Acicular particles (average length $\sim 0.2\mu$m, width $\sim 0.03\mu$m) were deduced to be SD since the ratio of remanence to saturation magnetization for a spatially-random distribution of particles was ~ 0.5. In contrast, cubic particles of side ~ 0.035 to 0.22μm were MD since the reduced remanence was only 0.103 to 0.257. The possibility that these low values were caused by superparamagnetic particles in the powder was rejected since the reduced remanence was almost unchanged on cooling the samples to 77 K. An interesting but unexplained result was that the anhysteretic remanence was just as high for the MD particles as for the SD ones.

At the other end of the single-domain range that is in particles smaller than a certain critical size, random thermal forces are large enough to cause the magnetization direction to reverse spontaneously. Néel (1949a, b) calculated the average time between reversals and found it to be an exponential function of the ratio of the particle volume to absolute temperature. A particle will spontaneously reverse its magnetization even in the absence of an applied field when the energy barrier to rotation is about $25\,kT$. Particles smaller than the critical size are said to be "superparamagnetic" (SP) since they show no measureable remanence but possess a magnetic moment that may be many orders of magnitude greater than that of a paramagnetic atom. Superparamagnetism was reviewed by Bean and Livingston (1959) and now is an important aspect of rock magnetism, (Dunlop 1973, Stacey and Banerjee 1974). The colloidal particles of ferric oxide which are used to develop domain patterns (or recorded patterns on tape) and so make them visible, are examples of superparamagnetic particles. Their diameters are usually in the range 20 to 200 Å and this is below the critical size for single domains in iron oxide. Since the direction of magnetization in SP particles is constantly changing, the particles cannot be used for recording.

McNab et al. (1968) measured the Mössbauer effect on very small magnetite particles (100–160 Å) and found that all were superparamagnetic.

The lower limit for SD behavior in acicular particles of γ-Fe$_2$O$_3$ was determined by Berkowitz et al. (1968) from observations of the dependence of coercivity on crystallite size (from X-ray line broadening). Their results are shown in fig. 9 which also shows the dependence of $H_c(83\text{ K})/H_c(293\text{ K})$ on crystallite size. They concluded that particles smaller than about 300 Å were SP and that at room temperature the range of stable SD particles centered on about

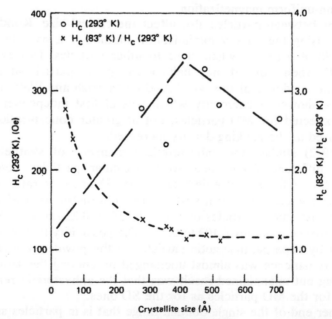

Fig. 9. Coercivity (o) and the ratio of the coercivity at 83 K to the coercivity at 293 K (x), vs crystallite size for acicular particles of γ-Fe$_2$O$_3$ (Berkowitz et al. 1968).

400 Å. It is noteworthy that while their particles having a crystallite size greater than 100 Å were standard magnetic recording oxides, the particles with crystallites smaller than 100 Å were prepared by the dehydration of lepidocrocite. Thus the points on figs. 8 and 9 really fall into two distinct groups and the functional dependence suggested by Berkowitz et al. (1968) must be viewed critically.

Dunlop (1968) suggested that the crystallites of Berkowitz et al. (1968) might behave as MD if the particles were sufficiently isolated.

Dunlop (1973) determined experimentally both the upper and lower critical sizes for SD behavior in cubic magnetite. His analysis was based on two facts: (1) SD particles showed a saturation remanence M_{rs} which is independent of grain size, (Kneller and Luborsky 1963), while MD and SP particles have low M_{rs}, (2) the SP threshold is strongly temperature dependent and so the SP fraction can be calculated from the difference in M_{rs} at room temperature and at a temperature low enough that all the superparamagnetic particles behave as single-domains. Finally the MD fraction was found from the difference between the measured M_{rs} and that calculated for SD particles. Dunlop's results were

Cubic particles of Fe_3O_4

Lower critical size, $d_s = 290$ to 360 ± 50 Å
Upper critical size, $d_0 < 480 \pm 50$ Å.

Radhakrishnamurty (1974) argued that the cation deficient magnetite phases present in Dunlop's samples could lead to a low value for the range of SD behavior.

A theoretical estimate of the critical sizes in equant particles of magnetite was given by Butler and Banerjee (1975). The lower critical size was found from Néel's (1955) relaxation equation and the upper critical size by an energy minimization method. The results were:

Cubic particles of Fe_3O_4

Lower critical size, $d_s \simeq 500$ Å
Upper critical size, $d_0 \simeq 760$ Å.

The calculations of Butler and Banerjee also included acicular magnetite particles for which the results were $d_s = 500$ Å, $d_0 = 1.4\mu m$.

The range of particle lengths over which $\gamma\text{-}Fe_2O_3$ or Fe_3O_4 particles show single domain behavior appears to be approximately from 300 Å to $1.5\mu m$. Superparamagnetic particles are undesirable in this application since they contribute no remanence and therefore no signal. Multi-domain particles are also undesirable if their coercivity is low; such particles would be too sensitive to erasing fields and to print-through. The aims of magnetic particle manufacturers then is not just to limit the lengths of their particles to the range of SD behavior but to restrict them still further in order to narrow the distribution function of switching fields. The coercivity and the remanence-coercivity (switching field) are length-dependent properties which help to determine the sharpness of the transition region on the recording surface and must therefore be confined between the narrowest possible limits.

Coercivity and remanence coercivity. Almost invariably when the magnetic properties of particles or recording surfaces are discussed the coercivity is reported. The reason for this is that it is easily measured from the major hysteresis loop using a vibrating sample magnetometer or an a.c. loop tracer. As we shall see in the discussion of the relationship between the output signal level of a magnetic tape and its magnetic properties, it is the remanence-coercivity (H_r, the field at which $M_r = 0$, see fig. 17 in section 3.6) which is found to be the more significant. Fortunately, as long as we are dealing with the properties of a random array of particles or with aligned particles close to the average direction of orientation, H_r exceeds H_c by no more than 10%. It is only when measurements are made with fields nearly perpendicular to the direction of orientation that large differences arise between H_c and H_r; this condition occurs during the writing process for particles close to the tape surface. In any case it is convenient to have a parameter which in some way expresses the average switching field of a material and its ability to withstand demagnetization and coercivity, H_c, meets that requirement.

Since almost all the iron oxide particles used in recording surfaces are acicular it is natural that the shape anisotropy should be considered as the principal source of their magnetic hardness. The magnetocrystalline anisotropy is not negligible in γ-Fe$_2$O$_3$ but as we have seen the anisotropy constant is small (it is a factor of ten smaller than that of iron and a factor of a hundred smaller than that of cobalt).

The critical field at which a SD particle reverses was calculated by Néel (1947b) and by Stoner and Wohlfarth (1948). The coercivity can be written $H_c = (N_b - N_a)M_s$, where N_b and N_a are the demagnetizing factor of the particle (assumed to be a prolate ellipsoid) along the minor and major axes, respectively. The maximum value that $(N_b - N_a)$ can take is 2π. Thus the maximum coercivity of an isolated acicular particle of γ-Fe$_2$O$_3$, assuming $M_s = 400$ emu/cc (0.50 Wb/m^2), is about 2500 Oe. Unfortunately this exceeds the experimental value by about a factor of ten. Allowing for the spatial randomness of the particles only reduces the theoretical value by a factor of 0.48.

Johnson and Brown (1958) measured the axial ratio of an assembly of acicular particles in the electron microscope and found it to be at least 5 but the particles possessed a coercivity which indicated a theoretical acicularity of about 1.6. They concluded that interactions between particles could not explain the disagreement. An alternative mechanism was proposed by Jacobs and Bean (1955). They postulated a chain-of-spheres model in which the magnetization on alternate spheres rotated in opposite directions. Using this model Wohlfarth (1959b) calculated that the effective magnetic axial ratio should have a maximum value of 1.4 rather than the observed value of 5. A different model for the "incoherent" rotation of magnetization during reversal was postulated by Frei et al. (1957). This is the "curling" mode in which the rotation of the spins is similar to the motion of a bundle of straws grasped at each end and twisted. Experimental evidence for curling came from the measurements of Luborsky and Morelock (1964) on iron whisters.

Further evidence for an incoherent mode of magnetization for particles lying within about 50° at the angle of orientation came from the angular dependence of coercivity of a partially-aligned array of γ-Fe_2O_3 particles (Bate 1961). The coercivity was found to increase at first with the angle between the average direction of particle-orientation and the direction of the applied field. For angles above about 50° the coercivity begins to decrease with increasing angle (as would be found at all angles for particles rotating coherently).

Flanders and Shtrikman (1962) measured the distribution of only those particles lying at about 90° to the applied field. Under these conditions only coherent rotation should occur and they found a close agreement between the axial ratio of the particles, as deduced from their magnetic measurements, and that observed in the electron microscope.

The phenomenon of rotational hysteresis has also been used to distinguish between the different modes of magnetization reversal. When a sample containing single-domain particles is rotated with respect to a steady magnetic field, the magnetization direction of the particles reverses discontinuously if the field is large enough. During the reversals, the particles absorb energy from the field and they do not return it when the direction of sample rotation is reversed. This energy (W) is the rotational hysteresis and can be measured on a torque balance as a function of the applied field. The rotational hysteresis integral

$$\int_0^\infty (W/M_s)\, d(1/H)$$

can then be calculated and compared with the prediction of the coherent and incoherent models (see table 5). The experiment indicates that the mechanism of reversal involves incoherent rotation but it fails to distinguish between those

TABLE 6

Values for the rotational hysteresis integral

Reversal mode	Rotational hysteresis integral
Coherent (Stoner and Wohlfarth 1948)	0.4
Chain-of-spheres fanning, incoherent (Jacobs and Bean 1955)	1.0–1.5
Curling, random infinite cylinders, incoherent (Shtrikman and Treves 1959)	0.4–π
Curling, aligned infinite cylinders, incoherent (Shtrikman and Treves 1959)	0.4–4
180° wall motion (Doyle et al. 1961)	4.0
γ-Fe_2O_3, experimental (Bate 1961)	1.6

modes. However, Bottoni et al. (1972) warn that the agglomeration of particles make it difficult to draw conclusions about particles from these measurements.

We have discussed above the origins of coercivity in iron oxide SD particles solely in terms of shape anisotropy. The role of magnetocrystalline anisotropy must now be treated, even though the magnitude of the anisotropy constants is small (particularly for γ-Fe_2O_3) and suggests that their contribution is not large. Magnetoelastic anisotropy is still smaller and will be neglected here but considered later (section 3.6.2) in the discussion of the stress sensitivity of magnetic tapes.

For spherical SD particles the coercivity is given by

$$H_c = \frac{a|K_1|}{M_s}$$

$a = 0.64$ for random arrays and $a = 2$ for aligned particles. Assuming a random assembly of particles of γ-Fe_2O_3 for which $|K_1| = 4.7 \times 10^4$ erg/cc (4.7×10^3 J/m^3) and $M_s = 400$ emu/cc (0.5 Wb/m^2), we find that $H_c = 75$ Oe which agrees with the experimental values of coercivity, which were found in early tapes made of equiaxed particles (Westmijze 1953).

The temperature dependence of coercivity has been used to distinguish between shape and crystalline anisotropies since shape anisotropy is independent of temperature and crystalline anisotropy is strongly temperature dependent. Furthermore, for SD particles in which shape predominates, $H_c = (N_b - N_a)M_s$ and M_s changes by only 10% between room temperature and absolute zero. Thus these particles should show a temperature-independent coercivity in contrast to the behavior of particles in which the dominant anisotropy is crystalline. Speliotis et al. (1965b) measured the coercivity of acicular particles of γ-Fe_2O_3 from 70–430 K and found a decrease from 380 Oe to 250 Oe (30.2–19.9 kA/m). They also measured samples of cobalt-doped Fe_2O_3 in the form of equiaxed particles. For these particles in which magnetocrystalline anisotropy predominates, the rate of change of coercivity with temperature was ten times as great. Similar measurements by Dunlop (1969) on both synthetic and natural magnetite and maghemite showed that the principal anisotropy in both materials was shape.

The relative contributions of the two anisotropies was determined quantitatively by Eagle and Mallinson (1967). They made use of the fact that at 130 K, $K_1 = 0$ (Bickford et al. 1957), and so at this temperature the coercivity of magnetite particles must be determined only by the particles shape (again assuming that magnetoelastic anisotropy can be neglected). Two samples consisting of particles having the same average axial ratio but different average lengths were measured over the temperature range from 93 to 573 K. The sample with the longer average particle length was then reduced to Fe_3O_4 and its coercivity and magnetization were measured at 130 K. From these measurements the "shape factor" of the sample could be calculated and used at other temperatures to determine the proportion of the coercivity due to shape at any temperature. They found that about 67% of the coercivity at room temperature could be attributed to shape and the remainder to crystalline anisotropy. Eagle and Mallinson also noted that the shape term was almost independent of the particle diameter and agreed well with the chain-of-spheres model.

Remanence. The saturation magnetization M_s of magnetic particles is an intrinsic property since normally it depends only on the chemical composition and the temperature (the dependence of M_s on crystallite size in γ-Fe_2O_3 is an exception found by Berkowitz et al. 1968). The remanent magnetization, M_r, is an extrinsic property since in addition to M_s it depends on the shape and size of the particles (and on their distributions), on the degree of interactions and their alignment, Stoner and Wohlfarth (1948) assumed an assembly of identical, non-interacting uniaxial particles and calculated the reduced remanence ($= M_r/M_s$ and usually called "squareness," S, in discussions of recording materials) as 1.0 for fully aligned particles measured along the alignment direction; 0, when the measurements where made perpendicular to that direction, and 0.50 for a random array of particles. Wohlfarth and Tonge (1957) considered particles having between two and ten equivalent easy directions of magnetization and showed that S increases monotonically as the average angle decreases through which the magnetization must rotate to find the nearest easy direction. The more complicated situation of particles having mixed anisotropies, uniaxial (e.g. from the shape of the particle), and cubic (from magnetocrystalline anisotropy) was treated by Tonge and Wohlfarth (1958). The calculation could be made exactly when the uniaxis was aligned with one of the cube axes. The values of S corresponding to pure cubic anisotropy are 0.83 for positive anisitropy and 0.87 for negative anisotropy (Lee and Bishop 1966). Heuberger and Joffe (1971) treated the case of assemblies of SD elliptical platelets having a uniaxial anisotropy in the plane of the particles. For a random assembly they found a value of $S = 0.638$ in contrast to $S = 0.5$ which Stoner and Wohlfarth (1948) found for a random assembly of prolate ellipsoids.

Experimentally the values of S obtained from diluted powders are in excellent agreement with these calculations. For example, Dunlop (1972) found $S = 0.464$ for powders of SD, and acicular γ-Fe_2O_3 particles. Kneller and Luborsky (1963) in a very different system obtained $S = 0.46$, at temperatures from 4 to 207 K, from small spherical particles of Fe–Co. These particles were also single-domain and uniaxial but the anisotropy was magnetocrystalline. Equiaxed SD particles of γ-Fe_2O_3 (K_1 negative, $\langle 110 \rangle$ probably the easy axis, and thus $S = 0.86$) were measured by Dunlop (1972) and found to have $S = 0.620$ i.e. between 0.5 and 0.86. The presence of MD or of SP causes a reduction in the values of S (and also of H_c) and must be avoided in recording materials.

The theoretical value for M_s for γ-Fe_2O_3 is 400 emu/cc (0.503 Wb/m^2) and since the density is 5.197 g/cc (5.197×10^{-3} kg/m^3) the saturation magnetization-per-unit-mass is 76.97 emu/g (96.67 μWb m/kg). Assuming that the powder samples consist of SD acicular particles the remanent magnetization should be $0.5 \times 76.97 = 38.49$ emu/g and this value is remarkably close to the values reported for the high-coercivity particles in the previous section. For example, Sato et al. (1976a) found $\sigma_r = 36.9$ emu/g (46.3 μWb m/kg); Horiishi and Fukami (1975) obtained $\sigma_s = 78$ emu/g $(98 \mu$ Wb m/kg); Leitner et al. (1973) measured $\sigma_r = 38$ emu/g $(47.7 \mu$ Wb m/kg).

Particle interactions; experimental results. Figure 10 shows a transmission electromicrograph of a cross-section of a magnetic tape; the magnetic particles occupy about 40% of the volume (and about 75% of the mass) of the coating. It is immediately evident that the extrinsic properties of a sample must be greatly affected by the proximity of and the interaction between neighboring particles. We cannot expect such an assembly to behave like a collection of isolated particles and it is not surprising that the subject of interactions has received a great deal of attention from both the theorist and the experimentalist. It is particularly unfortunate that their efforts have, as yet, born so little fruit.

Néel (1947a) considered the effect of interactions on the coercivity of an assembly of SD particles by calculating the magnetostatic interaction energy. He obtained the formula $H_c = H_c(0)(1 - p)$, where $H_c(0)$ is the coercivity at infinite

Fig. 10. Transmission electron micrograph of a thin section of tape (P.S. Laidlaw, IBM, Boulder, Colorado, unpublished).

dilution and p is the volume packing factor. Néel obtained the formula by calculating the magnetostatic energy (from a modified Lorentz-field approach) and relating this energy to the coercivity.

The derivation of the formula has been criticized by Wohlfarth (1963) and by others but acicular particles of γ-Fe_2O_3 and Fe_3O_4 were found by Morrish and Yu (1955) to obey the relation as is seen in fig. 11. Kondorskii (1952) argued that Néel's approach did not allow sufficiently for the effect of nearest-neighbor particles. He suggested an improved formula

$$H_c(p) = H_c(0)(1 - p/p_0) + C \quad \text{for} \quad p < p_0$$
$$H_c(p) = c \quad \text{for} \quad p > p_0$$

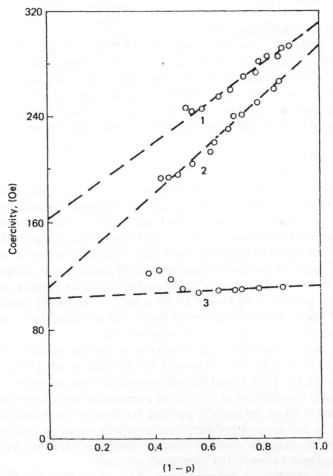

Fig. 11. Coercivity vs $(1 - p)$, where p is the volume packing fraction, for samples of iron oxide. (1) Acicular Fe_3O_4, (2) acicular γ-Fe_2O_3, (3) equiaxed Fe_3O_4 (Morrish and Yu 1955).

where

$$p_0^{-1} = \frac{4\pi}{3}\left(\frac{2N_a + N_b}{N_a N_b}\right).$$

Morrish and Watt (1957) measured the coercivity as a function of volume concentration for four samples of γ-Fe_2O_3 of different average particle shape and size. They found a rather qualitative agreement with Kondorskii's predictions.

Another formula was deduced by Martin and Carmona (1968) to describe the variation of the measured coercivity with packing fraction. They obtained good agreement between experiments with elongated single-domain (ESD) iron particles and the expression

$$H_c(p) = H_c(0) - \frac{\Lambda p}{1 - p}$$

where Λ was an empirical constant.

A related problem, considered by Weil (1951), was the effect of internal cavities in the particles on the coercivity. We have seen that cavities arise in particles of γ-Fe_2O_3, prepared by the standard method, during the dehydration and reduction steps. Weil considered that the dilution formula of Néel (1947a) should be applied to cavities rather than to the space between particles as appears to have been Néel's intention. The formula then becomes

$$H_c = H_c(0)\left(1 - \frac{\rho_a}{\rho_0}\right)$$

where ρ_a is the measured density of the powder and ρ_0 the bulk value of density.

Osmond (1953) saw the effect of cavities as reducing the magnetization from M_s to $M_s(1 - v)$ where v is the volume fraction of cavities. He estimated v by first assuming that the volume of an iron ion is roughly one eighth that of oxygen and thus the reduction to magnetite (α-$Fe_2O_3 \rightarrow Fe_3O_4$) involved a volume reduction from $12\frac{3}{4}$ units (equivalent oxygen ions) to $8\frac{3}{4}$ units; that is a loss of 31.4%. In the step, $Fe_3O_4 \rightarrow \gamma$-$Fe_2O_3$, oxygen is regained to a total of $9\frac{3}{4}$ units, so that for the whole process α-$Fe_2O_3 \rightarrow Fe_3O_4 \rightarrow \gamma$-$Fe_2O_3$ a net loss of 23.5% occurs. Using the latest ASTM figures for the densities of Fe_3O_4 and γ-Fe_2O_3, these volumes losses must be changed to 25.7% and 21.7% respectively.

The experimental dependence of coercivity on volume packing fraction in the range $p = 0.0003$ to 0.20 was examined by Bottoni et al. (1972) for acicular particles of γ-Fe_2O_3. They found that the coercivity was apparently independent of p (as did also Candolfo et al. (1970) but commented on a besetting problem of all measurements when the particle packing fraction is the independent variable. The problem is that magnetized particles invariably occur in clumps and that in attempting to increase the average distance between particles we may merely change the distance between the clumps of particles.

Bottoni et al. (1972) also examined the effect of packing fraction on squareness and found for particles of γ-Fe_2O_3 that a slight decrease occurred at first with increasing dilution but then no further changes were seen. They caution that "the formation of agglomerates is inevitable" and that measurements of S probably reflect more the properties of the agglomerates than the behavior of isolated particles.

Smaller and Newman (1970) also measured the coercivity and squareness as a function of packing factor of acicular particles of γ-Fe_2O_3. They found that the results depended on the method by which dilution was achieved. In general, the coercivity tended to increase and the squareness decreased slightly from $p = 0.4$ to $p = 0.004$. They also commented on the agglomeration of particles and noticed that the degree of agglomeration increased with increased dilution. The breaking up of agglomerates is an important step in the manufacture of recording coatings, since if the agglomerates were allowed to remain not only would the magnetic properties be inferior but, more importantly, the surface would be rough and the separation between it and the head would be increased. The effect of this would be to reduce the signal level and to increase the modulation noise. The particles are single-domains and this undoubtedly is a factor in their agglomeration but the sintering which occurs during the dehydration and reduction steps in the preparation of the particles is perhaps a more important factor. For this reason, as we have seen, many attempts have been made recently to isolate the particles and to reduce the temperatures at these stages.

Particle interactions; theoretical results. The theoretical problem of accounting for the interactions between single-domain particles is a very complex one. The approaches which have been taken are the two-particle models and the models involving large assemblies of particles.

As a first step, it is reasonable to treat the two-particle model as Néel (1958, 1959) did. There are then three remanent states for the pair; $+M_r$, 0, $-M_r$ and two critical fields, one to switch the magnetizations anti-parallel and a larger field to switch them parallel. Brown (1962) criticized the "local field" method of nucleation field calculations on the grounds that it tested the stability of one particle while holding all the others constant. Bertram and Mallinson (1969) avoided this objection in their calculation of the nucleating field for a pair of interacting anisotropic dipoles whose axes were parallel although the particles were not necessarily side by side. They calculated the switching fields for a range of interparticle distances and bond angles. In contrast to the independence of nucleating field and packing fraction found in previous "local field" models, they found a monotonically decreasing switching field with packing fraction in agreement with the experimental results of Morrish and Yu (1955), shown in fig. 11. Bertram and Mallison (1969) also found that as the bond angle decreased, the mode of magnetization reversal changed from coherent to incoherent which agreed with the conclusions of Bate (1961) based on experiments with partially-aligned samples of iron oxide particles.

Many modern calculations of the effect of interactions on switching fields treat large assemblies of particles. Moskowitz and Della Torre (1967) used an $n \times n \times n$ $(n > 6)$ array of aligned particles each of which was given a dipole moment and a switching field and they found that coherent switching occurred in strings of particles by three different processes. In type I, a given element switches and then so do its neighbors until the entire column has switched. In type II, the second element to switch may not be adjacent to the first but several particles removed and in type III randomly-positioned elements switch.

Straubel and Spindler (1969) described an even simpler model; an infinite linear chain with only nearest neighbor interactions within the chain. Their justification for this was that aligned particles in practice form chains in which the distance between chains is greater than the distance between particles within a chain. Hysteresis loops were plotted for a range of values of a parameter that was the ratio of the interaction field to the minimum switching field of the particles in the absence of interactions. Kneller (1969) considered a planar assembly of infinitely long chains of particles; within each chain the interactions were treated as a linear Ising model and for interactions between chains a local field was used. The total interaction behavior was found to depend largely on the weaker interaction between chains and hardly at all on the stronger interactions within the chains. Thus each chain behaved as a unit and could be replaced by an infinite cylinder. Kneller's model was applied to the calculation of anhysteretic and normal remanence curves. "Anhysteresis" is the name given to the process whereby an assembly of particles becomes magnetized by the simultaneous application of a steady field (or very slowly changing) and a decaying, alternating field which is initially large enough to saturate the sample but ultimately is zero. This process is believed to resemble the recording method in which the analog signal (slowly varying) is added to an ac bias signal of much higher frequency. Most models of anhysteresis obtain the result that the anhysteretic susceptibility is infinite. An exception was the result of Jaep (1969) who analyzed the process by a method that took account of the time dependence of particle switching. Each particle was assigned a switching probability which depended on the applied field and the switching time constants of the particle. In a second paper Jaep (1971) considered the effect of interaction between the particles and concluded that interactions do not provide a mechanism for example, for the reversal process, but they can greatly modify the size of the effects. The anhysteretic susceptibility calculated for non-interacting particles is about forty times the value found in practice. Jaep (1971) showed that the discrepancy could easily be accounted for by interactions. This work was extended to interacting particles of chromium dioxide by Bierlein (1973). He found that hysteresis and time-effects were only weakly dependent on interaction between particles and that these effects can be satisfactorily explained in terms of particle size distributions. Anhysteretic Remanent Magnetization (ARM) and Thermo-Remanent Magnetization (TRM), however, were found to be sensitive both to interactions and to size distribution. Narrow size distribution led to longer linear regions in the curve of anhysteretic susceptibility versus applied field "In" audio recording, it is most important to have this linear region as long as possible.

Bertram and Bhatia (1973) calculated a spatial mean interaction field and its effect on the saturation remanence of an assembly of identical, uniaxial, single-domain particles. The particles were assumed to be angularly arranged according to a known distribution function and the remanence was found by calculating the magnetization orientation function of the particle easy-axis orientation which minimized the energy of the assembly. The model yielded increased remanences due to interactions when the effect of the shape of the sample could be neglected. This agreed with the observation that the squareness of a tape increases when the coating is compressed or "densified" (Smaller and Newman 1970).

The interactions between large numbers of particles are usually treated by statistical methods that are based on the Preisach (1935) diagram. In this model, the particles are considered to have equal switching fields for positive and negative applied fields when the particles are isolated but unequal switching fields (caused by interaction fields) when interactions occur. The diagram is thus a two-dimensional distribution function that represents the number of particles, n (a, b), having a particular combination of positive (a) and negative (b) switching fields. The cross-section of the two-dimensional function along the diagonal $a = -b$ gives the distribution of particle switching fields in the absence of interactions and the cross-sections parallel to $a = b$ give the spectrum of spatially-fluctuating interaction fields (Dunlop 1969). The application of the Preisach diagram to particles of recording materials was first considered by Daniel and Levine (1960), by Woodward and Della Torre (1960) and by Schwantke (1961). One of the principal attractions of the approach was its ability to describe in a readily understood form the process of anhysteretic magnetization which is a first approximation to the description of ac bias recording (Mee 1964). In this process the recording medium is subjected simultaneously to a high-frequency field which is initially large and ultimately zero and a smaller (by a factor of about five) slowly varying signal field. In the limit the ratio of the moment so acquired to the signal field is the anhysteretic susceptibility which the model correctly predicts should be independent of particle magnetization, coercivity and packing fraction (Mallinson 1976).

The Preisach model was criticized by Brown (1962) on the grounds that it relies on a local interaction field. Brown showed that the true field between two identical particles was about half that given by the local field calculation. However, the introduction of a third particle reduces the difference by 30% and Dunlop (1969) suggested that in the limit of many interacting particles the interaction field result probably gives reasonable results. Brown (1971) appeared to overcome his earlier objections.

The magnitude and the direction of the interaction field at any given particle change as the magnetization of the assembly changes and thus it might appear that a different diagram would be needed at each point on the major and minor hysteresis loops. The principle of statistical stability of the diagram was invoked as a possible answer to this difficulty. Several experimenters have tested the principle, most (Bate 1962, Della Torre 1965, Dunlop 1968, Paul 1970) have found it to hold at least over most of the area bounded by the major hysteresis loop; some disagreed (Girke 1960 and Noble 1963). This is an important question since, as Dunlop (1969) observes only within the limits of statistical stability can the magnetic properties of an assembly of interacting particles be predicted with reasonable accuracy.

Waring (1967) attempted to formulate a model which combined the advantages of a statistically stable Preisach diagram with a microscopic description of the interactions between particles. His model involved bundles of particles between which the interactions are negligible and within which the interaction field can be replaced by a demagnetizing field. Support for the model was obtained in the measurements of Waring and Bierstedt (1969) on mixtures of CrO_2 and γ-Fe_2O_3 particles.

3.2. Iron–cobalt oxide particles

It appeared for many years that the highest coercivity achievable with iron oxide particles was about 300 Oe (23.9 kA/m). Yet progress in recording required that ever shorter wavelengths be recorded and, therefore, that particles be found with the higher coercivities needed to resist demagnetization. It was not easy to find ways of enhancing the acicularity of particles but Bickford et al. (1957) had shown that magnetocrystalline anisotropy of iron oxide could be greatly increased by substituting a small amount of cobalt in place of the iron atoms. They studied the magnetic properties of single crystals of $Co_xFe_{3-x}O_4$ where $x = 0$, 0.01, 0.04 and their results are shown in fig. 12. The starting materials were iron and cobalt nitrates mixed in aqueous solution, evaporated to dryness and converted directly to the oxide by slow heating to 700°C in a platinum crucible. They obtained the their values for K_1 and K_2 at 300 K (see table 7). The temperature dependence of K_1 (fig. 12) and of K_2 increased greatly with increasing amounts of cobalt. In most ferrites having the spinel structure, K_1 is negative and K_2 is of relatively minor importance at room temperature, indicating that the body diagonal of the cubic lattice, [111], is the easy direction of magnetization. Cobalt ferrite, $CoO \cdot Fe_2O_3$, is distinct in that the first-order anisotropy constant is larger than that of the other cubic ferrites by at least an order of magnitude. Furthermore, it is positive and [001] is the direction of easiest magnetization. Bickford et al. (1957) showed that, for their crystals of cobalt-substituted magnetite K_1 was positive at least up to a temperature of 450°C. In this material, then, there are six equivalent easy directions of magnetization.

TABLE 7

Anisotropy constants for $Co_xFe_{3-x}O_4$, for different values of x (after Bickford et al. 1957)

x	$K_1 \times 10^{-4}$ erg/cc ($K_1 \times 10^{-3}$ J/m³)	$K_2 \times 10^{-4}$ erg/cc ($K_2 \times 10^{-3}$ J/m³)
0	−11.0	−2.8
0.01	0	−3.3
0.04	+30.0	−18.0

Wohlfarth and Tonge (1957) calculated the dependence of M_r/M_s (squareness, S) on the number, n, of equivalent easy axes possessed by each particle, on the assumption of a completely random spatial distribution of identical SD particles. For $n = 6$ they found that $S = 0.825$ which is very close to the value found experimentally at low temperatures (77 K) for particles of cobalt substituted γ-Fe$_2$O$_3$. Lower values for S were measured at room temperature but the average particle size was about one tenth that of typical acicular particles of γ-Fe$_2$O$_3$, and so it is quite possible that some particles had crossed the threshold into the superparamagnetic region.

Clearly in order to increase both the coercivity of iron oxide particles the most promising element for substitution is cobalt.

3.2.1. Cobalt-substituted magnetite, $Co_xFe_{3-x}O_4$

The composition CoFe$_2$O$_4$ corresponds to the substitution of Co^{2+} for all the Fe^{2+} in magnetite. A method of preparing fine powders of this cobalt ferrite was given

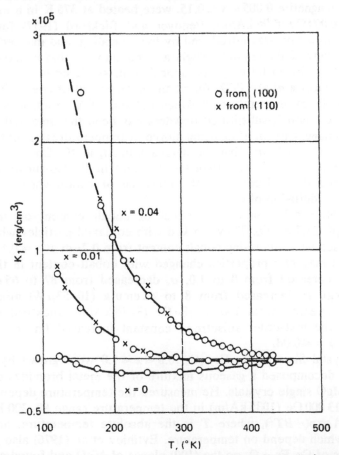

Fig. 12. Temperature dependence of first order anisotropy constant K_1 vs absolute temperature for magnetite and cobalt-substituted magnetite Co$_x$Fe$_{3-x}$O$_4$; $x = 0, 0.01, 0.04$ (Bickford et al. 1957).

by Schuele and Deetscreek (1961); the particles were precipitated from an aqueous solution of cobaltous and ferric chlorides by boiling sodium hydroxide solution and the mixture was vigorously stirred. The product was shown by X-ray diffraction to contain only Fe^{3+} $(Co^{2+}Fe^{3+})O_4$ and had the properties $\sigma_s = 70$ emu/g (88μ Wb m/kg) and $H_c = 750$ Oe (59.7 kA/m). Much higher coercivities (>2.100 Oe, 167 kA/m) were measured by Berkowitz and Schuele (1959) on particles of $CoFe_2O_4$ of rather larger size. Makino et al. (1974a, b) described a method for making cobalt–ferrite ($Co_xFe_{3-x}O_4$), $x < 1$, particles in which sodium hydroxide was used to precipitate the hydroxides from solutions of ferrous and cobaltous sulfates. Then potassium nitrate was used to oxidize the Fe^{2+} ions to Fe^{3+}. The precipitates were finally heated to 400°C for one hour and cooled in a rotating magnetic field of 5000 Oe (398 kA/m). The purpose of the last step was explained by Kamiya et al. (1970) and was to remove the substantial uniaxial anisotropy which tends to be induced by the heat-treatment of cobalt-substituted magnetite in the absence of an applied field. When single crystals of cobalt-substituted magnetite $0.005 < x < 0.15$, were heated at 375 K in a magnetic field of about 10 000 Oe (796 kA/m), Penoyer and Bickford (1957) found that an additional anisotropy was introduced by the annealing, and masked the original cubic anisotropy of the crystals. Sugiura (1960) showed that uniaxial anisotropy could also be induced by magnetic annealing in single crystals of composition $Co_{0.8}Fe_{2.2}O_4$. Ogawa et al. (1972) found an enhancement of coercivity on annealing the particles at temperatures between 100 and 400°C in nitrogen. A besetting problem of the cobalt-substituted powders has been the time and temperature dependent changes in M_r after being stored at temperatures of more than 40°C. Ogawa et al. (1972) found that the decrease in M_r in 4% Co-substituted Fe_3O_4 was usually considerably less than the decrease in 4% Co-substituted γ-Fe_2O_3 and attributed the difference to the presence of cation vacancy in γ-Fe_2O_3 permitting the diffusion of cobalt.

Sato et al. (1962) studied the effect of cobalt content on the magnetic properties of $Co_xFe_{3-x}O_4$. They worked with equiaxial particles whose average size decreased with increasing cobalt content from 0.2μm at $x = 0.1$ to 1.0μm at $x = 1.0$. The magnetic properties changed with cobalt content in the following way: as x increased from 0 to 1.0, σ_s decreased from 87 to 65 emu/g (109–82μ Wb m/kg), σ_r increased from 8 to 42 emu/g (10–53μ Wb m/kg), and the coercivity increased from 100 to 980 Oe (8–78 kA/m). Bickford et al. (1957) found that the first-order anisotropy constant increased linearly with cobalt content for $x = \leqslant 0.04$.

Single-crystal films of $Co_xFe_{3-x}O_4$, $0.8 \leqslant x \leqslant 1.0$ were grown by Kumashiro (1968) who decomposed a gaseous mixture of the metal bromides onto various planes of MgO single crystals. He measured the temperature dependence of K_1 in fields $\leqslant 13\,000$ Oe (1035 kA/m) in the temperature range 80–300 K and found that $K_1 = A\exp(-BT^2)$, where T is the absolute temperature, and A, B are constants which depend on temperature. Evtihiev et al. (1976) also grew single-crystal films of $Co_xFe_{3-x}O_4$ on the (100) planes of MgO and found an extremely large perpendicular anisotropy which had a maximum of 4 to 8×10^6 erg/cc ($4–8 \times 10^5$ J/m³) for values of x between 0.7 and 0.8.

It is common in commercial ferrites to find the two or more divalent cations are used to replace some of the Fe^{2+} in magnetite. Thus, Fukuda et al. (1962a) used manganese, nickel, and magnesium with nickel in addition to cobalt. They found that the coercivity and the remanence could be enhanced (by 20 to 40%) by magnetic annealing in a closed tube filled with nitrogen at 200–500°C for about 30 min in a field of about 1000 Oe (79.6 kA/m). The temperatures used were below that at which sintering occurs (600°C) but to reduce further the possibility of agglomeration an ultrasonic vibration (80 kHz) was applied during the annealing. Sato (1970) also used manganese with cobalt and found that the coercivity ratio of $MnFe_2O_4$ and $CoFe_2O_4$ is 1:50 and equal to that of the magnetocrystalline anisotropy constant, K_1, for the sintered ferrites. The additional elements are used in an attempt to overcome the undesirable loss of remanence (and thus, of output signal) which occurs when pure cobalt ferrite particles are kept at temperatures above room temperature. Makino et al. (1972) compared the behavior of cobalt ferrite particles and chromium dioxide particles and found that although the temperature depndence of H_c was about the same (for different reasons), the saturation magnetization, M_s, also decreased with increasing temperature for chromium dioxide but not for cobalt ferrite. Thus, the temperature independence of signal output in chromium dioxide resulted from an equal rate of change with temperature of the driving force for demagnetization (M_r) and the ability to withstand demagnetization (H_c). Their solution was to increase the temperature dependence of M_s (and of M_r) in cobalt ferrite so that it matched the temperature dependence of the H_c. They achieved this by the addition of small amounts of M in

$$(Co_xM_y)O_{x+y} \cdot Fe_2O_3$$

where M was zinc, magnesium or calcium. When zinc or magnesium were used the optimum amounts were $x = 0.3$ to 1, $y = 0.1$ to 0.25 and when cadmium or calcium were used the best results were found for $x = 0.05$ to 1, $y = 0.03$ to 0.1.

3.2.2. Cobalt-substituted iron oxides between Fe_3O_4 and Fe_2O_3

In Berthollide compounds the ratio of positive ions (e.g. iron) to negative ions (e.g. oxygen) is not fixed. Thus Berthollide compounds in the iron–oxygen system can show a degree of oxidation which is between that of Fe_3O_4 and Fe_2O_3. Sasazawa et al. (1975) and Sasazawa et al. (1976b) precipitated seeds of goethite in the usual way and then grew layers of goethite containing cobalt on the seeds. The crystals were dehydrated and reduced to cobalt-magnetite and, finally, were oxidized to give particles whose composition can be represented as MO_x where $M = Fe + Co$ and $x = 1.36$ to 1.47. Coercivities were reported between 780 and 4820 Oe (62–383 kA/m) and the latter value was increased to 5830 Oe (464 kA/m) by heating at 40–80°C for 3 days (possibly in a magnetic field). Itoh and Satou (1975a) also prepared Berthollide compounds of iron–oxygen and increased the coercivity of the particles by adding to them phosphorus in the form of P_2O_5 in water and heating the mixture. A method of stabilizing particles of cobalt-substituted Berthollide compounds of iron and oxygen was given by TDK (1976). The particles, after being coated onto tape, were exposed to a longitudinal field of more than 2000 Oe (159.2 kA/m) then briefly heated to a temperature of 80–120°C and finally stored at 60°C for one week. The process was said to work particularly well for particles whose coercivity exceeded 1000 Oe (79.6 kA/m).

3.2.3. Cobalt-substituted γ–Fe₂O₃

Iron oxide particles containing cobalt were first prepared from a mixture of ferrous and cobaltous sulfates by Jeschke (1954) and Krones (1960), by a method very similar to the standard one for γ-Fe$_2$O$_3$. However, the result was equiaxed rather than acicular particles. Except for the very earliest iron oxide particles, the ones used to make magnetic recording surfaces have been acicular. The reason for this is that particles that derive their magnetic hardness from shape anisotropy usually show only a slight temperature dependence of coercivity. Since the particle shape can be considered independent of temperature, the coercivity changes as the saturation magnetization and this is negligible for pure iron oxides at all practical temperatures. Equiaxed particles rely for their magnetic hardness on magnetocrystalline anisotropy which is often strongly temperature dependent, as seen in fig. 12.

However, there is another important difference; particles in which magnetocrystalline anisotropy dominates usually have six or eight equivalent easy directions of magnetization in contrast to two for acicular particles in which shape is the only anisotropy. Thus equiaxed particles are able to respond to the perpendicular as well as to the in-plane component of the writing field and the resulting pattern of magnetization is more faithful to the shape of the field. It is not clear just how important this distinction is. Figure 7c shows typical particles of cobalt-substituted γ-Fe$_2$O$_3$ and in this case the cobalt content was 3 wt%. The H_c for 10 at.% Co was 800 Oe (63.7 kA/m). Nobuoka et al. (1963) starting with an alkaline solution of ferric chloride and cobaltous nitrate under pressure, obtained particles containing 0.2 to 3% cobalt ($H_c \leqslant 600$ Oe, 47.8 kA/m, $S = 0.5$). Toda et al. (1973) claimed to overcome the disadvantages of low acicularity and non-uniform distribution of cobalt (it appears that much of the cobalt occurred on the surface of the particles rather than forming a solid solution) of the earlier cobalt-substituted particles by performing the precipitation and growth in a strongly basic solution (pH > 11) and by blowing the oxidizing gas through the mixture which was kept between 30 and 50°C. The result was acicular particles, length 0.2–0.5 μm and acicularity 10 : 1 in which the cobalt ions are dissolved to form a solid solution. The magnetic properties of the particles were: $H_c = 480$ Oe (38.2 kA/m), $\sigma_s = 80.2$ emu/g (101μ Wb m/kg), $S = 0.58$. Kawasaki (1973) used a very similar procedure but continued the reduction until metal particles (10 at.% Co, 90 at.% Fe) were formed, having a coercivity of 1250 Oe, (99.5 kA/m). Trandell and Fessler (1974) used cobaltous nitrate with ferrous sulfate and precipitated the oxyhydride which, in a departure from standard procedure, was brought to a pH of 8.5 by being heated with sodium hydroxide solution. The acicular particles contained 4.9% Co and had a coercivity of 1402 Oe

(111.6 kA/m). A variation in the precipitation procedure was due to Kugimiya (1975) who heated the solution of iron and cobalt salts in the presence of urea and subsequently an alkali solution. His particles had an average length of 0.2μm, an acicularity of 15:1 and a coercivity of 760 Oe (60.5 kA/m).

Ando and Wakei (1976a) treated the precipitated iron–cobalt oxyhydroxide particles with oxalic acid solution which was absorbed by the particles and helped to prevent sintering in the dehydration and reduction stages.

A detailed chemical, structural, and magnetic study of the solid solutions between γ-Fe$_2$O$_3$ and CoFe$_2$O$_4$ was made by Mollard et al. (1975). Their samples were prepared by decomposition of the oxalates and reduction to magnetite followed by a low-temperature oxidation. Differential thermal analysis during the last step showed an endothermic peak at 130°C corresponding to the release of absorbed water, and exothermic peaks at 180°C (oxidation) and at higher temperatures, 450–700°C (transformation $\gamma \rightarrow \alpha$). It is interesting that increasing amounts of substituted cobalt cause a monotonic increase in the transition temperature.

a. Stabilization of cobalt-substituted γ-Fe$_2$O$_3$

The marked decrease in remanent moment which was found in cobalt–magnetite after being held for a few hours at quite modest temperatures (50–100°C) also occurred in cobalt-substituted γ-Fe$_2$O$_3$. Clearly this is a most undesirable attribute for a material whose application is the long-term storage of information. The problem was recognized early in the development of the materials and Jeschke (1966) claimed that his method of substituting not only cobalt but manganese nickel and zinc, etc. led to a doped oxide of improved thermal stability. His method was to precipitate cobaltic ferric hydroxide (if cobalt was the additive) from a solution of ferrous sulfate and cobalt sulfate by sodium thiosulfate with potassium iodate added as an oxidizing agent. The precipitate was reduced to cobalt–magnetite with carbon monoxide and then re-oxidized in air to give cobalt-substituted γ-Fe$_2$O$_3$. It is not clear how the increased stability was achieved.

The method of Makino et al. (1972) to stabilize cobalt–magnetite was discussed above: it consisted essentially of substituting additional elements in order to achieve a temperature dependence of magnetization which matched the strong temperature dependence of coercivity. The same technique has been used with cobalt–γ-Fe$_2$O$_3$, Makino et al. (1973), and the additives were cadmium and zinc. The final product had the composition Co$_{0.4}$Cd$_{0.1}$Zn$_{0.05}$Fe$_2$O$_3$ and the coercivity was 440 Oe (35 kA/m). Bayer A-G (1975) similarly reported the stabilizing effect of zinc, titanium, chromium and also of phosphate ions when included in the particles of Co–γ-Fe$_2$O$_3$. For example ZnSO$_4$·7H$_2$O and Na$_3$PO$_4$·12H$_2$O were mixed with ferrous and cobaltous sulfates in the initial step of the preparation. The resulting particles of γ-Fe$_2$O$_3$ containing cobalt and coated with phosphate had a coercivity of about 550 Oe (43.8 kA/m) and the remanent magnetization, M_r, decreased by only 4–15% after the particles had been heated to 600–700°C for one hour.

Annealing the particles of [FeCo]O·OH at 680–700°C before dehydration, reduction and oxidation was reported to stabilize M_r by Woditsch and Leitner (1973). Finally the method of Matsumoto and Matsuo (1974) to stabilize M_r was to dehydrate and reduce the α-(FeCoO)OH in a chlorine-containing atmosphere at 200–450°C. The particles were then reoxidized to γ-(Fe, Co)$_2$O$_3$ and both the H_c and M_r were found to show a reduced dependence on temperature. No mechanisms were proposed for these stabilization procedures.

3.2.4. Cobalt-impregnated γ-Fe$_2$O$_3$

The method of forming oxide particles of iron and cobalt by starting with a mixture of their salts frequently led to equiaxed particles or at least to particles with a reduced acicularity. Furthermore, the magnetocrystalline anisotropy was strongly temperature dependent and the dependence increased with the amount of substituted cobalt and therefore, with coercivity. An appealing alternative was to make particles containing cobalt but also retaining the acicularity of γ-Fe$_2$O$_3$. If the two anisotropies were additive then to achieve a given coercivity the amount of cobalt (and the temperature dependence of coercivity) could be reduced.

Haller and Colline (1971) coated standard acicular γ-Fe$_2$O$_3$ particles with a cobalt compound (e.g. CoCl$_2$·6H$_2$O) which would decompose on heating to give CoO. Further heating at 245°C in an atmosphere of nitrogen caused the cobalt to impregnate the surface of the particle. The coercivity of the particles increased rapidly with cobalt content to a maximum of about 850 Oe (67.7 kA/m) at 17.5% cobalt. Further increase in coercivity to 1875 Oe (149.3 kA/m) could be achieved by partially reducing the particles until the amount of FeO represented about 10% of the weight of the particles. Hwang (1972) used a similar method except that his particles were coated at the α-Fe$_2$O$_3$ stage or even in the form of α-(FeO)OH, and then heated at 350°C for an hour to decompose the cobalt salt to the oxide Co$_3$O$_4$. Then the usual path of reduction and oxidation was followed and particles, which preserved the acicularity of the starting material, were obtained. As the percentage of cobalt in the end product was increased from 0% to 8% the coercivity increased from 235–560 Oe (18.7–44.6 kA/m) and the squareness increased from 0.35 to 0.54. Hwang (1973) cautioned that the cobalt salt used must decompose below 600°C since the acicularity is destroyed at higher temperatures. He recommended the nitrate, acetate, formate, or chloride and suggested that the temperature used to diffuse the cobalt or CoO into the surface of the iron oxide particle be kept between 250 and 450°C. Iron oxide particles were mixed with freshly precipitated cobaltous hydroxide gel which was uniformly adsorbed on the surface of the particles. After decomposing the hydroxide the material was partially reduced to (Fe, Co)O and cooled slowly in air to which nitrogen had been added. A slow cooling rate enhanced the coercivity as did increasing the amount of Fe^{2+} up to about 25% of the total iron content. Hwang (1973) found that by using these procedures high coercivity particles could be made with less cobalt and thus at lower cost.

Kaganowicz et al. (1975) discussed the effect of small amounts of zinc on stabilizing the time- and temperature-dependent properties M_r and H_c, in cobalt-impregnated particles of γ-Fe_2O_3. Their particles, also, were partially reduced toward magnetite (14.5–19.5% Fe^{2+}). The optimum ratio of zinc to cobalt depended on the amount of Fe^{2+} but generally the zinc content was equal to or greater than the cobalt content and more than 3.5% Zn was preferred. The instability of M_r and H_c in cobalt-containing particles is believed to be related to the pressure- and temperature-sensitivity of the magnetocrystalline anisotropy and the role of the zinc is to reduce that sensitivity. Matsumoto and Matsuo (1975) added cuprous chloride to the cobalt salts and obtained particles having enhanced thermal stability.

Particles of iron oxide surrounded by a shell of cobalt-doped iron oxide are being used increasingly in the preparation of recording surfaces for use at high densities (or short wavelengths). Not surprisingly there are a large number of variations which have been worked on this original theme. Wunsch et al. (1974) decomposed the carbonates of iron and cobalt onto particles of γ-Fe_2O_3 and obtained a coercivity of 620 Oe (49.4 kA/m). Kubota et al. (1974) recommended the use of complexing agents such as Rochelle salt to promote the absorption of the cobalt compound onto the surface of the particle. Horiishi and Takedoi (1975b, c, f) added up to 1% chromium to their acicular particles of α-(Fe, CoO)OH and followed the normal procedure to obtain cobalt, chromium doped iron oxide particles. A material which has been given the name Avilyn was dscribed by Umeki et al. (1974a) as an acicular iron oxide on whose surface cobalt compounds had been adsorbed. Similar particles were described by Umeki et al. (1974b) and were made by dispersing acicular particles of γ-Fe_2O_3 in water and adding solutions of cobaltous chloride and sodium hydroxide (to form the hydroxide). The mixture was heated at 100°C to give particles having $H_c = 550$ Oe (43.8 kA/m), $\sigma_s = 73.5$ emu/g (92.3μ Wb m/kg), $\sigma_r = 36.9$ emu/g (46.3μ Wb m/kg).

Imaoka et al. (1978) supplied the preparation details for the Avilyn particles. Well-dispersed particles of γ-Fe_2O_3 in cobalt chloride solution were added to sodium hydroxide solution. The mixture was stirred for several hours and the pH was kept above 12. The liquid was then kept at 100°C for several more hours by which time the adsorption of cobalt was completed. The coated particles were then filtered and dried at low temperature.

The prefered starting material was γ-Fe_2O_3 rather than Fe_3O_4 because of the superior thermal stability of the end product. The instabilities with respect to temperature and stress are related to the concentration of ferrous ion. In the latest material, this is held to 2–3% cf. 5–7% in previous cobalt-impregnated iron oxides. Imaoka et al. found that the coercivity of the particles increased with increasing cobalt content to about 2%; thereafter no increase in coercivity occurred but the moment density continued to decrease. It appeared that cobalt in excess of 2 wt% existed as the hydroxide in the particle. The change of moment with cobalt content was found to follow a curve that was intermediate between the curve predicted from substitution of cobalt for iron in the lattice and the curves desired by assuming mixtures of γ-Fe_2O_3 with $Co(OH)_2$ or Co_2O_3.

The surface nature of the change to the particles was shown by etching the particles with dilute hydrochloric acid, while monitoring the increase of cobalt and iron in the acid. Cobalt was preferentially dissolved and its appearance coincided with a sudden decrease in the coercivity of the particles. Furthermore, if the cobalt-impregnated particles were heat-treated at temperatures above 300°C the temperature-dependence of coercivity increased from about 1 Oe/°C to 23 Oe/°C; similar to that of cobalt-doped iron oxide.

The cobalt-impregnated particles of Imaoka et al. (1978) had $\sigma_s = 73$ emu/g $(92\mu$ Wb m/kg), $H_c = 550$ Oe (43.8 kA/m). The torque curves of oriented tapes made of the particles showed uniaxial behavior and a surface anisotropy of 1.5×10^6 erg/cm^3 was calculated. This was of the same order as the magnetocrystalline anisotropy constant K_1 of iron–cobalt ferrite.

Rather different procedures were adopted by Shimizu et al. (1975) who warmed particles of γ-Fe$_2$O$_3$ or of magnetite in sodium hydroxide solution containing FeSO$_4$ and CoSO$_4$ in an inert atmosphere. The result was magnetite particles having the surface modified with Fe^{2+} and Co^{2+}. The particles could be oxidized to modified γ-Fe$_2$O$_3$ by heating in air at 250–300°C and had coercivities (depending on the starting and finishing oxide) between 510 and 785 Oe (40.6–62.5 kA/m) and saturation magnetizations between 76 and 86 emu/g (95–108μ Wb m/kg). Finally an interesting variation to the method of applying the cobalt to the surface of the iron oxide particles was provided by Ando (1974) who sensitized the surface of the particles with stannous chloride and palladium chloride and then electrolessly-plated cobalt from a bath containing hypophosphite.

A problem which is referred to frequently is the loss of acicularity of the starting oxide by dissolution while the cobalt layer is being applied and decomposed. Once again a variety of solutions to the problem were proposed; Ando and Wakei (1975b) mixed goethite particles with the γ-Fe$_2$O$_3$ and Ando and Wakei (1976b) added a dicarboxylic acid; Ando (1976) added a phosphate solution to the particles of γ-Fe$_2$O$_3$.

The addition of both cobalt and manganese was recommended by Hirata and Wakei (1976) as a way of controlling the coercivity of the product more easily than when cobalt was used alone. Before coating the particles with cobalt salts Wakai and Suzuki (1976) dispersed γ-Fe$_2$O$_3$ particles in an aqueous solution of ferrous sulfate and sodium hydroxide with the objective of filling up the pores in the oxide particles and thus producing a smoother base for the later deposition of cobaltous sulfate.

After coating particles having a composition between Fe$_2$O$_3$ and Fe$_3$O$_4$ with cobaltous chloride Sasazawa et al. (1976a) annealed the particles for an hour at 350°C in nitrogen in a magnetic field of 800 Oe (63.7 kA/m). The treatment had the effect of increasing the coercivity to 500–600 Oe (39.8–46.9 kA/m).

All the methods discussed above for preparing iron oxide particles having a cobalt (oxide) coating use as starting material iron oxides or oxyhydroxides which were made by the standard process discussed earlier, of preparing seeds, then particles, of α-(FeO)OH. The principal alternative method of making Fe$_3$O$_4$

and Fe_2O_3 particles is to grow acicular particles of lepidocrocite, γ-(FeO)OH by the method of Bennetch et al. (1975). These particles were used as the basis for cobalt-coated iron oxide particles by Bennetch (1975) who first precipitated lepidocrocite at a pH of 2.6 to 4.1 and then added cobaltous chloride and enough sodium hydroxide to increase the pH to 7–9.5; thereby precipitating cobalt onto the lepidocrocite particles. The final steps were to mix the particles in a solution of coconut oil fatty acid and morpholine at 150°C to coat the particles and thus prevent sintering. This was followed by the reduction to cobalt-magnetite and re-oxidation to cobalt doped γ-Fe_2O_3. The resulting particles had a length up to 2μm and an average acicularity of 6.7. The magnetic properties as a function of cobalt content are given in table 8. It appears from these results that slightly more cobalt must be added to achieve a given coercivity when the surface impregnation method is used than when cobalt is in solid solution in the iron oxide. But the results obtained by either method depend strongly on the shape and size of the particle and not just on the amount of cobalt.

Both methods allow the coercivity of iron oxides to be increased greatly beyond that of the pure oxide but this enhancement was obtained at the risk of increasing the temperature sensitivity of H_c and M_r. As we have seen much work has been done in an attempt to decrease this sensitivity. Another disadvantage of cobalt-modified particles was alluded to by Bennetch (1975) and that was their tendency to high pH which made them incompatible with some binder materials, particularly the polyurethanes.

3.2.5. Iron oxides modified by other elements
In the hope of combining high anisotropy with stable properties other elements have been investigated as modifiers of the iron oxides.

The use of chromium has frequently been reported. Horiishi et al. (1974) precipitated seeds and grew particles of modified goethite by mixing solutions of $FeSO_4$ and $Cr_2(SO_4)_3$ with sodium hydroxide solution through which air was bubbled. The rest of the preparation was standard. Horiishi modified the procedure by alternately reducing the (chromium-modified) hematite to magnetite then re-oxidizing at 300°C in air back to hematite. The cycle was performed three times before the magnetite was fully oxidized at 270°C in air to chromium-modified γ-Fe_2O_3. The coercivity was 410 Oe (30.6 kA/m) and $\sigma_s =$ 79 emu/g (99μ Wb m/kg). The variation introduced by Matsui et l. (1975) was to

TABLE 8

Magnetic properties for cobalt-coated iron oxide particles (after Bennetch 1975)

Wt% Co	H_c		M_s		M_r	
	(Oe)	(kA/m)	(emu/cc)	(Wb/m²)	(emu/cc)	(Wb/m²)
3.18	475	37.8	380.9	0.479	2139	2.69
4.64	532	42.3	374.2	0.470	2274	2.86
5.92	796	63.4	349.3	0.439	2422	3.04

anneal the chromium-modified magnetite at 200–800°C in an atmosphere of nitrogen containing a controlled amount (80 ppm) of oxygen for three hours. On oxidizing to γ-(Fe, Cr)$_2$O$_3$ the coercivity was 450 Oe (35.8 kA/m) and $\sigma_r =$ 43.5 emu/g (54.6 μWb m/kg). The possibility of modifying the iron oxide particles with both cobalt and chromium was discussed by Horiishi and Takedoi (1975b) Uesaka (1974) used trivalent rare earth elements (e.g. Ce, Pr, Nd etc.) and claimed a high and temperature-independent coercivity. Pingaud (1974) added ferrous sulfate and cadmium sulfate to a solution of sodium hydroxide and washed the precipitate with water containing a slight amount of calcium. He used the standard method to obtain particles of γ-Fe$_2$O$_3$ modified with 3% Cd, 0.4% Ca and 0.055% Na. The particles had not only a high coercivity, 770 Oe (61.3 kA/m), but also the $\gamma \rightarrow \alpha$ transition temperature was increased to 760°C.

3.2.6. Magnetic properties of cobalt-modified iron oxides

The experimental values at 20°C for the intrinsic magnetic properties of magnetite and cobalt ferrite given by Smit and Wijn (1959) are given in table 9. Magnetite is an inverse ferrite and as cobalt ferrite is also inverse Bickford et al. (1957) assumed that the cobalt enters the octahedral sites and replaces the divalent ions. Thus the x in Co$_x$Fe$_{3-x}$O$_4$ represents the fraction of divalent ions replaced by cobalt. The effect on the anisotropy constant, K_1, of substituting a small amount of cobalt in magnetite is much greater than if cobalt is added to other ferrites. Bickford et al. (1957) suggested that the effect of cobalt is enhanced in magnetite by the presence of Fe^{2+} ions. For small cobalt concentrations ($x \sim 0.04$) they found that the changes in K_1 in the positive direction and of K_2 in the negative direction are linear with cobalt content. Thus the net effect of substituting cobalt is the additive effect of a contribution from the cobalt ions superimposed on the anisotropy of the magnetite itself. The results of Bickford et al. (1957) for the temperature dependence of K_1 in Co$_x$Fe$_{3-x}$O$_4$ for $x = 0, 0.01,$ 0.04 is shown in fig. 12. Slonczewski (1958, 1961) explained theoretically the large change in anisotropy in terms of a "one-ion" model involving the spin–orbit coupling of Co^{2+} ions placed in an electric field of trigonal symmetry in the lattice. The theory, while satisfactory for small amounts of cobalt, could not be used, for example, for Co$_{0.8}$Fe$_{2.2}$O$_4$ in which Sugiura (1960) examined the uniaxial and cubic anisotropies induced by magnetic annealing.

Kumashiro (1968) measured the temperature dependence of K_1 in single-

TABLE 9

Experimental values at 20°C for the intrinsic magnetic properties of magnetite and cobalt ferrite (after Smit and Wijn 1959)

| Ferrite | σ_s | | M_s | | θ_c |
	(emu/g)	(μWb m/kg)	(emu/cc)	(Wb/m^2)	(°C)
Fe$_3$O$_4$	92	116	480	0.603	585
CoFe$_2$O$_4$	80	100	425	0.534	520

crystal films of this composition which he prepared by the decomposition of gaseous mixtures of the metal bromides and found $K_1 = A \exp(-BT^2)$ where A and B are constants. Eagle and Mallinson (1967) found a very slight dependence of H_c on the cobalt content in the range 3–6% and concluded from the ferromagnetic resonance line width measurements of Schlömann (1956) that the anisotropy constant K_1, changes with cobalt content only up to 3%. Between 3 and 6% cobalt K_1 remained at $+10^6$ erg/cc ($+10^5$ J/m^3). Eagle and and Mallinson also found that M_s for their samples was within 90% of that of pure γ-Fe$_2$O$_3$. Köster's (1972) results for the crystalline anisotropy field of cubic cobalt-substituted γ-Fe$_2$O$_3$ particles as a function of cobalt content are shown in fig. 13. Here also the rate of change of H_c with cobalt content begins to decrease when more than 4% cobalt is included.

The critical size for single-domain behavior of CoFe$_2$O$_4$ was determined by Berkowitz and Schuele (1959) to be about 700 Å (larger than for Fe$_3$O$_4$). For particles of this size they measured coercivities between 1000 and 2000 Oe (79.6–159.2 kA/m) at 300 K. The squareness, S, was only slightly above 0.5 indicating that the anisotropy was principally uniaxial. Kamiya et al. (1970) also found that cubic particles of Co$_x$Fe$_{3-x}$O$_4$ had $S = 0.6$, whereas cubic particles of γ-(Fe, Co)$_2$O$_3$ gave $S = 0.8$ which was quite close to the theoretical value of 0.83 for K_1 positive.

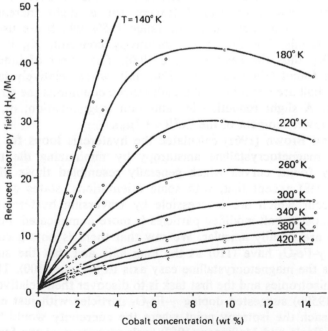

Fig. 13. Crystalline anisotropy field of cubic, cobalt-substituted γ-Fe$_2$O$_3$ particles vs the cobalt concentration (Köster 1972).

Khalafalla and Morrish (1974) examined the structure and magnetic properties of equiaxed particles of cobalt-substituted γ-Fe$_2$O$_3$ containing up to 9.9 mole percent Co and found that the lattice parameter increased linearly from 8.348 Å at 0% Co to 8.366 Å at 9.9% cobalt. They inferred that a significant amount of the cobalt was in the lattice and that it probably occupied some of the octahedral (B) sites and vacancies.

Mollard et al. (1975) confirmed that the cobalt which is incorporated in the lattice goes into the B sites by comparing the measured increase in lattice parameter with that calculated assuming three different possibilities; that the tetrahedral sites were occupied, that the octahedral sites were occupied, or that both sites were used. They also measured the γ–α transformation temperature and found a monotonic increase from 460°C for pure γ-Fe$_2$O$_3$ to 700°C for γ-Fe$_{0.4}$Co$_{1.6}$O$_3$. The substitution of Al in γ-Fe$_2$O$_3$ was used previously by Michel et al. (1950) as a device to increase the transformation temperature so that the Curie temperature could be measured. Morrish and Sawatzky (1970) using the Mössbauer effect, found an increase in the $\gamma \rightarrow \alpha$ temperature on substituting cobalt.

Both Khalafalla and Morrish (1974) and Mollard et al. (1975) found, as had earlier workers, that the increase in coercivity was proportional to the amount of cobalt incorporated, in keeping with the single-ion anisotropy model of Slonczewski (1958, 1961).

In summary it appears that least part of the added cobalt enters the lattice of γ-Fe$_2$O$_3$ and takes the place of Fe^{3+} ions on the B sites or occupies B vacancies. The structure remains unchanged (except for a slight increase in lattice parameter) for cobalt additions in the range 1–5% which are usually used in recording materials. The first order anisotropy constant, K_1, is positive and larger than K_2, thus the easy axes are $\langle 100 \rangle$. The effect of induced uniaxial anisotropy is found (Kamiya et al. 1970) only when relatively large amounts ($\sim 13\%$) of cobalt are substituted and under those conditions the structure is that of magnetite. A slight reduction in saturation magnetization, σ_s, occurs as a result of the lower moment of the Co^{2+}($< 3.7\mu_B$).

Johnson and Brown (1961) calculated the hysteresis loops for SD particles having only magnetocrystalline anisotropy by minimizing the magnetostatic energy. They found curves which generally resembled those of Stoner and Wohlfarth (1948) except that, with some orientations, states of soluble magnetization occur which are inaccessible by traversing hysteresis loops. The situation in most cobalt-modified particles is more complicated, however, since the particles are usually acicular. We saw that there is now a consensus that particles of γ-Fe$_2$O$_3$ have $\langle 110 \rangle$ as the long axis and that the substitution of cobalt causes the magnetocrystalline easy axis to become $\langle 100 \rangle$. Thus there are competing anisotropies and the first task is to discover their relative importance. Wohlfarth (1959b) suggested doping γ-Fe$_2$O$_3$ particles with just enough cobalt (fig. 12) to reach the isotropy point where the coercivity would be caused by shape alone. Eagle and Mallinson (1967) took advantage of the fact that, at the isotropy point (-143°C), the value of K_1 in pure magnetite is zero and any remaining anisotropy must be due to the shape of the particles. By converting γ-Fe_2O_4 to Fe_3O_4 without change of shape of the particles and then measuring the temperature dependence of H_c they were able to conclude that about two-thirds of the coercivity (~ 200 Oe, 15.9 kA/m) of γ-Fe_2O_3 is due to shape and the remainder to magnetocrystalline anisotropy (100 Oe, 7.96 kA/m). Köster (1973) observed that the latter contribution is in good agreement with Néel's (1947a) relation $H_c = 0.32H_k$ which gives 86 Oe (6.8 kA/m) using Birks (1950) value for K_1. Since shape anisotropy dominates in acicular γ-Fe_2O_3 particles for a random assembly we find $S = 0.5$. Köster (1972) determined the contribution of magnetocrystalline anisotropy directly in equiaxed particles of cobalt-substituted γ-Fe_2O_3. He measured the temperature dependence of the reversible initial susceptibility, χ_{rev}, and used these values to calculate the reduced anisotropy field H_k/M_s from:

$$H_k/M_s = \tfrac{2}{3}p(1/\chi_{rev} + N_b) - (N_b - N_a)$$

where p is the volume packing fraction and N_b, N_a are the demagnetizing factors along the major and minor axes. For equiaxed particles it is reasonable to set $N_b = N_a = \tfrac{4}{3}\pi$.

The results are shown in fig. 13. Köster (1972) also plotted the squareness, S, against temperature and found for each value of the cobalt content that there was a critical temperature, T_k, above which S was invariant and below which S increased greatly with decreasing temperature. Thus, he concluded that for $T > T_k$ for each composition shape anisotropy is dominant and for $T < T_k$ magnetocrystalline anisotropy is more important; for 2.5% Co, $T_k = 360$ K. T_k was found to be independent of the degree of particle alignment. At T_k, the ratio of uniaxial to magnetocrystalline anisotropy was found to be 0.86 for any cobalt concentration, in agreement with Wohlfarth and Tonge's (1958) prediction that with mixed anisotropies the squareness $S = 0.5$ for $H_m/H_k > 1$ and for $H_m/H_k < 1$, S increases until the maximum value for six easy directions $S = 0.83$ is obtained.

The measurements of Eagle and Mallinson (1967) on coercivity of cobalt-substituted equant particles of γ-Fe_2O_3 as a function of particle diameter are shown in fig. 14. The solid line was calculated from the curling equation

$$H_n = [2\pi(1.39)(d/d_0)^2 - \tfrac{4}{3}\pi]M_s + 0.64K_1/M_s$$

where d_0 is the critical diameter which was set at 150 Å. The agreement between theory and experiment argues strongly for an incoherent mode of magnetism reversal in cobalt-substitued γ-Fe_2O_3 particles.

The loss of remanence after storage at temperatures above ambient (40–100°C) has always been a cause for concern in cobalt-substituted iron oxide particles. Ogawa et al. (1972) compared the performance of particles of Fe_3O_4 and γ-Fe_2O_3 each containing 4% Co and found that the magnetite particles showed a smaller charge in both acicular and equant forms. They attributed the difference to the existence on the octahedral sites of γ-Fe_2O_3 of vacancies which permitted the movement of the Co^{2+} ions. Umeki et al. (1974a) found that the

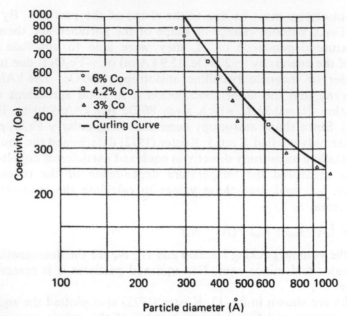

Fig. 14. Coercivity vs particle diameter for equiaxed particles of cobalt-substituted γ-Fe$_2$O$_3$ (Eagle and Mallinson 1967).

loss of remanence after storage and the related temperature-dependence of remanence were much lower in cobalt-modified particles of iron oxide i.e., particles which consisted of an iron oxide core around which was a cobalt-rich layer, than in cobalt-substituted γ-Fe$_2$O$_3$ particles.

Köster (1973) gave three reasons for a decreasing squareness with increasing temperature. They were: (1) competition between shape and crystalline anisotropies (2) thermal fluctuations and (3) self-demagnetization. In particles of cobalt-substituted γ-Fe$_2$O$_3$, Flanders (1976b) found an additional problem; the material suffers from an impact-induced demagnetization which is related to the saturation magnetostriction, λ_s. As cobalt is added to γ-Fe$_2$O$_3$, λ_s becomes more negative and the stress demagnetization increases. In contrast, when cobalt is added to Fe$_3$O$_4$, λ_s becomes less positive and the stress demagnetization is smaller. Losses up to 30% of M_r were measured on a tape containing 2.8% Co–γ-Fe$_2$O$_3$ with $\lambda_s = -15 \times 10^{-6}$. Tapes made of chromium dioxide particles, $\lambda_s = 1 \times 10^{-6}$ suffered only a 1% loss under the same impact conditions. The measured values of magnetostriction as a function of cobalt content for Co$_x$Fe$_{3-x}$O$_4$ and γ-Fe$_{2-x}$Co$_x$O$_3$ are shown in fig. 15.

Despite the high coercivities which could easily by achieved by adding cobalt to iron oxide particles the material was, until recently, only used for special applications e.g. master tapes for transfer copying. The reasons were the temperature and stress-sensitivity of the particles. It appears that after much work this problem may have been solved, and audio cassettes made of cobalt-

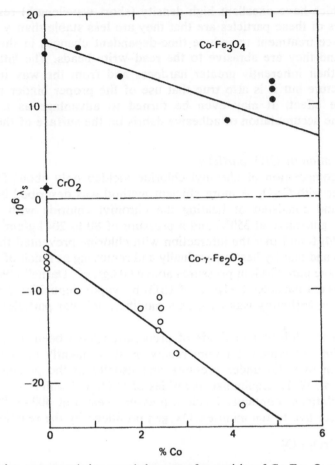

Fig. 15. Saturation magnetostriction vs cobalt content for particles of $Co_xFe_{3-x}O_4$ and γ-$(FeCo)_2O_3$ (Flanders 1976b).

modified iron oxide particles and having a high-density (short wavelength) response at least equal to that of chromium dioxide are now being made.

3.3. Chromium dioxide

The preparation of a ferromagnetic oxide of chromium, CrO_2, by the decomposition of chromyl chloride. CrO_2Cl_2 was first reported by Michel and Benard (1935). It is presently second in importance to γ-Fe_2O_3, as a material for magnetic recording surfaces but its position is now being threatened by the improved cobalt-modified iron oxides. Chromium particles are clearly-acicular as is apparent from fig. 7d (section 3.1.4) with none of the dendrites and voids which mar the appearance of the standard iron oxide. Consequently the CrO_2 particles are relatively easily dispersed and oriented and can be made into tapes and disks which have excellent high density (short wavelength) response. The disadvantages of these particles are that they are less stable than γ-Fe$_2$O$_3$, they require surface treatment to reduce time-dependent changes in their magnetic properties, and they are abrasive to the read–write heads. The latter property stems from their inherently greater hardness and from the way in which the particles fracture but it is also true that use of the proper binder material can minimize the effect. It may even be turned to advantage as a method of preventing the accumulation of adhesive debris on the surface of the head.

3.3.1. Preparation of CrO$_2$ particles

The first decomposition of chromyl chloride yielded only about 50% acicular CrO$_2$ together with Cr$_2$O$_3$. A more efficient method was described by Uchino et al. (1974a) and consisted of heating the chromyl chloride with CrO$_3$ under oxygen in an autoclave at 350°C and a pressure of 80 to 204 kg/cm^2. Funke and Kleinert (1974) found that the interaction with chlorine prevented the formation of pure CrO$_2$ and that by heating gradually and removing as much of the chlorine as possible gave pure CrO$_2$ at pressures above 60 kg/cm^2. Darnell (1961) obtained roughly equiaxed particles ($\sim 1\mu$m) of CrO$_2$ by vapor deposition from chromyl chloride. When antimony was used as a modifier, acicular particles, $1 \times 0.2\mu$m, were obtained.

More than ten different methods of preparation have been described in the literature. The one which is commercially most frequently used involves the decomposition of CrO$_3$ under pressure and usually in the presence of water (Swoboda et al. 1961). Equimolar quantities of CrO$_3$ and water were heated in a thin-walled platinum container inside a pressure vessel at 400–525°C and 500–3000 kg/cm^2 for five to ten minutes. Oxygen produced by the reaction

$$CrO_3 \rightarrow CrO_2 + O_2$$

was used to stabilize the CrO$_2$. The use of a double vessel is common; the inner one is thin and made of platinum or glass which does not react with CrO$_2$ and the outer vessel is of steel and thicker to withstand the high pressure. Terada et al. (1973) advised that the amount of water used, usually between 10 and 40% of the weight of CrO$_3$, was important in obtaining CrO$_2$ of uniform quality.

The thermal decomposition of anhydrous CrO$_3$ was also used by Ariya et al. (1953) and by Kubota (1960, 1961); again high temperatures (400–550°C) and high pressures (500–625 kg/cm^2) were necessary. It would clearly be an advantage to have a preparation method which could be carried out at atmospheric pressure and such a method has been described by Hirota et al. (1974) and by Sugimori (1974). It consisted of heating CrO$_3$ in air with hydrogen peroxide. Fukuda (1975) was also able to prepare CrO$_2$ particles at atmospheric pressure by dissolving CrO$_3$ and potassium nitrate in water and heating the mixture after drying, at 415°C for 40 min, Shimotsukasa (1972) added to CrO$_3$ solution a concentrated solution of nitric acid and oxalic acid in order to reduce about two-thirds of the Cr^{6+} to Cr^{3+}. The solution was dried and the residue was heated at 330°C for four hours under an atmosphere whose oxygen partial pressure was very carefully controlled, to give CrO_2.

Several workers combined CrO_3 with dichromate. Thus Balthis (1969) mixed CrO_3, Cr_2O_3 and sodium dichromate and heated at 878 kg/cm^2 at 265°C for four hours then at 400°C for six hours. Kawamata (1975) used ammonium dichromate. Morero et al. (1972) prepared particles of CrO_2 by heating chromium chromate to 350°C under 280 kg/cm^2 of oxygen. Shibasaki et al. (1970) precipitated chromium hydroxide from chromic nitrate and excess ammonium hydroxide solution. On heating at temperatures above 350°C and at pressures above 50 kg/cm^2 in oxygen, chromium dioxide particles were formed. Cox and Hicks (1969) prepared CrO_2 particles by heating CrO_3 with Cr_2O_3 and water at 200–350°C under oxygen at 0.5–5 kg/cm^2. The Cr_2O_3 was obtained by drying the chromium(III) hydroxide.

Other oxides of chromium have been used as starting materials. For example Amemiya et al. (1973a) used a mixture of Cr_2O_5 and 0.1 μm diameter particles of CrO_2. Uchino et al. (1974b) also used Cr_2O_5 which was heated in a double vessel at 370°C and 124 kg/cm^2 with sodium nitrate and water to form acicular particles of CrO_2.

Roger and Weisang (1973) prepared CrO_2 particles by the thermal decomposition of chromium tartrate, formate, or acetate which were heated in moist nitrogen at 250–300°C and then oxidized at a pressure of more than 90 kg/cm^2 and a temperature above 360°C.

3.3.2. Additives to chromium dioxide

In the preparation of particles of CrO_2 it is clearly advantageous to be able to avoid very high temperatures and pressures and to provide nuclei on which the particles can grow. Consequently the literature (and particularly, the patent literature) is rich with suggestions of elements and compounds which act as catalysts or promote particle growth in a favorable way. In fact the proposed additive elements amount to rather more than half of the naturally occuring elements in the periodic table. We shall deal only with a representative selection.

Ingraham and Swoboda (1960) found that by using Sb_2O_3 as an additive, the reaction temperature could be reduced to 300°C and the pressure to 50 kg/cm^2. X-ray diffraction photographs of the resulting particles showed only the tetragonal, rutile structure of CrO_2. Robbins (1973) found that the addition of Sb_2O_3 had the further effect of increasing the coercivity of the particles in which the antimony content was about 1%. Mihara et al. (1974) used antimony and tin in amounts up to 10%. Montiglio et al. (1975) added Sb_2O_3 to chromium chromate which was heated at 375°C under 350 kg/cm^2. The resulting CrO_2 particles were found to contain 0.29% Sb by weight. They had an average length of 0.2 μm, an axial ratio of 8:1 and the magnetic properties: $H_c = 440$ Oe (35 kA/m), $\sigma_s = 88$ emu/g (111 μWb m/kg) and $S = 0.51$.

After antimony, tellurium and iron are perhaps the most commonly used additives to chromium dioxide. Thus Mihara et al. (1975) mixed 1 at.% of tellurium in H_6TeO_6 with Cr_2O_5 in the presence of water and heated to 350°C at

400 kg/cm². More commonly tellurium is mixed with other additives, for example, Sn (Osmolovskii et al. 1976), Sb (Kawamata et al. 1975), W (Mihara et al. 1973) and Ca (Matsushita 1972). Particles prepared with 0.25% Ca had a length 0.2–0.5 μm, width 0.01–0.1 μm, $H_c = 470$ Oe (37.4 kA/m), $\sigma_s = 62.8$, $\sigma_r = 37.2$ emu/g (78.9, 46.8 μWb m/kg).

The modification of CrO_2 by iron in amounts of up to 10% was studied by Shibasaki et al. (1970) and by Mihara et al. (1970) who examined the effect of adding Te and Sn, Te + Fe and Sb + Fe on the magnetic properties of the CrO_2 particles. By means of these additions it was found possible to increase the Curie temperature to 160°C and the coercivity to 700 Oe (55.7 kA/m). Even higher coercivities of 800 Oe (63.6 kA/m) were obtained by Hirota et al. (1974) by combining Te, Sn and Fe as additives.

Iridium compounds were found to be more effective as additives (although rather costly ones) by Amemiya et al. (1973). The most effective compound was Ir (OH)₃, which was added to CrO_3 before the hydrothermal synthesis of CrO_2. The effect on coercivity is given in table 10.

The use of tungsten alone was proposed by Asada et al. (1974) and at 1.5% W a coercivity of 540 Oe (43 kA/m) was obtained. Tungsten at a level of 1.1% was combined with several other elements by Amemiya and Asada (1974, 1976) to give the following coercivities: 3% Ca (498 Oe, 39.6 kA/m), 4% Zn (500 Oe, 39.8 kA/m), 1.2% Mg (555 Oe, 44.2 kA/m), 1.5% Cu (570 Oe, 45.4 kA/m), Fe (more than 600 Oe, 47.8 kA/m). Uchino et al. (1975) added small amounts of lithium and sodium to produce highly acicular particles of CrO_2 which had $H_c = 560$ Oe (44.6 kA/m), = 82.3 emu/g (103.4 μWb m/kg) and $S = 0.42$.

It has been shown that copies from master tape may be made by the method of thermoremanent transfer (TRM) by allowing the slave tape to cool through its blocking temperature while in contact with the master. It is advantageous to have the blocking temperature (which is somewhat below the Curie temperature) to be as low as possible so that the amount of heating required is minimized. Thus additives are sought which reduce the Curie temperature and among these sulfur (Hirota et al. 1972) and fluorine (Chamberland et al. 1973) are the most common. The addition of sulfur has also been recommended as a method of reducing the abrasiveness of chromium dioxide particles. Phosphorus in

TABLE 10

Effect on the coercivity of addition of Ir(OH)₃ to CrO_3 before hydrothermal syntheses of CrO_2 (after Amemiya et al. 1973b)

Wt% Ir	H_c	
	(Oe)	(kA/m)
2	640	50.9
5	850	67.7
10	1050	83.6

amounts up to 5% was found by Kawamata et al. (1973) to reduce both the coercivity and the Curie temperature.

Chromium dioxide is not a naturally occurring compound and it is not surprising that it reacts somewhat with its environment. This is particularly true when the material is in the form of fine particles and undesirable changes have been noted as a result of interactions with the binder resins as well as with [OH] and [NH₂] groups. What is needed is a barrier layer covering each particle and there have been many suggestions on how this can be achieved. Bottjer and Ingersoll (1970) found that reducing the surface, for example, by agitating the CrO_2 particles in an aqueous solution of sodium bisulfite at 55°C for 15 h, decreased the reactivity. The passivation was accompanied by the appearance of X-ray diffraction lines which were intermediate in spacing between CrO_2 and $CrO(OH)$. Most of the stabilization procedures involve the precipitation of compounds which are sparingly soluble in water onto the surface of the CrO_2 particles. Thus Leutner et al. (1972, 1973) used nitrates of Pb, Ba, Ca or Hg or oxides of Sb or Pye (1972) recommended the use of Al_2O_3 and Mori et al. (1976) applied iron oxides. Montiglio et al. (1973) reported that stabilization could be achieved by coating CrO_2 particles with molybdates, tungstates or phosphates of sodium, zinc, lanthanum, etc., and Woditsch and Hund (1974) also coated their particles with sodium phosphate.

Schoenafinger et al. (1976) achieved stabilization of CrO_2 particles by treating them in an alcohol suspension with the hydrolysis products of alkoxyalkylsilanes and drying them in air at 100°C.

The use of sulfur, fluorine, and phosphorus as additives to decrease the Curie temperature of chromium dioxide for thermo-remanent recording was mentioned earlier. For conventional recording surfaces the normally low Curie temperature of CrO_2 has been considered a possible disadvantage and attempts have been made to increase it. Dezawa and Kitamoto (1973) reported that heating the particles in a reducing vapor stream of nitrogen and methyl alcohol for one hour at 200°C enhanced the Curie temperature to 130.4°C. In contrast, heating particles in an oxidizing atmosphere was found by Bottjer and Cox (1969) to cause an increase in σ_s and σ_r (to values close to 90 emu/g, 113 μWb m/kg and 45 emu/g, 56.5 μWb m/kg respectively) without changing the coercivity of the particles.

3.3.3. Structure and transformations of CrO_2

The formula CrO_2 was established by Michel and Benard (1935) who also found that the compound had the rutile structure. Wilhelmi and Jonsson (1958) compared their own diffraction measurements with those reported previously. They found the tetragonal structure with $a_0 = 4.423$ Å; $c_0 = 2.917$ Å and the density = 4.83 g/cc (4.83×10^3 kg/m³) in general agreement with earlier results. Cloud et al. (1962) prepared a single crystal of CrO_2 and found $a_0 = 4.4218$ Å and $c_0 = 2.9182$ Å. The space group is D_{4H}^{14}–P4₂/mnm (ASTM). The chromium ions are situated at the corners and at the body of the tetragonal unit cell giving two chromium per unit cell. The oxygen ions are found at $(u, u, 0)$, $(\bar{u}, \bar{u}, 0)$ $(u - \frac{1}{2}, \frac{1}{2} -$

$u, \frac{1}{2})$ and $(\frac{1}{2} - u, u - \frac{1}{2}, \frac{1}{2}))$ where where the value of u was determined by Siratori and Iida (1960) for diffraction measurements as 0.294 and by Cloud et al. (1962) as 0.301 ± 0.04.

Chromium dioxide is seldom made without additives and their effect for small concentrations is to increase the lattice constants. Thus, Hirota et al. (1972) investigated the effect of sulfur and found that up to 0.17 wt% the lattice parameter, a_0, increased from 4.427 Å to 4.433 Å and the X-ray diffraction pattern indicated a single phrase; c_0 remained constant. When $CrO_{2-x}F_x$ was prepared by Chamberland et al. (1973) both a_0 and c_0 increased up to $x = 0.30$ at which point two phases were found. The observed values were are given in table 11.

The presence of intermediate compounds having a rutile structure was also noticed by Mihara et al. (1970) when they used $Te + Sn$, $Te + Sn$ or $Sb + Te$ in large amounts.

Siratori and Iida (1960) first observed and Darnell and Cloud (1965) confirmed that the lattice constant, c_0, decreased with increasing temperature. Darnell and Cloud made their measurements on a single crystal of CrO_2 and found that the temperature dependence of c_0 was the same for increasing and decreasing temperatures and thus they eliminated the possibility of the anomaly being caused by a chemical change on heating.

Claude et al. (1968) investigated the formation of different oxides of chromium which were heating CrO_3 with 7% $CaCl_2 \cdot 2H_2O$ for 24 h at temperatures between 300 and 450°C in a sealed glass tube, followed by cooling in liquid nitrogen. At low temperatures, ~ 300°C, Cr_2O_5 was formed. At about 360°C, CrO_2 was formed and at the highest temperature they obtained Cr_2O_3. Similar results were found by Drbalek et al. (1973). In their search for the optimum temperature for the decomposition of CrO_3 they found that to obtain pure CrO_2 at 300 kg/cm², the temperature should be 350–400°C. Between 250–300°C, $CrO_{2.44}$ was formed and above 400°C, Cr_2O_3 appeared.

Rodbell and De Vries (1967) measured the transformation temperature, $CrO_2 \rightarrow Cr_2O_3$, for both single crystals and powders. They concluded that CrO_2 could be made (by the decomposition of CrO_3) at temperatures up to 400°C provided that nucleation of Cr_2O_3 was avoided. Malinin et al. (1971) found that the thermal

TABLE 11

Observed values of a_0 and c_0 for different x, after preparation of $CrO_{2-x}F_x$ (after Chamberland et al. 1973)

x	a_0 (Å)	c_0 (Å)
0.028	4.4225 ± 5	2.9179 ± 4
0.100	4.4360 ± 8	2.9210 ± 7
0.200	4.4480 ± 2	2.9267 ± 1
0.281	4.4578 ± 12	2.9274 ± 7

decomposition started at 430°C and measured the rate constant at 10^4 s and the activation energy as 30 ± 10 kcal/mole. Shannon (1967) found that the transformation was topotactic; the rhombohedral Cr_2O_3 developed an a-axis which was parallel to the CrO_2 c-axis and a c-axis parallel to the a-axis of CrO_2. The hydrothermal reaction $2CrO_2 + H_2O \rightarrow 2(CrO)OH + \frac{1}{2}O_2$ was studied by Shibasaki et al. (1973) and is of practical importance as a possible mechanism for the loss of magnetization with time of CrO_2 particles. Small single crystals of CrO_2 prepared by oxidizing $Cr(OH)_3$ in oxygen at 1800 kg/cm^2 and 450°C, were heated at 335°C under a water vapor pressure of 1500 kg/cm^2 for seven days. The product was the orthorhombic $(CrO)OH$ ($a = 4.861$ Å, $b = 4.292$ Å and $c = 2.960$ Å) which was found to be topotactic with the CrO_2; the c-axes were parallel and the a- and b-axes of $(CrO)OH$ were parallel to the a-axes of CrO_2. The reaction was found to be reversible. Alario-Franco et al. (1972) also found that the interconversion of $(CrO)OH \rightarrow CrO_2$ was topotactic and that only a small rearrangement of the lattice was required by the removal of hydrogen from the orthorhombic structure to give the tetragonal CrO_2.

De Vries (1966) showed that it was possible to grow epitaxially, films of CrO_2 on faces of a single crystal of TiO_2 (rutile). The structure of the films was identical in all respects to that of the bulk material. Acicular particles mostly have the tetragonal c-axis as the long axis. This was shown by Gustard and Vriend (1969) using dark-field electron microscopy. Shibasaki et al. (1970) used this method and selected area electron diffraction to show that Fe-modified CrO_2 particles also were single crystals of rutile structure having their long axis parallel to the c-axis.

3.3.4. Magnetic properties of CrO_2

a. Intrinsic magnetic properties of CrO_2

The Cr^{4+} ions have a magnetic moment per ion of $2\mu_B$ and hence, the expected saturation magnetization per unit mass at 0 K is $\sigma_{so} = 5585 n_B/M$, where n_B is the number of Bohr magnetons per ion ($= 2$) and M is the molecular weight ($= 84.01$); $\sigma_{SO} = 133$ emu/g (167 μWb m/kg).

Darnell and Cloud (1965) prepared pure CrO_2 and at 77 K measured the magnetization which was plotted versus the reciprocal field and extrapolated to $1/H = 0$. They found $n_B = 2.00 \pm 0.5$ Bohr magnetons per chromium atom. Pulsed field measurements at 77 K and 4.2 K gave respectively, 2.01 ± 0.4 and 2.03 ± 0.4 in fields up to 150 000 Oe (11.9 MA/m). Darnell and Cloud found CrO_2 difficult to saturate with the difficulty increasing with decreasing particle size. In fields of 10 000 Oe (0.796 MA/m), which are much greater than the anisotropy fields, the magnetization reached only 94 to 98% of saturation. The effect was attributed to the lower symmetry experienced by surface atoms, resulting in larger and randomly directed anisotropies, and the conclusion was confirmed when measurements on single crystals in the same field gave results closer to saturation. Swoboda et al. (1961) found $\sigma_S = 100 \pm 1$ emu/g (125.6 μWb m/kg) in single crystals at room temperature and 10 000 Oe (0.796 MA/m); using a density of 4.83 g/cc (4.83×10^3 kg/m^3) this gives $M = 483$ emu/cc (0.607 Wb/m^2).

Dickey and Robbins (1971) found that small amounts of sulfur can be substituted in the cation sites of CrO_2 and depress the Curie point. Hirota et al. (1972) reported measurements of the σ_s–T curves for CrO_2 containing up to 1 wt% sulphur. The saturation magnetization at room temperature was reduced to 67 emu/g; (84 μWb m/kg), mainly because the Curie temperature was reduced to 95°C. Chamberland et al. (1973) found that fluorine also depressed both the saturation magnetization and the Curie temperature of CrO_2. In contrast, Shibasaki et al. (1970) found that the inclusion of iron increased the Curie temperature and decreased σ_s. When 10% Fe was included in the starting material the values were: $\sigma_s = 73$ emu/g (91.7 μWb m/kg) and $\theta_C = 142$°C. Kouvel and Rodbell (1967) studied the magnetic critical-point behavior of pure CrO_2 and found that $\theta = 386.5$ K (113.5°C). Darnell and Cloud (1965) found $\theta_c = 119$°C from a graph of σ^2 versus H/σ at different temperatures. Measurements between 70° and 170°C have also been reported for modified CrO_2 by Waring (1970).

Measurements at room temperature of the magnetization curves of single crystal samples of CrO_2 along the c-axis and orthogonal to it by Cloud et al. (1962) gave the first order anisotropy constant $K_1 = 2.5 \times 10^5$ erg/cc (2.5×10^4 J/m³). They found the direction of easiest magnetization was in the (100) plane and at 30° to the c-axis. Darnell (1961) gave the angle as approximately 40° and Rodbell et al. (1967) found that the angle was 0°. Köster (1973) calculated the angle; ϕ between the direction of magnetization and the long axis of the particle for the case in which the magnetocrystalline easy axis was at 40° to the long axis of the particle. He found that $\phi = 10$° and concluded that the possibility was slight of distinguishing between the angles of 30° (Cloud et al. 1962), 40° (Darnell 1961), and 0° (Rodbell et al. 1967) by making magnetic measurements on powders.

At 103°C, $K_1 = 0$ and at temperatures between 103°C and the Curie temperature (119°C), K_1 is negative and the easy axes form a cone.

b. Extrinsic magnetic properties of CrO_2

Particles of CrO_2 prepared by the methods described above are highly acicular, they usually have lengths in the range 0.3-1.0 μm and the length to diameter ratio may be 10 or 20 to 1. Darnell (1961) calculated the upper limit for SD behavior, by balancing magnetostatic energy and wall energy, and found that for diameters less than 0.4 μm the particles should be single domains. His experimental study on a variety of particle sizes and shapes supported this and he concluded that particles smaller than 0.2 μm in diameter and having a length diameter 5 : 1 showed SD behavior.

Darnell (1961) deduced that the principal source of magnetic hardness in SD particles of CrO_2 was shape anisotropy by comparing the experimental dependence of H_c on temperature with that calculated on the assumption of pure shape anisotropy or pure magnetocrystalline anisotropy. For a spatially-random assembly of CrO_2 particles having $M_s = 483$ emu/cc (0.607 Wb/m²) and $K = 3 \times 10^5$ erg/cc (3×10^4 J/m²) the expected H_c is

$$\langle H_c \rangle = 0.94 \, K_1/M_s = 585 \text{ Oe } (46.6 \text{ kA/m})$$

Assuming only shape anisotropy for particles of axial ratio 5:1 and using Stoner and Wohlfarth's (1948) formula, the coercivity is

$$\langle H_c \rangle = 0.48(N_b - N_a)M_s = 1230 \text{ Oe } (97.9 \text{ kA/m}).$$

If the chain-of spheres model is used

$$\langle H_c \rangle = 0.18\pi I_s \, (6K_N - 4L_N)$$

where K_N, L_N are parameters that depend on the length and axial ratio of the particles. The equation gave $\langle H_c \rangle = 590$ Oe (47 kA/m). The measured dependence of H_c on temperature fitted quite closely the dependence calculated assuming shape anisotropy. This result, and the measurement of $H = 2200$ Oe (175 kA/m) for the maximum anisotropy field from rotational hysteresis, led Darnell to conclude "that the coercivity is determined almost entirely by shape."

Köster (1973) estimated the relative contributions of magnetocrystalline and shape anisotropies by taking advantage of the fact that at 103°C, $K_1 = 0$ (Rodbell et al. 1967), and the coercivity is attributable entirely to shape. The samples measured by Köster contained 0.15 wt% Sb and 0.25 wt% Fe and this had the effect of raising the Curie temperature (125°C) and raising the anisotropy point to 110°C. Köster found that the relative contributions at 20°C were 53% magnetocrystalline anisotropy and 47% shape anisotropy.

Darnell (1961) measured the rotational hysteresis integrals for different samples and found values between 1.10 and 1.36 which suggested that the mode of reversal is one of incoherent rotation.

To meet the need for recording materials to be used in high-density ($\lambda <$ 1.5 μm) helical video recorders chromium dioxide particles having high coercivity are now being made in 900 kg lots (Hiller 1978). The basic method of Ingraham and Swoboda (1960, 1962) was modified to obtain particles having both high coercivity and high moment with the results shown in table 12.

It is noticeable that the coercivity can be increased without any reduction in the remanent magnetization. Furthermore, it is possible to make very small particles (with very high surface areas) and still have high coercivity and fairly high remanent moments.

Chromium dioxide particles for magnetic recording invariably have additives, the purpose of which is: (1) to reduce the high temperatures and pressures involved in the decomposition of pure CrO_3, (2) to encourage favorable growth

TABLE 12

Magnetic properties of chromium dioxide particles made in 900 kg lots (after Hiller 1978)

Sample	H_c (Oe)	H_c (kA/m)	σ_r (emu/g)	σ_r (μW m/kg)	Surface area (m²/g)
Commercial CrO_2	485	38.6	35	44	23
a. High-coercivity CrO_2	654	52.1	35	44	26
b. High-coercivity CrO_2	581	46.2	36	45	26
c. High-coercivity CrO_2	595	47.4	33	42	41

habits in the particles, or (3) to modify the magnetic or other physical properties (e.g., Curie temperature, H_c, abrasiveness). The coercivity of a sample then is determined by the additives used as well as by the conditions under which the particles were grown. The effect of additives may be seen in fig. 16 in which the effects of Te + Sn, Te + Fe and Sb + Fe on the coercivity of CrO_2 particles is

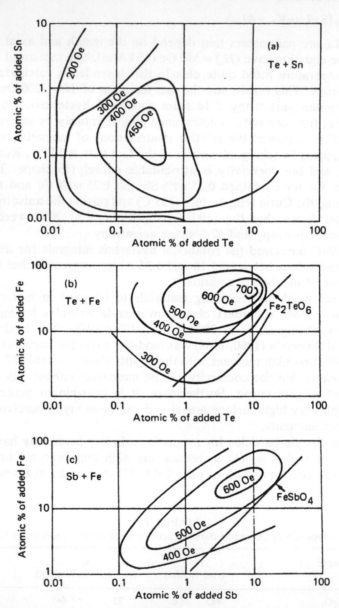

Fig. 16. Coercivity of CrO_2 particles containing (a) Te + Sn, (b) Te + Fe, (c) Sb + Fe (Mihara et al. 1970).

shown. With Te + Fe the maximum coercivity was about 700 Oe (55.7 kA/m). A higher coercivity of 800 Oe (63.7 kA/m) was reported by Hirota et al. (1974) for particles containing a combination of Te, Sn and Fe. Amemiya et al. (1973b) used iridium in the form of the hydroxide as an additive in the hydrothermal synthesis of CrO_2. Their results are given in table 13.

Particles of chromium dioxide prepared for use in recording surfaces are highly acicular and despite the slight uncertainty about the direction of the magnetocrystalline easy axis with respect to the long axis of the particle, there is no doubt about the uniaxial nature of the resultant anisotropy. Consequently, we expect for a random assembly of particles, a value of squareness ($S = M_r/M_s$) of 0.5 and this was found by Umeki, et al. (1974a). In addition to being highly acicular, particles of chromium dioxide show no dendrites in the electron microscope and are relatively easy to disperse and to align in a magnetic field so that values of $S = 0.82$ are not uncommon.

Darnell (1961) investigated the dependence of H_c on particle loading (p) and found that the dependence was much less than the Néel (1947a) relation

$$H_c = H_c(0)(1 - p).$$

Darnell attributed the disagreement to the presence of agglomerates and to the role of magnetocrystalline anisotropy. Bottoni et al. (1972) found that S was independent of p in the range 0.0003 to 0.2 in contrast to the behavior of particles of γ-Fe_2O_3 where in the vicinity of $p = 0.2$, S was found to increase with increasing p. Bierlein (1973) also found that the magnetic properties of assemblies of CrO_2 particles are only weakly sensitive to interactions between particles. He concluded that the properties are essentially explained by the particle size distributions which he found to be log normal.

The principal advantages of chromium dioxide particles are that they can be made of uniform size and are free of dendrites. They are readily dispersed and oriented and give excellent high-density (short-wavelength) recording performance. The disadvantages are that they are ten times as expensive as iron oxide particles, they can be very abrasive to heads, and their magnetic properties are time-dependent. The last property shows itself in two forms. First, a decrease in moment and in coercivity can be seen in tapes stored for 2 to 3 years. It is believed that this is caused by a chemical reaction between the particles and the

TABLE 13

Effect on the coercivity of CrO_2 particles, after using irridium hydroxide as an additive in the hydrothermal synthesis (after Amemiya et al. 1973b)

Wt% Ir(OH)₃	H_c	
	(Oe)	(kA/m)
2	640	50.9
5	850	67.7
10	1050	83.6

atmosphere or the binder. The second time-dependent effect takes place in minutes and is observed during the measurement of the hysteresis loop when an applied field about 80–90% of H_c is held constant. The measured moment steadily decreases according to the relation

$$\Delta M = P + Q \log t$$

where P and Q are constants.

The magnetic properties of chromium dioxide particles are summarized in table 5 (see section 3.1.4).

3.4. Ferrites with hexagonal structure

Ferrites having a complex, close-packed, hexagonal structure similar to that of the mineral magnetoplumbite have found application as permanent magnet materials because of their very high coercivities (over 1000 Oe, 79.6 kA/m) combined with a moderate saturation magnetization (σ_s is about 70 emu/g; 88 μWb m/kg). These materials have so far found only limited use in recording surfaces primarily because the high coercivity requires that the material of the writing head have a high saturation flux density. However, the high coercivity gives the recording surfaces a unique ability to withstand high demagnetizing fields or erasing fields and two kinds of application have so far appeared which take advantage of this. In the first application, high-coercivity master tapes are used to transfer information to copy tapes by bringing the two tapes together in the presence of a magnetic field. The transfer field should be high enough so that the combination of transfer-plus signal-fields are enough to switch the particles in the copy tape, yet the transfer field must not be so high that erasure of the master tape occurs. Clearly it would be advantageous to use very high coercivity master tapes. The second application is in magnetic stripes on credit cards and magnetic badges. The density of recorded information is only a few hundred bits per inch but the material must be capable of withstanding successfully accidental stray fields which could reach values of more than 100 Oe (7.96 kA/m) or in some applications, deliberately applied fields of 2000 Oe (159 kA/m) (Fayling 1977).

The most common example of these hexagonal ferrites is barium ferrite, $BaO \cdot 6Fe_2O_3$.

3.4.1. Barium ferrite, $BaO \cdot 6 Fe_2O_3$

a. Preparation

The usual method of preparing the hexagonal ferrites is similar to that described by Tenzer (1965). Barium carbonate was wet-mixed with ferric oxide and calcined at 1300°C for one hour. This temperature was higher than the sintering temperature and, since the high coercivities are found only in the powdered material, the last step was to grind with steel balls in water for about 120 h until the average particle size was about 1 μm.

Other methods of preparation include electrolytic coprecipitation, used by

Beer and Planer (1958) and a chemical coprecipitation technique used by Mee and Jeschke (1963). Haneda et al. (1974) also used the coprecipitation method starting with an aqueous solution of ferric and barium chlorides. The solution was poured into a solution of sodium hydroxide and sodium carbonate and the resulting precipitate was washed and dried at temperatures between 700°C and 1100°C. The saturation magnetization increased sharply with drying temperature from <5 emu/g (6.3 μWb m/kg) at 700°C to 56 emu/g (70.3 μWb m/kg) at 800°C, and then more slowly to a value of 63 emu/g (79 μWb m/kg). The coercivity showed a maximum of nearly 6000 Oe (478 kA/m) at a firing temperature of 925°C. The particle size was 200 Å after a firing temperature of 650°C and increased slowly to 1000 Å at 900° and thereafter very rapidly to diameters of 1 or 2 μm. The particles prepared by Mee and Jeschke (1963) had the form of very small platelets whose diameters were in the range 800 to 1500 Å and whose ratio of diameter to thickness was about 15:1.

b. Structure and properties

The structure of $BaO \cdot 6 Fe_2O_3$ is a complex one (see Smit and Wijn 1959, p. 182) based on a hexagonal lattice in which close-packed layers of oxygen atoms have, in every fifth layer, a mixture of Ba^{2+} and oxygen ions in the proportion of three to one. The smaller Fe^{3+} ions are arranged in five different kinds of interstitial sites. Two kinds are tetrahedral and two are octahedral and the fifth type of site is one in which the ferric ion is surrounded by five oxygen atoms forming a trigonal bipyramid. This gives five hyperfine Mössbauer spectra which were identified by Haneda et al. (1974). Ten oxygen layers are needed to form a unit cell consisting of two $BaO \cdot 6 Fe_2O_3$ formula units.

The magnetic moments of each formula unit are found by adding the moments of the seven octahedral Fe^{3+} ions and the Fe^{3+} ion in the trigonal site and then subtracting the moments of two octahedral and two tetrahedral Fe^{3+} ions. Since all the magnetic ions are Fe^{3+}, having a magnetic moment of 5 μ_B, the magnetization-per-formula-unit is then $(7 + 1 - 2 - 2)5 \mu_B = 20 \mu_B$. This corresponds to a value of $\sigma_{so} = 100$ emu/g (126 μWb m/kg).

At room temperatures this becomes 72 emu/g (90 μWb m/kg) or 380 emu/cc (0.477 Wb/m^2) which is very close to the saturation magnetization found for γ-Fe_2O_3. Mee and Jeschke (1963) measured $\sigma_s = 68$ emu/g for their 0.1 μm diameter particles. The Curie temperature of barium ferrite is 450°C, and the first order anisotropy constant, $K_1 = +3.3 \times 10^6$ erg/cc ($+3.3 \times 10^5$ J/m^3) and so the easy axis is the hexagonal c-axis.

To produce SD particles it is necessary to grind the sintered ferrite but it has long been recognized that prolonged grinding reduces the coercivity. This was investigated by Haneda and Kojima (1974) who observed that the decrease had been ascribed to an increase in the number of SP particles or to lattice defects induced by milling. They found, from measurements of the angular variation of H_c that milling lowers the nucleation field of the reverse domain or the effective anisotropy field or both. Haneda and Kojima concluded that lattice defects rather than superparamagnetic fine particles are the cause of reduction in H_c, in agreement with the earlier observations of Smit and Wijn (1959) who found that annealing the particles in air for one hour brought about an increase in H_c from 1600–3400 Oe (12.7–271 kA/m). Tenzer (1965) believed that SP particles were responsible for the reduction of H_c on prolonged grinding and that Smit and Wijn's results could by a sintering of the smaller (almost SP) particles to form stable SD particles. But Shtol'ts (1962) deduced that there was a more fundamental difference than particle size between high and low coercivity particles of barium ferrite. He produced particles of 1–2 μm by two methods. In the first, the material was sintered at 1400°C and had a coarse structure, with an average particle size of about 300 μm; the coercivity was about 20 Oe (1.59 kA/m). The second material was sintered at 1200°C; the particles were relatively fine-grained, having an average size of about 5 μm and had high coercivity, 2000 Oe (159 kA/m). When both materials were ground to the same average particle size, the coercivity of the first material increased for 20–350 Oe (1.59–27.9 kA/m) and that of the second decreased from 2000–1000 Oe (159–80 kA/m). By aligning the c-axes in a magnetic field, Shtol'ts was able to produce a pseudo-single crystal and measure its anisotropy. He found that the material which had been sintered at 1400°C was appreciably more anisotropic than that sintered at 1200°C and the difference did not appear to change on grinding. These results led Shtol'ts to conclude that there was a fundamental structural difference between the two materials. Ratnam and Buessem (1970) examined the nature of defects in barium ferrite by electron microscopy and concluded that the reduction in H_c was related to the introduction of stacking faults and deformation twins.

According to Stoner and Wohlfarth (1948) the theoretical coercivity for a random assembly of SD particles of barium ferrite would be

$$H_c = 0.96 \, K_1/M_s$$

substituting: $K = +3.3 \times 10^6$ erg/cc and $M_s = 390$ emu/cc gives $H_c = 8350$ Oe (66.5 kA/m), higher than any experimental value. However the graph showing the angular dependence of H_c obtained by Haneda and Kojima (1974) is very similar, for milling times of 36–48 hours, to the results obtained by Bate (1961) for aligned assemblies of γ-Fe_2O_3 particles. It appears, therefore, from this result, that the mode of magnetization reversal is an incoherent one. However, there is also evidence that the Stoner and Wohlfarth (1948) theory of coherent rotation applies to barium ferrite particles.

The platelets that Mee and Jeschke (1963) prepared had a very small average diameter, 0.1 μm, and a very high coercivity at room temperature, 5350 Oe (426 kA/m). The magnetization $\sigma = 60.3$ emu/g (75.7 μWb m/kg) was measured in a field of 10 400 Oe (828 kA/m) from which they concluded that $\sigma_s = 68.0$ emu/g (85.4 μWb m/kg). Assuming that the platelets have crystalline easy directions perpendicular to their planes and that the easy direction for shape anisotropy lies in the plane, the two anisotropies are opposed and the coercivity for a random array of particles becomes:

$$H_c = 0.48 \, (2K_1/M_s - 4\pi M_s).$$

The temperature dependence of M_s and K_1 was determined by Rathenau et al. (1952); hence the theoretical temperature dependence of H_c could be compared with that found experimentally. Mee and Jeschke (1963) found fairly good agreement between theory and experiment and Haneda et al. (1974) found excellent agreement except in the immediate vicinity of H_c between the measured hysteresis loop (in the first, second, and third quadrants) and that calculated for a random assembly of particles by Stoner and Wohlfarth (1948). Even there the disagreement was only 10%.

Since the contribution of the magnetocrystalline term $(2 K_1/M_s)$ is 3 to 4 times greater than the shape factor, it seems likely that the effects of interactions between particles would not be of great importance in barium ferrite. This question was investigated by Martin and Carmona (1968) on particles prepared by chemical precipitation followed by firing at a temperature of 920°C; the average particle diameter was 0.5 ± 0.05 μm. At a packing factor, $p = 0.5$ the sample coercivity was 4640 Oe (369 kA/m); on dilution the coercivity changed, only slightly at first and then (below $p \sim 0.15$) more rapidly, finally reaching 5200 Oe (413 kA/m) at $p = 0.03$. The experimental points fitted the curve

$$H_c(p) = 4.706 + 16.63 \, (1 - p)/p \text{ Oe}.$$

The anisotropy constants of higher order than K_1 are apparently negligible in barium ferrite. Furthermore, the magnetization M_s and K_1 change with temperature at almost the same rate (Smit and Wijn 1959) in the temperature range 20–500°C. Thus the dominant magnetocrystalline contribution to the coercivity, depending on K_1/M_s is practically independent of temperature in this range, and hence, so is the coercivity itself. We expect, therefore, that SD particles of barium ferrite, unlike those of the cobalt-substituted iron oxides, should be quite stable with respect to changes in temperature.

3.4.2. Substituted barium ferrites and related materials

The coercivity of barium ferrite is usually in the range 2000–3600 Oe (159–287 kA/m) and heads made of conventional materials (Permalloy, Alfesil, ferrites) are incapable of switching it. There exists a group of substituted barium ferrites in which the basal plane is the easy plane and the c-direction is the hard axis. The substitution elements are divalent metals such as Co, Ni, Mn, Fe, or Zn which is combined in the form of the oxide with BaO and Fe_2O_3. The structure and properties of these materials were discussed by Jonker et al. (1956/7), by Stuijts and Wijn (1957/8), and by Smit and Wijn (1959, p. 204). The anisotropy constant is usually much lower than that of barium ferrite ($K_1 + 2K_2 = -0.4$ to -2.6×10^6 erg/cc, -0.4 to -2.6×10^5 J/m³, whereas for barium ferrite $K_1 = +3.3 \times 10^6$ erg/cc, $+3.3 \times 10^5$ J/m³). Since the basal plane is preferred, there are six equivalent easy directions of magnetization. Thus, we expect the squareness to be higher and the coercivity to be lower than for barium ferrite. Furthermore, the plate-like particles may be oriented in a tape coating as described by Duinker et al. (1962).

Materials having the composition $MO \cdot 6 \, Fe_2O_3$ where $M = $ Pb or Sr, also show

hard magnetic properties, and the range of available oxides with the mag-
neto-plumbite structure can be increased still further by substituting up to 3%
BiO_2, TiO_2, Al_2O_3, SiO_2, etc. for part of the MO or Fe_2O_3. The magnetic
properties of these compounds was discussed by Kojima (1956, 1958).

The use of the hexagonal ferrites is, at present, limited to special recording
applications in which their high coercivities are needed to withstand strong
erasing fields. The magnetic properties are summarized in table 5 (see section
3.1.4).

3.5. Metal particles

The attraction of metal particles lies in their higher intensities of magnetization
and higher coercivities. The first permits high signal levels to be obtained from
thinner coatings and the second gives a resistance to demagnetization which
becomes increasingly important as the bit length or wavelength becomes shorter.
Unfortunately fine metallic particles are highly reactive and in many cases,
pyrophoric. Thus, something must be done to passivate the surface if the
materials are to have any practical importance in recording surfaces.

The principal methods of preparing metal particles in the SD range of sizes are
(1) reduction in hydrogen of oxides, hydroxides, oxalates, etc. (2) reduction in
aqueous solution by borohydrides, or hypophosphites (3) decomposition of
metallo-organic compounds (4) vacuum deposition and sputtering, and (5) electrodeposition.

3.5.1. Reduction in hydrogen

The key to this method of preparing fine metal particles is to find a compound
which can be reduced in a stream of hydrogen at a temperature low enough that
sintering of the product does not occur.

Acicular particles of iron (or iron–cobalt) oxides, prepared by any of the
methods described above, can be used as the starting material and reduced to
the metal by being heated in a stream of hydrogen. Carman (1955) found that a
temperature of about 350°C was optimum. At lower temperatures the process
was inefficient and higher temperatures caused sintering, some loss of acicularity
and a loss of M_r and H_c. Carman obtained coercivities greater than 600 Oe
(47.8 kA/m) and $M_r = 414$ emu/cc (0.52 Wb/m^2) at a packing density of 0.3. More
recently Van der Giessen and Klomp (1969) and Van der Giessen (1973)
prepared iron powders by the pseudo-morphic conversion of α-(FeO)OH which
contained a small amount of tin to minimize the change of shape and sintering
during reduction. The average length of the particles could be chosen between
0.1 and 1 μm. Particles having an average length of 0.2 μm gave $H_c = 1320$ Oe
(105 kA/m) and $\sigma_s = 159$ emu/g (200 μWb m/kg). The reduced remanence was 0.5
indicating the dominance of shape anisotropy. Van der Giessen and Klomp
calculated that in the temperature range 25–400°C the reaction

$$3 Fe_2O_3 + H_2 \rightarrow 2 Fe_3O_4 + H_2O$$

is possible if

$$P_{H_2O}/P_{H_2} < 10^4$$

and indeed, experimentally the reaction proceeds quickly. However the reaction

$$\tfrac{1}{4} Fe_3O_4 + H_2 \rightarrow \tfrac{3}{4} Fe + H_2O$$

can take place only at much lower partical pressures of water vapor. In fact, at 350°C P_{H_2O}/P_{H_2} must be less than 10^{-1}. They concluded that at higher temperatures the removal of water is the rate limiting factor while at lower temperatures it is probably the rate of formation of metallic nuclei. Aonuma, et al. (1975a), Köster et al. (1976a) and Schneehage et al. (1976) reduced the goethite particles in the presence of aliphatic barium compounds in methyl alcohol. After evaporation of the solvents the oxide was vacuum dried at 100°C and reduced in hydrogen at 300°C to give acicular iron particles having a length of 0.2–0.4 μm, length to width of 15–25: 1 and a coercivity of 1320 Oe (105 kA/m).

Kawasaki and Higuchi (1972) prepared goethite by the standard method and mixed a suspension of particles with cobalt hydroxide which was made from a solution of cobaltous chloride and ammonia. The $Co(OH)_2$ was absorbed on the surface of the goethite particles and heating at 700°C for two hours caused the cobalt to diffuse into particles. Thus far the process is similar to that described by Haller and Colline (1971) for the preparation of particles of cobalt-modified γ-Fe_2O_3. The final step was to reduce the particles to the metal by heating in hydrogen for 6 h at 380°C. Kawasaki and Higuchi investigated the magnetic properties of the product as a function of cobalt content and found the most attractive properties between 10 and 50% Co. In this range the coercivity exceeded 1100 Oe (87.6 kA/m) and the saturation magnetization was greater than 160 emu/g (201 μWb m/kg). The properties of particles of partially reduced oxides of Fe,Co and Ni prepared by co-precipitation of the oxyhydroxide and reduction in 90% N_2, 10% H_2, were described by Morrison, et al. (1969). Inagaki and Tada (1972) precipitated β-$(FeO)OH$ by the hydrolysis of ferric chloride and heated the precipitate in sodium hydroxide solution in order to remove trapped Cl^- ions. The removal of the ions before heat-treatment and reduction was claimed to help in preserving the acicularity of the particles. Cobalt particles may also be prepared from $Co(OH)_2$ by direction reduction in hydrogen at 350°C. (Swisher et al. 1971).

The methods just described use metal oxides, oxyhydroxides or hydroxides as starting materials. However, it is possible, starting with combinations of metal chloride solution to precipitate the oxalates and then to reduce them to the metal in hydrogen or in forming gas. Thus, Nakamura et al. (1975a) rapidly added a solution of ferrous, cobaltous and nickelous chlorite to a solution of acetone, toluene, and oxalic acid. The precipitate was reduced in hydrogen at 300–400°C to a powder which had $\sigma_s = 200$ emu/g (251 μWb m/kg). A narrower distribution of switching fields was claimed for this method. This usually implies a tighter control on particle size. Sueyoshi (1976) claimed that control of the reaction temperature was simpler and control of the oxalate particle size easier when the oxalic acid was dissolved first in dialkyl formamide.

The "Coballoy" particles described by Deffeyes et al. (1972) were made by reducing the oxalate. They found that carefully controlled pH and concentration during the precipitation of the oxalate particles and careful reduction gave uniform particles whose coercivity could be controlled in the range 600 to 1200 Oe (47.8–95.5 kA/m) to within ±50 Oe (3.98 kA/m). The particles had an average length of about 0.05 μm and length-to-diameter ratio of 2 or 3:1.

Deffeyes et al. (1973) believed that failure to control the lattice structure of the metal oxalate crystals explained the lack of reproducibility in earlier experiments to make recording materials by the oxalate method. They found that the control factors were temperature, composition, rate of addition of reagents and the use of a 30% alcohol–water mixture as solvent. The addition of Ce, Ti, or Ca as impurities allowed the coercivity to be chosen from 100–1200 Oe (7.96–95.5 kA/m). Deffeyes et al. (1973) found the cubic lattice parameter of their particles (64 Co, 18 Ni, 18 Fe) was 3.517 Å compared to 3.545 Å for pure fcc cobalt; a similar result was found by Gen et al. (1970) for pure Co particles prepared by evaporation.

The particles ranged in length from 0.1–0.5 μm and had an acicularity of about 5:1. The coercivity was found to be only slightly temperature dependent and thus shape anisotropy appeared to be dominant. The rotational hysteresis integral of 2.04 suggested an incoherent mode of reversal.

Takahashi et al. (1975) prepared metal particles containing Fe, Co, and Ni by a method similar to the one described by Nakamura et al. (1975a) above except that a small amount of aluminum chloride was added to the other chlorides. In this way oxalates and the alloy particles containing Fe, Co, Ni + 1% Al were obtained and showed $H_c = 880$ Oe (70 kA/m), $\sigma_s = 191$ emu/g. (240 μWb m/kg) and S = 0.34.

Takahashi et al. (1975) claimed that the incorporation of Al allowed them to find processing conditions which permitted H_c and σ_s to be maximized simultaneously. Ando and Wakai (1975a) combined both a gaseous reduction method by dispersing acicular particles of α-(FeO)OH in water and oxalic acid. The dried powder was reduced by heating in hydrogen for an hour at 300°C.

Other organic compounds which have been successfully reduced at low temperatures in hydrogen include iron formate Stewart et al. (1955) and iron phthalate which Rabinovici et al. (1977) prepared in the presence of a magnetic field of up to 1000 Oe (79.6 kA/m). The effect of the field was principally to reduce the average particle size and increase the coercivity and squareness.

3.5.2. Reduction in solution

It is also possible to reduce salts of the ferromagnetic metals directly in aqueous solution. For example, Oppegard et al. (1961) found it possible to make SD particles of iron, iron–cobalt and iron–nickel, by reducing the sulfates or chlorides of the metals with an aqueous solution of potassium borohydride in a magnetic field of about 5000 Oe (398 kA/m). The resulting particles had, in the case of iron, lengths of several micrometers and had a coercivity of 645 Oe (51.3 kA/m).

Measurements of the temperature dependence of coercivity in the range −200 to +200°C indicated contributions from both shape and magnetocrystalline anisotropies with the former dominating. The rotational hysteresis integral for a number of samples of iron–cobalt had an average of 1.42 indicating a mechanism involving incoherent rotation. Particles prepared without a magnetic field were much shorter and of lower coercivity. Similar particles of iron were prepared by Olsen (1974) who studied their corrosion behavior; this work will be discussed later.

The basic method has been varied by the use of additives or by changes in the processing conditions. Aonuma et al. (1975b), and Aonuma et al. (1975c) added an aqueous solution of chromium–potassium sulfate immediately after adding the borohydride solution to the metal salts. When these were ferrous and cobaltous chlorides the resulting metal powders had the composition Fe 72, B 5.6, Co 7.0, and Cr 4.1 parts by weight. It was claimed that the addition of a chromium-based surface layer inhibited oxidation of the particles which, when coated into a tape showed $H_c = 1000$ Oe (79.6 kA/m) and $S = 0.81$. The addition of lead acetate to the solutions was claimed by Aonuma and Tamai (1975b) to give improved high frequency response to the tape. Controlling the temperature during and after the reaction to within 20°C was found by Aonuma et al. (1975d) to help in reducing the distribution of particles sizes and Furukawa et al. (1976) added butyl acetate to the aqueous suspension of particles as a means of creating a minimum-boiling point mixture from which the particles could be dried at low temperature without oxidation.

The magnetic properties of iron–cobalt alloy particles produced by the reduction of their salts with sodium borohydride was discussed by Uehori et al. (1978). The particles were obtained by mixing a 1 M metal salt solution with 1 M sodium borohydride solution in a magnetic field of 1000 Oe (79.6 kA/m). As the atomic percentage of cobalt was increased from 0 to 60, the coercivity rose from 580 Oe (46.2 kA/m) to a maximum of 1250 Oe (99.5 kA/m) at about 30 at.% Co and then decreased to 650 Oe (51.7 kA/m) at 60 at.% Co. At the peak coercivity the mean particle diameter was 300 Å and particles of that size showed the smallest temperature coefficient of coercivity. From measurements on aligned samples of the variation of coercivity and of torque with angle the authors concluded that the mechanism of magnetization reversal was one of incoherent rotation, probably fanning.

The borohydride method of reducing metal salts is an effective but expensive process.

The autocatalytic process for the deposition of continuous metal films will be described in detail later. A catalytic substrate is immersed in an aqueous solution of metal chlorides or sulfates and on adding a reductant, e.g. hypophosphite or borohydride, plating begins spontaneously and continues since the deposited metal in turn catalyzes the reaction. However, if any of the three rate-controlling parameters, temperature, pH, or reactant concentration, are increased excessively, the homogenous reaction occurs between metal ions in the solution and the adjacent reductant ions to form metal atoms distributed throughout the solution. These atoms then catalyze the deposition of more metal and a metallic slurry is rapidly formed and collects on the bottom of the vessel. Afer filtering and drying, the particles are found to have a size in the range 200–700 Å and to have coercivities usually between 100–1000 Oe (7.96–79.6 kA/m).

A typical example of this process was provided by Ginder (1970) who prepared cobalt–nickel particles from a solution of their sulfates. Sodium citrate was used as a complexing agent to control the rate at which the metal ions were made available to the reductant (sodium hypophosphite). The reaction was initiated by the addition of a palladium chloride solution to provide the nuclei on which the first metal atoms could form. The reaction is often performed in a magnetic field of several hundred oersteds.

In order to control the distribution of particle sizes and shapes Akashi and Fujiyama (1971), Harada et al. (1972), Aonuma and Tamai (1975a), and Tsukanezawa et al. (1975) added proteins such as gelatin to the solution of salts in order to increase the viscosity. Harada, et al. (1972) kept the pH between 7.5 and 9.5 and added the palladium chloride solution at 90°C. The reaction was carried out in a field of 800 Oe (63.7 kA/m). They investigated the effects of variations in pH, hypophosphite concentration and interpreted the different particles properties which they obtained in terms of particle shape and size, crystal structure and phosphorous content. Spherical particles containing 82 Co–Ni, were hcp and between 300–500 Å in diameter had a coercivity of 1100 Oe (87.6 kA/m), $M_s =$ 880 emu/cc (1.11 Wb/m²). The squareness was 0.53 indicated that the dominant anisotropy was uniaxial (magnetocrystalline).

Parker et al. (1973) made similar particles without using a catalyst, the combined sodium hyposulfite and dimethylamine borane acted as reductants. It is not essential to use metal ions as nuclei in the autocatalytic deposition of metal particles. Fuji Co. (1974) used particles of goethite, or of attapulgus clay, as nuclei and thus obtained acicular particles having an average length of 1.2 μm, an acicularity of 8:1 and a coercivity of 1000 Oe (79.6 kA/m). Attapulgus clay was used as a base in the earlier work of Schuele (1959, 1962). The highly acicular clay particles were prevented from flocculating by giving them a negative charge from a dispersing agent such as sodium pyrophosphate which contains the $P_2O_7^{3-}$ radical. Scheule then deposited iron oxide on the nuclei using the standard method described earlier and finally reduced the oxide to metal in a stream of hydrogen at 180–200°C. The particles had coercivities between 950 and 1450 Oe (75.6–115.4 kA/m) and a squareness of 0.5.

To avoid the possibility of oxidation, during drying, of metal particles prepared from aqueous solutions, Aonuma and Tamai (1974) displaced the water first with acetone and then with butyl acetate.

3.5.3. Decomposition of metal carbonyls

Thomas (1966), Harle and Thomas (1966) and Thomas and Lavigne (1966) described a process whereby cobalt particles of closely controlled size may be made by the decomposition of dicobalt octacarbonyl. The starting material was dissolved in toluene containing a terpolymer and the solvent was refluxed until the theoretical quantity of carbon monoxide was evolved. The product was a stable colloid of metallic cobalt particles having a roughly-spherical shape and a very uniform diameter of about 200 Å. The particles were found to be predominantly (96–98%) fcc in structure and appeared in the electron microscope surrounded by a polymer film which prevented agglomeration. By varying the reagent conditions, temperature, and the composition of the polymer, the particle size could be controlled from about 20 Å to about 300 Å. Above diameters of about 100 Å the particles spontaneously joined together to form chains and the coercivity of the sample depended not only on the average size of the cobalt particles but also quite critically on the thickness of the polymer coating between the particles. Thus particles whose average diameter was 200 Å, and which had an initial separation of about 20 Å, had a coercivity of 500 Oe (39.8 kA/m) which changed to 1300 Oe (103.5 kA/m) when the particle separation was reduced to zero.

Thomas (1966) found that cobalt particles of approximately 80 Å in diameter behaved super-paramagnetically as long as the particles were randomly oriented. However, if the particles were dried in a field of several thousand Oersteds so as to form chains coercivities as high as 100 Oe (7.96 kA/m) and squareness of 0.8–0.9 developed.

Thomas and Lavigne (1966) showed that the coercivity of the particles could be substantially increased by heating in an inert atmosphere and preferably, at elevated pressures. Thus, particles consisting of 75% cobalt and 25% polymer, after heating at 110°C and 730 000 kg/m², showed an increase in coercivity from 560 Oe (44.6 kA/m) to 1150 Oe (91.5 kA/m).

3.5.4. Vacuum deposition

Morelock (1962) described a technique for growing whiskers of iron and nickel (diameter <1 μm, length ≤300 μm) by deposition from the vapor phase. The metal to be evaporated was put with the iron substrates in an evacuated tube which was heated to 1100°C. The evaporator filament was heated to 1150°C and so the whisker growth occurred in a supersaturated metal vapor; the pressure was about 10^{-6} mm Hg. Luborsky et al. (1963) and Luborsky and Morelock (1964) used the method to make iron, iron–cobalt, and cobalt whiskers and found by selected-area diffraction that the smaller diameter whiskers of Fe or Fe–Co were bcc with ⟨111⟩ orientation. The cobalt whiskers were hcp with ⟨11.0⟩ along the whisker length. Dragsdorf and Johnson (1962), however, found that a majority of their whiskers had a ⟨10.0⟩ orientation.

The coercivities of isolated whiskers and of planar-random arrays showed a marked dependence on particle diameter. For diameters less than about 0.05 μm, coercivities exceeding 2000 Oe (159 kA/m) were found; while for whiskers with a diameter of about 100 μm the coercivity was about 0.5 Oe (40 Å/m). The dependence of coercivity, remanence, and rotational hysteresis on particle diameter agreed very well with the incoherent, curling mode of magnetization reversal and in fact provided the first experimental evidence for that mode. The magnetic properties of iron and of iron–cobalt whiskers were very similar and since the only difference was the possible contribution of magnetocrystalline anisotropy, it was concluded that this anisotropy was negligible in these whiskers. In the cobalt whiskers, however, the crystalline anisotropy was not negligible and Luborsky and Morelock (1964) found that particles whose diameter was less than 250 Å the magnetization reversed coherently while in the larger particles an incoherent buckling mode occurred.

Kimoto et al. (1963) prepared fine particles of metals and alloys by evaporation in argon at 0.5–50 mm Hg. Tasaki et al. (1965) found that the size of such particles depended critically on the argon pressure in which they were formed. To prepare particles of about 300 Å diameter, a pressure of 3 mm Hg was needed; higher pressures gave larger particles and particles whose M_s approached that of the bulk ferromagnetic metals. The highest remanence and the highest coercivities (750 Oe, 50.7 kA/m for Fe and 1600 Oe, 127 kA/m for Co) were found for particles of 300 Å diameter. Even for the smallest particles (80 Å at 0.5 mm Hg of argon) no evidence of superparamagnetic behavior was found and this was attributed to the interaction between particles in chains which kept the moment stable even above the blocking temperatures of the individual particles. Similar results were given in section 3.5.3 for cobalt particles prepared by the decomposition of the carbonyl. Tanaka and Tamagawa (1967) prepared particles of Fe–Co particles by evaporation in argon and also found that the average particle size and the saturation magnetization both increased with increasing particle size. They also found that particles whose radius was as small as 50 Å did not behave super-paramagnetically as long as the particles were arranged in chains. From the observation that particles having the isotropy composition ($K_1 = 0$ at 60 Fe, 40 Co) did not differ appreciably in coercivity from similarly sized particles of different composition, they concluded that shape anisotropy was the principal cause of magnetic hardness.

Gen et al. (1970) found that cobalt particles prepared by evaporation had the fcc structure and a lattice parameter considerably smaller than for bulk cobalt. The lattice parameter increased and both coercivity and squareness, (<0.5) decreased with increasing particle size in the range 100 to 800 Å. Shtol'ts et al. (1972) studied similar particles of Ni, Co (fcc), Fe, Fe–Co and Fe–Ni and concluded that the coercivity was highly sensitive to the condition of the surface.

Not surprisingly, other alloy compositions and modification to the basic preparation technique have been reported. Iga and Tawara (1974) evaporated iron and palladium (40–65% Pd) in helium or in argon and after heating the resulting particles in vacuo for three hours at 610°C obtained $H_c = 1100$ Oe (87.6 kA/m) and $M_s = 358$ emu/cc (0.45 Wb/m^2). Particles of 90Co–10Si produced by evaporation in an inert gas at a pressure of 15 mm Hg. were found by IGA (1973) to be cubic with a crystallite size between 200 and 500 Å and to have a coercivity of only 80 Oe (6.4 kA/m). After being heated for 24 h at 510°C the structure became hexagonal, the crystallite size increased to 400–900 Å and the coercivity became 550 Oe (43.8 kA/m).

Metal particles (Fe–Co or Fe–Co–Ni) having a protective oxide coating were prepared by Ozaki et al. (1973) by evaporation in vacuo (10^{-5} torr) followed by treatment in an inert gas containing small amounts of oxygen. Acicular particles of Fe, Co, Fe–Co, were made by Osakawa and Harada (1974) by evaporation into a reactive gas mixture (hydrogen, methane, ethylene, oxygen, etc.) rather than the inert gases or vacuum used by other workers. Finally the inclusion of particles of Co–Fe prepared by evaporation in Ar directly into the binder polymers was described by Tamai and Shirahata (1976); their tape had the properties $H_c = 725$ Oe (57.7 kA/m), $\sigma_S = 132$ emu/g (166 μWb m/kg) and $S = 0.72$.

3.5.5. Electrodeposition

Elongated SD particles of iron or iron–cobalt may be prepared by electrodeposition from an aqueous electrolyte into a mercury cathode as described by Luborsky (1961). The current–density, time of deposition, temperature, and pH were the key processing parameters and were chosen to given particles 100–200 Å in diameter. The particles were then thermally grown at 100–200°C for 15 min in the mercury to remove the dendrites and to form the optimum particle shape. A reactive metal such as antimony or tin was added to the iron dispersion in mercury and serves to form a layer on the surface of each particle (Luborsky and Opie 1963). Finally the magnetic particles were removed from the mercury with a permanent magnet and any adhering mercury was removed by vacuum distillation. Particles having average diameters of 125–325 Å and acicularities greater than 10:1 were prepared by this method. The shape of the particles resembled a chain-of-spheres and in fact the model was conceived to explain the magnetic properties of these particles. Measurements reported by Kneller and Luborksy (1963) of coercivity and squareness showed only slight dependence on temperature in the range 4–207 K and suggested that the predominant anisotropy was that of shape. The values of the rotational hysteresis integral (1.2–1.3) and the absence of size dependence of the extrinsic properties support the incoherent fanning mode of reversal. Aligned, elongated SD particles gave coercivities in the range 700–1000 Oe (55.7–79.6 kA/m) and the saturation magnetization was between 700 and 760 emu/cc (0.88–0.95 Wb/m²).

Luborsky et al. (1963) found that the long axes of the particles were [111] which is a hard direction and explained their observed increase in coercivity with increasing temperature in terms of competition between shape and magnetocrystalline anisotropy. They inferred from the maximum in coercivity of 1000 Oe (79.6 kA/m) at a composition of 80Fe, 20Co, that $K_1 = 0$ at that composition (Tanaka and Tamagawa 1967).

Luborsky and Opie (1963) reported that reactive elements capable of forming intermetallic compounds with iron, increased the coercivity when added to the electrodeposited and thermally-grown particles. Since there was no change of magnetic moment of the particles (at least, with Mn, Zn, Sn, and Sb), they concluded that the additive was absorbed as a mono-molecular layer on the iron particle and thus no compound formation occurred. The increase in coercivity (as much as 15% with Sn or Zn) was ascribed to the decrease in magnetostatic interaction of the coated particles. An additional advantage of the coating was described by Yamartino and Falk (1961): the coating stabilized the particles against loss of acicularity during the removal of mercury by vacuum distillation at temperatures above 250°C.

3.5.6. Wire drawing
Levi (1959, 1960) reported a method of preparing elongated iron particles which was very different from any of those described above. The method consisted of taking a bundle of iron and copper wires, enclosing the bundle in a copper sheath, and drawing it down to a wire of diameter between 0.002″ and 0.030″. Repeatedly bundling, annealing, and drawing resulted in the formation of extremely fine and elongated particles of iron with diameters as small as 300 Å. The coercivities reported were between 100 and 410 Oe (7.96–32.6 kA/m) depending on the final diameter of the iron fibers and on the particle packing fraction. Luborsky (1961) concluded from measurements of the angular dependence of coercivity that the magnetization reversal process in these particles was an incoherent one. Nearly perfect alignment of the particles is to be expected in view of the method of preparation and so it is not surprising to find $S \sim 1.0$.

3.5.7. Stabilizing metal particles
Most of the methods described above for the preparation of metal particles gave a product whose saturation magnetization was, either by accident or design, significantly below that of the bulk material. Oxidation occurs during the preparation, or afterwards, and the magnetization of the oxide surface is generally lower than that of the bulk metal. For smaller particles, proportionately more atoms are at the surface and the effect is so pronounced, that the particles may burn spontaneously on exposure to the atmosphere. Such pyrophoric particles must be kept in an inert atmosphere or, more commonly, below the surface of a liquid until they can be protected. The loss of moment by oxidation or corrosion vitiates an important reason for choosing metals for recording materials and a considerable amount of work has been done to understand the problem and to control it.

Olsen (1974) identified two mechanisms of corrosion, chemical and electrolytic. In the first, SD iron particles held at 65°C and 85% relative humidity, developed a protective shell of amorphous iron oxide which protected the particles against further oxidation. When the iron particles were contained in a tape, a more serious, electrochemical, process occurred. The oxide shell no longer offered protection and electrolytic cells in the form of blisters were formed having cationic impurities within the blister and anionic impurities outside it. Deffeyes et al. (1973) observed that the binders used in tapes offer little protection to the particles. Polyurethanes, the most common binders in modern tapes, have high transmission rates for water vapor, and polyvinyl-polyvinylidene copolymers (also used in tapes) are excellent barriers against moisture but decompose when exposed to traces of iron, cobalt, or nickel.

Three principal methods of protection of metal particles against atmospheric corrosion have been proposed. They are (1) deliberate oxidation of a surface layer of the particles, usually involving about 10% of the mass of the particle (2) coating the particles with an inorganic compound (usually of chromium) or an organic compound, and (3) encapsulating the particles in a polymer film.

Deffeyes et al. (1972) carefully oxidized the surface of their metal particles (prepared by reduction of the oxalate) to form a coating which represented about 10% of the particles' mass. The oxidation was carried out at a temperature of 40°C and an oxygen partial pressure which began at $19 \, kg/m^2$ and was gradually increased to atmospheric pressure. Van der Giessen (1973) also used a thin oxide coating on acicular iron particles prepared by the pseudomorphic reduction of goethite particles. Particles having an iron core and an outer shell of Fe_4N were made by Haack and Beegle (1972) by nitriding acicular magnetite particles in a mixture of ammonia and hydrogen. Similarly, Nakamura et al. (1975b) prepared particles of 48Fe, 46.1Co, 1.9Ni, 4Al by coprecipitation of the oxalates followed by reduction in a stream of hydrogen mixed with ammonia. Commonly, metallic particles are kept under methyl alcohol or toluene which permits the formulation of a thin, stable oxide layer on the particles (Kawasaki and Higuchi 1972). Workers at Ampex (1974) mixed bismuth nitrate with hematite particles and dried the coated particles in nitrogen at 375°C to decompose the nitrate. The bismuth-modified hematite particles were then mixed with sodium silicate, dried, and reduced to give iron particles containing about 5% Bi and about 6% Si on the surface of the particles. The treatment was claimed to maintain the acicularity, reduce sintering, and yield a non pyrophoric product. The use of coatings of MgO or of ZrO_2 as protective layers on cobalt particles was described by Ushakova et al. (1976).

Metal particles based on Fe, Co, or Ni, and prepared by reduction in solution by the borohydride method may be coated with a chromium compound by the method described by Roden (1976). For example, a mixture of ferrous and cobaltous sulfates in solution were reduced with sodium borohydride solution and a solution of potassium dichromate was added to the slurry formed. The particles were thus coated with chromate ions and lost none of their remanent magnetization after being held to 38°C and 80% relative humidity for 21 days. Further protection for particles coated with chromium ions was claimed by Heikkinen and Kanten (1974) after adding tris (dimethylaminomethyl) phenol.

Metallic particles have been stabilized against atmospheric corrosion by being coated with organic compounds which are usually applied as a solution in toluene. Thus, Kawasaki (1974) applied sodium oleate, Aonuma and Tamai (1975c) used sodium phosphinate and an imide, and Uehori and Mikura (1976) used titanium tetraisopropylate.

Thomas (1966) described a method for preparing SD particles of cobalt in which dicobalt octacarbonyl was thermally decomposed in a hydrocarbon solvent containing a polymeric material. Thus, as the particles of cobalt formed they were enveloped in an overcoat of polymer and protected from the environment. Hess and Parker (1966) discussed more fully the choice of polymer and solvent and concluded that the secret to making particles having a narrow range of sizes by this method depended on a very delicate balance between the dispersant polymer, the solvent, and the growing metal particle. The best polymers had an average molecular weight of at least 100 000 and had polar groups attached to a relatively nonpolar backbone. The molecular weight should be high enough to allow absorption of the polar groups to the metal particle while leaving the bulk of the polymer to form a thick coating around the particle. The preferred polymer was an acrylic terpolymer of methyl methacrylate-ethyl acrylate and vinyl pyrrolidone and toluene. The solvent should be less polar than the most polar group of the polymer so that the solvent does not compete with the polymer at the metal surface.

In the methods just described, the encapsulating polymer was included in the bath in which the cobalt particles were prepared. Robbins et al. (1970) described a method of applying a polymer coating, as a separate step, to iron particles prepared by reduction of hematite or maghemite in hydrogen. Since the particles were pyrophoric it was necessary to keep them in hydrogen. The monomer used was tetraethylene glycol dimethacrylate diluted 1:6 in benzene. The benzene was removed in a vacuum oven at 60°C 110°C for four hours. No change in saturation magnetization could be detected after storing at 97°C in air for 67 h. The same coating technique was used by Swisher et al. (1971) on cobalt and iron–cobalt particles and by Robbins et al. (1971) on particles of iron–cobalt–nickel. Particles of $Co_{0.4}Fe_{0.6}$ had a saturation magnetization of 102 emu/g (128 μWb m/kg). When this value is compared with $\sigma_S = 243$ emu/g (305 μWb m/kg) for an alloy of composition $Co_{0.4}Fe_{0.6}$, it appears that rather more than 50% of the sample weight is actually polymer. The protection offered by the polymer coating was investigated at temperatures up to 166°C and it was found that the coating effectively stabilized the particles at least to 97°C for 67 h.

The most effective methods of protecting metal particles are probably (1) the careful oxidation of their surface and (2) the careful selection of binder materials.

Metal particles do indeed show some very desirable properties for high density recording materials. It is possible to combine higher moments and higher coercivities than can be achieved with any known oxide particle. However, the

TABLE 14
Magnetic properties of metal particles

Metal	Intrinsic				Extrinsic				Structure
	σ_s (emu/g)	θ (°C)	K_1 (ergs/cc)	λ_s	ρ (g/cc)	H_c (Oe)	σ_r (emu/g)	Shape	
Fe	178	770	$+4.4 \times 10^5$	$+4 \times 10^{-6}$	7.9	1030	89	Acicular	bcc $a = 2.86$ Å
70Fe–30Co	80.5	875	$+1 \times 10^5$	$+4 \times 10^{-6}$	8.0	1250	40	Acicular	bcc $a = 2.86$ Å

great price which has to be paid is the high reactivity of metals and particularly of metals in the form of fine particles. The magnetic properties of metal particles are summarized in table 14.

3.6. Magnetic properties and recording performance

The physics of magnetic recording has recently been the subject of two excellent and complementary tutorial articles by McCary (1971) and Mallinson (1976) but it is appropriate here to make a few observations about the narrower subject of the relationship between the magnetic properties of a recording surface and its recording performance. The theorist tries to steer a middle course between mathematical models that are complete enough to be believable yet not so complete that their solution is prohibitively expensive. The experimentalist would like to be able to design and carry out a set of one- or two-variable experiments and thence discover in turn the roles of each magnetic parameter. Unfortunately it is extremely difficult to make, say, a series of tapes in which only one or two variables change and everything else that matters stays the same. As we try to change just the coercivity for example by using particles of different average diameters, we may make a set of coating inks with very different dispersion characteristics and hence, prepare tapes of different roughnesses. It is impossible to over-emphasize the vital role in determining recording performance which is played by the non-magnetic parameters, roughness, compliance, dispersion, thickness, etc., and their importance becomes greater at an increasing rate as recording wavelengths decrease.

It is possible to study experimentally the dependence of recording performance of magnetic tapes on their magnetic properties over a limited range by taking advantage of the angular dependence of magnetic properties in oriented tapes (Bate 1961). By confining each correlation experiment to one oriented medium we can obtain a range of magnetic properties by cutting strips of tape at different angles to the orientation direction while the mechanical properties are practically the same for all samples. The magnetic parameters which were found to have the closest relationship to the signal output at recording densities below about 4000 flux reversals-per-inch were the remanence coercivity, H_r, and the quantity ϵ which is described in fig. 17; it represents the rate of change of remanent magnetization with applied field, H, at $H = H_r$. Of all magnetic properties ϵ shows the most pronounced change with angle, θ, between the measurement direction and the direction of orientation. The empirical relationship between signal output, V_0, and magnetic properties was found to be

$$V_0 \propto \epsilon H_r^{1.5-2.0}$$

That this is a reasonable relationship may be shown very simply. The signal output level at the reading head depends on the rate of change of flux with distance over the surface of the recorded medium, and ultimately on the rate of change of the remanent magnetization, M_r, with position within the tape, dM_r/dx. We can treat dM_r/dx as the product $dM_r/dH \times dH/dx$, both quantities being

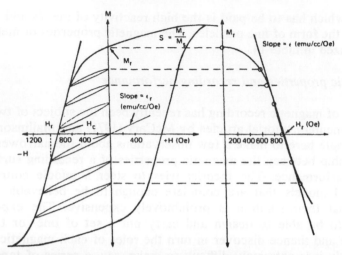

Fig. 17. Sketch of the hysteresis loop with recoil lines and of the descending remanence curve on which the parameters remanent magnetization M_r, saturation magnetization M_s, coercivity H_c, remanent coercivity H_r, ϵ and ϵ_r are shown.

evaluated at $H = H_r$ which marks the magnetic center of the recorded transition. By definition $\epsilon = (dM_r/dH)_{H_r}$ and to a first approximation the field from a writing head may be treated as coming from a current carrying wire running parallel to the head gap. Then $dH/dx)_{H_r}$ is proportional to H_r^2 and so we find that the output signal level should depend approximately on ϵH_r^2 in agreement with experiment. The relationship appears to hold as long as the magnetic pattern on the recorded medium can be regarded as a sequence of dipoles whose magnetization is M_r, separated by relatively narrow transition regions. However, at high densities, say above about 4000 bits-per-inch, this picture is no longer accurate (Bate and Dunn, 1976). As the length of a recorded dipole is decreased it becomes equal to, and ultimately smaller than, the length of the transition region. Because of the spatial extension of the writing-head field, a second dipole is written while the first is still within the range of influence of the head and consequently the first dipole is partially demagnetized. The phenomenon is known as "recording demagnetization" and it can be measured by a technique described by Bate and Dunn (1976). The effect can be minimized by using highly oriented recording surfaces.

Another effect which depends critically on the degree of particle orientation arises because of the strong perpendicular field component in the trailing edge of the writing field. This has two adverse effects. First, it induces a perpendicular component of magnetization which leads to asymmetry in the output pulse. Second, it creates at the surface closest to the head a layer of reduced in-plane magnetization, Bate and Dunn (1974). Unfortunately the output signal depends exponentially on separation/wavelength and so this reduction in magnetization occurs where it can do the most harm in high density recording. Improving the particle alignment also reduces the susceptibility to perpendicular fields.

The original reason for aligning tape particles was to increase the remanent magnetization in the direction of motion of the head and thereby to increase the signal output level. Orientation is achieved by applying a field of 1000–2000 Oe (79.6–159.2 kA/m) to the particles immediately after the ink has been applied to the webs and before appreciable drying has occurred. The field may be obtained from a permanent magnet or from a solenoid (driven by dc or ac). The degree of orientation is called the "orientation ratio" and is found by dividing the remanence measured along the orientation direction by the remanence measured in the plane of the coating but perpendicular to the direction of orientation. Orientation ratios of modern tapes range from 1.5 to 3.0 and are obtained by the application of the field at the right moment in the drying process to a highly dispersed assembly of particles. The direction of particle orientation is usually, but not always, parallel to the relative motion of head and tape. Bate et al. (1962) found that the interference (peak shift) between two adjacent magnetization transitions isolated from other transitions decreases as θ increases and goes through a minimum near $\theta = 60°$. The effect may be explained in terms of the angular dependencies of the driving force of demagnetization, M_r, and the ability to resist demagnetization, H_c.

It thus appears that there is no universal relationship linking magnetic properties and recording performance. A simple relationship may be found to hold at densities low enough that overlapping of adjacent reversal regions does not occur. However, at higher densities the relevant magnetic properties are much more complex and involve minor loop behavior in response to fields which change in direction as well as in magnitude and polarity. Moreover, as recording densities increase, variations in magnetic properties become of lesser importance than the non-magnetic properties of dispersion, roughness, and compliance. The latter properties largely determine the pre-eminent parameter which is the separation between the head and the recording medium.

3.6.1. Noise
As the performance of magnetic recording heads, mechanics, and electronics improves, the point is reached at which the factors limiting further improvement are noise and defects in the recording surface itself. The topic of the infrequent but gross irregularities in the coating, which are the principal cause of errors in the recorded signal, is of great practical importance but is too remote from the subject of magnetic materials to be considered further. However, medium noise is, at least partly, of magnetic origin and must be discussed briefly. A comprehensive recent discussion of tape noise was given by Daniel (1972) and fig. 18 taken from his paper shows the various kinds of noise which can occur. Equipment noise is that noise which comes from the head and the electronics and is observed when the tape is not moving over the head. Erased noise is also called "ac noise" and comes from the particularity of the recording material. It depends on the number-density of particles in the coating and can be minimized by using the highest possible particle packing density and the smallest possible particles. As we have seen, the danger in working with very small particles is that

G. BATE

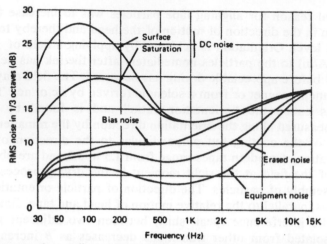

Fig. 18. Noise spectra of an audio cassette recorder, measured at 1.88 ips (48 mm/s) using standard playback equalization showing the contributions from equipment noise (head and electronics), the noise remaining after bulk erasure of the tape, the noise after applying ac bias to the writing head, the dc noise resulting from a writing current sufficient to saturate the medium (lower curve) and a writing current which maximizes the noise (upper curve) (Daniel 1972).

some will be small enough to be superparamagnetic and thus useless for recording. It has also been found that smaller particles are more susceptible to print-through which is the phenomenon by which an erased region on the tape may acquire a magnetization pattern simply by being close to a recorded part of the tape on the reel. Hayama et al. (1973) gave 0.046 to 0.1 μm as the critical range of particle size for high print-through for iron oxide particles. Daniel showed that particulate noise increased with the ratio of the standard deviation to the average length of the particles in an assembly. This is a further reason for striving to achieve the narrowest possible distribution of particle sizes.

Table 15 shows the dependence of the various noise contributions on particle length and we see that the noise does indeed decrease with decreasing particle

TABLE 15
Noise level as a function of particle length (after Hayama et al. 1973)

Mean particle length (μm)	H_c (Oe)	H_c (kA/m)	Axial ratio	Noise (dB) dc	Noise (dB) bias	Noise (dB) ac
γ-Fe$_2$O$_3$						
0.9	290	23.1	5:1	−45	−53	−57.5
0.4	340	27.1	8:1	−52.5	−56	−60
0.2	300	23.9	3:1	−53	−60.5	−61
CrO$_2$						
0.9	500	39.8	10:1	−52	−56	−59

size. It is interesting that particles of CrO_2 of average length 0.9 μm showed very similar noise figures to γ-Fe_2O_3 particles 0.4 μm long.

Daniel calculated that by achieving complete orientation of the particles, the signal output would increase by a factor of two but the noise would increase by $(\frac{3}{2})^{1/2}$ so that the signal to noise ratio (SNR) increases by $2(\frac{2}{3})^{1/2}$ or 4 dB over a completely disoriented tape. However, Hayama et al. (1973) found no increase in bulk-erased noise on orientation with iron oxide particles but they did remark that "excessive" orientation leads to rough surfaces and higher dc noise.

The next higher noise occurs when an erased tape is passed over a recording head driven by a high-frequency bias and is known as "bias noise." It was investigated by Ragle and Smaller (1965) who deduced two causes. First, if the bias frequency is too close to the signal frequency then bias noise can appear in the audio band as the lower of the amplitude-modulated side bands of the recorded bias signal. The second cause of this noise is that the bias field causes the particle noise flux from the tape which is imaged in the head, to be re-recorded anhysteretically on the tape. The correctness of this explanation was shown experimentally by Daniel (1972).

The largest contribution to tape noise is the modulation noise, the simplest case of which occurs when the head is driven with direct current. As we see in fig. 18 the noise has a maximum at a current that is less than that required to saturate the medium. Two departures from perfection are responsible for this noise. The first is the occurrence of agglomerates or voids in the coating. The effect of this contribution was shown by Hayama et al. (1973) who measured the changes which occurred in ac noise, bias noise, and dc noise with the degree of dispersion of the particles. The differences in noise level between their least dispersed and most dispersed samples were 1.5 dB for ac noise, 5 dB for bias noise, and 21.5 dB for dc noise. The second important cause of modulation noise is the presence on the surface of the tape of asperities which are high enough to cause intermittent separation between head and tape and yet rare enough to be difficult to detect. Many of these asperities can be removed or reduced by calendering or otherwise surface-finishing the tape, but Mills et al. (1975) showed that the success of this process depended critically on the composition of the soft roll of the calender. A roll of the correct composition was found to be capable of reducing modulation noise by 9 dB.

The bulk-erased and bias contributions to the noise can in principle be minimized by selecting small particles (but not so small that print-through becomes troublesome) having a very narrow size distribution and packing the largest possible number of them into the coating. Modulation noise is minimized by making sure that the particles are perfectly dispersed so that equal numbers of particles are found in equal and vanishingly small volumes sampled from different places in the coating. Clearly, agglomeration, sintering, and the occurrence of voids must be avoided. Finally the surface of the recording medium must be smooth so that a constant separation can be maintained between it and the head.

3.6.2. Stress- and impact-induced loss of magnetization

In considering the contributions to the total magnetic anisotropy energy we have
so far considered only shape and magnetocrystalline anisotropy. Flanders in a
series of papers (1974, 1975, 1976a,b) showed that stress- and impact-induced
losses in magnetic tapes are related to the magnetostriction constant λ_s of the
particles. Figure 19 gives the loss of remanent magnetization produced by
rubbing the back of magnetized tapes, backed by a metal plate, with a 3 mm
metal ball as a function of the magnetostriction constant. Clearly the most
serious losses of remanence occur in the magnetite and in the cobalt-substituted
γ-Fe_2O_3 tapes and the least losses were found for pure γ-Fe_2O_3 and for CrO_2
tapes. When Flanders (1976a) measured the hysteresis loops under compressive
and tensile stresses he found that the major changes occurred in the minor loops
obtained when the applied fields were just slightly greater than the sample's
coercivity. We must therefore expect that stresses will have a considerable
effect on the level of magnetization in the transition region between two

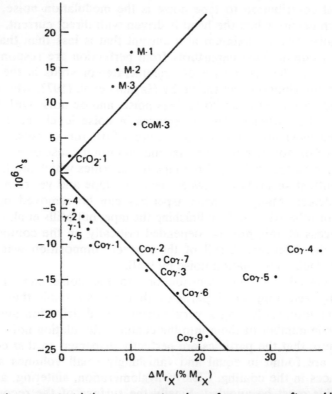

Fig. 19. Saturation magnetostriction vs percentage loss of remanence after rubbing magnetized
tapes, backed by a metal plate, with a 3-mm metal ball. The tapes were made of particles of Fe_3O_4
(M-1 to M-3), γ-Fe_2O_3 (γ-1 to γ-5), cobalt-substituted γ-Fe_2O_3 (Co-1 to Co-9), cobalt-substituted
Fe_3O_4 (CoM-3) and CrO_2 (Flanders 1976a).

oppositely magnetized regions in the tape since it is here that fields $\sim H_c$ or H_r were encountered during the writing process.

Reversible changes of magnetization were also found and they were roughly proportional to the product of the magnetostriction constant and the stress. These changes were generally much smaller than the irreversible changes and amounted to only $\pm 1\%$ of M_s.

The irreversible losses which occurred due to an impact of 1 kg/s on an area of 0.2 cm^2 of the tape samples were measured by Flanders (1976b) and are shown in fig. 20. Clearly the loss increases with $|\lambda_s|$ and it is thus important in minimizing stress-induced or impact-induced losses to reduce $|\lambda_s|$. Pure γ-Fe$_2$O$_3$ and CrO$_2$ have small magnetostriction (fig. 19) but if the coercivity of γ-Fe$_2$O$_3$ is increased by the substitution of cobalt, λ_s increases approximately linearly from -5×10^{-6} to -17×10^{-6} as the cobalt content is increased from 0 to 2.8%. The magnetostriction of magnetite, in contrast, decreases with the substitution of cobalt from $+17 \times 10^{-6}$ for Fe$_3$O$_4$ to $+7 \times 10^{-6}$ at Co$_{0.12}$Fe$_{2.88}$O$_4$. Flanders (1977) identified one composition for which $\lambda_s = 0$. It is in the γ-Fe$_2$O$_3$–Fe$_3$O$_4$ system when the amount of FeO reaches about 4%. He also suggested that if cobalt

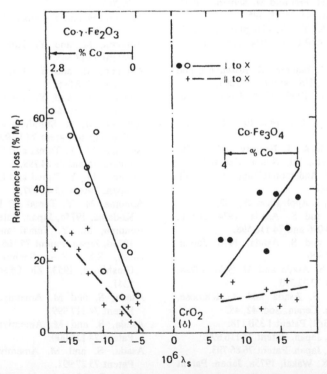

Fig. 20. Remanence loss due to a 1 kg/s impact on a 0.2 cm^2 area of tapes magnetized in the plane, parallel and perpendicular to their long dimension, vs saturation magnetostriction. The tapes were made of particles of cobalt-substituted γ-Fe$_2$O$_3$ (0–2.8% Co), cobalt-substituted Fe$_3$O$_4$ (0–4% Co) and CrO$_2$ (Flanders 1976b).

were added to partially-reduced $\gamma\text{-Fe}_2\text{O}_3$ in the following way: 0% Co for 4% FeO up to about 3.5% Co for 28.5% FeO ($= \text{Fe}_3\text{O}_4$), then the magnetostriction would be zero.

Magnetic particles for recording surfaces must not only have the right composition, size, and shape to give the correct remanent moment, remanent coercivity etc., but must also show the smallest possible change in these properties with their environment, stress, temperature, and time. As we have seen, the preparation of such particles can be a complex process and yet making the magnetic particles is only one part of the total job of making high quality recording surfaces. It is possible to waste the best magnetic materials by improper dispersion, coating, surface treatment, etc., but despite their critical effects the non-magnetic properties of inks and recording surfaces are not yet well understood.

References

Adams, G.M., J.I. Crowley and K.H. Larson, 1975, US Patent 3 894 970.

Aharoni, A., E.H. Frei and M. Schieber, 1962, J. Phys. Chem. Solids 23, 545.

Ahn, K.Y., K-N. Tu, A. Gangulee and P.A. Albert, 1973, AIP Conf. Proc. No. 18 (Pt. 2) 1103.

Ahn, K.Y., C.H. Bajorek, R. Rosenberg and K-N. Tu, 1976, US Patent 3 996 095.

Akad.d. Wiss. d. DDR, 1974, Japan. Patent 74 122 829.

Akashi, G. and M. Fujiyama, 1971, US Patent 3 607 218.

Alario-Franco, M.A., J. Fenerty and K.S.W. Sing, 1972, 7th. Int. Symp. Reactivity of solids, ed. J.S. Anderson (Chapman and Hall, London) p. 327.

Amar, H., 1957, J. Appl. Phys 28, 732.

Amemiya, M. and S. Asada, 1974, Japan. Patent 74 114 096 and 74 116 598.

Amemiya, M. and S. Asada, 1976, Japan. Patent 76 107 300.

Amemiya, M., S. Asada and M. Seki, 1973a, Japan. Patent 73 22 390.

Amemiya, M., S. Asada and Y. Ichinose, 1973b, J. Can. Ceram. Soc. 42, 45.

Ampex, 1974, Brit. Patent 1 358 118.

Ando, H., 1974, Japan. Patent 74 116 899.

Ando, H., 1976, Japan. Patent 76 86 793.

Ando, H. and K. Wakai, 1975a, Japan. Patent 75 161 697.

Ando, H. and K. Wakai, 1975b, Japan. Patent 75 136 699.

Ando, H. and K. Wakai, 1976a, Japan. Patent 76 20 595.

Ando, H. and K. Wakai, 1976b, Japan. Patent 76 20 596.

Aonuma, M. and Y. Tamai, 1974, Japan. Patent 74 130 864.

Aonuma, M. and Y. Tamai, 1975a, Japan. Patent 75 18 345.

Aonuma, M. and Y. Tamai, 1975b, Japan. Patent 75 78 896.

Aonuma, M. and Y. Tamai, 1975c, Japan. Patent 75 104 164.

Aonuma, M., R. Shirahata and T. Kitamoto, 1974, Japan. Patent 74 15 999.

Aonuma, M., Y. Tamai and G. Akashi, 1975a, Japan. Patent 75 24 799.

Aonuma, M., Y. Tamai and G. Akashi, 1975b, Japan. Patent 75 41 097.

Aonuma, M., Y. Tamai, T. Kitamoto and F. Kodama, 1975c, Japan. Patent 75 72 859.

Aonuma, M., Y. Tamai and E. Tadokoro, 1975d, Japan. Patent 75 106 198.

Ariya, S.M., S.A. Shchukarev and V.B. Glushkova, 1953, Zh. Obshchei Khimii 23, 1241.

Asada, S. and M. Amemiya, 1974a, Japan. Patent 74 117 999.

Asada, S. and M. Amemiya, 1974b, Japan. Patent 74 118 000.

Asada, S. and M. Amemiya, 1975, Japan. Patent 75 27 991.

Asada, S., M. Amemiya and M. Seki, 1974, Japan. Patent 74 69 999.

Aspland, M., G.A. Jones and B.K. Middleton, 1969, IEEE Trans. Mag. MAG-5, 314.

Ayers, J.W. and R.A. Stephens, 1962a, US Patent 3 015 627.

Ayers, J.W. and R.A. Stephens, 1962b, US Patent 3 015 628.

BASF A-G, 1974, Fr. Patent 2 180 802.

Balthis, J.H. Jr., 1969, US Patent 3 449 073.

Bando, Y., Y. Kiyama, T. Takada and S. Kachi, 1965, Japan J. Appl. Phys. 4, 240.

Banerjee, S.K. and H. Bartholin, 1970, IEEE Trans. Mag MAG-6, 299.

Baronius, W., F. Henneberger and W. Geidel, 1966, East Ger. Patent 48 590.

Bate, G., 1961, J. Appl. Phys. 32, 261S.

Bate, G., 1962, J. Appl. Phys. 33, 2263.

Bate, G. and L.P. Dunn, 1974, IEEE Trans. Mag. MAG-10, 667; IBM J. Res. Dev. 18, 563.

Bate, G. and L.P. Dunn, 1976, Int. Conf. Video and Data Recording, Proc. IRE 35, 21.

Bate, G., H.S. Templeton and J.W. Wenner, 1962, IBM J. Res. Dev. 6, 348.

Bate, G., D.E. Speliotis, J.K. Alstad and J.R. Morrison, 1965, Proc. Int. Conf. on Magnetism, Nottingham, 816.

Baudisch, O., 1933a, US Patent 1 894 749.

Baudisch, O., 1933b, US Patent 1 894 750.

Bayer, A-G., 1975, Belg. Patent 828 540.

Bean, C.P. and J.D. Livingston, 1959, J. Appl. Phys. 30, 120S.

Beer, H.B. and G.V. Planer, 1958, Brit. Comm. Electron. 5, 937.

Bennetch, L.M., 1975, Ger. Patent 2 520 379.

Bennetch, L.M., H.S. Greiner, K.R. Hancock and M. Hoffman, 1975, US Patent 3 904 540.

Berkowitz, A.E. and J.A. Lahut, 1973, AIP Conf. Proc. No. 10 (Pt. 2) 966.

Berkowitz, A.E. and W.J. Schuele, 1959, J. Appl. Phys. 30, 134S.

Berkowitz, A.E., W.J. Schuele and P.J. Flanders, 1968, J. Appl. Phys. 39, 126.

Bernal, J.D., D.R. Dasgupta and A.L. McKay, 1959, Clay Min. Bull. 4, 15.

Berndt, D.F. and D.E. Speliotis, 1970, IEEE Trans. Mag. MAG-6, 510.

Bertram, H.N. and A.K. Bhatia, 1973, IEEE Trans. Mag. MAG-9, 127.

Bertram, H.N. and J.C. Mallinson, 1969, J. Appl. Phys. 40, 1301.

Bickford, L.R., 1950, Phys. Rev. 78, 449.

Bickford, L.R., J.M. Brownlow and R.F. Penoyer, 1957, Proc. Inst. Elec. Eng. 104B, Supl. No. 5, 238.

Bierlein, J.D., 1973, Intermag Conf. Paper 5.4.

Birks, J.B., 1950, Proc. Phys. Soc. B63, 65.

Blackie, I., N. Truman and D.A. Walker, 1967, Int. Conf. on Magnetic Materials and their Applications, Conf. Pub. 33 (IEEE London) p. 207.

Bonn, T.H. and D.C. Wendell Jr., 1953, US Patent 2 644 787.

Bottjer, W.G. and N.L. Cox, 1969, Fr. Patent 2 001 806.

Bottjer, W.G. and H.G. Ingersoll, 1970, US Patent 3 512 930.

Bottoni, G., D. Candolfo, A. Cecchetti and F. Masoli, 1972, IEEE Trans. Mag. MAG-8, 770.

Bratescu, V. and P. Vitan, 1969, Rumanian Patent 51 610.

Braun, P.B., 1952, Nature 170, 1123.

Brown Jr., W.F., 1962, J. Appl. Phys. 33, 1308.

Brown Jr., W.F., J. de Phys. 32, Colloque C1. Suppl. to 2–3, 1037.

Brown Jr., W.F. and C.E. Johnson Jr., 1962, J. Appl. Phys. 33, 2752.

Brownlow, J.M., 1968, IBM Tech. Discl. Bull. 11, 238.

Butler, R.F. and S.K. Banerjee, 1975, J. Geophys. Res. 80, 4049.

Campbell, R.B., 1957, J. Appl. Phys. 28, 381.

Camras, M., 1954, US Patent 2 694 656.

Candolfo, D., A. Cecchetti and F. Masoli, 1970, IEEE Trans. Mag. MAG-6, 164.

Carman, E.H., 1955, Brit. J. Appl. Phys. 6, 426.

Chamberland, B.L., C.G. Frederick and J.L. Gillson, 1973, J. Sol. St. Chem. 6, 561.

Chen, S-L. and J.A. Murphy, 1975, US Patent 3 859 129.

Chevalier, R., 1951, J. Phys. Rad. 12, 172.

Claude, R., G. Lorthioir and C. Mazieres, 1968, Compt. Rend. Acad. Sci. Paris C 266, 462.

Cloud, W.H., D.S. Schreiber and K.R. Babcock, 1962, J. Appl. Phys. 33, 1193.

Clow, H., 1972, Brit. Patent 1 297 936.

Clow, H. and R.G. Gilson, 1976, Brit. Patent 1 427 724.

Colombo, U., G. Fagherazzi, F. Gazzarrini, G. Lanzavecchia and G. Sironi, 1964, Nature 202, 175.

Colombo, U., G. Fagherazzi, F. Gazzarini, G. Lanzavecchia and G. Sironi, 1967, Ind. Chim. Bel. XXXVI, Cong. Int. Chim. Ind. 32, 95.

Comstock, R.L. and E.B. Moore, 1974, IBM J. Res. Dev. 18, 555 and Intermag Conf. Paper 14.4.

Cox., N.L. and W.T. Hicks, 1969, US Patent 3 451 771.

Craik, D.J. and R. Lane, 1967, Brit. J. Appl. Phys. **18**, 1269.

Craik, D.J., 1970, Contemp. Phys. **11**, 65.

Creer, K.M., I.G. Hedley and W. O'Reilly, 1975, Magnetic oxides, ed. D.J. Craik, (Wiley, New York) p. 664.

Daniel, E.D., 1972, J. Audio Eng. Soc. **20**, 92.

Daniel, E.D. and I. Levine, 1960, J. Acoust. Soc. Am. **32**, 1.

Darnell, F.J., 1961, J. Appl. Phys. **32**, 1269.

Darnell, F.J. and W.H. Cloud, 1965, Bull. Soc. Chim. France **4**, 1164.

Dasgupta, D.R., 1961, Indian J. Phys. **35**, 401.

Daval, J. and D. Randet, 1970, IEEE Trans. Mag. MAG-6, 768.

David, I. and A.J.E. Welch, 1956, Trans. Farad. Soc. **52**, 1642.

Davies, A.V., B.K. Middleton and A.C. Tickle, 1965, Intermag Conf. Paper 12.2.

Deffeyes, R.J., C.E. Johnson Jr. and J.H. Judy, 1972, IEEE Trans. Mag. MAG-8, 465.

Deffeyes, R.J., C.E. Johnson Jr. and J.H. Judy, 1973, AIP Conf. Proc. No. 18 (Pt. 2) 1108.

Della Torre, E., 1965, J. Appl. Phys. **36**, 518.

Denisov, P.P., 1969, Phys. Met. Metallography **29**, 183.

De Pew, J.R. and D.E. Speliotis, 1967, Plating **54**, 705.

Derie, R., M. Ghodsi and R.J. Hourez, 1975, J. Appl. Chem. Biotechnol. **25**, 509.

De Vries, 1966, Mat. Res. Bull. **1**, 83.

Dezawa, S. and T. Kitamoto, 1973, Ger. Patent 2 325 132.

Dezsi, I. and J.M.D. Cooey, 1973, Phys. Stat. Sol.(a) **15**, 681.

Dickey, D.H. and H. Robbins, 1971, AIP Conf. Proc. No. 5, 826.

Doan, D.J., 1941, Magnetic separation of ores, ed. R.S. Dean and C.W. Davis (US Bur. Mines Bull., Govt. Printing Office, Washington, D.C.) No. 425, p. 113.

Dovey, D.M., R.C. Chirnside and H.P. Rooksby, 1954, US Patent 2 689 167.

Dovey, D.M. and W.R. Pitkin, 1954, US Patent 2 689 168.

Doyle, W.D., J.E. Rudisill and S. Shtrikman, 1961, J. Appl. Phys. **30**, 1785.

Dragsdorf, R.D. and R.J. Johnson Jr., 1962, J. Appl. Phys. **33**, 724.

Drbalek, Z., D. Rykl, V. Seidl and V. Sykora, 1973, Sb. Vys. Sk. Chem.-Technol. Praze. Mineral. **G15**, 43.

Duinker, S., A.L. Struijts, H.P.J. Wijn and W.K. Westmijze, 1962, US Patent 3 023 166.

Dunlop, D.J., 1968, Phil. Mag. **19**, 369.

Dunlop, D.J., 1969, Phil. Mag. Ser. 8 **19**, 329.

Dunlop, D.J., 1972, IEEE Trans. Mag. MAG-8, 211.

Dunlop, D.J., 1973, J. Geophys. Res. **78**, 1780.

Eagle, D.F. and J.C. Mallinson, 1967, J. Appl. Phys. **38**, 995.

Elder, T., 1965, J. Appl. Phys. **36**, 1012.

Engin, R., A.G. Fitzgerald, 1973, J. Mat. Sci. **8**, 169.

Evtihiev, N.N., N.A. Economov, A.R. Krebs, N.A. Zamjatina, V.D. Baurin and V.G. Pinko, 1976, IEEE Trans. Mag. MAG-12, 773.

Fagherazzi, G., G. Cocco and L. Schiffini, 1974, Atti. Accad. Peloritana Pericolanti Cl. Sci. Fis. Mat. Nat. **5**, 385.

Fayling, R.E., 1977, Intermag Conf. Paper 22.5.

Feng, J.S-Y., C.H. Bajorek and M-A. Nicolet, 1972, IEEE Trans. Mag. MAG-8, and US Patent 3 860 450.

Ferguson, G.R. Jr. and M. Hass, 1958, Phys. Rev. **112**, 1130.

Fisher, R.D., 1962, J. Electrochem. Soc. **109**, 479.

Fisher, R.D., 1966, IEEE Trans. Mag. MAG-2, 681.

Fisher, R.D., 1972, Symp. on Electrodeposition of Metals, (Battelle, Memorial Inst., Colombus, Ohio) (MCIC 72-05) p. 38.

Fisher, R.D., E.W. Jones, 1972, Intermag Conf. Kyoto, Paper 27.5.

Fisher, R.D., R.P. Williams and F.J. Harsacky, 1963, US Patent 3 116 159.

Flanders, P.J., 1974, IEEE Trans. Mag. MAG-10, 1050.

Flanders, P.J., 1975, Intermag Conf. Toronto, Paper 22.2.

Flanders, P.J., 1976a, IEEE Trans. Mag. MAG-12, 348.

Flanders, P.J., 1976b, IEEE Trans. Mag. MAG-12, 770.

Flanders, P.J. and S. Shtrikman, 1962, J. Appl. Phys. **33**, 216.

Francombe, M.H. and H.P. Rooksby, 1959, Clay Minerals Bull. **4**, 1.

Frei, E.H., S. Shtrikman and D. Treves, 1957, Phys. Rev. **106**, 446.

Freitag, W.O. and J.S. Mathias, 1964, Electroplating and Metal Finishing, p. 42.

Frieze, A.S., R. Sard and R. Weil, 1968, J. Electrochem. Soc. **115**, 586.

Fuji Photo Film Co., 1974, Brit. Patent 1 348 457.

Fukuda, S., T. Miyake, G. Akashi and M. Seto, 1962a, US Patent 3 026 215.

Fukuda, S., H. Goto, G. Akashi and T. Miyake, 1962b, US Patent 3 046 158.

Fukuda, Y., 1975, Japan. Patent 75 05 395.

Funke, A., P. Kleinert, 1974, Z. Anorg. Allg. Chem. 403, 156.

Furukawa, M., H. Ogawa, M. Aonuma, Y. Tamai, N. Nakahara and M. Ingarashi, 1976, Japan. Patent 76 80 998.

Gălugăru, Gh., 1970, Thin Solid Films 5, 365.

Gazzarini, F. and G. Lanzavecchia, 1969, Reactivity of Solids, ed. J.W. Mitchell (Wiley, New York) p. 57.

Gen, M.Ya., Ye V. Shtol'ts, I.V. Plate and Ye.A. Fedorova, 1970, Phys. Metals Metallog. 30, 200.

Gerkema, J.T., J.L. Melse and J. DeVries, 1971, IEEE Trans. Mag. MAG-7, 415.

Ginder, P., 1970, US Patent 3 494 760.

Girke, H., 1960, Z. Angew. Phys., 11, 502.

Glemser, O. and E. Gwinner, 1939, A. Anorg. Allgem. Chem. 240, 161.

Gorter, E.W., 1954, Philips Res. Rep. 9, 245.

Goto, H. and G. Akashi, 1962, US Patent 3 047 428.

Greenblatt, S. and F.T. King, 1969, J. Appl. Phys. 40, 4498.

Greiner, J., 1970, US Patent 3 498 748.

Gustard, B. and W.J. Schuele, 1966, J. Appl. Phys. 37, 1168.

Gustard, B. and H. Vriend, 1969, IEEE Trans. Mag. MAG-5, 326.

Gustard, B. and M.R. Wright, 1972, IEEE Trans. Mag. MAG-8, 426.

Haack, G.C. and L.C. Beegle, 1972, US Patent 3 700 499.

Hägg, G., 1935, Z. Phys. Chem. B29, 88.

Halaby, S.A., N.S. Kenny and J.A. Murphy, 1975, US Patent 3 892 888.

Haller, W.D. and R.M. Colline, 1971, US Patent 3 573 980.

Haneda, K. and H. Kojima, 1974, J. Am. Ceram. Soc. 57, 68.

Haneda, K., C. Miyakawa and H. Kojima, 1974, J. Am. Ceram. Soc. 57, 354.

Harada, S., T. Yamanashi and M. Ugaji, 1972, IEEE Trans. Mag. MAG-8, 468.

Harle, O.L. and J.R. Thomas, 1966, US Patent 3 228 882.

Hartmann, J.W., H. Roller, H.J. Hartmann, J. Hack and H. Hartmann, 1975, Can. Patent 962 536.

Hayama, F., H. Miyagawa and Y. Sensai, 1973,

Audio Eng. Soc. Preprint 918(6–3).

Hayashida, H., K. Shinjo, M. Hirose, M. Ohno, F. Soeda, M. Endo, K. Yokoyama, K. Ito and R. Tsunoi, 1975, 1976, Japan, Patents 75 33 934, 75 38 098, 76 11 023, 76 11 197, 76 11 198.

Healey, F.H., J.J. Chessick and A.V. Fraoli, 1956, J. Phys. Chem. 60, 1001.

Heikkinen, D.G. and T.M. Kanten, 1974, Fr. Patent 2 212 597.

Heller, J., 1974, US Patent 3 829 372.

Heller, J., 1976, IEEE Trans. Mag. MAG-12, 396.

Henry, W.E. and M.J. Boehm, 1956, Phys. Rev. 101, 1253.

Herczog, A., M.M. Layton and D.W. Rice, 1975, US Patent 3 900 593.

Hess, P.H. and P.H. Parker, Jr., 1966, J. Appl. Polymer Sci. 10, 1915.

Heuberger, R. and I. Joffe, 1971, J. Phys. C. 4, 805.

Hiller, D.M., 1978, J. Appl. Phys. 49, 1821.

Hirata, J. and K. Wakai, 1976, Japan. Patents 76 96 097 and 76 96 098.

Hirota, E., 1976, Japan. Patent 76 28 700.

Hirota, E., T. Kawamata, T. Mihara and Y. Terada, 1972, Intermag Conf. Digests, Paper 23.3.

Hirota, E., T. Mihara, T. Kawamata and Y. Terada, 1974, Nat. Tech. Rep. Matsushita Elec. Ind. 20, 1974.

Horiishi, N., 1974, Japan. Patent 74 43 900.

Horiishi, N. and K. Fukami, 1975, Japan. Patent 75 155 500.

Horiishi, N. and A. Takedoi, 1975a, Japan. Patent 75 21 996.

Horiishi, N. and A. Takedoi, 1975b, Japan. Patent 75 99 995.

Horiishi, N. and A. Takedoi, 1975c, Japan. Patent 75 99 996.

Horiishi, N. and A. Takedoi, 1975d, Japan. Patent 75 99 997.

Horiishi, N. and A. Takedoi, 1975e, Japan. Patent 75 101 299.

Horiishi, N. and A. Takedoi, 1975f, Japan. Patent 75 118 299.

Horiishi, N., A. Takedoi and F. Kurata, 1974, Japan. Patents 74 23 199 and 74 23 200.

Hoyama, M. and Y. Suzuki, 1973, Japan. Patent 73 70 899.

Hrynkiewicz, A.Z. and D.S. Kulgawchuk, 1963, Acta Phys. Polonica 24, 6.

Hurt, J., A. Amendola and R.E. Smith, 1966, J. Appl. Phys. 37, 1170.

Hwang, P.Y., 1972, US Patent 3 671 435.

Hwang, P.Y., 1973, US Patent 3 748 270.

Iga, T., 1972, Japan. Patent 72 04 131.

Iga, T., 1973, Japan. Patent 73 42 780.

Iga, T. and Y. Tawara, 1974, Japan. Patent 74 15 320.

Imaoka, Y., 1968, J. Electrochem. Soc. Japan 36, 15.

Imaoka, Y., S. Umeki, Y. Kubota and Y. Tokuoba, 1978, IEEE Trans. Mag. MAG-14, 649.

Inagaki, M. and T. Tada, 1972, Japan. Patent 72 39 477.

Inagaki, N., S. Hattori, Y. Ishii and H. Katsuraki, 1975a, IEEE Trans. Mag. MAG-11, 1191.

Inagaki, N., S. Hattori, Y. Ishii and M. Asahi, 1975b, Japan. Patent 75 51 517.

Inagaki, N., S. Hattori and Y. Ishii, 1975c, Denki Tsushin Kenkyujo Jitsuyoka Hokoku 24, 2765.

Inagaki, N., S. Hattori, Y. Ishii, A. Terada and H. Katsuraki, 1976, IEEE Trans. Mag. MAG-12, 785.

Ingraham, J.N. and T.J. Swoboda, 1960, 1962, US Patents 2 923 683, 3 034 988.

Ishii, K. and S. Uedaira, 1974, Japan. Patent 74 89 896.

Ito, K., 1974, Japan. Patent 74 46 424.

Ito, K., K. Yokoyama and S. Muramatsu, 1970, NHK Lab. Note No. 133.

Itoh, F. and M. Satou, 1975a, Japan. Patent 75 151 397.

Itoh, F. and M. Satou, 1975b, Japan J. Appl. Phys. 14, 2091.

Itoh, F. and M. Satou, 1976, Nippon Kagaku Kaishi 6, 853.

Iwasaki, S. and Y. Nakamura, 1977, IEEE Trans. Mag. MAG-13, 1272.

Iwasaki, S. and H. Yamazaki, 1975, Proc. 7th Ann. Conf. on Magnetism, 4pA-7.

Jacobs, I.S. and C.P. Bean, 1955, Phys. Rev. 100, 1060.

Jaep, W.F., 1969, J. Appl. Phys., 40, 1297.

Jaep, W.F., 1971, J. Appl. Phys., 42, 2790.

Jeschke, J.C., 1954, E. Ger. Patent 8 684.

Jeschke, J.C., 1966, US Patent 3 243 375.

Johnson, C.E. Jr., 1969, J. Phys. C. Ser. 2, 2, 1996.

Johnson, C.E. Jr. and W.F. Brown Jr., 1958, J. Appl. Phys. 29, 1699.

Johnson, C.E., Jr. and W.F. Brown, Jr., 1961, J. Appl. Phys. 32, 243S.

Jonker, G.H., H.P.J. Wijn and P.B. Braun, 1956/7, Philips Tech. Rev. 18, 145.

Judge, J.S., 1972, Ann. N.Y. Acad. Sci. 189, 117.

Judge, J.S., J.R. Morrison and D.E. Speliotis, 1965a, J. Appl. Phys. 36, 948.

Judge, J.S., J.R. Morrison, D.E. Speliotis and G. Bate, 1965b, J. Electrochem Soc. 112, 681.

Kachi, S., N. Nakanishi, K. Kosuge and H. Hiramatsu, 1970, Proc. Int. Conf. on Ferrites, Kyoto, p. 141.

Kaganowicz, G., E.F. Hockings and J.W. Robinson, 1975, IEEE Trans. Mag. MAG-11, 1194 (and Ger. Patent 2 343 701).

Kajitani, K. and K. Saito, 1975, Japan. Patent 75 131 896.

Kamberska, Z. and V. Kambersky, 1966, Phys. Stat. Sol. 17, 411

Kamiya, I., Y. Makino and Y. Sugiura, 1970, Proc. Int. Conf. on Ferrites, Kyoto, p. 483.

Kanbe, T. and K. Kanematsu, 1968, J. Phys. Soc. Japan 24, 1396.

Kawai, S., 1973a, J. Japan Inst. Light Metals 23, 143.

Kawai, S., 1973b, J. Japan Inst. Light Metals 23, 151.

Kawai, S., 1975, J. Electrochem. Soc. 122, 1026.

Kawai, S. and I. Ishiguro, 1976, J. Electrochem. Soc. 123, 1047.

Kawai, S. and R. Ueda, 1975, J. Electrochem. Soc. 122, 32.

Kawai, S., H. Sato and T. Sakai, 1975, Brit. Patent 1 418 933.

Kawamata, T., 1975, Japan. Patent 75 146 599.

Kawamata, T., Y. Terada and E. Hirohota, 1973, Japan. Patent 73 65 500.

Kawamata, T., E. Hirota and Y. Terada, 1975, Ger. Patent 2 435 874.

Kawasaki, M., 1973, Ger. Patent 2 305 153.

Kawasaki, M., 1974, Japan. Patent 74 97 738.

Kawasaki, M. and S. Higuchi, 1972, IEEE Trans. Mag. MAG-8, 430 and US Patent 3 702 270.

Kay, E., 1961, J. Appl. Phys. 32, 99S.

Kefalas, J.H., 1966, J. Appl. Phys. 37, 1160.

Keitoku, S., 1973, J. Sci. Hiroshima Univ. Ser. A. Phys. Chem. 37, 167.

Khalafalla, D. and A.H. Morrish, 1974, J. Appl. Phys. 43, 624.

Kimoto, K., Y. Kamiya, M. Nonoyama and R. Uyeda, 1963, Japan J. Appl. Phys. 2, 702.

Klimaszewski, B. and J. Pietrzak, 1969, Bull.

Acad. Pol. Sci. Ser. Math Astron. Phys. 17, 51.

Kneller, E.F., 1969, Magnetism and metallurgy, ed. Berkowitz and Kneller (Academic Press, New York) Ch. 8.

Kneller, E.F. and F.E. Luborsky, 1963, J. Appl. Phys. 34, 656.

Kojima, H., 1954, Sci. Rep. Res. Inst. Tohoku Univ. A6, 178.

Kojima, H., 1956, Sci. Rep. Res. Inst. Tohoku Univ. A8, 540.

Kojima, H., 1958, Sci. Rep. Res. Inst. Tohoku Univ. A10, 175.

Kondorskii, E., 1952, Izvest, Akad. Nauk. SSSR Ser. Fiz., 16, 398.

Kondorsky, E.I. and P.P. Denisov, 1970, IEEE Trans. Mag. MAG-6, 167.

Kordes, E., 1935, Z. Krist. A91, 193.

Koretzky, H., 1963, Proc. 1st Austral. Conf. on Electrochemistry, 417.

Koshizuka, K., Y. Sakamoto, T. Tamura, M. Shimazaki and T. Kazama, 1975, Japan. Patent 75 80 999.

Köster, E., 1970, J. Appl. Phys. 41, 3332.

Köster, E., 1972, IEEE Trans. Mag. MAG-8, 428.

Köster, E., 1973, Conf. on Video and Data Recording IRE Conf. Proc. 26, 213.

Köster, E., G. Wunsch, P. Deigner, W. Stumpfi and H. Schneehage, 1976a, Ger. Patent 2 434 058.

Köster, E., G. Wunsch, E. Schoenafinger, H.H. Schneehage and J. Jakusch, 1976b, Ger. Patent 2 447 386.

Kouvel, J.S. and D.S. Rodbell, 1967, J. Appl. Patent Phys. 38, 979.

Krones, F., 1955, Mitt. Forschungslab. Agfa 1, 289.

Krones, F., 1960, Technik des Masnetspeicher (Springer-Verlag, Berlin) p. 479.

Kubota, B., 1960, J. Phys. Soc. Japan 15, 1706.

Kubota, B., 1961, J. Am. Cer. Soc. 44, 239.

Kubota, Y., S. Umeki and Y. Tokuoka, 1974, Ger. Patent 2 413 430.

Kugimiya, K., 1975, Japan. Patent 75 88 598.

Kumashiro, Y., 1968, Denki Kagaku 36, 680.

Lazzari, J.P., I. Melnick and D. Randet, 1967, IEEE Trans. Mag. MAG-3, 205.

Lazzari, J.P., I. Melnick and D. Randet, 1969, IEEE Trans. Mag. MAG-5, 995.

Lee, E.W. and J.E.L. Bishop, 1966, Proc. Phys. Soc. 89, 661.

Leitner, L., F. Hund and J. Rademachers,

1973, Ger. Patent 2 162 716.

Leutner, B. and E. Schoenafinger, 1975, Ger. Patent 2 352 440.

Leutner, B., M. Schwarzmann and M. Ohlinger, 1972, Ger. Patents 2 120 206 and 2 119 932.

Leutner, B., M. Schwarzmann and M. Ohlinger, 1973, US Patent 3 769 087.

Levi, F.P., 1959, Nature 183, 1251.

Levi, F.P., 1960, J. App. Phys. 31, 1469.

Luborsky, F.E., 1961, J. Appl. Phys. 32, 171S.

Luborsky, F.E., 1970, IEEE Trans. Mag. MAG-6, 502.

Luborsky, F.E. and J.D. Opie, 1963, J. Appl. Phys. 34, 1317.

Luborsky, F.E., E.F. Koch and C.R. Morelock, 1963, J. Appl. Phys. 34, 2905.

Luborsky, F.E. and C.R. Morelock, 1964, J. Appl. Phys. 35, 2055.

MacLeod, D.B., 1971, J. Motion Pic. Tel. Engs. 80, 295.

Makino, Y., S. Higuchi and Y. Masuya, 1972, US Patent 3 654 163.

Makino, Y., S. Higuchi, I. Kamiya and Y. Masuda, 1973, Japan. Patent 73 22 270.

Makino, Y., I. Kamiya and G. Sugiura, 1974a, Japan. Patent 74 15 756.

Makino, Y., I. Kamiya and G. Sugiura, 1974b, Japan. Patent 74 15 757.

Malinin, G.V., Yu. M. Tolmachev and V.B. Yadrintsev, 1971, Kinet. Katal. 14, 234.

Mallinson, J.C., 1976, Proc. IEEE 64, 196.

Marcot, G.C., 1974, US Patent 3 845 198.

Marcot, G.C., W.J. Cauwenburg and S.A. Lamanna, 1951, US Patent 2 558 302.

Martin, A. and F.C. Carmona, 1968, IEEE Trans. Mag. MAG-4, 259.

Matsui, G., K. Toda, S. Shimizu, N. Horikoshi and A. Takedoi, 1975, Japan. Patent 75 118 298.

Matsumoto, K. and Y. Matsuo, 1974, Japan. Patent 74 44 298.

Matsumoto, K. and Y. Matsuo, 1975, Japan. Patent 75 36 354 and Ger. Patent 2 345 375.

Matsushita Ltd., 1972, Fr. Patent 1 604 939.

Maxwell, L.R., S. Smart and S. Brunauer, 1949, Phys. Rev. 76, 459.

McCary, R.O., 1971, IEEE Trans. Mag. MAG-7, 4.

McNab, T.K., R.A. Fox and A.J.F. Boyle, 1968, J. Appl. Phys. 39, 5703.

Mee, C.D., 1964, The physics of magnetic recording (Wiley, New York).

Mee. C.D. and J.C. Jeschke, 1963, J. Appl. Phys. **34**, 1271.

Melezoglu, C., 1972, US Patent 3 703 411.

Michel, A., J. Bénard, 1935, Compt. Rend. Acad. Sci. Paris **200**, 316.

Michel, A., G. Chaudron and J. Bénard, 1950, Colloq. Int. du Ferromagnetisme et Antiferromagnetisme, Grenoble, p. 51.

Mihara, T., T. Kawamata, Y. Terada and E. Hirota, 1970, Proc. Int. Conf. on Ferrites, Kyoto, p. 476.

Mihara, T., H. Kawamata, Y. Terada and E. Hirota, 1973, Japan. Patent 73 18 318.

Mihara, T., H. Kawamata, Y. Terada and E. Hirota, 1974, Japan. Patent 74 24 754.

Mihara, T., H. Kawamata, Y. Terada and E. Hirota, 1975, Japan. Patent 75 37 640.

Mills, D., H. Kristensen and V. Santos, 1975, Aud. Eng. Soc. Preprint No. 1084 (D-3).

Mizushima, K., T. Yamase and O. Saito, 1975a, Japan. Patent 75 90 997.

Mizushima, K., T. Yamase and O. Saito, 1975b, Japan. Patent 75 95 798.

Mizushima, K., T. Yamase and O. Saito, 1975c, Japan. Patent 75 95 799.

Mizushima, K., T. Yamase and O. Saito, 1975d, Japan. Patent 75 98 500.

Mizushima, K., T. Yamase and O. Saito, 1975e, Japan. Patent 75 98 700.

Mollard, P., A. Collomb, J. Devenyi, A. Rousset and J. Paris, 1975 IEEE Trans. Mag. MAG-11, 894.

Montiglio, U., P. Aspes, G. Scotti and G. Basile, 1973, Ger. Patent 2 325 719.

Montiglio, U., G. Basile, P. Aspes and E. Gallinotti, 1975, Ger. Patent 2 520 030.

Morelock, C.R., 1962, Acta Metallurgica **10**, 161.

Morero, D., U. Montiglio, P. Aspes and G. Basile, 1972, Ital. Patent 922 283.

Mori, M., I. Osada and S. Takatsu, 1976, Japan. Patent 76 21 200.

Morrish, A.H. and P.E. Clark, 1974, Tr. Mezhdunar Konf. Magn. (Moscow) **2**, 180.

Morrish, A.H. and G.A. Sawatzky, 1970, Proc. Int. Conf. on Ferrites, Kyoto, p. 144.

Morrish, A.H. and E.P. Valstyn, 1962, J. Phys. Soc. Japan **17**, Suppl. Bl. 392.

Morrish, A.H. and L.A.K. Watt, 1957, Phys. Rev., **105**, 1476.

Morrish, A.H. and S.P. Yu, 1955, J. Appl. Phys. **26**, 1049.

Morrison, J.R., 1968, Electrochem. Tech. **6**, 419.

Morrison, J.R., D.E. Speliotis and J.S. Judge, 1969, J. Appl. Phys. **40**, 1309.

Morton, J.P. and M. Schlesinger, 1968, J. Electrochem. Soc. **115**, 16.

Morton, V. and R.D. Fisher, 1969, J. Electrochem. Soc. **116**, 188.

Moskowitz, R. and E. DellaTorre, 1967, J. Appl. Phys. **38**, 1007.

Münster, E., 1971, IEEE Trans. Mag. MAG-7, 263.

Nakamura, M., N. Takahashi, S. Ozaki, I. Fukushima and H. Isono, 1975a, Japan. Patent 75 03 948.

Nakamura, M., N. Takahashi, S. Ozaki, H. Isono, T. Naruse and T. Nisihara, 1975b, Japan. Patent 75 84 430.

Nakano, S., 1974, Japan. Patent 74 28 600.

Nakao, K., C. Shibuya and T. Ikeda, 1975, Japan. Patent 75 133 127.

Néel, L, 1947a, Comptes Rend. Acad. Sci. Paris **224**, 1488.

Néel, L., 1947b, Compt. Rend. Acad. Sci. Paris **224**, 1550.

Néel, L., 1949a, Compt. Rend. Acad. Sci. Paris **228**, 664.

Néel, L., 1949b, Ann. de Geophys **5**, 99.

Néel, L., 1955, Adv. Phys. **4**, 191.

Néel, L., 1957, Compt. Rend. Acad. Sci. Paris **244**, 2668.

Néel, L., 1958, Compt. Rend. Acad. Sci. Paris **246**, 2313.

Néel, L., 1959, J. Phys. Radium **20**, 215.

Noble, R., 1963, Intermag Conf. Paper 4.2.

Nobuoka, S., T. Ando and F. Hayama, 1963, US Patent 3 081 264.

Nose, H., I. Tashiro, M. Hashimoto and R. Kimura, 1970, Trans. Nat. Res. Inst. Met. **12**, 1.

Ogawa, S., S. Nagase, K. Ogasa and Y. Uehara, 1972, Intermag. Conf. Kyoto, Paper 23.2.

Ohiwa, T., S. Yamashita, T. Adachi, 1975, Japan. Patent 75 78 898.

Okamoto, S. and T. Baba, 1969a, Japan. Patent 69 09 730.

Okamoto, S. and T. Baba, 1969b, Japan. Patent 69 09 731.

Olsen, K.H., 1974, IEEE Trans. Mag. MAG-10, 660.

Olsen, K.H. and J.W. Cox, 1971, Proc. Electron Microscope Soc. America Conf. 152.

Oppegard, A.L., F.J. Darnell and H.C. Miller, 1961, J. Appl. Phys. **32**, 184S.

Osakawa, S., S. Harada, 1974, Japan. Patent

74 52 134.

Osmolovskii, M.G., M.P. Morozova, G.P. Kostikova, V.A. Bogolyubskii and L. Yu. Ivanova, 1976, USSR Patent 533 922.

Osmond, W.P., 1953, Proc. Phys. Soc. (London) 66B, 265.

Osmond, W.P., 1954, Proc. Phys. Soc. (London) 67B, 875.

Ōuchi, K. and S. Iwasaki, 1972, IEEE Trans. Mag. MAG-8, 473.

Ozaki, S., H. Isono and I. Fukushima, 1973, Japan. Patent 73 25 663 and 73 25 664.

Panter, G.B., I.I. Eliasberg and N.K. Yakobson, 1971, Tr. Vses, Nauch.-Issled. Inst. Telev. Radioveshchaniya 1, 205.

Parker, C.C., R.W. Polleys and J.S. Vranka, 1973, US Patent 3 726 664.

Parker, R., 1975, Magnetic oxides, ed. D.J. Craik, (Wiley, London) p. 457.

Parfenov, V.V., Yu.A. Lobanov and I.V. Ivanova, 1966, Fiz. Met. Metalloved 22, 569.

Parry, L.G., 1975, Aust. J. Phys. 28, 693.

Paul, M., 1970, Z. Angew. Phys. 28, 321.

Penniman, R.S., Jr. and N.M. Zoph, 1921, US Patent 1 368 748.

Penoyer, R.F. and L.R. Bickford Jr., 1957, Phys. Rev. 108, 271.

Pingaud, B.J., 1972, Fr. Patent 2 129 841.

Pingaud, B.J., 1974, Ger. Patent 2 318 271.

Piskacek, Z. and J. Vavra, 1968, Thin Solid Films 2, 487.

Preisach, F., 1935, Z. Phys. 94, 277.

Pye, D.G., 1972, Ger. Patent 2 151 471.

Rabinovici, R., Gh. Gălugăru and E. Luca, 1977, IEEE Trans. Mag. MAG-13, 931.

Radhakrishnamurty, C., 1974, J. Geophys. Res. 79, 3031.

Ragle, H.U. and P. Smaller, 1965, IEEE Trans. Mag. MAG-1, 105.

Rathenau, G.W., J. Smit and A.L. Stuitjts, 1952, Z. Phys. 133, 250.

Ratnam, D.V. and W.R. Buessem, 1970, IEEE Trans. Mag. MAG-6, 610.

Robbins, H., 1973, US Patent 3 778 373.

Robbins, M., J.H. Swisher, H.M. Gladstone and R.C. Sherwood, 1970, J. Electrochem. Soc. 117, 137.

Robbins, M., J.H. Swisher and R.C. Sherwood, 1971, J. Appl. Phys. 42, 1352.

Robinson, J.W. and E.E. Hockings, 1972, RCA Rev. 33, 399.

Rodbell, D.S. and R.C. De Vries, 1967, Mat. Res. Bull. 2, 491.

Rodbell, D.S. and D.M. Lommel, 1972, US Patent 3 700 500.

Rodbell, D.S., R.C. De Vries, W.D. Barber and R.W. De Blois, 1967, J. Appl. Phys. 38, 4542.

Roden, J.S., 1976, US Patent 3 932 293.

Rogalla, D., 1969, IEEE Trans. Mag. MAG-5, 901.

Roger, B. and E. Weisang, 1973, Fr. Patent 2 179 501.

Sallo, J.S. and J.M. Carr, 1962, J. Electrochem. Soc. 109, 1040 and J. Appl. Phys. 33, 1316.

Sallo, J.S. and J.M. Carr, 1963, J. Appl. Phys. 34, 1309.

Sallo, J.S. and K.H. Olsen, 1961, J. Appl. Phys. 32, 203S.

Sard, R., 1970, J. Electrochem. Soc. 117, 864.

Sasazawa, K., M. Shimizu and T. Kitamoto, 1975, Ger. Patent 2 510 799.

Sasazawa, K., S. Komine, T. Kitamoto and G. Akashi, 1976a, Ger. Patent 2 526 363.

Sasazawa, K., Y. Yamada, T. Kitamoto and G. Akashi, 1976b, Japan. Patent 76 02 998.

Sata, T., N. Tsuchiya, K. Amari, K. Ohta and N. Suzuki, 1975, Japan. Patent 75 09 229.

Sato, M., S. Yokoyama and Y. Hosino, 1962, Kogyo Kagaku Zasshi 65, 1336.

Sato, T., 1970, IEEE Trans. Mag. MAG-6, 795.

Sato, S., T. Mudo and N. Iwata, 1976a, Japan. Patent 76 20 098.

Sato, S., T. Saito and K. Ishii, 1976b, Japan. Patent 76 20 100.

Sato, S., S. Iizuka and M. Suzuki, 1976c, Japan. Patent 76 20 597.

Schlömann, E., 1956, Proc. Conf. on Magnetism and Magnetic Materials (AIEE) Spectrum Publ. T-91, p. 600.

Schneehage, H., G. Wunsch, E. Köster and E. Schoenafinger, 1976, Ger. Patent 2 434 096.

Schneider, J. and A.M. Stoffel, 1973, Tr. Mezhdunar, Konf. Magn. 5, 206 and Ger. Patent 2 146 008.

Schneider, J.W., A.M. Stoffel and G. Trippel, 1973, IEEE Trans. Mag. MAG-9, 183.

Schoenafinger, E., P. Deigner, E. Köster, M. Ohlinger, B. Schäfer, W. Stumpfi, J. Amort, C.D. Seiler, H. Nester and O. Ambros, 1976, Ger. Patent 2 459 766.

Schräder, R. and G. Büttner, 1963, Z. Anorg. Algem. Chem. 320, 205.

Schuele, W.J., 1959, J. Phys. Chem. 63, 83.

Schuele, W.J., 1962, US Patent 3 042 543.

Schuele, W.J. and V.D. Deetscreek, 1961, J. Appl. Phys. 32, 235S.

Schwantke, G., 1961, J. Audio Eng. Soc. 9, 37.

Sellwood, P.W., 1956, Magnetochemistry (Interscience, London) p. 307.

Shannon, R.D., 1967, J. Am. Ceram, Soc. 50, 56.

Shibasaki, Y., F. Kanamaru and M. Koizumi, 1970, Proc. Int. Conf. on Ferrites, Kyoto, p. 480.

Shibasaki, Y., F. Kanamaru and M. Koizumi, 1973, Mat. Res. Bull. 8, 559.

Shimizu, S., N. Umeki, T. Uebori, N. Horiishi, Y. Okuda, Y. Yuhara, H. Kosaka, A. Takedoi and K. Yaguchi, 1975, Japan. Patents 75 37 667 and 75 37 668.

Shimotsukasa, J., 1972, Japan. Patent 72 46 675.

Shirahata, R., T. Kitamoto, M. Shimizu, A. Tasaki and M. Suzuki, 1975a, US Patents 3 898 592 and 3 929 604.

Shirahata, R., T. Kitamoto and M. Suzuki, 1975b, Japan. Patent 75 116 330.

Shtol'ts, Ye.V., 1962, Bull. Acad. Sci. USSR Phys. Ser. 26, 318.

Shtol'ts, Ye.V., M.Ya. Gen, A.N. Mertemyanov, I.V. Plate and Ye.A. Fedorova, 1972, Phys. Metals Metallog. 34, 77.

Shtrikman, S. and D. Treves, 1959, J. Phys. Radium 20, 286.

Siakkou, M., D. Effenburger, E. Muenster and K. Willaschek, 1973, J. Signalaufzeichnungsmaterialen 2, 157.

Siratori, K. and S. Iida, 1960, J. Phys. Soc. Japan 15, 210.

Skorski, R., 1972, Nature 240, 15.

Slonczewski, J.C., 1958, J. App. Phys. 29, 448.

Slonczewski, J.C., 1961, J. App. Phys. 32, 253S.

Smaller, P. and J.J. Newman, 1970, IEEE Trans. Mag. MAG-6, 804.

Smit, J. and H.P.J. Wijn, 1959, Ferrites (Wiley, New York) p. 157, 182, 204.

Smith, D.O., 1956, Phys. Rev. 102, 959.

Speliotis, D.E., G. Bate, J.K. Alstad and J.R. Morrison, 1965a, J. Appl. Phys. 36, 972.

Speliotis, D.E., J.R. Morrison and G. Bate, 1965b, Proc. Int. Conf. on Magnetism, Nottingham, p. 623.

Stacey, F.D. and S.K. Banerjee, 1974, The physical principles of rock magnetism (Elsevier, Amsterdam) Ch. 2, p. 35.

Steck, W., G. Wunsch, P. Deigner, W. Ostertag and K. Uhl, 1976, Ger. Patent 2 429 177.

Stewart, E.W., G.P. Conard II and J.F. Libsch, 1955, J. Met. 203, 152.

Stone, E. and P.K. Patel, 1973, US Patent 3 738 818.

Stoner, E.C. and E.P. Wohlfarth, 1948, Phil. Trans. Roy. Soc. A240, 599.

Straubel, R. and S. Spindler, 1969, IEEE Trans. Mag. MAG-5, 895.

Strickler, D.W. and R. Roy, 1961, J. Amer. Ceram. Soc. 44, 225.

Stuijts, A.L. and H.P.J. Wijn, 1957/8, Philips Tech. Rep. 19, 209.

Sueyoshi, T., 1976, Japan. Patent 76 96 100.

Sugimori, M., 1974, Japan. Patent 74 101 295.

Sugiura, Y., 1960, J. Phys. Soc. Japan 15, 1461.

Suzuki, S. and M. Yazaki, 1975, Japan. Patent 75 140 899.

Swisher, J.H., M. Robbins, R.C. Sherwood, E.O. Fuchs, W.H. Lockwood and J.P. Keilp, 1971, IEEE Trans. Mag. MAG-7, 155.

Swoboda, T.J., P. Arthur, N.L. Cox, J.N. Ingraham, A.L. Oppegard and M.S. Sadler, 1961, J. Appl. Phys. 32, 374S.

Sykora, V., 1967, Industrie Chemique Belge 32, Spec. No. Pr. 2, 623.

TDK, 1976, Brit. Patent 1 422 098.

Tadokoro, E. and M. Aonuma, 1974, Japan. Patent 74 72 699.

Tadokoro, E., M. Suzuki, M. Aonuma and T. Kitamoto, 1975, Japan. Patent 75 14 326.

Takahashi, N., M. Nakamura, S. Ozaki, H. Isono and I. Fukashima, 1975, US Patent 3 925 114.

Takao, M. and A. Tasaki, 1976, IEEE Trans. Mag. MAG-12, 782.

Takei, H. and S. Chiba, 1966, J. Phys. Soc. Japan 21, 1255.

Tamai, Y. and R. Shirahata, 1976, Japan. Patent 76 54 064.

Tanaka, T. and N. Tamagawa, 1967, Japan J. Appl. Phys. 6, 1096.

Taniguchi, T., 1975, Japan. Patent 75 92 499.

Tasaki, A., S. Tomiyama, S. Iida, N. Wada and R. Uyeda, 1965, Japan J. Appl. Phys. 4, 707.

Tenzer, R., 1965, J. Appl. Phys. 36, 1180.

Terada Y., T. Mihara, T. Kawamata and E. Hirota, 1973, Japan. Patents 73 69 795 and 73 69 796.

Thomas, J.R., 1966, J. Appl. Phys. 37, 2914.

Thomas, J.R. and J.B. Lavigne, 1966, US Patent 3 248 358.

Toda, H., S. Shimizu and H. Ihara, 1973, US Patent 3 720 618.

Tonge, D.G. and E.P. Wohlfarth, 1958, Phil. Mag. 3, 536.

Trandell, R.F. and R.G. Fessler, 1974, Ger. Patent 2 419 800.

Tsukanezawa, K., T. Kobayashi, Y. Nishizawa and S. Harada, 1975, Japan. Patent 75 86 694 and Ger. Patent 2 504 995.

Uchino, T., K. Sato, M. Takeuchi and H. Kijimuta, 1974a, Japan. Patent 74 00 198.

Uchino, T., K. Sato, M. Takeuchi and Y. Ota, 1974b, Japan. Patents 74 127 196 and 74 127 197.

Uchino, T., K. Sato, M. Takeuchi and Y. Ota, 1975, Japan. Patent 75 118 992.

Uehori, T. and C. Mikura, 1976 Japan. Patent 76 109 498.

Uehori, T., S. Umeki and M. Moteki, 1974, Japan. Patent 74 08 496.

Uehori, T., A. Hasaka, Y. Tokuoka, T. Izumi and Y. Imaoka, 1978, IEEE Trans. Mag. MAG 14, 852.

Uesaka, Y., 1974, Japan. Patent 74 107 010.

Umeki, S., 1975, Ger. Patent 2 455 158.

Umeki, S., S. Saitoh and Y. Imaoka, 1974a, IEEE Trans. Mag. MAG-10, 655.

Umeki, S., T. Uebori and M. Motoki, 1974b, Japan. Patent 74 113 199.

Umeno, S., 1975, Japan. Patent 75 80 499.

Ushakova, S.E., M.N. Raevskaya and Yu.A. Ovchinnikov, 1976. USSR Patent 532 127.

Uyeda, R. and K. Hasegawa, 1962, J. Phys. Soc. Japan 17, Suppl. B-1, 391.

Valstyn, E.P., J.P. Hanton and A.H. Morrish, 1962, Phys. Rev. 128, 2078.

Van der Geissen, A.A., 1973, IEEE Trans. Mag. MAG-9, 191.

Van der Giessen, A.A., 1974, Rev. de Phys. Appl. 9, 869.

Van der Giessen, A.A. and G.J. Klomp, 1969, IEEE Trans. Mag. MAG-5, 317.

Van Oosterhout, G.W., 1960, Acta Cryst. 13, 932.

Van Oosterhout, G.W., 1965, Proc. Int. Conf. on Magnetism, Nottingham, p. 529.

Van Oosterhout, G.W. and C.J.M. Rooijmans, 1958, Nature 181, 44.

Verwey, E.J.W., 1935, Z. Krist. 91, 65.

Wakai, K. and Y. Suzuki, 1976, Japan. Patent 76 23 697.

Waring Jr., R.K., 1967, J. Appl. Phys. 38, 1005.

Waring Jr., R.K., 1970, US Patent 3 535 688.

Waring Jr., R.K. and P.E. Bierstedt, 1969, IEEE Trans. Mag. MAG-5, 262.

Watanabe, S., 1970, Ferrites, Proc. International Conf. p. 473.

Weil, L., 1951, J. Phys. Radium 12, 437.

Welch Jr., R.H. and D.E. Speliotis, 1970, J. Appl. Phys. 41, 1254.

Welo, L.A. and O. Baudich, 1925, Phil. Mag. 50, 399.

Welo, L.A. and O. Baudisch, 1933, Naturwissenschaften 21, 659.

Wenner, J.W., 1964, US Patent 3 150 939.

Westmijze, W.K., 1953, Philips Res. Rep. 8, 245.

Wilhelmi, K.A. and O. Jonsson, 1958, Acta. Chem. Scand. 12, 1532.

Woditsch, P. and F. Hund, 1974, Ger. Patent 2 249 273.

Woditsch, P. and L. Leitner, 1973, Ger. Patent 2 221 218, 1975, US Patent 2 897 354.

Woditsch, P., K. Hill, F. Hund and F. Rodi, 1972, Ger. Patent 2 122 312.

Woditsch, P., G. Buxbaum, F. Hund and V. Hahnkamm, 1975a, Ger. Patent 2 347 486.

Woditsch, P., P. Hund, G. Buxbaum, V. Hahnkamm and I. Pflugmacher, 1975b, Ger. Patent 2 339 142.

Wohlfarth, E.P., 1959a, Adv. Phys. 8, 87.

Wohlfarth, E.P., 1959b, J. Appl. Phys. 30, 1465.

Wohlfarth, E.P., 1963, Magnetism, eds. Rado and Suhl 3 (Academic Press, New York) p. 35.

Wohlfarth, E.P. and D.G. Tonge, 1957, Phil. Mag. 2, 1333.

Wohlfarth, E.P. and D.G. Tonge, 1958, Phil. Mag. 3, 536.

Wolff, N.E., 1964, US Patent 3 139 354.

Woodward, J.G. and E. DellaTorre, 1960, J. Appl. Phys. 31, 56.

Wunsch, G., D. Graw, P. Deigner and E. Köster, 1974, Ger. Patent 2 242 500.

Yada, Y., S. Miyamoto and H. Kawagoe, 1973, IEEE Trans. Mag. MAG-9, 185.

Yamamoto, N., 1968, Bull. Inst. Chem. Res. Kyoto Univ. 46, 28.

Yamanaka, K., T. Yamashita, S. Yamamura and Y. Ichinose, 1975, Japan. Patent 75 18 326.

Yamartino, E.J. and R.B. Falk, 1961, US Patent 2 888 777.

Yoshimura, F., Y. Maeda and T. Manabe, 1976, Japan. Patent 76 87 196.

chapter 8

FERROMAGNETIC LIQUIDS

S.W. CHARLES AND J. POPPLEWELL

School of Physical and Molecular Sciences
University College of North Wales
Bangor, Gwynedd
UK

Ferromagnetic Materials, Vol. 2
Edited by E.P. Wohlfarth
© North-Holland Publishing Company, 1980

CONTENTS

1. Introduction*

Intrinsic ferromagnetism arising from a ferromagnetic alignment of the moments on atoms in the liquid state is not known to exist. Its existence, however, is not excluded on theoretical grounds and both Gubanov (1960) and Handrich and Kobe (1970) point out that the ferromagnetic state is mathematically feasible in liquid and amorphous structures. Though non-crystalline solids may be ferromagnetic (see volume 1, chapter 6 of this handbook) reports of ferromagnetism in liquid structures (Busch and Guentherodt 1968) most probably relate to liquids containing ferromagnetic particles rather than to intrinsic liquid ferromagnetism (Nakagawa 1969, Wachtel and Kopp 1969). The analysis of the ferromagnetic state in liquids requires a calculation of an exchange field in a structure with no long range periodicity. As a consequence, a fluctuating component of the exchange integral has to be included in any determination of the magnetisation or Curie temperature. Hence the Curie temperature T_c of a ferromagnetic liquid (Handrich and Kobe 1970), is given by

$$T_c = S(S + 1)N\langle J_{12}\rangle(1 + \Delta^2)/3k \tag{1a}$$

where $N\langle J_{12}\rangle = \langle \Sigma_{j(i \neq j)} J_{ij}\rangle$ is the mean exchange integral and S is the total spin. Δ^2 measures the mean square fluctuation of the exchange integral due to the lack of order in the liquid state.

The reduced magnetisation $M_0 = M/M_s$ is given by

$$M_0 = \tfrac{1}{2}\{B_s[x(1 + \Delta)] + B_s[x(1 - \Delta)]\} \tag{1b}$$

where B_s is the Brillouin function and

$$x = 3M_0 T_c S/(S + 1)T. \tag{1c}$$

It follows (Handrich 1969) from eqs. (1b) and (1c) that for ferromagnetism Δ should be less than unity, and the melting point must be less than the Curie temperature. As the theory stands, therefore, there is no mathematical reason for the non-existence of intrinsic ferromagnetism in the liquid state provided the fluctuations are below a critical value. The absence of observations of ferromagnetism, however, does seem to suggest that there are other factors which

---

* Cgs units are the primary system used. SI units are also given secondarily.

have not been taken into account and which make its existence impossible or unlikely. The production of a ferromagnetic liquid is, therefore, restricted at the present time to the fabrication of liquids containing highly stable colloidal suspensions of single domain ferromagnetic particles. These two phase systems have characteristics which would be expected to be associated with single phase ferromagnetic liquids; they have relatively large magnetisations and respond readily to magnetic fields (Rosensweig 1969).

The first reported attempts at producing ferromagnetic liquids by dispersing ferromagnetic particles in a carrier fluid were by Gowan Knight in 1779 (Wilson 1779. He attempted to produce a ferromagnetic liquid using essentially the same technique as that used at the present time, that is by dispersing ferromagnetic particles in carrier liquids, in this case iron filings in water. After several hours mixing, the water contained a suspension of small particles but the mixture did not have long term stability. Bitter (1932) produced a colloid for magnetic domain studies which consisted of a colloidal suspension of magnetite in water. The particle size was around 10^4 Å. Suspensions containing smaller particles (200 Å) have been prepared by Elmore (1938a) and more recently by Craik and Griffiths (1958). Such colloids closely resemble ferrofluids but refinements involving the removal of the larger particles are necessary to produce ultra-stable systems of high magnetisation and high particle concentrations.

In the 1940's, liquids containing micron-sized ferromagnetic particles were used as clutch fluids. In these fluids the particles migrate in the presence of a magnetic field gradient producing a solid plug as the particles congeal. This is a feature not observed in ferrofluids where the liquid characteristics remain even in large field gradients.

TABLE 1

Some properties of magnetic materials which can be used in the preparation of ferromagnetic liquids

			Magnetisation				Anisotropy constants [$\times 10^{-3}$($J\,m^{-3}$), $\times 10^{-4}$ (erg cm^{-3})]			
			cgs (G)		SI (kA m^{-1})					
			$M_s = 4\pi I_s$				0 K		290 K	
	Crystalline structure	T_c(K)	0 K	290 K	0 K	290 K	K_1	K_2	K_1	K_2
Fe(α)	bcc	1043	22016	21450	1752	1707	52	~9.5	48	20
Co(α)	hcp	~1130[a]	18171	17593	1446	1400	670	175	450	150
Co(β)	fcc	1404					23[b]	4.3[b]	0.9[c]	3.6[c]
Ni(α)	fcc	631	6409	6095	510	485	−126	−46	−50	−2
Gd[d]	hcp	289	24881	13697	1980	1090	−75	240	20	~0
Fe$_3$O$_4$	Inverse spinel	858	6400	6000	509	484	4[e]	4[e]	−11	−2.8

[a] By extrapolation
[b] Value at 773 K
[c] Value at 1173 K
[d] Values of K_3 and K_4 are given by Smith et al. (1977).
[e] Value at 130 K.

Ultrastable ferromagnetic liquids were first prepared by Papell (1965) for the National Aeronautics and Space Administration (NASA). These liquids were similar to the ferromagnetic liquids now commercially available and consisted of ferrite particles in a non-conducting liquid carrier. Other ferromagnetic liquids containing the more magnetic metal particles of iron, cobalt and nickel dispersed in similar non-conducting carriers are currently being developed. The preparation of ferromagnetic liquids containing iron, nickel–iron and gadolinium particles in a metallic carrier liquid have been investigated by Shepherd et al. (1970) and Shepherd and Popplewell (1971). The metallic ferromagnetic liquids present additional problems of stability and though long term stability can be achieved in zero magnetic field (Popplewell et al. 1976), the problem of producing similar stability in large field gradients is still currently under investigation.

Table 1 gives a list of the basic and relevant properties of magnetic materials which can be used in the preparation of ferromagnetic liquids.

1.1. Applications

Ferromagnetic liquids were developed in the 1960's by NASA as a means for controlling fuel flow under conditions of weightlessness in space. More recently they have been developed for use in rotating seals (Moskowitz and Ezekiel 1975); the ferromagnetic liquid providing the seal between the rotating shaft and its static housing. The performance of such a seal (fig. 1) is impressive. They may sustain a differential pressure of 600 psi at a continuous rating of 10 000 rpm or 100 000 rpm for a shorter period. This compares most favourably with more conventional O-ring seals which have a limited rotational speed of approximately 300 rpm. Further applications of ferromagnetic liquids are found in magnetic

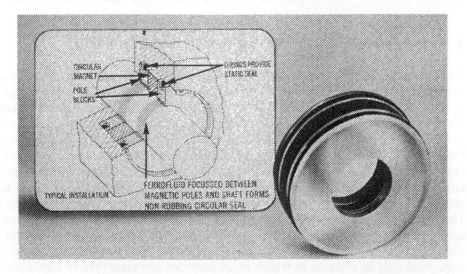

Fig. 1. Ferrofluid Modular Seal (by kind permission of Ferrofluidics Corp.).

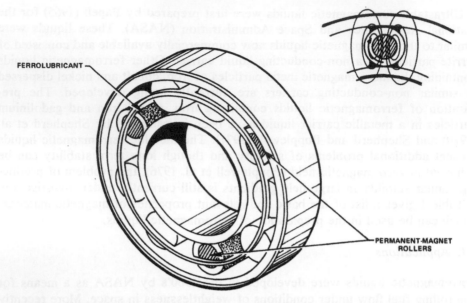

FERROLUBRICANT

PERMANENT-MAGNET
ROLLERS

Fig. 2. Bearing containing Ferrolubricant (by kind permission of Ferrofluidics Corp.).

lubricants (fig. 2), magnetic damping (stepping motors and loudspeaker coils) (Ezekiel 1975), magnetic separation of scrap metals (Nogita et al. 1977) and for energy conversion. The latter application is considered in detail in section 7. Many other applications covering a wide field of interest from the medical to the engineering fields are being seriously considered and it is to be expected that in the near future devices employing ferromagnetic liquids will not be uncommon.

2. Essential parameters in the preparation of ferromagnetic liquids

Ferromagnetic liquids may be conveniently divided into two categories, those consisting of ferrite particles or ferromagnetic metal particles dispersed in non-metallic liquids and ferromagnetic metal particles dispersed in liquid metals. The mechanism by which the small single-domain particles are produced invariably leads to particle size distributions which play a profound role in determining the magnetic properties (see section 4.4). For a dispersion to remain stable, it is essential that the particles be sufficiently small so that Brownian motion opposes any tendency for particle agglomeration. If particle agglomeration does take place the desirable characteristics of the ferromagnetic liquid will be destroyed. The particle interactions in a ferromagnetic liquid which promote agglomeration arise from magnetic and London-type Van der Waals forces. The process of diffusion whereby large particles grow at the expense of the smaller particles is important only in systems in which metal particles are dispersed in liquid metals (Greenwood 1956).

2.1. Magnetic interactions

The minimum potential energy arising from the magnetic interaction of two particles is given by

$$U = -2\mu_1 \cdot \mu_2/r^3 \tag{2}$$

where the moments μ_1 and μ_2 are considered parallel and separated by a distance r (Maxwell 1954, Brown 1962).

Assuming $\mu_1 = \mu_2 = \mu$, where $\mu = \pi D^3 I_s/6$ and I_s and D are the saturation magnetisation/unit volume and diameter of the single domain particle, respec-

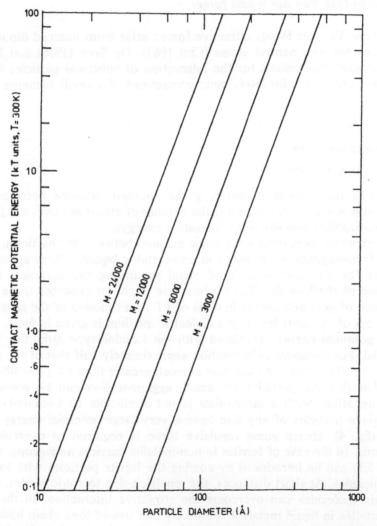

Fig. 3. Contact magnetic potential energy plotted against particle diameter (Rosensweig et al. 1965).

tively, then

$$U = -2(\pi D^3 I_s/6)^2/r^3. \tag{3}$$

The potential energy is a minimum for surfaces in contact ($r = D$). If this potential energy is equated to the thermal energy kT, an upper particle diameter is obtained below which it is expected that thermal agitation alone would be sufficient to prevent agglomeration. From fig. 3, it is seen that for iron having an $I_s = 1707$ emu cm^{-3} [$M_s = 4\pi I_s = 21450$ G (1707 kA m^{-1})] the upper limit of the particle diameter at 20°C is approximately 30 Å. For ferrite particles with $I_s \sim 300$ emu cm^{-3} (300 kA m^{-1}) at 20°C the upper limit is approximately 100 Å.

2.2. London-type Van der Waals forces

London-type Van der Waals attractive forces arise from induced dipole–dipole interactions between neutral atoms (Chu 1967). De Boer (1936) and Hamaker (1936) obtained expressions for the interaction of spherical particles based on the London expression for interatomic interactions. For small distances between spheres

$$U = -AD/24s \tag{4}$$

and for large distances

$$U = -16A(D/2s)^6/9 \tag{5}$$

where D is the particle diameter, s the shortest distance between sphere surfaces and $A = \pi^2 n^2 \lambda'$, where n is the number of atoms per cm^3, $\lambda = \frac{3}{4} h \nu_0 \alpha^2$, ($\alpha$ is the polarizability and $h\nu_0$ is the ionisation energy).

These expressions assume a dielectric medium between the particles, as is the case for ferromagnetic suspensions in non-metallic liquids. There appears to be little data for the case of a liquid metal separating the particles. However attenuation of the Van der Waals interaction might be expected to be appreciable. A plot of potential energy in units of kT as a function of the separation of the surfaces of the particles, s, in a dielectric medium is given in fig. 4. It shows that the potential energy associated with the London-type attractive forces is substantial. For distances of separation approximately half that of the radius of the particles, the energy of attraction is much greater than kT ($T = 300$ K).

Provided that the particles are small, agglomeration can be prevented by thermal agitation. Such a mechanism is not applicable to London-type interactions since particles of any size have a very large potential energy when in contact (fig. 4). Hence some repulsive force is required to overcome these interactions. In the case of ferrites in non-metallic carriers an entropic repulsion (section 5.1) can be introduced by coating the ferrite particles with long chain polar molecules. At short distances, the repulsion due to steric hindrance of the long chain molecules can overcome the attractive interaction. In the case of metal particles in liquid metals it is unlikely that use of long chain hydrocarbon molecules to overcome these attractions is possible since such molecules are not

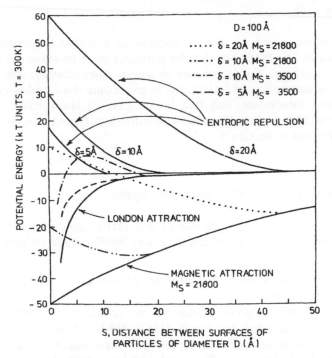

Fig. 4. Potential energy plotted against particle separation.

compatible with liquid metals. However, the differences in the workfunction and Fermi energy of the metal particle and liquid metal carrier may give rise to an effective repulsive interaction which is sufficient to overcome the short range London-type forces (section 5.2).

2.3. Diffusional growth of particles

Apart from particle agglomeration due to inter-particle interactions, particle growth can occur by diffusion. Such a growth mechanism is expected to be important for metallic particles dispersed in liquid metals but of no consequence in the non-metallic systems. A theoretical study of particle growth in a liquid medium has been carried out by Greenwood (1956). Atoms diffuse from the smallest particles with higher surface energy, to the larger particles with lower surface energy. Greenwood predicted that the fastest growing particles are those with a radius a_{2m} twice that of the mean, and that the growth of these particles is governed by the relationship $a_{2m}^3 \propto t$. Experimental studies of the growth of iron and cobalt particles in mercury are consistent with this relationship (Luborsky 1957). The addition of certain metal additives to the dispersion results in a reduction of the rate of diffusion due to a coating of an atomic monolayer of the additive about the particles (Windle et al. 1975).

2.4. Summary

The conditions necessary for the production of a stable suspension of ferro-magnetic particles in a liquid are that the particles must be small (so that thermal agitation can overcome the long range magnetic interactions), that some short range repulsive interaction be introduced to overcome the London-type Van der Waals attractive interaction, and that in the case of the metallic systems the particles be coated to reduce the effects of diffusion. How these conditions may be met is discussed in section 5.

3. Methods of preparation of ferromagnetic liquids

This section deals with the production of fine ferromagnetic particles of metals and ferrites and the means by which they may be dispersed in liquid media.

3.1. Methods of preparation of fine particles of ferrites

The synthesis of ferromagnetic liquids by wet grinding of ferrites in ball mills was originally investigated by Papell (1963, 1965). Rosensweig et al. (1965) developed the technique further and produced ferromagnetic liquids consisting of ferrite particles in kerosene. This work has been extended to include a large number of carrier fluids including fluorocarbons, water, silicone liquids and esters (Kaiser and Rosensweig 1969).

Ferromagnetic liquids containing Fe_3O_4 particles are now manufactured commercially under the name "Ferrofluids" by Ferrofluidics Corporation*. They are produced by ball-milling the ferrite in a liquid in the presence of a suitable surfactant. The grinding needs to be continued for several weeks to produce the ferrite in the colloidal state. Centrifuging is used to remove the larger particles. The mean diameter of the particles produced in this way by Kaiser and Miskolczy (1970a) varied from 50 to 90 Å. The surfactant is included to prevent agglomeration due to London-type Van der Waals and magnetic forces. The surfactant must apparently contain one or more polar groups to adhere to the ferrite surface and another group, e.g. a hydrocarbon chain, to render the surfactant-coated particle compatible with the carrier. Oleic and linoleic acids have been amongst the most successful surfactants for ferrite particles.

Grinding ferromagnetic materials is complicated by interparticle magnetic flocculation arising from the attraction of one particle for another. This causes the process to be extremely inefficient. Khalafalla and Reimers (1973) found that by grinding a non-magnetic precursor compound to produce a stable colloidal suspension, followed by conversion of the precursor to a ferromagnetic form, the grinding time was reduced to a few percent of that used in conventional

* Ferrofluidics Corp., Middlesex Turnpike, Burlington, Mass. USA.

methods. Wustite was the favoured non-magnetic precursor. It can be readily converted to metallic iron and magnetite while in stable suspension by refluxing the liquid from $1\frac{1}{2}$ to 10 h, at temperatures between 200 and 300°C.

Magnetite particles can also be prepared by precipitating the magnetite from a solution of ferric and ferrous ions using excess of an alkali hydroxide solution (Elmore 1938a, Bibik 1973). To obtain a stable suspension of magnetite particles of 80 Å diameter in a hydrocarbon carrier, the precipitate is washed with water, acetone and then toluene. Finally, oleic acid is added to the precipitate and the resulting paste ball-milled for 6–12 h in the appropriate carrier liquid.

Ferromagnetic liquids containing Fe_3O_4 particles can be prepared with particle concentrations up to $\sim 10^{17}$ cm^{-3} before viscosity effects limit their application. Liquids with this concentration would have a magnetisation $\bar{M}_s \sim 600$ G (47 kA m^{-1}).

3.2. Methods of preparation of fine metal particles

For a number of years fine metal particles of diameter less than 10^4 Å have been produced by evaporation of the metal in an atmosphere of an inactive gas, such as nitrogen (Harris et al. 1948), argon (Kimoto et al. 1963, Kimoto and Nishida 1967, Granqvist and Buhrman 1976a) and helium (Yatsuya et al. 1973).

The enormous interest that has been shown in this method arose because of the industrial application of ferromagnetic particles particularly their use for recording tapes.

The growth mechanism of the particles has been considered by several groups of workers (Fritsche et al. 1961, Mizushima 1961, Kimoto et al. 1963, Wada 1967, 1968). They considered that on evaporation metal atoms collide frequently with gas atoms with the result that the metal vapour is cooled. Nucleation then takes place with the production of particles in the gas. There is strong evidence to show that particles also grow by coalescence (Uyeda 1964, Yatsuya et al. 1973, Granqvist and Buhrman 1976b). The ultimate size of the particles, so produced, is dependent on the pressure of the inactive gas and the temperature at which the metal is evaporated. Increase of the gas pressure at a fixed evaporation temperature more effectively supresses the expansion of the metal vapour with the result that a higher supersaturation is obtained. Thus nucleation starts at a higher vapour density leading to a larger particle size. For higher evaporation temperatures, the vapour pressure increases rapidly with a resultant increase in particle size.

Small single domain particles of ferromagnetic metals (Fe, Co, Ni, Gd) and alloys of these metals with themselves and non-ferromagnetic constituents (Cu, Si, Cr, Ho) have been prepared by the method outlined above (Tasaki et al. 1965, 1974, Kusaka et al. 1969). In most of these systems, argon was used as the inactive gas at a pressure of 3 torr and below. Tasaki et al. (1965) reported that the particles so produced are very stable in that only a slight oxidation is observed after the lapse of a year. The particle size distribution of the particles produced by this method is in general narrow.

The method just outlined has been used by Shepherd et al. (1970) to produce ferromagnetic liquids. The stream of particles formed, in this case gadolinium, were directed on to a mercury surface agitated by a stream of argon. The agitation was achieved by passing the argon through a capillary tube which terminated below the surface of the mercury. In this way the incoming argon maintained a condensing atmosphere and simultaneously agitated the mercury. This agitation appeared to be sufficient to prevent the growth of large particles at the mercury surface since the particle diameters obtained at a given argon pressure agree with those obtained by Kimoto et al. (1963).

A convenient way of producing small single domain ferromagnetic metal particles and introducing them into a liquid metallic carrier involves the use of electrodeposition techniques. Salts of the ferromagnetic metals dissolved in water or alcohol are used with the metallic liquid carrier as cathode in an electrolytic cell. In order to promote the production of small single domain particles in the carrier and prevent dendritic growth agitation of the carrier must be maintained throughout deposition. This can be achieved by both mechanical and magnetic agitation. Iron (Mayer and Vogt 1952, Paine et al. 1955, Luborsky 1958a) nickel and cobalt (Meiklejohn 1958, Deryugin and Sigal 1962) and alloys of iron/cobalt (Luborsky and Paine 1960) nickel/cobalt (Falk and Hooper 1961) and nickel/iron (Shepherd and Popplewell 1971) have been deposited into mercury and mercury alloys. In order to produce alloys of uniform composition it is important that the cell current density be maintained constant during the electrodeposition.

There is difficulty in producing ferromagnetic liquids consisting of iron particles in mercury with a high saturation magnetisation because the liquids become highly viscous at low iron particle concentrations. The concentration at which the liquids become slurries depends on particle size (Hoon et al. 1978). For particle sizes with $D \sim 40$ Å this concentration is approximately 0.8 wt%. A liquid with this concentration of particles has a saturation magnetisation $\bar{M}_s \sim$ 300 G (24 kA m^{-1}).

Gallium, low melting point alloys ("Cerro" alloys), tin, "Ingas" (an indium–gallium–tin alloy) and bismuth alloys have also been considered as carrier liquids (Kagan et al. 1970).

In the case of metals which oxidise readily in the presence of water, such as the rare-earth elements, the electrodeposition must be performed on solutions of anhydrous salts in anhydrous alcohol under a blanket of anhydrous inert gas.

Stable suspensions of metallic cobalt particles have been prepared by the thermal decomposition of dicobalt octacarbonyl in solutions containing polymer (Thomas 1966a, Harle and Thomas 1966). The formation of stable suspensions of single domain cobalt particles in organic liquids is dependent on a very delicate polymer–solvent–metal balance. The product is a black, stable colloid of metallic cobalt particles, predominantly of face-centred cubic crystal structure (Thomas 1966b). The presence of an appropriate polymer produces discrete particles coated with the polymer. If no dispersant polymer is present during the decomposition reaction, large multidomain particles are formed. The average particle size may be varied over the range 10–1000 Å by varying the polymer composition, average molecular weight and the solvent used. It has been found (Thomas 1966a, Harle and Thomas 1966) that about 85 wt% of the particles lie within a factor of two of the average particle size.

An investigation of a variety of polymeric materials as stabilizers for these cobalt particles has been made (Hess and Parker 1966). The most successful dispersant polymers, as judged by cobalt particle size distribution and particle magnetic properties, are the addition polymers with average molecular weight of 100 000 or more. The polymer consists of a relatively unreactive hydrocarbon chain with at least one very polar group spaced every 200 backbone atoms. The polar groups adhere to the metal surface leaving the bulk of the polymers to form a thick coating and thus protect the metal particles from agglomeration. Nonpolymer dispersants such as Aerosol O.T., Sarkosyl, etc., have been used successfully in the preparation of stable suspensions of cobalt particles (Martinet 1978, Chantrell et al. 1978). Ferromagnetic liquids containing cobalt particles with $M_s \sim$ 370 G (29 kA m^{-1}) and having a viscosity of a few centipoise can be readily prepared.

A method similar to that used in the production of plated materials using standard electroless plating techniques has been used to prepare fine ferromagnetic particles. The solutions used in the preparation contain the salts of ferromagnetic metals, a reducing agent such as sodium hypophosphite or dimethyl amine borane, complexing agents and buffers to control the pH during reaction. The formation of fine particles may be induced by raising the temperature of the solution and adding small quantities of PdCl$_2$ to the solution. The Pd ion is reduced to Pd metal upon which the ferromagnetic metal is deposited and which continues to grow by autocatalytic action. Metallic palladium is the catalyst most commonly used for initiation of electroless plating. Sub-micron particles of Co, Co–Ni alloys and Fe have been prepared by this method (Akimoto et al. 1972, Akashi and Fujiyama 1971, Judge et al. 1966).

The addition of water-soluble viscosity increasing materials, such as gelatin, water glass, carboxymethyl cellulose, etc. in the pH range where they do not precipitate or gel has been shown (Akashi and Fujiyama 1971) to be effective in reducing the particle size of the magnetic powder produced. It is possible to produce particles of sufficiently small size for them to remain in stable suspension. However, there are problems of preventing oxidation in aqueous media.

In the production of particles using electroless plating techniques, phosphorus or boron is invariably present, the concentration of which depends upon the conditions of preparation. However, by controlling the pH of the solution the concentration of these non-magnetic species can be kept to a sufficiently low level so that the magnetic properties of the particles are not severely affected.

The methods outlined for preparing ferromagnetic particles are those which have been used with some success in the preparation of stable ferromagnetic liquids.

4. Magnetic properties

4.1. Superparamagnetism in ferromagnetic liquids

The magnetic properties of ferromagnetic liquids are well described by Langevin's classical treatment of paramagnetism. Since the ferromagnetic particles within the liquid have a volume some 10^5 greater than that of atoms the magnetic properties are much enhanced and the particles exhibit "superparamagnetic" behaviour in that they have a large saturation magnetisation but no remanence or coercivity. A typical magnetisation curve for such a ferromagnetic liquid is shown in fig. 5.

Assuming a system of non-interacting spherical particles the magnetisation of the liquid $\bar{M}$ may be written as

$$\bar{M} = \epsilon M_s(\coth \alpha - 1/\alpha) \tag{6}$$

where $\alpha = D^3 M_s H/24kT$ and ϵ is the volume fraction of magnetic particles of saturation magnetisation M_s. In the case of the measurements by Elmore (1938b) on colloidal magnetite, good agreement between the Langevin curve and experimental observations was not obtained unless the particle size distribution

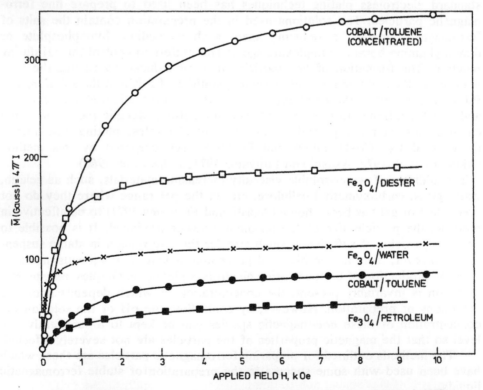

Fig. 5. Magnetisation curves for some ferromagnetic liquids (Chantrell et al. 1978).

(considered to be Gaussian) was taken into account. More recently Chantrell et al. (1978) have shown that good agreement may be obtained using a log-normal volume fraction distribution. This is illustrated by the good agreement between the theoretical curve and experimental results shown in fig. 5. The technique for determining the form of the size distribution is considered further in section 4.4.

There are two mechanisms by which the magnetisation of the ferromagnetic liquid may relax after the applied field has been removed. In the first instance the particle magnetisation may relax by particle rotation in the liquid. This mechanism is characterised by a Brownian rotational diffusion time τ_B given by

$$\tau_B = 3V\eta/kT \tag{7}$$

where V is the particle volume and η the viscosity of the liquid. In the frozen or polymerised state, however, this mechanism is effectively quenched and the relaxation process is then determined by the rotation of the magnetic vector within the particle in the presence of magnetic anisotropies. This process which is associated with a different relaxation time τ_N was first conceived by Néel (1949).

For uniaxial materials this relaxation time is given by

$$\tau_N = \tau_0 X^{-1/2} \exp X \tag{8}$$

where

$$X = KV/kT = \pi D^3 K/6kT \tag{9}$$

and K is the anisotropy constant. τ_0 is given by

$$\tau_0 = M_s/8\pi a\gamma K \tag{10}$$

where γ is the gyromagnetic ratio and a is a damping parameter $(\sim 10^{-2})$ which has been calculated by Skrotskii and Kurbatov (1961) and Anderson and Donovan (1959). Either relaxation process can take place in the liquid state though only the Néel mechanism is possible in solid systems.

In the liquid state the magnetisation of small particles is more likely to relax by the Néel mechanism whereas that of the large particles is more likely to relax by Brownian rotation. The transition from Néel to Brownian relaxation may be considered to take place for particles with a diameter D_s obtained by equating τ_B and τ_N. Hence $X^{-3/2} \exp X = 24a\eta\gamma M_s^{-1}$ where X depends on the diameter according to eq. (9).

Shliomis (1974) calculated the diameter D_s as 85 Å for iron at 300 K, assuming $K = 5 \times 10^5$ ergs cm^{-3} $(5 \times 10^4$ J m$^{-3})$, $M_s = 18000$ G $(1400$ kA m$^{-1})$, $\eta = 10^{-2}$ P $(10^{-3}$ N s m$^{-2})$, $\gamma = 1.7 \times 10^7$ s^{-1} G^{-1} $(1.7 \times 10^{11}$ s^{-1} T$^{-1})$ and $a = 10^{-2}$ $(\sim 10^{-6}$ SI units). A similar calculation for hexagonal cobalt with $K = 4.5 \times 10^6$ ergs cm^{-3} $(4.5 \times 10^5$ J m$^{-3})$ gives $D_s \sim 40$ Å. X for these calculations would be approximately equal to 4. The calculations of the "Shliomis" diameter D_s are based on the assumption that the only anisotropy is that due to crystalline anisotropy. This is probably not the dominant anisotropy in the case of cubic iron, where shape anisotropy is likely to be more important. Furthermore, cobalt particles of 40 Å diameter prepared by the method of Hess and Parker (1966) have a fcc structure and thus the estimate of D_s made by Shliomis is unlikely to correspond to that found experimentally. The form of magnetic relaxation is important when discussing viscosity measurements in magnetic fields. This effect is described more fully in section 6.2.

For a theoretical treatment where the Néel relaxation is dominant the particle anisotropy is considered to be uniaxial. The problem of determining the magnetic properties is then one of considering thermal affects on uniaxial particles. Further, in any real situation the magnetic properties of the ferromagnetic liquid should only be interpreted by considering a system containing particles of different sizes. It is possible in such a system to make use of magnetic granulometry (Bean and Jacobs 1956), to determine the size distribution since the initial susceptibility is given by the expression

$$\bar{\chi}_i = \epsilon M_s^2 V / 48\pi^2 kT = \epsilon M_s^2 D^3 / 288\pi kT \tag{11}$$

and the approach to saturation by

$$\bar{M} = \epsilon M_s (1 - 24kT/M_s D^3 H) \tag{12}$$

where M_s is the saturation magnetisation of the particles, $\bar{M}$ that of the ferromagnetic liquid and D the particle diameter.

No account has been taken of the possibility of the particles clustering, but for magnetic fields large compared with the interaction field (12) is applicable and thus the size of the small particles within the cluster can be determined. Equation (11) however, gives estimates of particle size only in the absence of clusters.

A more complete treatment of the magnetisation of ferromagnetic liquids has been made by Chantrell et al. (1978) who derived expressions equivalent to eqs. (11) and (12) which included a log-normal distribution of particle sizes. The equations then become

$$\bar{\chi}_i = \epsilon M_s^2 D_V^3 \exp(4.5\sigma^2)/288\pi kT \tag{13}$$

and

$$\bar{M} = \epsilon M_s (1 - 24kT \exp(4.5\sigma^2)/M_s D_V^3 H. \tag{14}$$

D_V is the median diameter and σ the standard derivation of the log-normal distribution. By considering eqs. (13) and (14) D_V and σ can be determined.

For a system of aligned uniaxial particles in a solid matrix magnetised along the easy symmetry axis the remanent magnetisation will be time dependent according to

$$M_r = M_s \exp(-t/\tau) \tag{15}$$

where M_s is the saturation magnetisation and τ is the relaxation time given by

$$\tau^{-1} = f_0 \exp(-KV/kT) \tag{16}$$

where f_0 is a frequency factor ($f_0 = 10^9 \text{ s}^{-1}$) and K is the anisotropy constant. For meaningful measurements the experimental observation time should be short compared with τ. Because of the exponential dependence of τ on the particle size (eq. (16)) the transition to non-superparamagnetic behaviour occurs over a narrow particle size range. The very rapid increase in τ with increase in particle diameter and temperature is illustrated in table 2. For spherical iron particles with only crystalline anisotropy a particle of 230 Å diameter has a calculated relaxation time of 10^{-1} s at room temperature and can, therefore, be considered to reach thermal equilibrium very quickly. In contrast, a particle of 200 Å diameter has a relaxation time of 10^9 s. A relaxation time of 10^2 s has been adopted as the criterion for the onset of superparamagnetic behaviour. This arises when the barrier height to rotation KV is close to $25\,kT$. The onset of superparamagnetic behaviour is, therefore, temperature dependent; the temperature at which it occurs is defined as the "blocking temperature". Particles with diameter of 80 Å for hcp cobalt, 280 Å for fcc cobalt and 250 Å for iron have room temperature blocking temperatures. For the case of an aligned system of particles in a reversing field H the barrier to rotation is field dependent being reduced by an amount μH. When the barrier height becomes small ($\sim 25kT$) thermal activation provides magnetisation reversal. The coercive force at temperature T is then given by

$$H_c = 2KV[1 - 5(kT/KV)^{1/2}]/\mu \tag{17}$$

where μ is the magnetic moment of the particle (Bean and Livingston 1959). Ideal superparamagnetic behaviour occurs when $H_c = 0$, i.e.

$$T = KV/25k. \tag{18}$$

The value of D_p as calculated from the volume V given in this last expression is the diameter below which a system of equal sized particles will exhibit no remanence and coercivity and, therefore, will be superparamagnetic. The analysis considered so far has been applied to particles in a solid matrix.

TABLE 2

Relaxation time τ (s) for different values of temperature T (K) and particle volume V (Å³) calculated for a crystalline anisotropy constant $K = 10^5$ erg cm⁻³ (10^4 J m⁻³)

$V \times 10^{-5}$ (Å)³	D (Å)	Relaxation time τ (s)				
		($T = 5$ K)	($T = 10$ K)	($T = 15$ K)	($T = 20$ K)	($T = 25$ K)
3.4	70	280	6×10^{-4}	8×10^{-6}	9×10^{-7}	2×10^{-7}
5.1	80	1×10^8	4×10^{-1}	6×10^{-4}	2×10^{-5}	3×10^{-6}
7.3	90	1×10^{15}	1×10^3	1×10^{-1}	1×10^{-3}	8×10^{-5}
10	100	1×10^{24}	4×10^7	1×10^2	2×10^{-1}	5×10^{-3}
13.3	110	1×10^{35}	1×10^{13}	6×10^5	1×10^2	8×10^{-1}

[3] Using the criterion that particles with a relaxation time greater than 10^2 s may be considered "blocked", then particles with relaxation times to the right of the stepped line are considered superparamagnetic.

For ferromagnetic particles in a liquid carrier the superparamagnetic state is not so clearly defined. For normal liquids the Shliomis diameter as calculated earlier from $KV_s/kT \sim 4$ is always less than D_p as calculated from $KV_p/kT \sim 25$. To make $KV_s/kT \sim 25$ the liquid viscosity would have to be exceptionally large ($>10^6$ P), hence in the normal liquid environment non-superparamagnetic behaviour would be unexpected on this treatment. Non-superparamagnetic behaviour has been observed in some metallic ferromagnetic liquids but this is possibly due to the effects of magnetic sedimentation or particle clustering. The effect of magnetic interactions between particles has not been included in the eqs. (11) and (12) derived by Bean and Jacobs (1956). Hence these cannot describe the experimental observations precisely. Kneller (1965) obtained an empirical formula, stated to include the effect of particle interactions, by replacing V in eq. (11) by an effective volume

$$V' = V \exp{(B'M_s^2/T)} \qquad (19)$$

where B' is a constant.

Hahn (1965) questioned Kneller's interpretation in terms of magnetic interactions and writes the initial susceptibility with higher order terms as

$$\chi_i = \beta(C_0 + C_1\beta + C_2\beta^2 + \ldots) \qquad (20)$$

with $\beta = 1/kT$, $C_0 = \epsilon M_s^2 D^3/288\pi$ and $C_1 = 4\pi C_0^2/3$. However, calculations by Hahn based on experimental data on iron–mercury magnetic liquids show $C_1 \gg 4\pi C_0^2/3$. Later Brown (1967) considered in more detail the effect of particle interactions and concluded that at high temperatures Hahn's results could be interpreted by assuming an interaction between spherical particles grouped in pairs; the higher order terms (eq. (20)) being attributed to particle interactions. A more detailed analysis could be expected to include aggregates of particles grouped in the form of long chains.

The question of particle interactions has been discussed by De Gennes and Pincus (1970). In zero field some clustering may be expected if the particle sizes are greater than the critical size for stability as calculated from eq. (3). Whether the larger particles form chains or rings in zero field remains in some doubt, although Brown (1962) considers clusters of four particles to form rings and clusters of three to form chains. Pincus (1978) considers that this may also depend on particle shape. In magnetic fields, however, chain formation takes place leading to optical effects and magnetic anisotropies.

De Gennes and Pincus (1970) and Pincus (1978) have considered particle chaining processes in the presence of particle interactions. If the coupling constant λ measured as the ratio of the magnetic to thermal energy $\lambda = \mu^2/kTD^3$ is much less than unity little clustering is expected. However it would seem that because it is impossible to exclude some large particles from the particle size distribution some clustering would be inevitable, particularly in an applied field. For $\lambda \gg 1$ which would not be the case for a stable suspension of particles, chains should form with a length given by

$$\nu = (1 - 2|B|n)^{-1} \tag{21}$$

where ν is the number of particles per chain, n is the number of particles per unit volume and

$$B = -\pi D^3 \exp{(2\lambda)}/18\lambda^2.$$

The chain width perpendicular to the field direction is $\sim \nu D^2/\lambda$.

Experimental evidence for the aggregation of particles in uniform magnetic fields has been put forward by Peterson et al. (1974) who measured settling times in concentrated liquids in uniform fields up to 230 G (18.4 kA m⁻¹). Micrographs of water based liquids frozen in magnetic fields of 2000 G (160 kA m⁻¹) (Goldberg et al. 1971) show evidence of particle alignment in chains. Micrographs by Hess and Parker (1966) on cobalt particles in zero field clearly show the clustering effect to be size dependent as would be expected from simple theory; clustering having a minimal effect for cobalt particles less than 100 Å diameter.

The effect on the magnetic properties of particles clustering into chains has been examined further by Martinet (1978) (see section 4.2).

4.2. Ferrite particles in a non-conducting carrier

The variation of the relative magnetisation with magnetic field as obtained by Kaiser and Miskolczy (1970a) is shown in fig. 6 for a ferromagnetic liquid containing ferrite particles in a hydrocarbon carrier. The magnetisation curve does not vary with the particle concentration suggesting that these magnetic measurements are not being influenced by particle interactions. The experimental points shown in fig. 6 can be interpreted in terms of Langevin paramagnetism but it is clear that the saturation value of $\bar{M}/\epsilon M_s$ is substantially less than unity. This indicates that there is less magnetic material in the ferromagnetic liquid

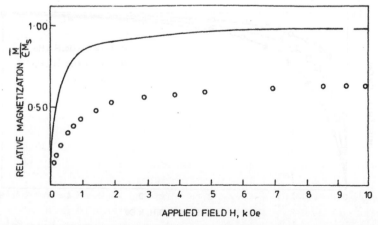

Fig. 6. Observed and calculated magnetisation curves for Fe_3O_4 particles in a hydrocarbon carrier liquid (Kaiser et al. 1970a).

than predicted by simply determining the solid content. Kaiser and Miskolczy (1970a) explained this by considering that the surface atoms on the ferrite particles react chemically with the absorbed molecules of the dispersant coating thus forming a non-magnetic layer. Hence the particle diameter should be replaced by a smaller effective diameter in expressions relating to magnetic properties.

A typical value for the saturation magnetisation, $\bar{M}_s$, for the ferrite based liquids would be of the order of 400 G (32 kA m^{-1}). The magnetisation curves are not dependent on viscosity for these ferromagnetic liquids, confirming that magnetisation changes occur by means of the rotation of the magnetic vector rather than particle rotation (Kaiser and Miskolczy 1970a). The Shliomis diameter, D_s, is therefore greater than the particle diameter.

Magnetic measurements have also been made by Martinet (1978) on Fe_3O_4 particles in kerosene and a liquid containing 65 Å diameter cobalt particles in toluene. Measurements were made in both static and rotating fields in both liquid and solid states. The latter, obtained by polymerisation of the carrier liquid, were made in order to investigate particle chaining effects and the magnetisation process. Martinet (1978) showed that for elongated clusters the magnetisation parallel and the perpendicular to the field varies as

$$\bar{M}_{\parallel} = \bar{\chi}_i H / (1 - 4\alpha' \bar{\chi}_i)$$
$$\bar{M}_{\perp} = \bar{\chi}_i H / (1 + 2\alpha' \bar{\chi}_i) \tag{22}$$

for weak fields and

$$\bar{M}_{\parallel} = \bar{M}_s - \bar{M}_s kT / \mu (H - 2\alpha' \bar{M}_s)$$
$$\bar{M}_{\perp} = \bar{M}_s - \bar{M}_s kT / \mu (H + 4\alpha' \bar{M}_s) \tag{23}$$

for $H \gg kT/\mu$; α' is an unknown factor reflecting the size of the particle interactions which has to be determined from experiment and $\bar{\chi}_i$ is the initial

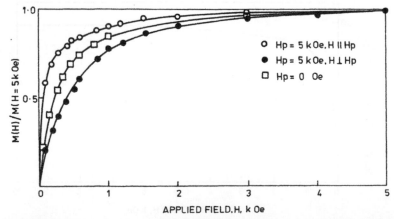

Fig. 7. Magnetisation curve of a ferromagnetic liquid containing Fe_3O_4 particles polymerized in a magnetic field H_p (Martinet 1978).

susceptibility. The observed magnetisation curves shown in fig. 7 can be considered consistent with eqs. (22) and (23). However, in order to make a complete and unambiguous interpretation of the magnetisation curves eqs. (22) and (23) should include contributions which arise from the magnetic anisotropy (Chantrell 1977).

Because of the different relaxation processes in the ferrite and cobalt ferromagnetic liquids, Néel relaxation for 100 Å diameter ferrite particles and Boltzmann relaxation for 65 Å cobalt particles, the anisotropy of the magnetisation of the liquids in a rotating field is different for the two systems (see section 6.3).

4.3. Systems containing ferromagnetic metal particles in a metallic carrier

Figure 8 shows a room temperature magnetisation curve for a ferromagnetic liquid containing iron particles in mercury. Typical superparamagnetic behaviour is observed as characterised by the absence of a coercivity and remanence. It should be noted that in the liquids containing iron much higher saturation magnetisations are possible in principle since the magnetisation of iron is substantially greater than that of ferrites. The difficulties in achieving high magnetisations are referred to in section 3.2.

Measurements of the magnetic properties of frozen metallic ferromagnetic liquids exhibit a remanence and coercivity (fig. 9). In this case the Shliomis diameter is less than the particle diameter, contrary to what is observed for ferromagnetic liquids containing ferrite particles in non-conducting carriers in the solid state. When a coercivity and remanence can be measured, much useful information regarding the properties of the ferromagnetic liquid in the liquid state can be inferred. The coercivity of a single domain particle follows a well documented behaviour with particle size. This is shown for spherical iron

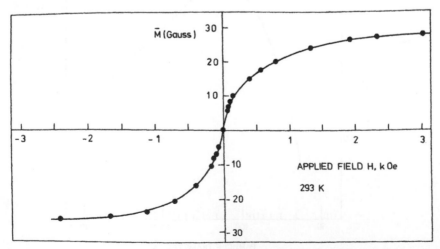

Fig. 8. Magnetisation curve for a ferromagnetic liquid containing iron particles in mercury at 293 K (Windle et al. 1975).

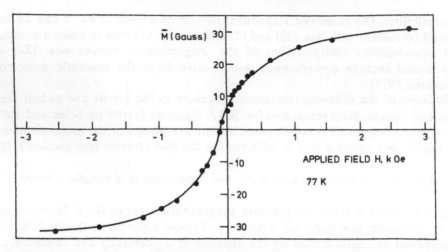

Fig. 9. Magnetisation curve for a ferromagnetic liquid containing iron particles in mercury at 77 K (Windle et al. 1975).

particles at 77 K in fig. 10 (Luborsky 1962a). The coercivity initially rises as the particle diameter increases reaching a maximum determined by the magnetic anisotropy, before falling again as the particle moves from single to polydomain size. In principle, therefore, estimates of particle diameter may be obtained by measuring the particle coercivity at 77 K. Further, the degree of particle elongation can be inferred from the maximum of the particle diameter–coercivity

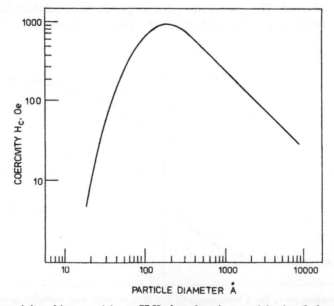

Fig. 10. Coercivity of iron particles at 77 K plotted against particle size (Luborsky 1962a).

curve. Small departures from sphericity make large changes in the shape anisotropy and though the maximum in the coercivity is given by $H_c \sim 2K/I_s (M_s = 4\pi I_s)$ for spherical particles with crystalline anisotropy, values greater than this would be observed for elongated particles. The value of the maximum for a system consisting of iron particles in frozen mercury indicates that such particles are generally elongated. These particles, however, can appear spherical on electron micrographs when they have a low aspect ratio.

The measurement of magnetic properties in the solid state is important in monitoring size changes in metallic ferromagnetic liquids where particle growth takes place and direct observation of particle growth by optical methods is impossible.

4.4. The effect of particle size distributions on magnetic properties

The theoretical treatment of the magnetisation of a system of small ferromagnetic particles with a certain size distribution $f(D)$ has been undertaken by Chantrell et al. (1977) who show that for non-interacting spherical particles in a solid matrix the reduced magnetisation is given by

$$\bar{M}/\epsilon M_s = \int_0^\infty F_H(D) f(D) \mathrm{d}D \qquad (24)$$

where $F_H(D)$ is the Langevin expression as given in eq. (6) which applied only for those particles less than a particular superparamagnetic diameter D_p. For particle sizes greater than D_p, $F_H(D)$ becomes time dependent as given by

$$F_H(D) = 2 \exp(-t/\tau_N) - 1. \qquad (25)$$

In order to write eq. (24) in terms of the particle size the form of $f(D)$ must be determined. It has been pointed out by Kaiser and Miskolczy (1970a) that a log-normal distribution is not appropriate for the ferrite particles in hydrocarbon liquids. These conclusions were based on the evidence of electron microscopy. However, Chantrell et al. (1977) have shown that the log-normal distribution is applicable to metallic systems; the conclusion following from a detailed comparison of theoretical and experimental values for the coercivity and remanence of fig. 11 and fig. 12. Good agreement was only observed for a log-normal volume fraction distribution (fig. 12). Further evidence supporting the log-normal distribution has been given by Luborsky (1961). For metallic systems containing iron particles electron microscopy has only limited usefulness because the liquids are opaque and thus particles must be removed for observation. Substantial changes may take place in the size distribution if attempts are made to remove the carrier fluid. Attempts to remove particles directly by placing microscope grids in contact with the fluid can be criticised on the grounds that it is difficult to ensure that one is observing a representative sample of particles since particles are removed only from the surface. A comparison of this method with others suggests that in fact the samples are not representative and

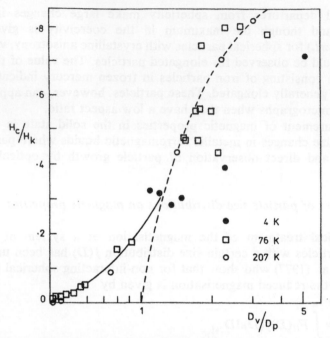

Fig. 11. Reduced coercivity plotted against reduced particle diameter for a log-normal volume
fraction distribution (Chantrell et al. 1977).

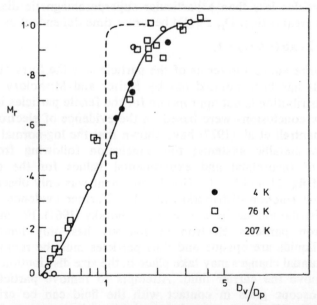

Fig. 12. Reduced remanence plotted against reduced particle diameter for a log-normal volume
fraction distribution (Chantrell et al. 1977).

TABLE 3

Particle diameters determined from magnetic measurements and electron microscopy for iron particles in mercury (Charles and Popplewell 1978).

Method of determination	Particle diameters for samples:		
	1	2	3
Magnetisation curve (293 K)	40–50Å	38–70Å	30–60Å
Coercivity (77 K)	50Å	44Å	44Å
Microscopy	80–800Å	120–240Å	60–200Å

contain a preponderance of the larger particles in the distribution (Charles and Popplewell 1978). Table 3 compares the particle sizes of iron particles in mercury obtained by the different methods.

4.5. Optical properties of ferrofluids

Optical studies have been made on dilute systems containing ferrite particles in water and other non-metallic liquids. Initial studies were made by Heaps (1940) and Elmore (1941) on micron-sized particles. They were able to relate the optical data as measured by the change in transparency of the fluid in a magnetic field to the fluid magnetisation and particle shapes. More recently Neitzel and Bärner (1977) used a similar light scattering technique to determine the particle size and magnetic moment of a ferromagnetic liquid containing 100 Å diameter particles in heptane. Optical studies such as those made by Bibik et al. (1966, 1969) have been generally concerned with the observation of cluster formation in magnetic fields and field gradients. Bibik et al. relate changes in fluid transparency to aggregation processes according to an expression

$$t = t_0 \exp(-\Delta h) \tag{26}$$

where Δh is a time dependent absorbance and t_0 is the transmission in zero field. Further studies by Goldberg et al. (1971) have been concerned with the polarising properties of ferromagnetic liquids when placed in a magnetic field. They consider the polarisation to be related to particle chaining thereby producing optical anisotropy in the liquid. The formation of needle-like chains has also been observed by Hayes (1975). The chains dissociate after the field is removed.

Martinet (1974, 1978) and Davies and Llewellyn (1979) have observed strong birefringence and linear dichroism when the ferrite non-conducting ferromagnetic liquids are subject to a magnetic field. Typically the difference in refractive index, parallel and perpendicular to the field is 1.2×10^{-3} in a field of 1 kOe (80 kA m^{-1}) (see fig. 13) for Fe_3O_4 particles in kerosene. The dichroism and birefringence is quasi-linear for small fields (< 300 Oe, 24 kA m^{-1}). The optical properties of these fluids are such as to make them attractive possibilities for use in light shutters (Martinet 1971).

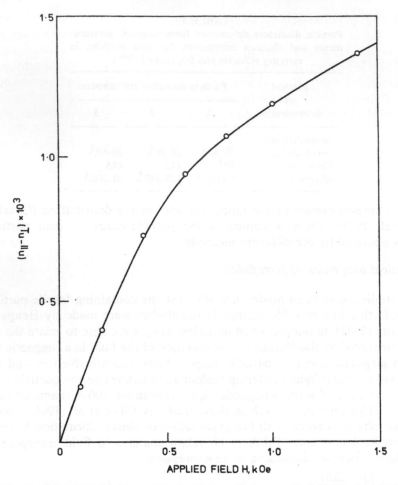

Fig. 13. Optical birefringence plotted against applied magnetic field (Martinet 1978).

5. Stability criterion

5.1. Stability of ferromagnetic particles in non-metallic carrier liquids

Magnetic and London-type Van der Waals interactions may lead to aggregation of particles and hence instability of the ferromagnetic liquid (see section 2). The London-type interactions are such that reducing the size of the particles will not prevent aggregation, unlike the situation for magnetic interactions. It follows from eqs. (4) and (5) that the interaction energy is the same for two similar spheres with the same ratio of diameter to separation. The London potential as plotted in fig. 4 is seen to exceed kT in magnitude for a separation of less than about 0.7 of the radius of the particle. To overcome the attractive interaction some repulsive interaction is required. Rosensweig et al. (1965) have shown that by coating ferrite particles with long chain molecules (e.g. oleic acid) the stability of the colloid is enhanced markedly. This effect is attributed to an increase in stability due to entropic repulsion resulting from steric interaction of the adsorbed molecules which is sufficient to overcome the attractive potential. The repulsion is due to the decrease in the number of possible configurations of the adsorbed long chain molecules as two particles approach one another (Mackor 1951).

Rosensweig et al. (1965) have derived the following expression for the repulsive potential for the interaction between two spheres of equal size

$$E_R = \pi D^2 NkT[b - \ln(1 + b)]/b \tag{27}$$

where E_R is the total repulsion energy, N is the number of adsorbed molecules per unit area, D is the diameter of the uncoated particle and $b = 2\delta/D$, where δ is the thickness of the coating. The cross sectional area of an oleic acid molecule has been determined as 46 Å^2 and the length as 11 Å in studies of surface spreading (Taylor and Glasstone 1951). N can be determined from the cross-sectional area of the molecule and δ from its length.

In fig. 4 the potential energy curves for entropic repulsion and London attraction have been plotted for ferrite particles of 100 Å diameter. The magnetic potential energy is negligible in the case of 100 Å diameter ferrite particles. Figure 4 shows that for a 10 Å adsorbed layer, a potential barrier of approximately $6kT$ is produced which is comparable to the barrier which Verwey and Overbeek (1948) consider suitable to prevent aggregation.

Also shown in fig. 4 is the situation to be expected for a 5 Å adsorbed layer. In this case no barrier is produced. As the size of the particle is reduced, the number of adsorbed molecules becomes insufficient to produce a repulsive force to overcome that of the attractive force. Rosensweig et al. (1965) have shown that for particles of diameter 20 Å with a coating of 10 Å thickness the net potential energy curve hardly differs from that due to London attraction.

The magnetic potential energy is small for ferrite particles up to approximately 100 Å in diameter. However, the magnetic potential energy cannot be ignored when considering larger ferrite particles or ferromagnetic metal particles having large saturation magnetisations. In these situations the net potential energy curve can possess a pronounced minimum. The presence of such a minimum has been put forward as a possible explanation of the transient gelling observed during grinding of ferrite powders. It is clear from fig. 4 that in the case of iron or cobalt particles which possess high saturation magnetisations the presence of an adsorbed layer of 10 Å or even 20 Å will not prevent the presence of a deep potential minimum and hence not prevent aggregation. This is consistent with the observations of Hess and Parker (1966) who observed that for the production of a colloidal suspension of cobalt particles, a polymer was needed which possessed a molecular weight of $100\,000$ and had at least fifty polar groups spaced along the chain which were capable of being adsorbed on the metal particles. Assuming that all polar groups were bonded to the particle such a coating would result in polymer loops emanating from the particle which could extend up to 100 Å, i.e. a distance far in excess of that required to stabilize comparably sized ferrite particles. Equation (27) predicts that an adsorbed layer of 30 Å thickness with $N = 2 \times 10^{18}$ molecules m^{-2} would be sufficient to prevent a potential minimum and hence prevent aggregation of cobalt particles of 100 Å diameter. For larger diameter particles it would appear that the adsorbed layer is not capable of preventing a minimum in the potential energy curve. Because of the occurrence of such a minimum the magnetic attraction leads to the formation of spatial structures such as chains, rings and clusters, Bibik and Lavrov (1965).

Most of the particles shown in fig. 14A have diameters less than 100 Å and show no evidence of aggregation as predicted. Formation of chains (fig. 14B) occur at larger average particle diameters. Cobalt particles shown in figs. 14C and D are in the 200–800 Å range and have chained and aggregated extensively.

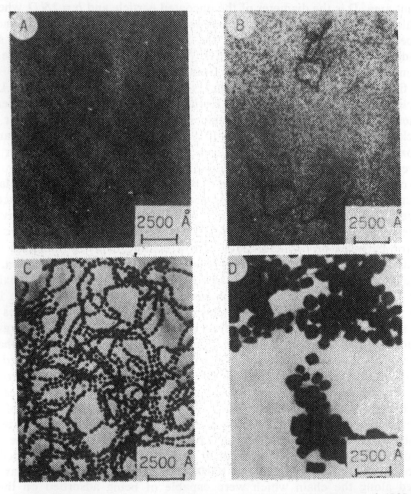

Fig. 14. Electron micrographs of a suspension of cobalt particles (Hess et al. 1966).

5.2. Stability of ferromagnetic particles in liquid metal carriers

The use of polymers and long chain molecules to prevent aggregation of ferromagnetic particles is not appropriate in the case of liquid metal carriers because of the incompatability of hydrocarbon chains and liquid metals. Clearly a different technique needs to be used to introduce a repulsion term in the potential in order to prevent aggregation and hence obtain an ultra-stable suspension.

A further problem, which is not encountered in ferrite and metal suspensions in non-metallic liquid carriers, arises because of diffusion. When dispersed particles of a second phase exist in a saturated solution the larger particles of low surface energy grow at the expense of the smaller particles of high surface energy by the diffusion of atoms through the metal carrier. This arises because of the consequent reduction in total interfacial area. Studies of the growth of small particles of iron (Falk and Luborsky 1965, Luborsky 1957, Windle et al. 1975), cobalt (Luborsky 1958b), cobalt–iron alloys (Luborsky and Paine 1960) and nickel–iron alloys (Shepherd et al. 1972) in mercury have been made by thermally "ageing" the systems at approximately 420 K. The size of the particles in the carrier liquid has been estimated indirectly by adsorption studies and magnetic measurements of the coercivity and remanence of the particles "in situ" and directly for the larger particles by electron microscopy (Luborsky 1962a). Luborsky found that there was good agreement between these three methods. The estimation of particle size and size distribution using magnetic measurements is the subject of section 4. Studies of iron particles in mercury have shown that rapid growth of the particles occurs on "ageing" at 420 K (Windle et al. 1975). This growth is shown in fig. 15.

It has been shown (Luborsky and Opie 1963, Falk and Luborsky 1965, Shepherd et al. 1972, Windle et al. 1975), that the addition of certain metals to such suspensions results in increased stability. A monolayer of adsorbed atoms

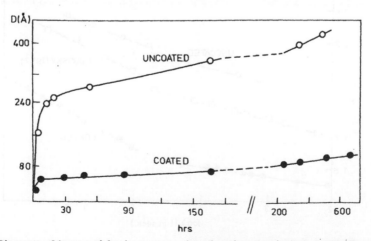

Fig. 15. Diameter of iron particles in mercury plotted against "ageing" time (Windle et al. 1975).

on the iron particles is produced with the effect of substantially reducing the growth rate (Luborsky 1962b). The most effective coating agents are those metals which can form intermetallic compounds with the iron. Addition of these metals over and above the concentration required to form a monolayer leads to an increased layer thickness over a period of time which is long compared to the time for the formation of the first monolayer. This growth was observed by Hoon et al. (1978) who measured the resistivity and specific heat capacity of such systems. The effect on the growth of the iron particles by the addition of tin to the mercury prior to the deposition of iron particles is shown in fig. 15 (Windle et al. 1975). The figure shows the much greater stability of the coated particles although growth is still occurring after many weeks.

Greenwood (1956) has analysed the diffusional growth of a system of particles. He obtained an expression for the rate of growth of the fastest growing particles in the distribution by making use of the Thomson–Freundlich equation (Freundlich 1922),

$$(d\mathscr{D}/dt)_{max} = 4\mathscr{D}S\mathscr{M}\sigma/RT\rho^2 D_m^2 \tag{28}$$

which relates solubility to particle size; D_m is the arithmetic mean diameter of the particles, $\mathscr{D}$ is the diffusion coefficient from the Stokes–Einstein equation, S is the solubility of the metal of the particles in the carrier, σ is the interfacial tension between the particles and carrier, $\mathscr{M}$ is the gram molecular weight of the particles, R is the gas constant and ρ is the density of the material of the particles. This expression leads to a cubed dependence of D_m with time which is in reasonable agreement with experimental observation of the growth of uranium particles in liquid sodium and in liquid lead (Greenwood 1956).

In the case of iron particles (coated and uncoated) in mercury, the results shown in fig. 15 plotted on a log–log scale (fig. 16) seem to point to two regions

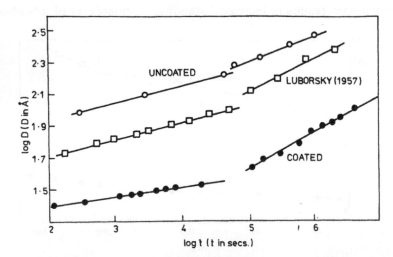

Fig. 16. Plots of log(diameter) against log("ageing" time in seconds) (Windle et al. 1975).

TABLE 4
Growth of particle diameters as a function of ageing time for uncoated and tin coated iron particles in mercury (Windle et al. 1975)

Uncoated	$D^{(9\pm2)} \propto t \ (<10^5 \text{ s})$	$D^{(5.6\pm0.2)} \propto t \ (>10^5 \text{ s})$
Coated	$D^{(20\pm5)} \propto t \ (<10^5 \text{ s})$	$D^{(4.2\pm0.8)} \propto t \ (>10^5 \text{ s})$

of growth (Windle et al. 1975). The relationships for growth in the two regions are shown in table 4. The results of Luborsky (1957) have also been plotted and exhibit a similar trend. There is evidence from electron micrographs (Windle et al. 1975), that growth of individual particles has occurred. There is also evidence of clustering but this must be treated with caution because the particles are studied out of the liquid environment. The results of magnetic measurements of Kneller and Luborsky (1963) on iron particles in mercury after freezing in zero magnetic field are consistent with the presence of a log-normal particle size distribution (Chantrell et al. 1977) (see section 4.4). Thus there is no evidence from these magnetic measurements that clustering takes place in the liquid state in zero magnetic field. It is evident from table 4 that the diameter of the tin-coated particles increases at a slower rate than for uncoated particles in the initial stages of growth. This may be due to a slower diffusional growth expected for coated particles (Falk and Luborsky 1965). A study of the form of the particle size distribution as a function of ageing time should, however, shed light on the mechanism of growth (Granqvist and Buhrman 1976b). London-type Van der Waals forces may in time lead to particle aggregation. Thus a mechanism whereby a particle–particle repulsion can be introduced into the system needs to be found to produce an ultrastable suspension. Such a mechanism may arise because the particle and carrier liquid are metallic with different Fermi energies. It is inevitable therefore that electron charge transfer will take place across the particle–metallic liquid interface. The situation is identical to the formation of a contact potential at the interface between two dissimilar metals. The transfer rapidly reaches an equilibrium situation and the quantity and extent of the charge distributed across the boundary depends on the respective Fermi energies and workfunctions of the two metals. The "space charge" resides around the particle on the metallic liquid and would oppose particle contact if the extent of the charge distributions were appreciable. Calculations of the magnitude and extent of the charge distribution by Popplewell et al. (1976) have been made, based on work by Ehrenberg (1958) and Lang (1971). These calculations indicate that for the "space charge" to be significant the difference in the workfunctions of the particles and carrier liquid must be as large as possible. For a system of tin-coated iron particles in mercury the "space charge" is small because of the workfunctions of the metals involved are similar. Thus a coating of a material with a workfunction very different from mercury could be effective in preventing particle contact.

The addition of metallic sodium to a ferromagnetic liquid of iron in mercury

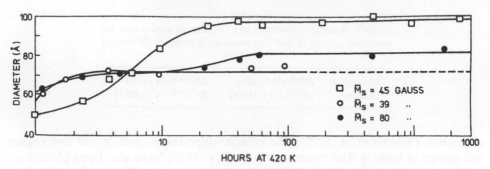

Fig. 17. Sodium stabilization of tin-coated iron particles in mercury (Popplewell et al. 1976).

containing tin has produced a ferromagnetic liquid in which the size of the particles remains unchanged when aged at 420 K over a period of a year (see fig. 17).

5.3. Stability of ferromagnetic liquids in the presence of a gravitational field

For a system of single sized spherical particles suspended in a liquid, undergoing Brownian motion in a gravitational field, the drift of particles downwards must be balanced by the diffusion upwards at equilibrium. In this situation the number of particles as a function of height z is given by

$$n = n_0 \exp - (\pi D^3 (\rho - \rho_c) gz/6kT) \qquad (29)$$

where ρ is the density of the material of the particles with its coating, ρ_c is the density of the carrier liquid and D is the diameter of the particles. The concentration gradient is thus given by

$$dn/dz = -\pi D^3 (\rho - \rho_c) gn/6kT. \qquad (30)$$

For the narrow particle size distributions as found in stable ferromagnetic liquids the above equations represent an adequate description of the concentration gradient. The "usefulness" of a liquid as defined by Rosensweig et al. (1965) is set by the value of $n^{-1}dn/dz$. Above a certain value the concentration of particles would be sufficient to cause the viscosity to be so large that the liquid would have limited use. A value of $n^{-1}dn/dz = 1$, i.e. a 100% change in concentration in 1 cm which leads to an upper value of 400 Å for the diameter of the particles for any useful liquid. The results of studies by Peterson et al. (1974) on a water based ferrofluid are in agreement with eq. (29) and are consistent with the absence of any large aggregates.

5.4. Stability of ferromagnetic liquids in the presence of a magnetic field gradient

A similar expression to that in eq. (29) may be derived for the concentration gradient produced in the presence of a magnetic field gradient by replacing $(\rho - \rho_c)g$ by $M_s dH/dz$ where M_s is the saturation magnetisation and dH/dz is the field gradient. The gravitational contribution may be neglected for $M_s dH/dz \gg (\rho - \rho_c)g$. Thus

$$dn/dz = -D^3 M_s n (dH/dz)/24kT$$

where D remains the overall diameter of the particle and coating.

Figure 18 indicates the upper limit of particle diameter as a function of saturation magnetisation and applied field gradient using the criterion of stability that $n^{-1} dn/dz \leq 1$. Thus in a field gradient of $10^3 \, \text{Oe cm}^{-1}$ ($8000 \, \text{kA m}^{-2}$), for which the magnetic force is very much greater than the gravitational force, the diameter of Fe_3O_4 particles should be less than 60 Å to satisfy this criterion.

The actual time required for the system to achieve equilibrium has been ignored in this treatment and that in section 5.3. Liquids containing particles with diameters greater than 60 Å may be considered stable in much larger gradients provided that the time to reach equilibrium is much longer than the time that the system is being subjected to the magnetic or gravitational force. This would also apply to a system that is disturbed mechanically. Fe_3O_4 particles with diameters as large as 200 Å would require approximately one hour, to reach the equilibrium state. The time to reach equilibrium as a function of particle diameter and magnetisation is shown in fig. 19. Hoon et al. (1978) show that for iron particles in mercury equilibrium is reached almost instantaneously whereas times of hours are predicted from fig. 19. Studies by Peterson et al. (1974) on a water based ferrofluid indicate that in uniform magnetic fields (2.5–200 Oe, 0.2–16 kA m^{-1}) where there are no external magnetic forces present to induce the particles to move, the particles settle much more quickly than in a gravitational field alone. The rate at which sedimentation occurs is consistent with the presence of aggregates of about 5×10^4 Å in diameter. Martinet (1978) has also shown from electron microscopy that aggregates of particles occur for cobalt particles in toluene and for Fe_3O_4 particles in water in a uniform magnetic field. These

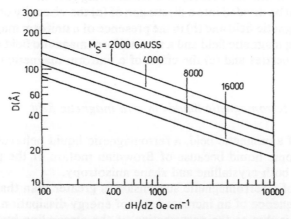

Fig. 18. Upper limit of particle diameter for stability plotted against particle saturation magnetisation and applied magnetic field gradient for $n^{-1} dn/dz < 1$ (Rosensweig et al. 1965).

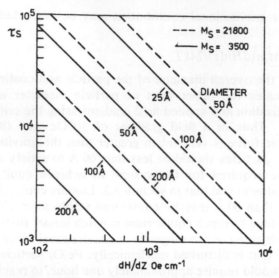

Fig. 19. Time for magnetic particles to achieve an equilibrium concentration distribution in a magnetic field gradient plotted against particle diameter and saturation magnetisation (Rosensweig et al. 1965).

aggregates were needle shaped ($\sim 10^5$ Å). Thus the rapid sedimentation observed by Hoon et al. (1978) can be accounted for by the presence of large aggregates of iron particles.

6. Ferrohydrodynamics of ferromagnetic liquids

The ferrohydrodynamic properties of ferromagnetic liquids have been reviewed by Bertrand (1970), Shliomis (1972) and Rosensweig (1978).

This section will be divided into a discussion of (a) the viscosity of ferromagnetic liquids in zero magnetic field and (b) in the presence of a uniform magnetic field, the effect of a rotating magnetic field and a non-uniform magnetic field on the physical and magnetic properties and (c) the effect of a uniform magnetic field on surface stability.

6.1. Viscosity of ferromagnetic liquids in zero magnetic field

In the absence of a magnetic field, a ferromagnetic liquid behaves macroscopically like an isotropic liquid because of Brownian motion of the particles which averages to zero both crystalline and shape anisotropy.

The viscosity of a ferromagnetic suspension is greater than that of the carrier liquid as a consequence of an increased rate of energy dissipation during laminar shear flow. This is due to the perturbation of the streamline by the suspended particles. A rotation is imparted to particles situated in a Newtonian liquid undergoing homogeneous shear flow due to the velocity difference of the layers of liquid forming the flow. In this situation, the angular velocity of the particle ω is equal to $\frac{1}{2}L$ perpendicular to the plane of shear, where L is the velocity gradient or shear rate. Its translational velocity is equal to the liquid velocity at a plane in the shear field coincident with the centre of the particle. For dilute suspensions Einstein (1906, 1911) showed that the dependence of the viscosity of a suspension on the volume fraction may be represented by

$$\eta = \eta_0(1 + 2.5\epsilon) \tag{31}$$

where ϵ is the volume fraction and η_0 the viscosity of the carrier liquid. Einstein assumed that the particles were spherical, uncharged, small compared with the dimensions of the measuring apparatus and large compared with molecules of the carrier liquid. These assumptions describe the situation to be found for most ferromagnetic suspensions. At volume fractions greater than 0.01, higher order terms in ϵ have to be introduced into eq. (31). This leads to an increase of viscosity of the suspension which is attributed to the formation of clusters enhancing the rate of energy dissipation. A review of the large number of empirical and theoretical equations that have been used to describe viscosities at higher volume fractions is not appropriate here. A recent review of the subject has been presented by Goodwin (1975) but no data is included for magnetic suspensions. However, some data on the viscosity of ferrite suspensions in liquids has been published by Rosensweig et al. (1965), Kaiser and Miskolczy (1970b) and Ferrofluidics Corporation.

6.2. Viscosity of ferromagnetic liquids in a uniform magnetic field

On application of a magnetic field, an anisotropy is introduced which is revealed by studies of the optical, magnetic and viscous properties of the liquid.

McTague (1969) measured the viscosity of a dilute suspension of polymer-coated cobalt particles in toluene in a magnetic field using a capillary viscometer. Measurements of the effect of the magnetic field parallel to the direction of flow through the capillary, i.e. parallel to the shear planes, and also perpendicular to the flow were made. The results of these measurements, which are given in fig. 20, show that in both situations the viscosity increases with increasing field approaching an asymptotic value at 6 kOe (480 kA m^{-1}) at 300 K for 60 Å diameter particles. In the case of the magnetic field parallel to the shear plane, the increase in viscosity is approximately twice that for the perpendicular field orientation. The latter case is complicated in that the angle between the field direction and shear planes assumes all values from 0 to 2π. A simple qualitative explanation can be used to explain the results. The introduction of a magnetic field introduces a torque $\mu \times H$ which hinders the rotation of the particle about axes perpendicular to the magnetic field. As a result an additional frictional coupling between fluid layers is introduced thereby increasing the viscosity. For this effect to be significant the mechanism of relaxation of the magnetisation must be determined by the Brownian rotational diffusion time τ_B which is a characteristic of the viscosity of

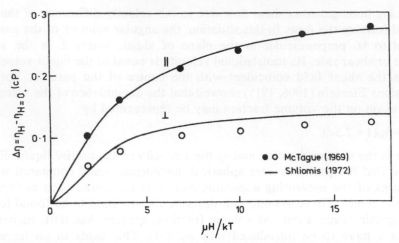

Fig. 20. Change of viscosity of a ferromagnetic liquid plotted against applied magnetic field.

the liquid. Thus the relaxation time of the magnetisation due to Brownian motion must be significantly less than the Néel relaxation time τ_N. This condition is met in the ferromagnetic liquids used by McTague (1969) where the mean diameter of the particles is greater than the critical diameter D_p for intrinsic super-paramagnetism. The increase in viscosity should be most marked when the magnetic field is parallel to the direction of flow. At this orientation the torque on the particles is most effective in hindering the rotation imparted to the particles by the homogeneous shear flow. The torque N producing this additional frictional coupling is given by

$$N = \pi D^3 \eta (\tfrac{1}{2} L - \omega).\tag{32}$$

This coupling is greatest when $\omega = 0$ i.e. the particles are no longer rotating. This should occur according to Hall and Busenberg (1969) when

$$\mu \times H > \tfrac{1}{2}\pi D^3 \eta L.\tag{33}$$

This conclusion is not in accord with the experimental results of McTague (1969) who found that magnetic fields of the order of 10^3 Oe (80 kA m^{-1}) were needed to achieve the condition above compared with the predicted value from eq. (33) of approximately 1 Oe (80 A m^{-1}). The explanation for this is that the orienting effect of the magnetic field is opposed by the hydrodynamic forces and the rotational Brownian motion. In order to overcome the Brownian motion, and achieve a saturation viscosity, the magnetic field must be such that $\mu H \gg kT$.

Brenner (1970), Brenner and Weissman (1972), Levi et al. (1973) and Shliomis (1972) have carried out a theoretical analysis of this magnetoviscous effect in dilute suspensions and have derived independently similar expressions for the viscotity as a function of external field. The analyses are valid for all values of the ratio $D^3 \eta L/2kT$.

Shliomis (1972) has obtained expressions for the relaxation times of the transverse and longitudinal components of the magnetisation as a function of field intensity. The relaxation time upon which the viscosity depends is $\tau_\perp$, the relaxation time of the magnetisation perpendicular to the field.

For $\mu H \ll kT$, i.e. in weak fields $\tau_\perp \approx \tau_B$. In this situation the contribution to the viscosity η_r due to the rotational component is given by

$$\eta_r = \tfrac{1}{4}\eta\epsilon(\mu H/kT)^2 \sin^2 \theta \tag{34}$$

where ϵ is the volume fraction, θ is the angle between H and L and η is the viscosity of the suspension in zero field. For $\mu H \gg kT$

$$\tau_\perp \approx 2\tau_B(\mu H/kT)^{-1}. \tag{35}$$

In this case η_r achieves the limiting value of

$$\eta_r = \tfrac{3}{2}\eta\epsilon \sin^2 \theta. \tag{36}$$

By including the rotational component in the Einstein formula and considering a Poiseuille flow as in the experiment of McTague (1969), Brenner and Weissman (1972) and Shliomis (1972) showed that the saturation values of the viscosity parallel ($\eta_\parallel$) and perpendicular ($\eta_\perp$) to the field direction are given by

$$\eta_\parallel = \eta(1 + \tfrac{3}{2}\epsilon), \qquad \eta_\perp = \eta(1 + \tfrac{3}{4}\epsilon) \tag{37}$$

where η is given by eq. (31). Thus $(\eta_\parallel - \eta)/(\eta_\perp - \eta) = 2$, which is consistent with the results obtained by McTague (1969). Shliomis (1972) obtained good agreement between theory and the experimental observations of McTague (1969) as illustrated in fig. 20.

Rosensweig et al. (1969) have studied the viscosity of commercial ferromagnetic suspensions as a function of applied magnetic field. The viscosity of the liquid in a magnetic field η_H is predicted by dimensional analysis to be a function of the ratio of the hydrodynamic stress to the magnetic stress $L\eta_0/\bar{M}H$, where L is the shear rate, η_0 is the viscosity of the carrier liquid, $\bar{M}$ is the magnetisation of the fluid and H the applied field. The prediction was verified by measurements of viscosity using a cone-and-plate viscometer placed in a solenoid so that the magnetic field was perpendicular to the thin layer of ferromagnetic liquid. The results of this investigation are shown in fig. 21. For values of $L\eta_0/\bar{M}H$ greater than 10^{-4} there is little change of viscosity with field. For values of $L\eta_0/\bar{M}H$ between 10^{-6} and 10^{-4} a transition occurs where the viscosity is a function of both magnetic field and shear rate. Theoretically, for high values of applied field, the viscosity again becomes independent of field and shear rate. The scatter of points in fig. 21 has been attributed by Rosensweig et al. (1969) to the fact that in their analysis no account has been taken of the solvated stabilizing sheath which influences the hydrodynamic behaviour of the particle. The solid curve in fig. 21 represents the predicted relationship for particles which have an effective sheath thickness to particle diameter of 0.35. On increasing the ratio to 0.45, the transition occurs when $L\eta_0/\bar{M}H$ lies between 10^{-6} and 10^{-8}; a significant shift of the curve from the experimental data.

Measurements of the magnetic properties of liquids similar to those used by

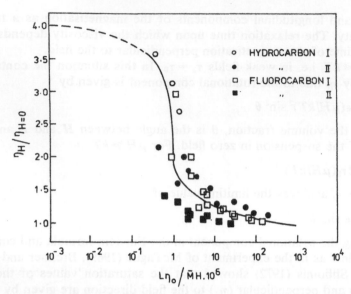

Fig. 21. Effect of magnetic field and shear rate on the viscosity of ferromagnetic liquids containing Fe$_3$O$_4$ particles (Rosensweig et al. 1969).

Rosensweig et al. (1969) have been made by Kaiser and Miskolczy (1970a). In addition, measurements were made on a suspension of the same Fe$_3$O$_4$ particles in paraffin wax, where particle rotation cannot occur. The same magnetisation curves are observed for a given assemble of particles dispersed in kerosene and paraffin wax. Thus the magnetisation of the wax dispersion is due to Néel rotation. However, the measurements do not allow the mode of the relaxation, whether Néel or Brownian, in the liquid state to be determined. Radiofrequency measurements by Curtis (1968) and Mossbauer spectroscopy by Winkler et al. (1977) seem to indicate that Néel relaxation is dominant in the liquid state.

The factors giving rise to the field dependence of the viscosity in the concentrated suspension studied by Rosensweig et al. (1969) are more complicated than those in the systems studied by McTague (1969). In concentrated suspensions there is evidence (Bibik and Lavrov 1965, Peterson et al. 1974) that clustering takes place in the presence of a magnetic field. The situation is further complicated by the fact that all shear planes are subjected to an effective rotating magnetic field, the frequency of which varies from zero to a maximum through the thin ferromagnetic sample in the viscometer.

Clusters containing a large number of particles would be expected to have a loose structure so that entrapped liquid would enhance the particle volume and thereby increase the viscosity. Increase in the field strength is expected to lead to an increase in clusters and a corresponding increase in viscosity. Clustering is probably responsible for the non-Newtonian behaviour of the liquid, particularly at high field strengths where a decrease in viscosity occurs on increasing the shearing rate. This behaviour is not predicted by Brenner and Weissman (1972) in their theoretical treatment even though the magnetoviscous effects of dilute suspensions are adequately explained. This discrepancy probably arises because clustering due to magnetic interactions has been ignored. Even the presence of clusters consisting of two particles leads to a behaviour in a shear field quite different from that of isolated spherical particles (Krieger 1959, Goldsmith and Mason 1967).

6.3. *Ferromagnetic liquids in rotating magnetic fields*

The effect of a rotating magnetic field on a ferromagnetic liquid has been investigated by several groups of workers (Moskowitz and Rosensweig 1967, Jenkins 1971, Laithwaite 1968, Brown and Horsnell 1969, Mailfert and Martinet 1972, Shliomis 1974, Martinet 1978, Bibik and Simonov 1972, Bibik and Skobochkin 1972).

The macroscopic motion of a ferromagnetic liquid under the influence of a uniform rotating magnetic field depends on the relaxation mechanisms of the magnetisation as described in section 4.1.

In the limiting case of $\tau_B \ll \tau_N$, the relaxation time will be governed by the Brownian rotational motion and each particle will begin to rotate at the same angular velocity Ω as the rotating field but lagging in phase by an angle ξ. For $\mu H/kT \ll 1$, the angle ξ is given by

$$\tan \xi \approx \Omega \tau_B \tag{38}$$

and for $\mu H/kT \gg 1$

$$\tan \xi \approx 2\Omega \tau_B kT/\mu H. \tag{39}$$

The liquid in the immediate vicinity of the particle will itself rotate so that each particle of the suspension will become a centre of a microscopic vortex motion. By averaging the vortex motion of all particles, each particle rotating with the same angular velocity, macroscopic immobility results. Only a thin layer of the carrier liquid between the outermost particles and the liquid/container boundary will have a resultant motion.

In the other limiting case of $\tau_B \gg \tau_N$, a situation in which the magnetic moment can move freely within the domain, the external field will influence the macroscopic motion only when the magnetic field is inhomogeneous.

Mailfert and Martinet (1972) have made a study of the macroscopic properties of two ferromagnetic liquids in the presence of a magnetic field of about 100 Oe (8 kA m^{-1}) rotating at frequencies up to several hundred Hz. One liquid consisted of a suspension of CrO_2 particles acicular in shape about 100 Å wide and a few 1000 Å long in water, toluene or alcohol. The other consisted of a suspension of cobalt particles in toluene, prepared by the thermal decomposition of cobalt octacarbonyl (Thomas 1966b, Hess and Parker 1966). It is expected from eqs. (7) and (8) that $\tau_B \ll \tau_N$ in these systems. It was found that for a uniform rotating field no net bulk rotation of the liquid was observed, as predicted. For highly concentrated suspensions, a marked rotation of the outer layers of the liquid occurred which rapidly decreased towards the centre of the liquid. This effect has been attributed to a concentration gradient formed as a result of interparticle attractive forces which increases the concentration at the centre of the liquid at the expense of that at the boundary. The rotation was in the positive sense, i.e. the rotation was observed to be in the same direction as the rotation of the field. A similar type of explanation may account for the results of Moskowitz and Rosensweig (1967), Bibik and Simonov (1972), and Bibik and Skobochkin (1972), who observed bulk rotation of fluids consisting of high volume fractions of ferrite particles. In such systems clustering due to attractive forces will lead to inhomogeneities in the local field configuration resulting in bulk rotation. Whenever there is a spatial variation of parameters, such as concentration, viscosity and magnetic field, bulk rotation of the liquid is likely to occur.

Martinet (1978) has made measurements of the components of the magnetisation in the plane of rotation of the magnetic field, perpendicular ($\bar{M}_\perp$) and parallel ($M_\parallel$) to the field direction. The measurements have been made on systems for which it has been calculated from eqs. (7) and (8) that $\tau_B \ll \tau_N$ for cobalt particles in toluene and $\tau_B \gg \tau_N$ for ferrite particles in kerosene.

For $\tau_B \ll \tau_N$ and $\mu H \ll kT$, the components of magnetisation are given by Shliomis (1972, 1975) as follows

$$\bar{M}_\parallel = 4\pi\bar{\chi}_i H/[1 + (\Omega\tau_B)^2] \tag{40}$$

$$\bar{M}_\perp = 4\pi\bar{\chi}_i H\Omega\tau_B/[1 + (\Omega\tau_B)^2] \tag{41}$$

with

$$\Omega\tau_B = \bar{M}_\perp/\bar{M}_\parallel = \tan \xi \ll 1. \tag{42}$$

For $\tau_B \gg \tau_N$ and $\mu H \ll kT$ the same expressions hold except that τ_B is replaced by τ_N.

For $\tau_B \ll \tau_N$ and $\mu H \gg kT$ the viscous torque exerted by the carrier liquid is counteracted by the magnetic torque giving

$$M_s \times H = 24\pi\eta\Omega \tag{43}$$

with the magnetisation lagging behind the rotating field. For cobalt particles ($M_s = 18\,172$ G, 1400 kA m^{-1}) in toluene ($\eta = 10^{-2}$ P) in a rotating field of frequency 1 kHz and 10^4 Oe (800 kA m^{-1}) the angle of lag ξ is very small ($\sim 10^{-6}$ rad). Thus a very small component of the magnetisation perpendicular to the field is to be expected. As the field strength is increased ξ becomes smaller and therefore the component of the magnetisation parallel to the field increases at the expense of the perpendicular component. Figure 22 shows the variation of $\bar{M}_\perp$ with the field strength from zero field up to high fields. For high fields the $1/H$ dependence of $M_\perp$ is clearly shown. A linear variation of the ratio $\bar{M}_\perp/\bar{M}_\parallel$ as a function of Ω was observed for low and high fields as predicted from eqs. (42) and (43). In the low field case the ratio $\bar{M}_\perp/\bar{M}_\parallel$ was observed to be independent of Ω although it is predicted from eqs. (40) and (41) that it should be proportional to Ω. In the high field case the value of the viscosity was calculated to be

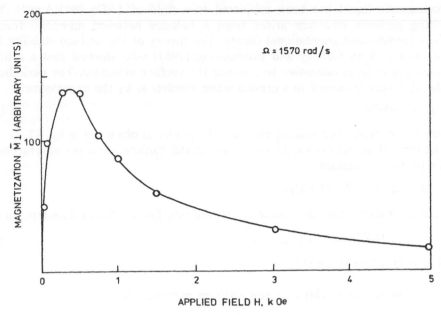

Fig. 22. Perpendicular component of the magnetisation plotted against magnetic field for a ferromagnetic liquid under rotation (Martinet 1978).

300 cP from experiment instead of 1 cP as expected. Martinet (1978) suggested
that these discrepancies were a result of the presence of clusters and surfactant.

6.4. Surface stability of a ferromagnetic liquid in a uniform field

In addition to the usual properties of liquids, ferromagnetic liquids have a ready
response to magnetic fields. The interface between the ferromagnetic liquid and
some other liquid or gas has a characteristic structure when subject to a uniform
magnetic field normal to the interface. This is shown in fig. 23 where a

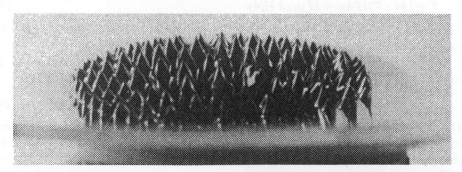

Fig. 23. Characteristic structure of the surface of a ferromagnetic liquid consisting of cobalt particles
in toluene, when subject to a uniform magnetic field normal to the surface.

ferromagnetic liquid has been subjected to a field of 1 kOe (80 kA m^{-1}). The resulting periodic structure arises from a balance between magnetic forces, surface tension and gravitational forces. The theory of the surface stability has been discussed by Cowley and Rosensweig (1967) who showed that a critical magnetisation $\bar{M}_c$ is necessary to produce the surface structure. The periodicity of the structure is related to a critical wave number k_c by the expression

$$l_c = 4\pi/3^{1/2}k_c \tag{44}$$

where l_c represents the spacing between the peaks as observed in fig. 23.

The critical magnetisation $\bar{M}_c$ below which the surface remains unperturbed is given by the expression

$$\bar{M}_c^2 = 8\pi(g\Delta\rho S)^{1/2}(1 + 1/r) \tag{45}$$

where $\Delta\rho$ is the density difference across the interface, S the surface tension and

$$r = (B_0/H_0)^{1/2}(\partial B/\partial H)_{H=H_0}^{1/2} \tag{46}$$

and k_c is given by $(g\Delta\rho/S)^{1/2}$.

6.5. Ferromagnetic liquids in a non-uniform magnetic field

Magnetic particles acquire an increased velocity when exposed to a field gradient. The terminal velocity is determined by the value of the field gradient and the viscosity of the fluid. It is small ($\sim 10^{-5}$ cm s^{-1}) in a field gradient of 10^3 Oe cm^{-1} (8000 kA m^{-2}) assuming a liquid viscosity of a few centipoise. Thus the relative motion of the particles with respect to the liquid motion can be ignored for most calculations. In addition to the translational motion of the particles, the particles experience a torque which can, however, be neglected if $KV \ll kT$.

The forces acting on the particles in a field gradient are transmitted to the liquid carrier. Hence the force F per unit volume on a non-conducting ferromagnetic liquid is found by summing the forces on the individual particles, thus giving in terms of the magnetisation of the suspension $\bar{M}$ and the field gradient ∇H

$$F = (\bar{M} \cdot \nabla)H/4\pi = \bar{M} \text{ grad } H/4\pi. \tag{47}$$

For most calculations on the macroscopic behaviour of ferromagnetic liquids the liquid may be considered a true continuum.

The magnetic term is written into the hydrodynamic equation for an incompressible ferromagnetic liquid (Neuringer and Rosensweig 1964), to give

$$\rho(\partial v/\partial t + (v \cdot \nabla)v) = -\nabla p - \nabla\psi + \eta\nabla^2 v + \bar{M}\nabla H/4\pi. \tag{48}$$

The first term on the right hand side of eq. (51) represents the pressure gradient, the second term the gravity force where $\psi = \rho gh$, the third term the viscous force and the final term the magnetic force.

The continuity of mass equation gives

$$\nabla \cdot v = 0. \tag{49}$$

An associated energy equation can also be derived but this need not be considered for the derivation of a generalised Bernouilli's equation. For irrotational flow eq. (49) reduces to the general Bernouilli's equation

$$\rho \partial \phi / \partial t + p + \tfrac{1}{2}\rho v^2 + \rho g h - \int_0^H \bar{M} \nabla H / 4\pi = g(t) \tag{50}$$

where $\nabla \times v = 0$, $v = -\text{grad } \phi$ and ϕ is a velocity potential. For time invariant flow $\partial \phi / \partial t = 0$ and $g(t) = \text{constant}$. $\bar{M}$ is given by eq. (6). Hence

$$p + \tfrac{1}{2}\rho v^2 + \rho g h - 24\epsilon k T \ln(\sinh \alpha / \alpha) / 4\pi D^3 = \text{const.} \tag{51}$$

Equation (51) is important in determining velocity and pressure distributions in non-uniform magnetic fields and thus is relevant in determining pressure differences in ferrofluid seals and the pressures produced in the type of energy convertor described in section 7.

Equation (51) also indicates that a non-magnetic body immersed in a ferromagnetic liquid would experience a magnetic force, in addition to that due to buoyancy, in a direction towards minimum field. Any non-magnetic body may, therefore, be made to float in a magnetisable liquid provided the magnetisation and field gradient are of the correct magnitude. A detailed analysis of the levitation force F' acting on a spherical body in a uniform field gradient is given by Rosensweig (1978). For a spherical body of volume V in a fluid of constant susceptibility $\bar{\chi}$

$$F' = -3V(1 + \bar{\chi})\bar{M} \text{ grad } H / 4\pi(3 + 2\bar{\chi}) \tag{52}$$

and for $\bar{\chi} \to 0$

$$F' = -V\bar{M} \text{ grad } H / 4\pi. \tag{52b}$$

Bodies may be levitated under this force in a magnetic field thereby circumventing Earnshaw's theorem (1842) which states that stable levitation of bodies in static fields is impossible.

7. Energy conversion

Ferromagnetic liquids may be used for energy conversion in a system such as that described diagrammatically in fig. 24. The magnetic liquid is heated in a magnetic field gradient to a temperature T_2 greater than the Curie temperature T_c of the ferromagnetic particle content thereby providing a temperature difference $T_2 - T_1$ within the field gradient. T_1 is the sink temperature which is less than T_c. Under these conditions the particles, with magnetisation M, experience a force $M\nabla H$ which is transmitted to the liquid causing it to flow. Heat for regeneration is removed in the heat exchanger and the overall cycle converts heat energy to mechanical energy as the liquid flows.

The details of the system shown in fig. 24 have been examined by Resler and

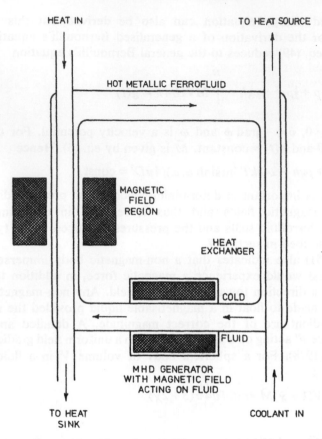

Fig. 24. A diagrammatic representation of an energy conversion system employing a metallic ferromagnetic liquid.

Rosensweig (1964) and Rosensweig et al. (1965). They showed that the power to weight ratio, an important factor in the design of a portable device, is given by

$$W/P = [\eta'/(\eta'_c - \eta')](d_t/H)^2/Z. \tag{53}$$

In eq. (53) W refers to the weight of magnetic liquid, H is the applied magnetic field, d_t the diameter of the tube, η'_c the Carnot efficiency, η' the overall efficiency and Z a parameter called the "figure of merit" which in turn is given by

$$Z = (\partial M/\partial T)^2 \epsilon^2 \kappa T_c/16\pi^2 C_0^2 \rho \tag{54}$$

where ϵ is the packing fraction, κ the thermal conductivity, T_c the Curie temperature, C_0 the volume specific heat, ρ the density and $\partial M/\partial T$ is the pyromagnetic coefficient. For space applications the ratio W/P should be small and Z large. It follows from an analysis of the parameters involved in eq. (54) that an attractive ferromagnetic liquid for energy conversion would consist of ferromagnetic particles of gadolinium in liquid sodium–potassium alloy, fig. 25. The thermal conductivity of the carrier should be kept high to maintain a small W/P ratio. In addition, the performance of the heat exchanger would be improved by using a liquid with a high thermal conductivity. Metallic ferromagnetic liquids with large κ enable power to be generated using magnetohydrodynamical principles. A low voltage, high current generator then becomes feasible and a single magnet or solenoid can be used for both MHD generation and magnetic pumping.

The efficiency of a system without regeneration such as that shown in fig. 24 would be

$$\eta' = [(\partial M/\partial T)\epsilon H(T_2 - T_1)]/[4\pi C_0(T_2 - T_1) - (\partial M/\partial T)\epsilon HT_2]. \tag{55}$$

However, in order to use regeneration a temperature difference must be maintained between the coolant in the heat exchanger and the hot ferromagnetic liquid. If the same ferromagnetic liquid flows through a counter current heat exchanger as in fig. 24, an extra amount of heat $2C_0\Delta T$ is required to maintain

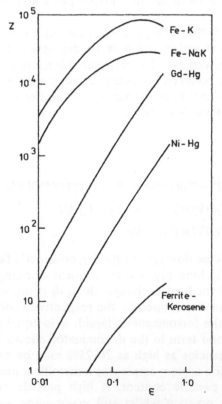

Fig. 25. Z (figure of merit) plotted against packing fraction ϵ for various ferromagnetic liquids (Rosensweig et al. 1965).

the heat flow and the efficiency now becomes

$$\eta' = \eta'_c/(1 + 8\pi C_0 \Delta T/(\partial M \partial T) \epsilon H T_2). \tag{56}$$

It is also necessary to take into account friction losses between the fluid and the tube wall for turbulent flow. The frictional force is given by

$$F_t = f\rho v^2 \pi d_t l \tag{57}$$

where f is the friction coefficient and l is the length of the regenerative section of the tube and v is the liquid velocity. If the fractional power necessary for pumping is known the overall length of the regenerative section can be determined for a specific value of W/P. It was shown by Resler and Rosensweig (1964) that for an iron based ferromagnetic liquid, a W/P ratio of 11 lbs/kW is possible when $\eta'_c = 25\%$, $\eta' = 15\%$, $H = 50$ kOe (4000 kA m^{-1}) and $d_t = 0.1$ cm. With $f = 0.01$, $l = 141$ cm, $W = 382$ lbs the power output would be 35 kW. It is possible by tolerating lower overall efficiencies to improve on the figure of 11 lb/kW thus making the comparison with turbo-alternators ($W/P \sim 20$ lbs/kW) attractive. Further, such an energy conversion system as described in fig. 24 is extremely reliable since there are no moving parts and thus is ideal for space applications.

The possible use of this energy conversion system for other than space applications is a subject which will now be discussed. Franklin (private communication, AERE, Harwell, UK) has considered in a more complete treatment the efficiency of an energy conversion system utilising a flat rectangular duct rather than circular tubes in the heat exchanger to minimise frictional losses. The efficiency of such a system is given by

$$\eta' = A/(B + C)$$

where

$$A = (\partial M/\partial T)\epsilon H(T_2 - T_1)(\partial V/\partial t) - 2\pi\rho(\partial V/\partial t)^3 fNl/A^2 d_t$$
$$B = (\partial M/\partial T)\epsilon H T_2(\partial V/\partial t) + 4\pi\alpha\kappa(T_2 - T_1)A/l \tag{58}$$
$$C = 4\pi C_0^2(T_2 - T_1)(\partial V/\partial t)^2 d_t^2/3Al\kappa$$

where $\partial V/\partial t$ is the volume flow rate of the ferromagnetic liquid, N is the ratio of the length of the liquid loop (fig. 24) to the heat exchanger length l, A is the cross-sectional area of the heat exchanger duct, d_t is the small dimension of the duct and α is the factor determined by the respective conductivities of the heat exchanger liquid and the ferromagnetic liquid. α is equal to 4 when the liquids are identical. The second term in the denominator allows for conduction losses along the liquid. Efficiencies as high as 20–25% may be expected from eq. (58) for an energy conversion system employing a metallic liquid of high conductivity and a high magnetic particle content. A high particle concentration may be difficult to achieve in practice whilst still maintaining a reasonable viscosity. High magnetic fields ($\sim$100 kOe, 8000 kA m^{-1}) are required to achieve these efficiencies. Thus, the use of superconducting solenoids and liquid helium is essential. The reliability and low weight to power ratio offered by the device are overriding advantages.

In the preceding paragraphs an outline of the use of metallic ferromagnetic liquids in an energy conversion system has been given. However, it should be pointed out that this application would only be possible if ultra-stable metallic ferromagnetic liquids could be prepared. These liquids need to possess high stability at high temperatures and in high magnetic field gradients. A further problem to be considered is that of the growth of the ferromagnetic particles by diffusion, a process which would ultimately give rise to large particles which would settle out of the liquid by means of their buoyancy. This problem, discussed in section 5.2, has been tackled by using particle coatings of tin (Luborsky and Opie 1963, Windle et al. 1975). The problem of preventing aggregation due to magnetic attraction has been discussed by Popplewell et al. (1976) (see section 5.2).

Studies of the stability of these liquids at high temperatures and in high magnetic fields have yet to be made. The chemical reactivity of liquid metals could present serious problems in the design of an energy convertor. However, problems of this nature already form part of nuclear reactor technology and should not, therefore, be insoluble. A problem not yet resolved but discussed in section 3.2 is that concerned with producing magnetic liquids with a high concentration of ferromagnetic particles. Mercury containing 1 wt% of iron in mercury may be liquid but at a 2.5 wt% concentration the liquid can become a viscous slurry.

The question of liquid stability in strong magnetic fields can be directly related to the problems discussed in the previous paragraph since if the particle concentration is likely to change on the application of a magnetic field the concentration limit may be reached causing the liquid to form a slurry. In a relatively low field gradient $10^3 \, \text{Oe cm}^{-1}$ ($8000 \, \text{kA m}^{-2}$) magnetic separation becomes evident (Hoon et al. 1978), so that the use of these liquids in an energy conversion system is not yet possible.

In concluding, it would appear that there are a great number of problems to be examined before ferromagnetic liquids can be used in devices employing high magnetic fields. The problems outlined in this section are concerned with energy conversion systems but will be relevant in any device utilising metallic liquids.

List of symbols

a	damping parameter	D_s	Shliomis particle diameter
B_s	Brillouin function		
C_0	volume specific heat	D_V	particle median diameter of log normal distribution
$\mathscr{D}$	diffusion coefficient		
D	particle diameter		
D_p	critical diameter for onset of superparamagnetism	F	force per unit volume on non-conducting ferromagnetic liquid

F' levitation force

f_0 frequency factor

g acceleration due to gravity

H applied magnetic field

H_c coercive force

I_s saturation magnetisation per unit volume

J exchange integral

K anisotropy energy per unit volume

k Boltzmann constant

k_c critical wavenumber for surface instability

L velocity gradient

l_c spacing between peaks resulting from surface instability

$\bar{M}_s$ saturation magnetisation per unit volume for a ferromagnetic liquid

$M_s = 4\pi I_s$ saturation magnetisation per unit volume of solid content

$M_0 = M/M_s$ reduced magnetisation

M_r remanent magnetisation per unit volume

$\mathcal{M}$ gram molecular weight

n number of particles per unit volume

N number of atoms adsorbed per unit area

N torque

P power (kW)

p pressure

R gas constant

S spin Q.N., surface tension and solubility

T temperature

T_c Curie temperature

U potential energy

V particle volume

V_s Shliomis particle volume

V_p particle volume for onset of superparamagnetism

W weight

Z "figure of merit" for energy conversion system

γ gyromagnetic ratio

δ particle coating thickness

ϵ packing volume fraction

η viscosity of suspension in zero field

η_0 viscosity of carrier liquid

η_r rotational contribution to viscosity

η'_c Carnot efficiency

η' overall efficiency

θ angle between applied field and shear flow

κ thermal conductivity

λ coupling constant

μ magnetic moment

ν number of particles per chain

ν_0 ionisation frequency

ξ phase angle

ρ density

σ standard deviation of log normal volume distribution and interfacial tension between particles and carrier liquid

τ_B Brownian rotational diffusion time

τ_N Néel relaxation time

ϕ velocity potential

χ_i initial susceptibility

ω velocity of particles in suspension

Ω angular velocity of rotating field

References

Akashi, G. and M. Fujiyama, 1971, US Patent 3 607 218.

Akimoto, Y., M. Sato and Y. Hoshino, 1972, Bull. Tokyo Inst. Technol. 108, 141.

Anderson, J.C. and B. Donovan, 1959, Proc. Phys. Soc. Lond. 73, 593.

Bean, C.P. and I.S. Jacobs, 1956, J. Appl. Phys. 27, 1448.

Bean, C.P. and J. Livingston, 1959, J. Appl. Phys. 30, 120S.

Bertrand, A.R.V., 1970, Rev. Inst. Fr. Petrole Ann. Combust. Liquides 25, 16.

Bibik, E.E., 1973, Kolloidn Zh. 35, 1141.

Bibik, E.E. and I.S. Lavrov, 1965, Kolloidn Zh. 27, 652.

Bibik, E.E. and A.A. Simonov, 1972, IFZh 22, 343.

Bibik, E.E. and V.E. Skobochkin, 1972, IFZh 22, 687.

Bibik, E.E., I.S. Lavrov and O.H. Merkushev, 1966, Kolloidn Zh. 28, 631.

Bibik, E.E., I.S. Lavrov and O.M. Merkushev, 1969, Chem. Abstr. 70, no. 6843u.

Bitter, F., 1932, Phys. Rev. 41, 507.

Brenner, H., 1970, J. Coll. Int. Sci. 32, 141.

Brenner, H. and M.H. Weissman, 1972, J. Coll. Int. Sci. 41, 499.

Brown, R. and T.S. Horsnell, 1969, Electrical Rev. 183, 235.

Brown, W.F. Jr., 1962, Magnetostatic principles in ferromagnetism (Interscience Publisher, New York) p. 112.

Brown, W.F. Jr., 1967, J. Appl. Phys. 38, 1017.

Busch, G. and H.J. Guentherodt, 1968, Phys. Lett. 27A, 110.

Chantrell, R.W., 1977, Thesis (University of Wales).

Chantrell, R.W., J. Popplewell and S.W. Charles, 1977, Physica 86–88B, 1421.

Chantrell, R.W., S.W. Charles and J. Popplewell, 1978, IEEE Trans. Mag. 14, 975.

Charles, S.W. and J. Popplewell, 1978, Proc. Int. Advanced Course and Workshop on Thermomechanics of Magnetic Fluids (Hemisphere Publ., Washington D.C.).

Chu, B., 1967, Molecular forces (Wiley and Sons, New York–London) Ch. 4,5.

Cowley, M.D. and R.E. Rosensweig, 1967, J. Fluid Mech. 30, 271.

Craik, D.J. and P.M. Griffiths, 1958, Brit. J. Appl. Phys. 9, 276.

Curtis, R.A., 1968, Thesis (Cornell University).

Davies, H.W. and J.P. Llewellyn, 1979, J. Phys. D: Appl. Phys. 12, 1357.

De Boer, J.H., 1936, Trans. Farad. Soc. 32, 10.

De Gennes, P.G. and P.A. Pincus, 1970, Phys. Kondens. Mater. 11, 189.

Deryugin, I.A. and M.A. Sigal, 1962, Sov. Phys. Solid State 4, 359.

Earnshaw, S., 1842, Trans. Cambridge Phil. Soc. 7, 97.

Ehrenberg, W., 1958, Electrical conduction in Semiconductors and metals (Clarendon Press, Oxford) Ch. 11.

Einstein, A., 1906, Ann. Physik 19, 289.

Einstein, A., 1911, Ann. Physik 34, 571.

Elmore, W.C. 1938a, Phys. Rev. 54, 309.

Elmore, W.C. 1938b, Phys. Rev. 54, 1092.

Elmore, W.C. 1941, Phys. Rev. 60, 593.

Ezekiel, D., 1975, Am. Soc. Mech. Eng., paper no. 75-DE-5.

Falk, R.B. and G.D. Hooper, 1961, J. App. Phys. 32, 2536.

Falk, R.B. and R.E. Luborsky, 1965, Trans. AIME 233, 2079.

Freundlich, H., 1922, Kapillarchemie (Leipzig).

Fritsche, L., F. Wolt and A. Schaber, 1961, Z. Naturforsch. 16a, 31.

Goldberg, P., J. Hansford and P. G. van Heerden, 1971, J. Appl. Phys. 42, 3874.

Goldsmith, H.L. and S.G. Mason, 1967, Rheology: Theory and applications, Vol. 4 (Academic Press, New York) pp 85–220.

Goodwin, J.W., 1975, Colloid science, Vol. 2 (Chem. Soc., London) p. 246.

Granqvist, C.G. and R.A. Buhrman, 1976a, J. Appl. Phys. 47, 2200.

Granqvist, C.G. and R.A. Buhrman, 1976b, J. Catalysis 42, 477.

Greenwood, G.W., 1956, Acta Met. 4, 243.

Gubanov, A.I., 1960, Sov. Phys. Solid State, 2, 468.

Hahn, A., 1965, Phys. Kondens. Mater, 4, 20.

Hall, W.F. and S.N. Busenberg, 1969, J. Chem. Phys. 51, 137.

Hamaker, H.C., 1936, Rec. Trav. Chim. 55, 1015.

Handrich, K., 1969, Phys. Stat. Sol. 32, K55.

Handrich, K. and S. Kobe, 1970, Acta Physica Polonica A33, 819.

Harle, U.L. and J.R. Thomas, 1966, US Patent 3 228 882.

Harris, L., D. Jeffries and B.M. Siegel, 1948, J. Appl. Phys. 19, 791.

Hayes, C.P., 1975, J. Coll. Int. Sci. 52, 239.

Heaps, C.W., 1940, Phys. Rev. 57, 538.

Hess, P.H. and P.H. Parker Jr., 1966, J. Appl. Polymer Sci. 10, 1915.

Hoon, S.R., J. Popplewell and S.W. Charles, 1978, IEEE Trans. Mag. 14, 981.

Judge, J.S., J.R. Morrison and D.E. Speliotis, 1966, IBM Tech. Discl. Bull. 9, 320.

Jenkins, J.T., 1971, J. de Phys. 32, 931.

Kagan, I. Ya., V.G. Rykov and E.I. Yantovskii, 1970, Magnetohydrodynamics USA 6, 441 [Magnit. Gidrodin. 6, 155].

Kaiser, R. and G. Miskolczy, 1970a, J. Appl. Phys. 41, 1064.

Kaiser, R. and G. Miskolczy, 1970b, IEEE Trans. Mag. 6, 694.

Kaiser, R. and R.E. Rosensweig, 1969, NASA Report NASA CR-1407.

Khalafalla, S.E. and G.W. Reimers, 1973, US Patent 3 764 540.

Kimoto, K. and I. Nishida 1967, Japan J. Appl. Phys. 6, 1047.

Kimoto, K., Y. Kamiya, M. Nonoyama and R. Uyeda, 1963, Japan J. Appl. Phys. 2, 702.

Kneller, E., 1965, Proc. Int. Conf. on Magnetism, Nottingham University (Inst. of Phys. and Phys. Soc., London) p. 174.

Kneller, E. and F.E. Luborsky, 1963, J. Appl. Phys. 34, 656.

Krieger, I.M. 1959, Trans. Soc. Rheol. 3, 137.

Kusaka, K., N. Wada and A. Tasaki, 1969, Japan J. Appl. Phys. 8, 599.

Laithwaite, E.R., 1968, Electrical Rev. 182, 130.

Lang, N.D., 1971, Phys. Rev. 4, 4234.

Levi, A.C., R.F. Hobson and F.R. McCourt, 1973, Can. J. Phys. 51, 180.

Luborsky, F.E., 1957, J. Phys. Chem. 61, 1336.

Luborsky, F.E., 1958a, Phys. Rev. 109, 40.

Luborsky, F.E. 1958b, J. Phys. Chem. 62, 1131.

Luborsky, F.E. 1961, J. Appl. Phys., 32, 171S.

Luborsky, F.E. 1962a, J. Appl. Phys. 33, 1909.

Luborsky, F.E. 1962b, J. Appl. Phys, 33, 2385.

Luborsky, F.E. and J.D. Opie, 1963, J. Appl. Phys. 34, 1317.

Luborsky, F.E. and T.O. Paine, 1960, J. Appl. Phys. 31, 665.

Mackor, E.L., 1951, J. Coll. Sci. 6, 492.

Mailfert, R. and A. Martinet, 1972, J. de Phys. 34, 197.

Martinet, A., 1971, Nouv. Rev. D'Optique Appliquee 2, 283.

Martinet, A., 1974, Rheol. Acta. 13, 260.

Martinet, A., 1978, Proc. Advanced Course and Workshop on Thermomechanics of Magnetic Fluids, (Hemisphere Publ., Washington D.C.).

Maxwell, J.C., 1954, A treatise on electricity and magnetism, Vol. II, (Publications Inc., New York) p. 12.

Mayer, A. and E. Vogt, 1952, Z. Naturforsch. 7a, 334.

McTague, J.P. 1969, J. Chem. Phys. 51, 133.

Meiklejohn, W.H., 1958, J. Appl. Phys. 29, 454.

Mizushima, Y., 1961, Z. Naturforsch. 16a, 1260.

Moskowitz, R. and D. Ezekiel, 1975, Instruments and Control Systems, 1.

Moskowitz, R. and R.E. Rosensweig 1967, Appl. Phys, Lett. 11, 301.

Nakagawa, Y. 1969, Phys. Lett. 28A, 494.

Néel, L. 1949, Compt. Rend. Acad. Sci. 228, 664.

Neitzel, U. and K. Bärner, 1977, Phys. Lett. 63A, 327.

Neuringer, J.L. and R.E. Rosensweig, 1964, Phys. Fluids 7, 1927.

Nogita, S., T. Ikeguchi, K. Muramori, S. Kazama and H. Sakai, 1977, Hitachi Rev. 26, 139.

Paine, T.O., L. I. Mendelsohn and F.E. Luborsky, 1955, Phys. Rev. 100, 1055.

Papell, S.S., 1963, Space Daily, 211 (Nov.).

Papell, S.S., 1965, US Patent 3 215 572.

Peterson, E.A., D.A. Krueger, M.P. Perry and T.B. Jones, 1974, AIP Conf. Proc. 24, 560.

Pincus, P.A., 1978, Proc. Int. Advanced Course and Workshop on Thermomechanics of Magnetic Fluids (Hemisphere Publ., Washington D.C.).

Popplewell, J., S.W. Charles and S.R. Hoon, 1976, IEE Conf. Publication No. 142, Adv. Mag. Mat. and their Appl. pp. 13–16.

Resler, E.L. and R.E. Rosensweig, 1964, AIAA J. 2, 1418.

Rosensweig, R.E., 1969, Int. Sci. and Technology, 1 (Oct.).

Rosensweig, R.E. 1978, Proc. Int. Advanced Course and Workshop on Thermomechanics of Magnetic Fluids (Hemisphere Publ., Washington D.C.).

Rosensweig, R.E., J.N. Nester and R.S. Timmins, 1965, AIChE-I Chem. E. Joint Meeting London, Symp. Ser. no. 5, 104.

Rosensweig, R.E., R. Kaiser and G. Miskolczy, 1969, J. Coll. Int. Sci. 29, 680.

Shepherd, P.G. and J. Popplewell, 1971, Phil. Mag. 23, 239.

Shepherd, P.G., J. Popplewell and S.W. Charles, 1970, J. Phys. D 3, 1985.

Shepherd, P.G., J. Popplewell and S.W. Charles, 1972, J. Phys. D 5, 2273.

Shliomis, M.I., 1972, Sov. Phys. JETP 34, 1291.

Shliomis M.I., 1974, Sov. Phys. Usp. 17, 153.

Shliomis, M.I., 1975, Sov. Phys. Dokl. 19, 686.

Skrotskii, G.V. and L.V. Kurbatov, 1961, Ferromagnetic resonance, ed. S.V. Vonsovskii (M. Fizmatgiz).

Smith, R.L., W.D. Corner, B.K. Tanner, R.G. Jordon and D.W. Jones, 1977, Proc. Int. Conf. on Rare Earths, Durham University (Inst. of Physics, Bristol) 215.

Tasaki, A., S. Tomiyama, S. Iida, N. Wada and R. Uyeda, 1965, Japan J. Appl. Phys. 4, 707.

Tasaki, A., M. Takao and H. Tokunaga, 1974, Japan J. Appl. Phys. 13, 2.

Taylor, H.S. and S. Glasstone, 1951, A treatise on physical chemistry Vol. II, (Van Nostrand, New York) p. 588.

Thomas, J.R. 1966a, US Patent 3 228 881.

Thomas, J.R. 1966b, J. Appl. Phys. 37, 2914.

Uyeda, R., 1964, J. Cryst. Soc. Japan 6, 106.

Verwey, E.J.W. and J.Th.G. Overbeek, 1948, Theory of the stability of lyophobic colloids (Elsevier, Amsterdam) p. 123.

Wachtel, E. and W.U. Kopp, 1969, Phys. Lett. 29A, 164.

Wada, N., 1967, Japan J. Appl. Phys. 6, 553.

Wada, N., 1968, Japan J. Appl. Phys. 7, 1287.

Wilson, B. (FRS), 1779, Philosophical transactions account of Dr. Knight's method of making artificial Lodestone, p. 480.

Windle, P.L., J. Popplewell and S.W. Charles 1975, IEEE Trans Mag. MAG-2, 1367.

Winkler, H., H.J. Heinrich and E. Gerdau, 1977, J. de Phys. C10-5, 340.

Yatsuya, S., S. Kasukabe and R. Uyeda, 1973, Japan. J. Appl. Phys. 12, 1675.

AUTHOR INDEX

Francombe, M.H. 423, 424
Frankel, S., see Caspari, M.E. 31
Frederick, C.G., see Chamberland,
  B.L. 468, 470, 472
Frei, E.H. 440
Frei, E.H., see Aharoni, A. 34,
  422, 432, 434, 435
Freitag, W.O. 385
Freundlich, H. 538
Frey, J., see Kornetzki, M. 225
Frey, J., see Löbl, H. 220
Frieze, A.S. 391
Frischmann, P.G., see Luborsky,
  F.E. 72, 75
Fritsche, L. 519
Fuchs, E.O., see Swisher, J.H. 65,
  68, 69, 70, 71, 481, 490
Fudge, A.D., see Enoch, R.D. 151,
  152
Fujiwara, S., see Esho, S. 350
Fujiyama, M., see Akashi, G. 484,
  521
Fukami, K., see Horiishi, N. 413,
  416, 443
Fukuda, B. 104, 105
Fukuda, B., see Irie, T. 104, 105
Fukuda, I., see Shibaya, H. 123
Fukuda, S. 414, 453
Fukuda, Y. 466
Fukushima, I., see Nakamura,
  M. 481, 482
Fukushima, I., see Ozaki, S. 487
Fudashima, I., see Takahashi,
  N. 482
Fullerton, L.D., see Nesbitt,
  E.A. 138
Funahashi, T., see Shimanaka,
  H. 100, 101, 104
Funatogawa, Z., see Miyata, N. 212
Funatogawa, Z., see Usami, S. 212
Funke, A. 466
Furukawa, M. 483

Gallagher, P.K., see Hoekstra,
  B. 144
Gallaher, L.E., see Ault, C.F. 161
Gallinotti, E., see Montiglio, U. 467
Gălugăru, Gh. 391
Gălugăru, Gh., see Rabinovici,
  R. 482
Gambino, R.J. 355
Gambino, R.J., see Argyle,
  B.E. 372
Gambino, R.J., see Chaudhari,
  P. 348, 349, 353, 362, 374
Gambino, R.J., see Cuomo, J.J. 350,
  373
Gambino, R.J., see Hasegawa,
  R. 376

Gambino, R.J., see Malozemoff,
  A.P. 372
Gambino, R.J., see Mizoguchi,
  T. 372
Gangulee, A. 351, 352, 353, 355,
  356, 357, 360
Gangulee, A., see Ahn, K.Y. 397
Gangulee, A., see Kobliska,
  R.J. 368
Gangulee, A., see Taylor,
  R.C. 351, 352, 355, 356, 357, 377
Gans, R. 231
Ganz, D., see Assmus, F. 77
Gazzarini, F. 420, 421
Gazzarrini, F., see Colombo,
  U. 420, 421
Geidel, W., see Baronius, W. 408
Geisler, A.H., see Martin,
  D.L. 170
Gelderman, P., see Wijn,
  H.P.J. 231, 237
Geller, S. 3, 4, 5, 6, 7, 8, 10, 11,
  12, 13, 14, 15, 16, 19, 20, 21, 22,
  23, 24, 28, 29, 30, 31, 32, 34, 35,
  36, 38, 40, 42, 50, 257
Geller, S., see Gilleo, M.A. 9, 14,
  21, 22, 23, 24, 25, 26, 34, 38, 39
Gen, M.Ya. 482, 486
Gen, M.Ya., see Shtol'ts,
  Ye.V. 486
Gerdau, E., see Winkler, H. 546
Gerkema, J.T. 387
Gersdorf, R. 62
Geschwind, S. 10, 34
Getting Jr., M.P. 108
Geusic, J.E., see Bonner,
  W.A. 319, 329
Ghez, R., see Giess, E.A. 320
Ghodsi, M., see Derie, R. 418
Gibbs, G.V., see Novak, G.A. 6
Gibson, B., see Scholefield,
  H.H. 151
Giess, E.A. 320, 324, 326, 336
Giess, E.A., see Davies, J.E. 318,
  319, 321, 324
Giese, E.A., see Hu, H.L. 330, 336
Gilleo, M.A. 9, 14, 16, 21, 22, 23,
  24, 25, 26, 27, 34, 35, 37, 38, 39,
  41, 45
Gilleo, M.A., see Caspari, M.E. 31
Gilleo, M.A., see Geller, S. 3, 4, 5,
  7, 8, 10, 11, 19, 20, 22, 31, 34, 257
Gillson, J.L., see Chamberland,
  B.L. 468, 470, 472
Gilson, R.G., see Clow, H. 399
Ginder, P. 484
Girke, H. 449
Gladstone, H.M., see Robbins,
  M. 490
Glass, H.L. 338

Glasstone, S., see Taylor, H.S. 535
Gleiser, M., see Hultgren, R. 77
Glemser, O. 416
Globus, A. 209, 220
Glushkova, V.B., see Ariya,
  S.M. 466
Gödecke, T., see Köster, W. 77
Goldberg, P. 527, 533
Goldsmith, H.L. 547
Goldstein, J.I. 124
Goman'kov, V.I. 125, 126
Goodenough, J.B. 232, 234, 235,
  236, 238
Goodwin, J.W. 543
Gordon, D.I. 136
Gordon, D.I., see Sery, R.S. 136
Gorter, E.W. 199, 231, 280, 435
Gorter, E.W., see Wijn,
  H.P.J. 231, 237
Goss, N.P. 76, 85, 115
Goto, H. 414
Goto, H., see Fukuda, S. 414
Goto, I. 98, 100
Gotoh, T., see Goto, I. 98, 100
Gould, H.L.B. 168, 174, 175, 176
Graczyk, J.F. 350
Graham Jr., C.D. 62, 134, 136
Graham Jr., C.D., see Chikazumi,
  S. 139
Granqvist, C.G. 519, 539
Graw, D., see Wunsch, G. 457
Green, J.J. 256, 286
Green, J.J., see Maguire, E.A. 34
Green, J.J., see Saunders,
  J.H. 257, 259, 262
Green, J.J., see Schlömann, E. 253
Greenblatt, S. 415
Greenwood, G.W. 514, 517, 538
Greenwood, T.S., see Ault,
  C.F. 161
Greifer, A.P. 191, 236, 238
Greiner, H.S., see Bennetch,
  L.M. 410, 412, 413, 417, 418,
  419, 427, 459
Greiner, J. 410
Grenoble, H.E. 98, 101, 107
Griffiths, P.M., see Craik, D.J. 512
Grodkiewicz, W.H., see Heilner,
  E.J. 17
Grodkiewicz, W.H., see Van Uitert,
  L.G. 338
Grodkiewicz, W.J., see Gyorgy,
  E.M. 325
Grupen, W.B., see Chin, G.Y. 159,
  161, 162
Guarnieri, C.R., see Heiman,
  N. 350, 353, 358
Guarnieri, C.R., see Minkiewicz,
  V.J. 375
Gubanov, A.I. 511

SUBJECT INDEX

MATERIALS INDEX

Printed and bound by CPI Group (UK) Ltd, Croydon, CR0 4YY

03/10/2024

01040330-0012